Handbook of Parallel Computing and Statistics

STATISTICS: Textbooks and Monographs

Recent Titles

Handbook of Parallel Computing and Statistics

Edited by

Erricos John Kontoghiorghes

University of Cyprus and
Birkbeck College, University of London, UK

Chapman & Hall/CRC
Taylor & Francis Group
Boca Raton   London   New York

Published in 2006 by
Chapman & Hall/CRC
Taylor & Francis Group
6000 Broken Sound Parkway NW, Suite 300
Boca Raton, FL 33487-2742

© 2006 by Taylor & Francis Group, LLC
Chapman & Hall/CRC is an imprint of Taylor & Francis Group

No claim to original U.S. Government works
Printed in the United States of America on acid-free paper
10 9 8 7 6 5 4 3 2 1

International Standard Book Number-10: 0-8247-4067-X (Hardcover)
International Standard Book Number-13: 978-0-8247-4067-2 (Hardcover)

Library of Congress Cataloging-in-Publication Data

Catalog record is available from the Library of Congress

Taylor & Francis Group
is the Academic Division of Informa plc.

Visit the Taylor & Francis Web site at
http://www.taylorandfrancis.com

and the CRC Press Web site at
http://www.crcpress.com

Dedication

———

This book is dedicated to Laurence, Louisa, Ioanna, and Emily

Preface

Over the last decades a wealth of parallel algorithms has been discovered for solving a wide range of problems that arise in diverse applications areas. Effort has been concentrated mainly on the solution of large scale industrial and engineering problems. Although some of these application areas, such as signal processing and pattern recognition, involve significant statistical computing components, the development of parallel algorithms for general use in statistics and econometrics remains comparatively neglected. This is, to some extent, due to a lack of strong interaction between the parallel computing and statistical communities.

A number of current parallel numerical libraries provide subroutines that are useful to statisticians. For example, most of them offer routines to solve constrained least squares problems and matrix problems that arise in statistical modelling and estimation. However, these routines have been constructed as numerical tools for general use and are unsuitable for the efficient solution of statistical problems that exhibit special properties and characteristics. The design of specifically targeted parallel numerical libraries and tools to facilitate the solution of computationally intensive statistical problems requires close collaboration between statisticians and parallel computing experts [1–3].

The aim of this handbook is twofold: first, to provide an overview of the state-of-the-art in parallel algorithms and processing from a statistical computing standpoint; and, second, to contribute toward the development and deepening of research in the interface between parallel and statistical computation.

REFERENCES

[1] E.J. Kontoghiorghes. *Parallel Algorithms for Linear Models: Numerical Methods and Estimation Problems*. Advances in Computational Economics, Vol. 15. Kluwer Academic Publishers, Boston, MA, 2000.

[2] E.J. Kontoghiorghes. Parallel processing and statistics. Special issue, *Computational Statistics & Data Analysis*, 31(4), 373–516, 1999.

[3] E.J. Kontoghiorghes, A. Nagurney, and B. Rustem. Parallel Computing in economics, finance, and decision-making. Special issue, *Parallel Computing*, 26(5), 507–676, 2000.

Editor

Erricos John Kontoghiorghes received a B.Sc. and Ph.D. from Queen Mary College, University of London, United Kingdom. He is a faculty member of the Department of Public and Business Administration at University of Cyprus and holds a visiting professorship at Birkbeck College, University of London, United Kingdom. Previously, he held faculty positions at City University Business School, London, United Kingdom and University of Neuchatel, Switzerland, and held various visiting appointments at INRIA-IRISA, France.

Dr. Kontoghiorghes has published over 50 technical papers in journals and books in the interface research area of parallel computing, statistics, econometrics, and numerical linear algebra. He has authored or edited 7 books, and edited 11 special issues of journals. He is a co-editor of the *Journal of Computational Statistics & Data Analysis* (Elsevier), editor-in-chief of the *Handbook Series of Computing and Statistics with Applications* (Elsevier), and a member of the editing committees of various journals. He was elected vice-president of the International Association for Statistical Computing (IASC) in April 2005.

Contributors

Michael W. Berry
Department of Computer Science
University of Tennessee
Knoxville, Tennessee

Yair Censor
Department of Mathematics
University of Haifa
Haifa, Israel

Maurice Clint
School of Computer Science
Queen's University
Belfast, Ireland

Marco D'Apuzzo
Department of Mathematics
Second University of Naples
Caserta, Italy

Jurgen A. Doornik
Nuffield College
University of Oxford
Oxford, United Kingdom

Lars Eldén
Department of Mathematics
Linköping University
Linköping, Sweden

Paraskevas Evripidou
Department of Computer Science
University of Cyprus
Nicosia, Cyprus

Janis Hardwick
University of Michigan
Ann Arbor, Michigan

Markus Hegland
Centre for Mathematics and its Applications
Mathematical Sciences Institute
Australian National University
Canberra, Australia

David F. Hendry
Nuffield College
University of Oxford
Oxford, United Kingdom

Elias S. Manolakos
Electrical and Computer Engineering
    Department
Northeastern University
Boston, Massachusetts

Marina Marino
Department of Agricultural Engineering
    and Agronomy
University of Naples Federico II
Portici, Italy

Dian I. Martin
Department of Computer Science
University of Tennessee
Knoxville, Tennesse

Dani Mezher
ESIB
Université Saint-Joseph
Beyrouth, Lebanon

Athanasios Migdalas
Department of Production, Engineering,
    and Management
Technical University of Crete
Chania, Greece

Marcin Paprzycki
SWPS
Computer Science Institute
Warsoue, Poland

Panos M. Pardalos
Department of Industrial and Systems
    Engineering
University of Florida
Gainesville, Florida

Haesun Park
College of Computing
Georgia Institute of Technology
Atlanta, Georgia

R. Perrott
School of Computer Science
Queen's University
Belfast, Ireland

Bernard Philippe
INRIA/IRISA
Campus Universitaire de Beaulieu
Rennes, France

C. Phillips
School of Computer Science
Queen's University
Belfast, Ireland

Ahmed Sameh
Department of Computer Science
Purdue University
West Lafayette, Indiana

Neil Shephard
Nuffield College
University of Oxford
Oxford, United Kingdom

T. Stitt
School of Computer Science
Queen's University
Belfast, Ireland

Quentin F. Stout
University of Michigan
Ann Arbor, Michigan

Przemysław Stpiczyński
Department of Computer Science
Marie Curie-Sklodowska University
Lublin, Poland

Gerardo Toraldo
Department of Agricultural Engineering and
    Agronomy
University of Naples Federico II
Portici, Italy

Pedro Trancoso
Department of Computer Science
University of Cyprus
Nicosia, Cyprus

Edward J. Wegman
Center for Computational Statistics
George Mason University
Fairfax, Virginia

Darren J. Wilkinson
School of Mathematics and Statistics
University of Newcastle
Newcastle upon Tyne, United Kingdom

Stavros A. Zenios
Department of Public and Business
    Administration
University of Cyprus
Nicosia, Cyprus
and
The Wharton School
University of Pennsylvania
Philadelphia, Pennsylvania

Table of Contents

General — Parallel Computing

1 A Brief Introduction to Parallel Computing

Marcin Paprzycki and Przemysław Stpiczyński

CONTENTS

ABSTRACT

A concise overview of the fundamentals of parallel computing is presented. It is intended for readers who are familiar with the general aspects of computing but are new to high-performance and parallel computing for solving large computationally intensive problems. A general overview of the field and a requisite background for the issues considered in subsequent chapters is provided.

1.1 INTRODUCTION

Over the past 20 years, parallel computing has emerged from the enclaves of research institutions and cutting edge technology firms and entered the mainstream. While multiprocessor machines were once a marvel of technology, today most personal computers arrive prebuilt with multiple processing units (i.e. main processor/CPU, video processor, sound processor, etc.), with computational power and memory substantially larger than those of the top of the line, high-performance workstations of just a few years ago. More importantly, the increased availability of inexpensive components has made "desktop" systems with two and four processors available for the price ranging from US$5,000 to $15,000, while the proliferation of eight and 16 processor systems has been delayed only by the recent economic downturn.

In this chapter, a concise overview of the fundamentals of parallel computing is presented. It is intended for readers who are familiar with the general aspects of computing but are new to high-performance and parallel computing and provides a general overview of the field and a background for the issues considered in following chapters. Specialists in high-performance or parallel computing may wish to skip directly to the subsequent material.

1.2 WHY PARALLEL COMPUTING?

Why parallel computing? The answer is simple: because there exists a need to solve large problems — the kinds that often arise, among others, in the application of statistical methods to large data sets.

While the ongoing applicability of Moore's Law promises uninterrupted increase in (single) processor "power," the size of the problems that researchers are interested in solving is increasing even faster. Parallel computing becomes a viable bridge across this gap, when it is possible to harness multiple processors in a straightforward way, allowing nonspecialists to take advantage of available computational power. This does not imply that the procedures involved are trivial (if this were the case, volumes like the present one would not be needed), but merely that the economics of computing is relatively simple. If by combining four processors it is possible to cut the execution time of a large program by half, this is most likely worth the effort. Of interest also is the trend in the overall growth of the world's most powerful computers, which follow the pattern described by Moore's Law (doubling approximately every 18 months) [1]. Accordingly, the size of problems that can be feasibly solved also nearly doubles every 18 months. However, to be able to efficiently solve problems on machines with hundreds and thousands of processors, knowledge must be accumulated and experience built across the spectrum of available architectures.

Large problems from many areas of scientific endeavors are often used to illustrate the need for large-scale parallel computing. Some examples include (this list is certainly not exhaustive, but rather is presented here to indicate the breadth of the need for parallel computing):

- Earth environment prediction
- Nuclear weapons testing (the ASCI (Advanced Simulation and Computing Initiative) program)
- Quantum chemistry
- Computational biology
- Data mining for large and very large data sets
- Astronomy and cosmology
- Cryptography
- Approximate algorithms for $\mathcal{NP}$-complete problems

The first two problems on the list led to the creation of the two largest supercomputers in the world, which achieve 35.86 and 7.72 Tflops (10×10^{12} floating-point operations per second), respectively. Unfortunately, their performance still does not reach the level necessary to accomplish the original scientific goals set forth by the researchers. For instance, the estimated necessary sustained speed of computations for the ASCI program is of the order of 10^{15} floating-point operations per second. Obviously, even with the exponential growth of computational power promised by Moore's Law, a single processor will not be able to achieve this level of performance any time soon (if ever); therefore combining multiple computational units is the only possible solution.

In summary, parallel computing and the construction of large multiprocessor systems has come to the forefront because of the need to solve large computationally intensive problems. While it is relatively easy to combine a few (2–16) processors to work on a problem on a small scale (it is in fact much easier and cheaper than to acquire a single processor that would be as powerful on its own), multiprocessing is the only way to reach the necessary performance on a large scale.

1.3 ARCHITECTURES

In this section, some important issues in computer architectures are briefly described (a more detailed treatment of hardware issues can be found in Chapter 2). For the purposes of this book, parallel computing is not considered a goal in its own right but rather a means of solving large computationally intensive problems. This being the case, performance needs to be considered on all "levels" of a multiprocessor system to assure that its overall efficiency is optimal. Starting from the processor itself, the following issues need to be addressed in order to indicate how its performance can be maximized: (1) what is happening inside of a processor to make them fast and (2) how can data be efficiently supplied to these fast processors.

1.3.1 INSIDE OF A PROCESSOR

One of the important features that drive the performance of modern processors is *pipelining*, that can be traced back to the CDC 7600 computer, in which it was for the first time successfully applied in a commercial processor. Pipelining is based on a very simple observation: in a classical von Neumann architecture, processors complete consecutive operations in a fetch-execute cycle, i.e., each instruction is fetched, decoded, the necessary data is fetched, the instruction is executed, the results are stored and only then is the processing of the next instruction initiated. This means that a single instruction is executed in five steps and, after the start-up period, each instruction is completed every cycle (see Figure 1.1). Assuming that there is no direct dependency between consecutive instructions, while one instruction is being decoded, another may be fetched. So, when

Stage	Cycle					
	1	2	3	4	5	6
Fetch operands	A[1], B[1]	A[2], B[2]	A[3], B[3]	A[4], B[4]	A[5], B[5]	A[6], B[6]
Fetch operands		A[1], B[1]	A[2], B[2]	A[3], B[3]	A[4], B[4]	A[5], B[5]
Execute multiplication			A[1], B[1]	A[2], B[2]	A[3], B[3]	A[4], B[4]
Nomalize result				A[1], B[1]	A[2], B[2]	A[3], B[3]
Store result					A[1], B[1]	A[2], B[2]

FIGURE 1.1 Partial representation of pipelined floating-point multiplication of elements of a vector.

the data for the first instruction is fetched (third step), yet another instruction is introduced. In this way, after a start-up period of five cycles, the pipeline fills in (five instructions are in various stages of execution) and an instruction is "completed" every step (instead of every five steps). In the case when there is a direct dependency between consecutive operations, the compiler can often rearrange the instructions in such a way as to keep these operations separated in order to maintain independence and prevent the disruption of the pipeline. Finally, when branch instructions are encountered, an educated guess is made (and there exists a large body of research about how to make a correct branch prediction) and "the most likely" of the available branches is selected. In modern architectures, the success rate of branch prediction is above 90%.

This general idea of pipelining has been further extended in the following directions. First, it was observed that scientific computations very often involve vectors and matrices. Accordingly, special pipelines have been developed for floating-point vector operations. In addition, the multiplication of a vector by a constant is often followed by vector addition: operations of the type $ax + y$ (where x and y are vectors). For this class of operations, it is possible to further improve processor performance by sending the results from the multiplication pipeline directly into the addition pipeline, without storing the intermediate results. This process is called *chaining*. Second, operations on integers differ from operations on floating-point numbers in the amount of steps required for their completion. For instance, when two integers are added, since both of them are objects belonging to the same range and represented as two strings of 32 bits, the only operation to be performed is the actual addition. In the case of floating-point addition, the two have to be assumed to belong to different ranges of available numbers and have to be aligned to be added. This involves operations on both their mantissas and exponents. After the operation is completed, the result needs to be "rescaled" to the desired final representation, which involves further operations on the mantissa and the exponent of the result. This difference opens up the possibility of developing processors with separate pipelines for integer and floating-point operations. Finally, since it is often possible to reorganize the instruction stream to sufficiently separate dependent operations and since many operations on long vectors can be divided into independent operations on subvectors (i.e., vector addition, vector scaling etc.), it is also possible to introduce multiple integer and floating-point pipelines. Currently, all modern processors — including products from Intel and AMD — are built this way.

While processors are constantly increasing their speed, an important problem arises: how to provide them with data sufficiently fast. The memory subsystem is typically either too slow to service the processor or the implementation of a memory service that is fast enough is too expensive to be economically feasible, which results in a bottleneck in the processing capabilities of the computer. It becomes impossible to feed the ever-faster processors with data at an appropriately fast rate. Attempts to minimize the effects of this bottleneck have led to the development of hierarchical memory.

1.3.2 Memory Hierarchies

There are two ways of addressing the limitations of the memory subsystem. One way is to consider only the performance of the memory and to install the fastest subsystem available at a time. To further increase its speed, the memory can be divided into separately accessible and refreshable memory "banks." Given this setup, if the data is laid out and accessed in an optimal way, each consecutive element is retrieved from a separate memory bank and this memory bank is ready before the next access is requested. Unfortunately, if consecutive elements are retrieved from the same memory bank, a memory bank conflict occurs (the next element cannot be retrieved until the memory bank is refreshed). This can result in the reduction of performance even by a factor of seven [2–5]. This approach to the memory bottleneck problem was a staple of Cray Research and their architectures i.e., Cray Y-MP, which while characterized by very high memory throughput, was chiefly responsible for the prices of the order of US$ 20 million (in 1990, with an academic discount) for a complete system.

Extremely fast uniform-speed memory cannot be afforded in quantity except by a few power users, i.e., very large companies, the government, its military and their research laboratories, etc. Thus the need for an alternative approach — cache memory. Here, a relatively small, high-grade, and fast memory subsystem is inserted between the processor and the main memory, which is of a lower grade but proportionally larger. Cache memory stores both the instructions that are to be utilized by the processor and the data, which is predicted to be used next. Since cache memory is faster than the main memory, it can increase the data throughput (assuming that it contains the correct data that is to be used in subsequent steps). Modern systems typically have at least two levels of cache memory and the general rule that characterizes such systems is: the further "away" from the processor, the larger and slower the available memory and the longer it takes for the data to reach the processor. There are downsides to this approach, of course. Since each level of cache has a different speed for every level of cache in the system, different data latencies are introduced. To utilize the hierarchical memory system to its fullest extent, algorithms have to be oriented toward *data locality*. First, groups of data elements that are to be worked on together should be moved up the memory hierarchy (closer to the processor) together. Second, all necessary operations on these data elements should be performed while they are stored in the closest cache to the processor and the algorithm should not return to these elements in subsequent steps. For example, in the context of numerical linear algebra, this is achieved through the application of block-oriented algorithms based on the level three BLAS kernels (see Section 1.6.7.1 for more details) and careful selection of blocksizes (utilizing, for instance, tools developed by the ATLAS project [6]).

1.3.3 Flynn's Taxonomy

Now that some of the techniques used to optimize the performance of single processor systems have been presented, a look at possible configurations for multiple processors is in order (for more detailed treatment of these topics, see Chapter 2). The old but still useful general computer taxonomy introduced by Flynn in 1972 [7] provides a good starting point. By considering the fact that information processing, which takes place inside of a computer, can be conceptualized in terms of interactions between data streams and instructions streams, Flynn was able to classify computer architectures into four types:

- SISD — single instruction stream/single data stream, which includes most of the von Neumann type computers
- MISD — multiple instruction streams/single data stream, which describes various special computers but no particular class of machines
- SIMD — single instruction stream/multiple data streams, the architecture of parallel processor "arrays"
- MIMD — multiple instruction streams/multiple data streams, which includes most modern parallel computers

The last two types of machines, SIMD and MIMD, are at the center of interest in this book since most of the currently used parallel systems fall into one of the two categories.

1.3.4 SIMD COMPUTERS

SIMD-based computers are characterized by a relatively large number of relatively weak processors, each associated with a relatively small memory. These processors are combined into a matrix-like topology, hence the popular name of this category: "processor arrays". This computational matrix is connected to a controller unit (usually a top of the line workstation), where program compilation and array processing management takes place (see Figure 1.2). For program execution, each processor performs operations on separate data streams; all of the processors may perform the same operation, or some of them may skip a given operation or a sequence of operations. In the past, the primary vendors producing such machines were ICL, MasPar and Thinking Machines. Currently, this class of machines is no longer produced for the mainstream of parallel computing. One of the main advantages of SIMD computers was the fact that the processors work synchronously, which enables relatively easy program tracing and debugging. Unfortunately, this advantage comes at a price. In experiments with actual SIMD computers, users realized that it is relatively difficult to use them for unstructured problems and, in general, all problems that require a high level of flexibility of data manipulation and data transfer between processing units. Currently, most of these machines have

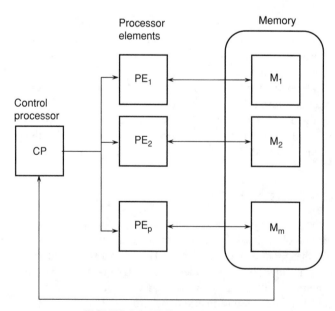

FIGURE 1.2 SIMD multiprocessor.

disappeared from the market, although there are researchers, who claim that SIMD machines are about to make a comeback [8].

1.3.5 MIMD COMPUTERS

The class of MIMD-based parallel computers can be divided into two subcategories: shared memory and distributed memory systems. This division is based on the way that the memory and the processors are connected (see Figure 1.3).

1.3.5.1 Shared Memory Computers

Computers of this type consist of a number of processors that are connected to the main (global) memory. The memory connection is facilitated by fast bus technology or a variety of switch types (i.e., omega, butterfly, etc.) These machines were initially developed with practically memory-less processors but were later equipped with cache memory, which resulted in relatively complicated data management/cache coherency issues. Many companies have produced shared memory computers over the years, including: BBN, Alliant, Sequent, Convex, Cray, and SGI. While these systems were relatively easy to program (with loop parallelization being the simplest and the most efficient means of achieving parallelism), computational practice exposed some important hardware limitations in their design. The most important of them is the fact that it always was and still is almost impossible to scale shared memory architectures to more than 32 processors and to simultaneously avoid saturating the bandwidth between the processors and the global memory. Nevertheless, one can observe a resurgence of shared memory computers in the form of: multiprocessor desktops, which usually contain two or four processors; midrange systems with four or eight processors, primarily used as servers; and high-performance nodes for the top of the line parallel computers, which usually contain 16, 32, or even 64 processors (in the latter case special switching technology is applied).

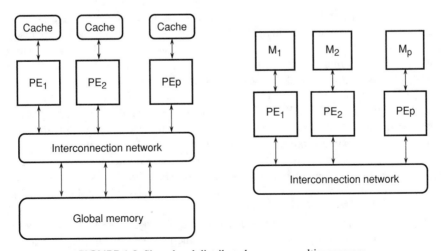

FIGURE 1.3 Shared and distributed memory multiprocessors.

1.3.5.2 Distributed Memory Computers

Since the early days of parallel computing, the second most popular architecture, is distributed memory technology, supported over time among others by Intel, NCube, Inmos, Convex, Cray, and IBM. In this configuration, processing units are composed of a processor and a large local memory and are interconnected through a structured network. There is no global memory. Typical topologies for these structured networks are meshes, toruses and hypercubes. While code for distributed memory computers is relatively more difficult to write and debug, the architecture can be scaled to a large number of processors. Even some of the early distributed memory machines successfully employed up to 1024 processors.

1.3.6 ARCHITECTURE CONVERGENCE

Over time the experiences of early adopters led to an evolution in the design of parallel computers architectures, a process which became even more accelerated with the end of the Cold War and the substantial drop in military funding for high-performance computing, which resulted in bankruptcies or product line changes (moving away from producing parallel computer hardware) for a number of parallel computer vendors (among others, Kendall Square Research, Alliant, BBN, and Thinking Machines). SIMD machines have been all but eliminated, although many of their conceptual and technological advances have been incorporated into modern processor design (video processors in particular). The development of shared memory machines had almost become dormant, until the rebirth of the architecture in desktop PCs and servers. Distributed memory architectures have been by far the most successful. This success has been further substantiated in 1995 when NASA scientists successfully completed the Beowulf project, which paved the way toward cluster computing [9].

1.3.7 CLUSTERS

While the largest computers in the world are still custom built for the highest performance, they cost tens of millions of dollars as well. Clusters made high-performance parallel computing available to those with much smaller budgets. The idea is to combine commodity-off-the-shelf (COTS) components to create a parallel computer. The Beowulf project was the forerunner in this approach.

It is now possible to network 16 top of the line PCs using a fast switch to form a parallel computer that is substantially more powerful than the fastest machine in the world in 1990 for no more than US$50,000 (a 400-fold price decrease). What is even more interesting is the possibility of combining higher-end shared memory PCs and servers into clusters. One can still use the same 16-port switch but with each node having four processors, for example, a resulting parallel computer has 64 processors. Notice however, that in the computational practice, it is the connectivity between the cluster nodes that is the weakest point of many clusters. Low-throughput switches can result in imbalanced systems and become a major performance bottleneck, especially when more powerful nodes are used in the cluster. Regardless of possible drawbacks and limitations, due to their excellent price–performance ratio, clusters have been successfully applied in the industry to solve practical problems (see for instance Ref. [10]).

It is important to stress here that clusters are not a different architectural category of parallel computers in Flynn's taxonomy; they are merely a practical way of building efficient low cost distributed memory MIMD computers.

1.3.8 THE GRID

The most recent developments in parallel computing have been in the area of *grid computing*. The idea is based on the metaphor of the electric power grid. If one plugs-in an appliance to the electric grid, one does not care about where the electricity comes from (as long as it is available). In the same fashion, one could plug a computer into the computer grid, request that a computational task be completed, and not have to care about where the actual computation takes place [11]. It can be said that the first successful grid-like computing endeavor was the SETI@home research project, in which the computers of more than 4,400,000 users work "together" in sifting through data acquired by the Arecibo radio telescope, searching for extraterrestrials. At the time of this writing, the SETI@home network of machines delivered approximately 54.7×10^{12} floating-point operations per second. Most of this power is gained by utilizing the unused otherwise cycles on users' computers. In the same way, computers connected to a grid are to be able to combine their computational power/unused cycles to serve their customers. While a very large amount of research still remains to be completed before the Grid becomes a reality outside special academic and industrial projects, this is one of the most interesting and most fashionable areas in parallel computing today. This latter fact is signified by a flurry of publications devoted to grid computing that appear outside of the realm of peer reviewed journals and conference proceedings (see for instance [12–16]). As an interesting footnote illustrating the interest in this approach, it can be noted that the PC maker Gateway tries to use its in-store demo computers as a computational grid and plans on making profit in this way to support company's bottom line [17, 18].

Summarizing, most of modern computer architectures used to solve large computationally intensive problems involve multiple levels of parallel computing: from the processors with multiple integer and floating-point pipelines and special vector processing units through multiple processors combined into shared memory computers followed by multiple processing units tightly integrated into distributed memory parallel computers to multiple computers (single processor and parallel) working together in a computational grid.

1.4 LEVELS OF PARALLELIZATION

It will be useful to consider as an example a modern parallel computer that consists of a number of state-of-the-art processors (vector, RISC, x86, IA-64, etc.). These processors are combined into shared memory modules. The machine consists of a number of such modules combined into distributed memory parallel computer (possibly a cluster). The computer needs to be examined from the "inside out" — what is happening inside a single processor, what is happening inside a single module, and what is happening in the whole machine. Finally, it will be assumed that a number of computers of this type are combined together into a loosely connected network. For the purposes of this discussion, all issues related to memory banks, cache memory, and memory hierarchies are omitted. They have been addressed previously and while they are crucial to the overall performance of the system they do not affect the available parallelism directly.

1.4.1 INSIDE OF A SINGLE PROCESSOR

As mentioned above (Section 1.3.1), modern day desktop processors such as those developed by Intel, AMD, IBM, etc., are already highly parallelized. Such processors have multiple pipelines for integer and floating-point operations [19], so two different levels of parallelization can be considered. First, the depth of the pipeline: if a pipeline of depth k is used then k operations can be executed at the same time. Second, number of pipelines: assuming that there are l integer pipelines of depth k_1 and m floating-point pipelines of depth k_2 in a given processor, and that all pipelines are operating at maximum capacity at a given moment, then $k_1 \times l + k_2 \times m$ operations are executed by the processor concurrently in every cycle.

Availability of this level of parallelism, which is often called *microparallelism*, is a function of dependencies inside a stream of machine language operations. These dependencies are analyzed and microparallelism is supported: (1) by the logic unit inside of the processor on the hardware level and (2) by the compiler and compiler supplied optimization on the software level. Typically, the user does not have to and does not need to have any control over microparallelization as it should be furnished automatically by the system.

1.4.2   Shared Memory Parallelism

While multiple operations are executed in parallel inside each processor at every step of program execution, parallelism occurring at higher levels is of real interest. In the case of an architecture, in which multiple processors are connected to a global (logically and physically) shared memory, the *typical* way of introducing parallelization is to perform similar operations on subsets of data. The most natural algorithmic level of achieving such parallelization is by dividing between multiple processors work performed by a loop. In particular, a given loop (simple or the outermost loop of a nested loop structure) is divided into as many parts as there are processors (while this is not required and a loop could be divided into a different number of parts, assuming that the number of parts equals the number of processors, is most natural and can be done without loss of generality). Subsequently, each part is executed independently by a separate processor. This kind of parallelization is often called *medium-grain* parallelization and is supported either through a set of special directives (see Section 1.6.3) or through high-level language extensions (see Chapter 3).

A large body of research has been and continues to be devoted to developing compilers capable of automatically generating this level of parallelization. Unfortunately, the results have been disappointing thus far. Parallelizing compilers are relatively successful in generating parallel code with simple loops, addition of two vectors, matrix multiplication, etc. However, in more complicated cases, i.e., when functions are called inside of loops, the code must still be manually divided into parallel units.

1.4.3   Distributed Memory Parallelism

The most typical approach to distributed memory parallelization is to create independent programming units that will execute separate work units (which may or may not be similar to each other), with these units communicating with each other via message passing. This latter functionality, although necessary, is expensive (in terms of the time required for messages to reach their destination and thus in terms of the latency introduced to the system). Accordingly, minimizing the number of messages passed between components becomes an important goal of program design. One must seek to divide a distributed parallel program into large computational units that are as independent from each other as possible and only rarely communicate. While it is possible that each work unit is completely different from others (i.e., some of them advance the solution of a differential equation, others perform interpolation, while others still use the result of interpolation to generate on-the-fly visualization of the solution), this is rarely the case. Most often each work unit is a derivative of the main program and performs the same subset of operations as the other work units but on separate data sets. This type of an approach is called SPMD (single program, multiple data) and often referred to as *coarse level parallelization*. Obviously, this level of parallelization must be implemented manually by the programmer as the division of work is based on a *semantic* analysis of the algorithms used to solve the problem (and possibly their interdependencies). One of the more important problems for distributed memory parallelization is also the question of load balancing.

Since each computational unit is relatively independent (which is especially the case when they perform completely different functions), it may require different time to complete its work. This may, in turn, lead to a situation when all processors but one remain idle as they wait for the last one to complete its job. (A theoretical model, which analyzes and illustrates the negative effects of such a situation, can be found in Section 1.5.3.) The last example illustrates the fact that while the work units should be large and independent, they should also complete their tasks within a similar time, which makes the SPMD approach somewhat more attractive than the division of work into completely independent units.

It should be noted that although these are the principal methods of shared and distributed memory parallelization, there are alternative ways as well. The approaches described above try to make the software "match" the underlying hardware. However, it is also possible to treat a shared memory machine as a distributed memory computer and apply approaches based on message passing. Although message passing does introduce certain overhead, this approach often results in good performance if implemented efficiently [20, 21]. The converse approach is to treat a distributed memory computer as a shared memory system and rely on the mechanisms provided by the vendor or third party software to emulate logically shared memory implemented on physically distributed memory hardware. Unfortunately, such an approach is impractical for real applications; not a single existing shared memory emulator is efficient enough to support "virtual" shared memory in a production environment. Finally, one needs to consider the hybrid hardware, where shared memory nodes have been combined into a distributed memory configuration, which results in a distributed shared memory computer. Here, again, treating such a machine as a distributed memory computer and applying appropriate parallelization techniques is usually more successful than treating it as a shared memory environment. Nevertheless, it is the combined approach of distributed and shared memory techniques that can be expected to be the most efficient in using the underlying architecture.

1.4.4 GRID PARALLELIZATION

In grid computing, a number of computers (irrespective of their individual architectures) are loosely connected via a network. In the most general case, each machine (including the properties of connections between them) is assumed to be different. This makes for an extremely heterogeneous system, which requires the coarsest level of parallelization since the work must be divided into independent units that can be completed on different computers at different speed and returned to the main solution coordinator at any time and in any order, without compromising the integrity of the solution. Although there are tasks that are naturally amenable to this level of parallelization, a broader applicability of this approach requires much further research and infrastructure development. Examples of successfully tested tasks include the analysis of very large sets of independent data blocks, in which the problem lies in the total size of data to be analyzed (such as in the SETI@home project [22]).

In summary, there are multiple levels at which parallelization can occur. The simplest *microparallelization* takes place inside a single processor and usually does not require the intervention of the programmer to implement. *Medium-grain* parallelization is associated with language supported or loop level parallelization. While some headway has been made in automating this level of parallelization with optimizing compilers, the results of these attempts are only moderately satisfactory. *Coarse-grain* parallelization is associated with distributed memory parallel computers and is almost exclusively introduced by the programmer. Finally, grid-level parallelization is currently the focus of intensive research and, while it is a very promising model for solving large problems, its applicability in the foreseeable future will probably continue to be limited to certain classes of computational problems, viz. those that belong to the "large-scale embarrassingly parallel" category.

1.5 THEORETICAL MODELS FOR PERFORMANCE ANALYSIS

After looking at the basic issues in the design of parallel systems, the theoretical analysis of parallel algorithms executed on these systems will be considered. There are a number of models that can be employed to predict the parallel usability of a given algorithm. While the approach advocated here is not uncontroversial, it could be argued that, for a user in the context of scientific computing, the most important criterion defining the usability of a given approach is its execution time (the speed of execution of an algorithm on a given machine). Obviously, this assumes that a particular algorithm, after parallelization, has to correctly solve the problem (under appropriate conditions of correctness).

Some further assumptions need to be made explicit. Since this book is concerned with computational statistics, it can be assumed that floating-point operations are the most important (time consuming) operations that will be performed during algorithm execution. It can also be assumed that data most often will be stored/represented as vectors and matrices. In this context, a parallel algorithm can be characterized by a number of parameters that influence its performance, i.e., size of vectors and matrices, data layout, number of processors of the parallel computer, etc. Algorithm analysis with a particular computational model in mind will allow initial optimization of these parameters without requiring testing on a real machine or at least a preliminary assessment of possible algorithm efficiency.

1.5.1 BASIC RAW PERFORMANCE

The processing speed of computers involved in scientific calculations is usually expressed in terms of a number of floating-point operations completed per second, a measure used above to describe the computational power of the world's largest supercomputers. For a long time, the basic measure was Mflops expressed as:

$$r = \frac{N}{t} \text{ Mflops}, \tag{1.1}$$

where N represents a number of floating-point operations executed in t microseconds. Obviously, when N floating-point operations is executed with an average speed of r Mflops, the execution time of a given algorithm can be expressed as:

$$t = \frac{N}{r}. \tag{1.2}$$

Due to the geometric increase in available speeds of computer hardware, the Mflop measure has been superseded by higher-order measures: Gflops (gigaflops), Tflops (teraflops), and even Pflops (petaflops) = 10^{15} floating-point operations per second. The floating-point operations rate can be used to characterize an algorithm executing on a given machine independently of the particular characteristics of the hardware, on which the algorithm is executed, as well as to describe the hardware itself. Many vendors of parallel computers advertise a theoretical peak performance for their machines; this is the maximum speed with which any algorithm can be potentially executed on their hardware. Of course, in computational practice (outside of special simplified cases, such as matrix multiplication), this performance is unattainable. At the same time, however, it indicates what performance can potentially be expected from a given machine.

1.5.2   HOCKNEY'S AND JESSHOPE'S MODEL FOR VECTOR PROCESSING

Although this chapter is focused on parallel computers and their performance, it is useful to start with a model for vector processors. While this model is relatively dated, its applicability has recently been revived with the development of the Earth Simulator, the largest supercomputer in the world, which was built by the NEC Corporation out of proprietary vector processors. In addition, since most modern processors consist of multiple pipelines, performance of each such pipeline can be also conceptualized in terms of the vector performance model presented here. For vector computers, the performance r_N of a vector-processing loop of length N can be expressed in terms of two parameters r_∞ and $n_{1/2}$, which are specific for a given type of a loop and a vector computer [23]. The first parameter represents the performance in Mflops for a very long loop, while the second the loop length for which a performance of about $r_\infty/2$ is achieved. Then

$$r_N = \frac{r_\infty}{n_{1/2}/N + 1} \text{ Mflops.} \tag{1.3}$$

This model can be applied to predict the execution time of different loops. For example, the vector update in the form

$$x \leftarrow x + \alpha y$$

(i.e. the _AXPY operation) can be expressed as a loop of the length N in which each repetition consists of two floating-point operations. Thus, the execution time of _AXPY is

$$T_{\text{AXPY}}(N) = \frac{2N}{10^6 r_N} = \frac{2 \times 10^{-6}}{r_\infty}(n_{1/2} + N) \text{ seconds.} \tag{1.4}$$

This model can be applied not only to predict the execution time of vectorized programs but also to develop optimal vector algorithms [23, 24]. Indeed, several algorithms (especially *divide-and-conquer* algorithms) consist of loops of different lengths, which are related to each other and can be treated as parameters of a vectorized program. Then Equation (1.3) can be used for finding optimal values of these parameters that minimize the execution time of a program. For instance, in Ref. [24], the Hockney–Jesshope model of vector processing was applied to find a very fast vector algorithm for solving linear recurrence systems with constant coefficients.

1.5.3   AMDAHL'S LAW

One of the most important methods of analyzing the potential for parallelization of any algorithm is to observe how the algorithm can be divided into parts that can be executed in parallel and into those that have to be executed sequentially. More generally, different parts of an algorithm are executed with different speeds and use different resources of a computer to a different extent. It would be naive to predict the algorithm's performance by dividing the total number of operations by the average speed of the computer. Such a calculation would be at best a very crude estimate for single processor machines with a very simple memory hierarchy. To find a quality performance estimate, one should separate all parts of the algorithm that utilize the underlying computer hardware to a different extent. Significant initial work in this area was done by Amdahl [25, 26]. He established how slower parts of an algorithm influence its overall performance. Assuming that a given program consists of N floating-point operations, out of which a fraction f is executed with a speed of V Mflops, while the remaining part of the algorithm is executed with a speed of S Mflops, and assuming further that the speed V is close to the peak performance while the speed S is substantially slower ($V \gg S$), then the total execution time can be then expressed using the following formula:

$$t = f\frac{N}{V} + (1 - f)\frac{N}{S} = N\left(\frac{f}{V} + \frac{1 - f}{S}\right) \tag{1.5}$$

which can be used to establish the total execution speed of the algorithm as

$$r = \frac{N}{t} = \frac{1}{\dfrac{f}{V} + \dfrac{(1-f)}{S}} \text{ Mflops.} \tag{1.6}$$

From formula (1.5), it follows that

$$t > \frac{(1-f)N}{S}. \tag{1.7}$$

If the whole program is executed at the slower speed S, its execution time can be expressed as

$$t_s = \frac{N}{S}. \tag{1.8}$$

If the execution speed of the part f of the program can be increased to V then the performance gain can be represented as

$$\frac{t_s}{t_v} < \frac{N}{S} \cdot \frac{S}{N(1-f)} = \frac{1}{1-f} \tag{1.9}$$

This last formula is called Amdahl's Law and can be interpreted as follows: *The speedup of an algorithm that results from increasing the speed of its fraction f is inversely proportional to the size of the fraction that has to be executed at the slower speed.*

In practice, Amdahl's Law *provides an estimate of the overall speed at which the algorithm can be executed.* Figure 1.4 illustrates the effect of the size of f, the fraction of calculation executed in vector speed, on the total performance of an algorithm to be executed on a vector computer, where $V = 1000$ Mflops (peak performance using a vector unit) and with $S = 50$ Mflops (performance of scalar calculations).

It can be observed that a relatively large $f = 0.8$ results in an average speed of only 200 Mflops. In computational practice, this result illustrates an intuitive fact: to reach the highest possible level of performance of a given program, the most important parts are those that are the slowest. Moreover, even a relatively small fraction of a slow code can substantially reduce the overall speed achieved by the whole program — see the rapid decrease of the performance graph from $f = 1.0$ to 0.8. A typical example of how the speed of the slowest parts of the program influences the overall performance of the vector computer would be a notorious case of recurrent formulas [27, 28].

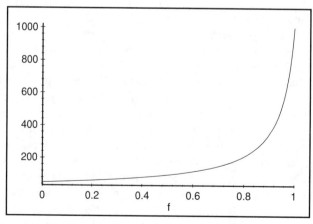

FIGURE 1.4 Amdahl's Law (performance in Mflops) for $V = 1000$ and $S = 50$ (increasing fraction f).

Formula (1.6) can be easily generalized to the case of a program consisting of n separate parts $(A_1, \ldots, A_n)$ that are executed at different speeds. Here, N_j operations of part A_j are executed with speed r_j Mflops, and $N = \sum_{j=1}^{n} N_j$. Thus the average speed of the whole program can be expressed as:

$$r = \frac{N}{\sum_{j=1}^{n} \dfrac{N_j}{r_j}} \text{ Mflops,} \tag{1.10}$$

and again, the average performance will be determined primarily by the speed of its slowest parts.

For a parallel program, the total execution time, understood as the sum of execution times on all processors, is usually larger than the total time used by the same algorithm executed on a single processor (not considering special cases, where the effects of job splitting among multiple processors with large memories affect the total performance by freeing the program from the influence of memory-related bottlenecks, which occur on a single processor machine). However, as specified above, the main goal of parallel computing is to reduce the humanly observable (wall-clock) execution time of the algorithm. The increase of the total time is the price to be paid for the reduction of the wall-clock time. If an algorithm can be divided into p equal parts that can be executed concurrently on p processors then it is conceivable that the parallel execution time will be $1/p$ of the single processor time. Such a situation is practically impossible for all nontrivial algorithms. Since the sequential parts of an algorithm are the "slowest," Amdahl's Law dictates that these parts have the most serious negative impact on the overall performance.

More precisely, in the case of a parallel machine with p homogeneous processors, the speedup s_p over the sequential algorithm achieved due to p processors can be expressed as:

$$s_p = \frac{t_j}{t_p}. \tag{1.11}$$

where t_j denotes time of execution of the algorithm on j processors. Assuming that fraction f of the algorithm can be divided into p parts and ideally parallelized (executed at exactly the same time t_1/p on p processors), the remaining $1 - f$ of operations cannot be parallelized and thus have to be executed on a single processor. The total execution time of this algorithm on p processors can be expressed as:

$$t_p = f\frac{t_1}{p} + (1-f)t_1 = \frac{t_1(f + (1-f)p)}{p}.$$

Therefore the speedup s_p is equal to

$$s_p = \frac{p}{f + (1-f)p}. \tag{1.12}$$

Obviously, $f < 1$, and therefore the following inequality is true:

$$s_p < \frac{1}{1-f}. \tag{1.13}$$

This inequality (1.13) is known as *Amdahl's Law for parallel computing*. It states that the *speedup achievable through parallel computing is bound by the value that is inversely proportional to the fraction of the code that has to be executed sequentially.*

Similarly to the general Amdahl's Law [25] the above considerations provide a simple way to initially assess the expected parallel performance of an algorithm. If $f = 0.9$, so that 90% of an algorithm can be ideally parallelized, and if $p = 10$, the formula (1.12) gives the result that the speedup cannot exceed 5.

While the situation presented here is obviously highly idealized, it allows to conceptualize the effects of load imbalance. In the case of a program executing on two processors, where at a certain moment processor 1 has completed its work while processor 2 is still executing its part of the parallel code, the still-to-be-executed part of the code on processor 2 becomes a serial part of the parallel program and formula (1.12) applies. In general, in the case of a parallel program executing on p processors, due to the load imbalance at various times t_j, different numbers p_j of processors are working on the code while the remaining $p - p_j$ processors are idle, therefore:

$$S_p = \frac{1}{\sum_{i=1}^{k-1} \frac{f_i}{p_i} + f_k} \qquad \text{where} \quad \sum_{i=1}^{k} f_i = 1. \qquad (1.14)$$

This case can be treated as the generalized case of Amdahl's Law, where various fractions of the code can be executed by a different number of processors. Obviously, each situation, when only a number of processors smaller than p is used, leads to the degradation of overall performance, which indicates how important load balancing is for the parallel program performance.

An obvious idealization lies in Amdahl's Law because it only takes into account a fixed problem size. More often, the problem size is expected to scale with the number of processors (one of the important reasons for applying a parallel computer with distributed memory is the fact that with every processor additional memory is added to the system thus allowing solution to a larger problem). Thus, Gustafson [29], proposed an alternative to Amdahl's Law. *Rather than asking how fast a given serial program would run on a parallel machine, he asks how long a given parallel program would have taken to run on a serial processor* [30]. Let t_s and t_p denote the serial and the parallel time spent on a P processor parallel machine, respectively, and let $t_s + t_p = 1$, then *the scaled speedup* of the parallel program will be given by

$$s = \frac{t_s + Pt_p}{t_s + t_p} = t_s + Pt_p = P + (1 - P)t_s.$$

This model is somewhat more realistic in its predictive power. Its usability was presented in Ref. [30].

Theoretical considerations presented here as well as in the literature show that, in computational practice, a very large number of factors influence the parallel execution time of an algorithm. Most of these factors are much more likely to degrade the performance gains of parallelization rather than to augment them. Some of the factors that should be taken into consideration are listed below:

- The algorithm itself must be parallelizable and the data set to which it is to be applied must be such that an appropriately large number of processors can be applied.
- Overheads related to synchronization and memory access conflicts will lead to performance deterioration.
- Load balancing is usually rather difficult to achieve and the lack of it results in performance deterioration.
- Creation of algorithms that can be used on multiple processors often leads to the increase of computational complexity of the parallel algorithm over the sequential one.
- Dividing data among multiple memory units may reduce the memory contention and improve data locality, resulting in performance improvement.

In summary, there are a number of theoretical models that can be applied to predict the performance of parallel algorithms. Unfortunately, all of them are highly idealized and thus are limited to supplying general performance expectations as well as to pointing out the most important issues that should be taken into account when a parallelization of a sequential algorithm and an existing sequential code is attempted. When a more detailed performance prediction is required, a particular model for a given problem and its characteristics as well as the hardware, on which it is to be executed, have to be taken into account.

1.6 PROGRAMMING PARALLEL COMPUTERS

Moving from theoretical considerations to the computational practice, one must consider a number of issues involved in programming parallel computers. This introduction to them will follow the above described levels of parallelization and start from the parallelization inside of a single processor as well as a brief note about language-based parallelization. It will be followed by the discussion of writing codes for shared memory and distributed memory computers, parallelization of existing codes, and library-based parallelization.

1.6.1 OPTIMIZING COMPILERS

Optimizing compilers [31, 32] offer several optimization levels. They transform a code according to the specified option(s). These transformations are cumulative: each higher level retains the transformations/optimizations of the previous level. The available optimization levels are typically as follows (while the levels described below are not specific to any particular hardware architecture and to any computer vendor, similar levels can be found across most of them):

1. *Machine-dependent scalar optimization, usually a default*, which fully exploits the machine's scalar functional units and registers.
2. *Basic block machine-independent scalar optimization* works at the local basic-block level. A basic block is a branch-less sequence of statements ending with a conditional or unconditional branch. At this level a compiler uses such techniques as assignment substitution, elimination of common subexpressions, constant propagation, and folding.
3. *Program block machine-independent scalar optimization* works at the global program-unit level (a subroutine, function, or main section) using eliminations of redundant assignments, dead-codes, and hoisting as well as sinking scalar and array references.
4. *Vector optimization* (if applicable) improves the performance of programs that manipulate arrays. Consider a loop adding the corresponding elements of two arrays. Within this level of optimization, the vector CPU (if available) can add groups of array elements utilizing a single machine instruction. For example, the following simple loop

```
do j=1,100
z(j) = x(j) + y(j)
end do
```

will be transformed to a vector instruction

```
z(1:100)=x(1:100)+y(1:100)
```

When the loop length is greater than the vector length or when it is unknown, the compiler will generate a loop that repeats the above vector instruction.

5. *Parallel optimization* allows to spread work across multiple CPUs and typically analyzes the loop structure of the code. In case of parallel machines with vector processors, the inner loops are vectorized while the outer loops are parallelized.

All optimization levels but the last one should generate code optimal for a given processor and the memory structure of a given machine. The last level of optimization is typically available on shared memory parallel computers (most vendors of parallel computers provide Fortran and C compilers capable of code parallelization). Early successful work on optimizing compilers for high-performance (parallel) computers can be traced to the Cray optimizing compilers for Fortran, and Kuck and Associates parallelizing compilers for Fortran and C [33]. While this statement is open to discussion, there is little question that Cray's Fortran optimizing compiler was one of the best (in terms of the quality of generated code) products existing at the times of Cray Y-MP and Cray C-90 supercomputers. Unfortunately, its success can be traced to the relative simplicity of the Cray architecture, few (up to 64, but typically 8 or 16) vector processor and a large global memory. As soon as the modern workstations with hierarchical memory are considered, the performance of optimizing compilers becomes less satisfactory. Experiments show a substantial performance degradation of a simple matrix multiplication operation on an MIPS and Alpha processor-based computers from SGI and DEC, when the matrix size increased past $n = 1000$ [34, 35]. These results indicate that optimizing codes for modern architectures with multiple levels of data latency and multiple sizes of intermediate memory layers is a complicated endeavor. Projects like ATLAS [6] attempt to remedy these problems for linear algebraic operations (see also Section 1.6.7.1) but, in general, a lot more work is required. The situation is even worse for code parallelization. While parallelizing compilers can deal with microparallelization and matching the code with multiple execution pipelines of modern processors, as well as with parallelization of simple loops, they have problems with parallelization of complicated structures of the program. The situation becomes particularly complicated when parallelization requires a serious restructurization of the program. It is arguable that while the compiler-based parallelization will play an increasingly important role in the implementation of algorithms (it is, for instance, claimed that with the increasing power of computers an even larger "window" of the code can be considered at once thus increasing compilers' ability to analyze and optimize the machine code), it will be mostly responsible for low-level parallelization and, at least for some time to come, cannot be relied on as a method for building parallel programs. In other words, implementers have to do the work themselves and the parallelizing/optimizing compiler can fine-tune the results of their work.

1.6.2 LANGUAGE-BASED PARALLELIZATION

The second approach to parallelization is based on the language constructs. Two scenarios can be distinguished. First, constructs inside of the language support various possible levels of parallelism, i.e., various versions of Fortran (primarily High-Performance Fortran) and SISAL (Streams and Iteration in a Single Assignment Language). Second, language constructs are geared toward high level of parallelism, i.e., Ada and Java. Since Fortran was one of the early languages applied in large-scale scientific computing, efforts were undertaken to extend the Fortran 77 definition to add more constructs supporting high-performance and parallel computing. These additions varied from the vector- and matrix-oriented operations in Fortran 90/95 standards [36, 37] to a more thorough support of parallelism in the High-Performance Fortran [38]. Interestingly, thus far neither of these approaches gained widespread popularity among parallel program developers (for more details about High-Performance Fortran, see Chapter 3). SISAL is an example of an attempt at bringing functional programming, which is said to be one of the better approaches to parallel program design, to scientific computing [39, 40]. The most interesting feature of SISAL is the fact that it combines the imperative and functional programming paradigms. Although initial results were quite encouraging [41, 42], SISAL has practically disappeared after a few years of development.

Ada was originally designed to support concurrency and thus included support for most functions necessary to develop and implement parallel programs. However, for a variety of reasons summarized in Ref. [43], it has never been seriously considered for scientific parallel computing application development. For instance, although until recently the US Department of Defense required all of its computing to be done in Ada, numerically intensive codes, i.e., ocean modeling applications, were developed and implemented in Fortran and translated into Ada by a separate group of programmers when ready to be turned into a production environment. Finally, there is Java, which also can be used to naturally develop parallel programs (i.e., through the application of multi-threading). Since it is a relatively new language, it is still unclear how much popularity it will gain in the scientific computing community. The main disadvantage of Java seems to be its widely perceived lack of efficiency (which should not be viewed as a problem with the language itself since it was not designed for that purpose). Java Grande "community" (among others) attempts to remedy this problem [44, 45] and time will tell how successful their efforts will be (for more detailed treatment of Java as a language for parallel computing, see Chapter 3).

1.6.3 SHARED MEMORY PARALLEL ALGORITHMS AND OPENMP

Parallel architectures based on a shared memory have now become commonplace and usually offer more than just a few processors. Until quite recently, each vendor has provided its own set of commands to support writing parallel programs. All these approaches were quite similar and consisted of directives for managing parallel code execution; i.e., loop parallelizing directives, locks, barriers, and other synchronization primitives etc., inserted into codes written in Fortran or C. It was only recently that OpenMP [46] emerged as a standard for code parallelization for shared memory parallel computers. While OpenMP provides support for three basic aspects of parallel computing:

- Specification of parallel execution
- Communicating between multiple thread
- Expressing synchronization between threads

and could be potentially used to support parallelism on any computer architecture, it is best suited for shared memory environments. OpenMP directives satisfy the following format:

```
!$omp directive name    optional clauses
```

Such approach allows to write the same code for both single-processor and multiprocessor platforms. Simply, compilers which do not support OpenMP directives or those that are working in a single-processor mode treat them as comments.

The OpenMP uses the fork-join model of parallel execution. A program starts execution as a single process, called the *master thread* of execution, and executes sequentially until the first parallel construct is encountered. Then the master thread spawns a specified number of "slave" threads and becomes a "master" of the team. All statements enclosed by the parallel construct are executed in parallel by each member of the team. Several directives accept clauses that allow a user to control the scope attributes of variables for the duration of the construct (i.e., shared, private, reduction, etc.). The following code can serve as an example of a simple parallel program [47]:

```
print *,'#procs='
```

```
      read *,p

      call omp_set_num_threads(p)

!$omp parallel shared(x,npoints) private(iam,np,ipoints)

      iam=omp_get_thread_num()

      np=omp_get_num_threads()

      ipoints=npoints/np

      call work_on_subdomain(x,iam,ipoints)

!$omp end parallel
```

Each thread in the parallel region determines what part of the global array x to work on. The code contains calls to OpenMP routines: omp_set_num_threads, which sets the number of threads in a parallel construct, omp_get_thread_num, which returns the number of a calling thread, and omp_get_num_threads, which returns the number of threads in a parallel region. In this code, each processor will execute the work_on_subdomain routine.

In the case of shared memory machines, the most common work-sharing construct within a parallel region is the *do-construct*, which distributes iterations of the do-loop among available working threads.

```
!$omp parallel do

      do j=1,n

          . . . . . . . .

      end do

!$omp end parallel do
```

The OpenMP compiler *do-construct* divides the iterations of a do-loop into subranges of the j index and hands them to different threads, which execute them in parallel.

Sometimes, parallelism can be expressed by means of the *sections* construct. In the example below, functions proc_one() and proc_two() are executed in parallel, which allows-high level parallelization based on heterogeneous tasks and enables the OpenMP written codes to be applied also beyond the simple loop parallelization.

```
!$omp parallel sections

!$omp section

      call proc_one()

!$omp section

      call proc_two()

!$omp end parallel sections
```

Finally, numerical integration can serve as a slightly more complicated example of utilizing OpenMP for parallel computing:

$$\int_a^b f(x)dx \approx h\left(\frac{f(x_0)}{2} + f(x_1) + \ldots + f(x_{n-1}) + \frac{f(x_n)}{2}\right) \tag{1.15}$$

where $h = (b - a)/n$ and $x_i = a + ih, i = 0, \ldots, n$. An OpenMP code to complete this task would look like this:

```
      h=(b-a)/n
 !$omp parallel do private(x) reduction(+:sum)
      do j=1,n-1
        x=a+j*h
        sum=sum+f(x)
      end do
 !$omp end parallel do
      sum=h*(sum+0.5*(f(a)+f(b)))
```

In this example, the `reduction(+:sum)` clause is used. It instructs the compiler that the variable sum is the target of a sum reduction operation.

The examples presented above as well as other examples are included to give the reader a general feeling for what a code written for a given environment looks like. For further details on OpenMP, consult Refs. [46–48]. Similarly, references should be consulted for the details concerning the remaining tools and environments.

1.6.4 DISTRIBUTED MEMORY PARALLELIZATION

As indicated above (Sections 1.3.5.2 and 1.4.3), parallelization for distributed memory computers typically consists of dividing the program into separate computational units that work independently and communicate by exchanging messages. To illustrate the basics of such an approach, consider a simple example of distributed computation of the vector norm

$$\|x\|_2 = \sqrt{\sum_{i=1}^{n} x_i^2}.$$

Assume that there exist integers p and q such that $pq = n$ and that the code is to be executed on a parallel computer. One of possible approaches to parallelizing the problem is based on the master–slave approach. Here, the program consists of two types of tasks: *master task* and *slave tasks* $T_i, i = 1, \ldots, p$. The following pseudo-code can illustrate this approach:

MASTER

1. Get n, x and choose p and q.
2. Spawn p slave tasks $T_i, i = 1, \ldots, p$.
3. Send q to all slave tasks T_i.
4. For $i = 1, \ldots, p$: send numbers $x_{(i-1)q+1}, \ldots, x_{iq}$ to T_i.
5. Set $sum \leftarrow 0$; for $i = 1, \ldots, p$: receive "partial sum" from a slave task and assign it to a variable s and $sum \leftarrow sum + s$.
6. Assign $sum \leftarrow \sqrt{sum}$

SLAVE

1. Receive from "MASTER" the value of q.
2. Receive from "MASTER" q numbers $y_1, \ldots, y_q$.
3. Calculate

$$s = \sum_{i=1}^{q} y_i^2.$$

4. Send s to "MASTER".

While the master–slave-based computing is only one of the possible approaches to distributed memory parallel computing, it illustrates the general idea of dividing work into independent tasks and coordinating computations through message passing. The two most popular environments supporting this mode of computing will now be considered.

1.6.4.1 PVM and MPI

Parallel virtual machine (PVM) has been developed in 1991 by a group of researchers at the University of Tennessee. It is a distributed-memory tool [49] designed to develop parallel applications on networks of heterogeneous computers. It allows to use such a heterogeneous environment as a single computational resource. PVM consists of a demon software that should be run on each node of the PVM, a console, and a library, which provides subroutines for process creation and message passing. Currently this library supports APIs for Fortran and C/C++. It should be noted that in its philosophy the PVM environment can be considered an "interactive" one. In its typical mode of operation, the PVM user works from the console, initializes the heterogeneous environment, instantiates the PVM daemons on all machines that participate in the virtual computer, and starts the execution of the main program. Experiments show that PVM sometimes do not work well in batch processing environments [20, 21, 50].

What follows is an example of the PVM code that calculates the integral (1.15) using the master–slave approach.

```
main() {
 int myid, nprocs, howmany;
 int tids[10];
 float a,b,h,lsum,tol;
 tids[0]=pvm_mytid();
 myid=pvm_joingroup("integral"); /* join the group */
 if (myid==0) /* i'am the first task - the master task*/
  {a=0.0;
   b=3.141569;
   printf("How many processors ? ");
   scanf("%d",&nprocs);
   /* spawn nprocs-1 tasks */
   howmany=pvm_spawn("int01",(char**)0,0,"*",nprocs-1,&tids[1]);
```

```
      /* wait for other processes */
      while(pvm_gsize("integral")!=nprocs)

         ;

      /* send data */
      pvm_initsend(PvmDataDefault);
      pvm_pkint(&nprocs,1,1);
      pvm_pkfloat(&a,1,1);
      pvm_pkfloat(&b,1,1);
      pvm_bcast("integral",10);

   }
else /* I'm a slave */
   { pvm_recv(-1,10);
     pvm_upkint(&nprocs,1,1);
     pvm_upkfloat(&a,1,1);
     pvm_upkfloat(&b,1,1);

   }
/* find the approximation */
 h=(b-a)/nprocs;
 a=a+h*myid;
 b=a+h;
 lsum=0.5*h*(f(a)+f(b));
/* gather partial results */
 pvm_reduce(PvmSum,&lsum,1,PVM_FLOAT,20,"integral",0);
 if (myid==0)
    { printf("%f\n",lsum);
    }
 pvm_barrier("integral",nprocs);
 pvm_exit();
}
```

It should be noted that in this as well as in the previous example of calculating the norm of the vector, the master is performing tasks that are completely different from the tasks of the slave processes. This does not need to be the case. If the computational workload of each of the slave processes is substantially larger than that of the master, the master should also perform some work (possibly a smaller overall amount than the slaves), while waiting for the slaves to complete their tasks. This approach, while slightly more difficult to implement, helps to improve the load balancing (the master is not idle) and thus results in a better overall performance (see Section 1.5.3).

Message passing interface (MPI) has been developed in 1993 by researchers from Argonne National Laboratory [51]. Over time, it has become a *de facto* standard for message passing parallel computing (superseding PVM, which is slowly becoming extinct). MPI provides an extensive set of communication subroutines including point-to-point communication, broadcasting and collective communication. It has been implemented on a variety of parallel computers including massively parallel computers, clusters, and networks of workstations. Due to its popularity, a number of open source and commercial tools and environments have been developed to support MPI-based parallel computing [52].

As an example, consider again the integration problem. However, this time, it will be solved using the SPMD model. Here each processor calculates its own part of the integral and then all of them exchange partial sums. At the end of the process, each processor calculates and contains its own local copy of the integral. While it may seem unreasonable to assume that each processor needs to calculate its own copy of the integral, one can assume that this is just a part of a larger code and these integral values are used by each processor independently in subsequent calculation.

```fortran
      integer ierr, myid, numprocs, rc, j
      real*8 x, sum, global_sum, a, b, my_a, my_b, my_h, h
*
      read *,a,b
      call MPI_INIT( ierr )
      call MPI_COMM_RANK( MPI_COMM_WORLD, myid, ierr )
      call MPI_COMM_SIZE( MPI_COMM_WORLD, numprocs, ierr )
      h=(b-a)/numprocs
      my_a=a+myid*h
      my_b=my_a+h
      sum=h*(sum+0.5*(f(my_a)+f(my_b)))
* collect all the partial sums
      call MPI_REDUCE(sum,global_sum,1,MPI_DOUBLE_PRECISION,
     MPI_SUM,0, $ MPI_COMM_WORLD,ierr)
* node 0 prints the answer.
      if (myid .eq. 0) then
        print *,'Result is ',global_sum
      endif
      call MPI_FINALIZE(rc)
      stop
      end
```

1.6.5   SHARED–DISTRIBUTED MEMORY ENVIRONMENTS

As indicated above (Section 1.3.5), there are a number of approaches to implementing parallel algorithms on shared–distributed memory computers. Such a parallelization can be done by treating such a computer as a virtual shared memory environment (and use, for instance, OpenMP), as a distributed memory machine (and use MPI). However, to be able to use the machine to the fullest extent, one may want to consider a mixed approach. A modification of the above code for the calculation of the integral that utilizes jointly MPI and OpenMP can illustrate such an approach.

```fortran
      parameter (n=1000)
      integer ierr, myid, numprocs, rc, j
      real*8 x, sum, global_sum, a, b, my_a, my_b, my_h, h
*

      read *,a,b
      call MPI_INIT( ierr )
      call MPI_COMM_RANK( MPI_COMM_WORLD, myid, ierr )
      call MPI_COMM_SIZE( MPI_COMM_WORLD, numprocs, ierr )
      my_h=(b-a)/numprocs
      my_a=a+myid*my_h
      my_b=my_a+my_h
      h=(my_b-my_a)/n
!$omp parallel do private(x) reduction(+:sum)
      do j=1,n-1
        x=my_a+j*h
        sum=sum+f(x)
      end do
!$omp end parallel do
      sum=h*(sum+0.5*(f(my_a)+f(my_b)))
* collect all the partial sums
      call MPI_REDUCE(sum,global_sum,1,MPI_DOUBLE_PRECISION,
     MPI_SUM,0, $ MPI_COMM_WORLD,ierr)
* node 0 prints the answer.
      if (myid .eq. 0) then
        print *,'Result is ',global_sum
      endif
      call MPI_FINALIZE(rc)
```

```
stop

end
```

Here, the division of the workload is initially done in the MPI environment. Then, on each shared memory computer, the OpenMP directives are used to perform loop parallelization. Obviously, such an approach is slightly more complicated and more tedious to implement but, as a result, the code matches the underlying hardware architecture, which may lead to further performance gains.

1.6.6 PARALLELIZATION OF EXISTING CODES

Consider the situation when an existing sequential code is to be parallelized. As stated above, at the present time one cannot count on the parallelizing compiler (or any other tools) to mechanize and substantially simplify the process of restructuring an old code for parallel processing. Most of the necessary work must be done manually and must be based on a clear understanding of the inner workings of the code. While each case has to be treated separately and will involve its own particular challenges, there are some general rules that follow from what has been said above.

First, the selection of the approach: while an OpenMP loop parallelization may be the best solution (and may be the easiest to implement) in the case of the code that is to run on a shared memory parallel computer, this approach may not be the best in the long run. It seems clear that in the near future most of the really large computers are going to be based on the distributed memory paradigm and may or may not consist of shared memory nodes. If so then the rational advice is to seriously consider an MPI-based parallelization. The MPI-based approach may be combined with the OpenMP when a shared–distributed memory hardware is to be used, but this may be an overkill not worth the effort. While there exist a number of other tools and environments that can be utilized, and each one of them has its own merits, it is arguable that if the goal of the project is to develop and apply software then researchers should stay with proven technology, which represents the state of the art at the time. For this and other reasons presented above, at this time Java does not seem to be ready in solving computationally intensive tasks.

Second, an important issue is the analysis of the existing code. Here the lessons learned from the application of Amdahl's Law play an important role. It follows from it that attention must be paid to the most computationally intensive and therefore most time-consuming parts of the code. It makes no difference whether a part of the code, which takes only 10% of the total time, is perfectly parallelized, if the remaining parts of the code are not. Therefore, the early stages of the code analysis should consist of benchmarking and time-profiling. As soon as a well-developed profile is created, it becomes clear where to focus further efforts.

Finally, one more area needs to be seriously considered. While an old code may be used for many years, there exists a possibility that this code, or its parts, can be replaced by the existing modules stored, among others, in the Netlib repository [53], the ACM TOMS [54] library, or announced, among others, on the NA Digest forum. Recent years have witnessed a rapid development of codes for the efficient parallel solution of a wide variety of mathematical problems. These codes have been implemented and tested on a number of parallel computers and, very often, they are very high quality both from the point of view of numerical properties as well as parallel performance. This approach can be called *library-based parallelization*. (In the next section, a discussion of software available from one of the more successful projects in the area of numerical linear algebra for dense matrices is presented.) It needs to be stressed, that the above discussion contains only a fragment of what is available; see the Chapter 5 for more examples of existing software. Overall, before an attempt is made to write a parallel code to solve a given problem, a thorough search should be conducted for existing software because chances are that a ready-to-use routines are already available.

1.6.7 LIBRARY-BASED PARALLELIZATION

There are a great many applications for which software has been implemented and which can be used when solving a particular computational problem. This can be done on two levels. First, there exist complete software packages that can be utilized to solve problems in parallel, i.e., parallel PDE solver ELLPACK [55, 56]. There exist also several packages designed for supporting parallel computing involving sparse matrices [57–59]. While some of these environments are definitely state-of-the-art and using them is preferable to developing the code oneself, there are problems that may not fit well enough into the existing software. There are also new algorithms that have to be implemented. In this case, there may exist libraries of "building blocks" that should be used in the process. What follows is a brief introduction to one of the more robust libraries supporting the development of high-performance and parallel codes involving matrix operations.

1.6.7.1 BLAS and LAPACK

In the area of linear algebraic computations for dense and band-structured matrices, there exists a *de facto* standard for writing high-performance software. More precisely, there exists a collection of interdependent software libraries that became the standard tool for dense and banded linear software implementation (even though the latter has been recently challenged by the work of Gustafson [60]).

The first step in the general direction took place in 1979, when the BLAS (basic linear algebra subroutines) standard was proposed [61]. Researchers realized that linear algebra software (primarily for dense matrices) consists of a number of basic operations (i.e., vector scaling, vector addition, dot product, etc.). These fundamental operations have been defined as a collection of Fortran 77 subroutines. The next two steps took place in 1988 and 1990, respectively, when the collection of matrix–vector and matrix–matrix operations have been defined [62, 63]. These two developments can be traced to the hardware changes happening at this time. The introduction of hierarchical memory structures resulted in the increasing need for the development of algorithms that would support data locality (move the *block* of data once, perform all the necessary operations on it and move a data back to the main memory and proceed with the next data block). It was established that to achieve this goal, one should rewrite linear algebra codes in terms of block operations and such operations can be naturally represented in terms of matrix–vector and matrix–matrix operations.

The BLAS routines were used in the development of linear algebra libraries that solved a number of the standard problems. Level 1 BLAS (vector-oriented operations) was used in the development of the LINPACK [64] and EISPACK [65] libraries devoted to the solution of linear systems and eigenproblems. The main advantages of these libraries were: the clarity and readability of the code, its portability as well as the possibility of hardware-oriented optimization of the BLAS kernels.

The next step was the development of the LAPACK library [66], which "combined" the functionalities available in the LINPACK and EISPACK libraries. LAPACK was based on utilizing level 3 BLAS kernels, while the BLAS 2 and 1 routines were used only when necessary. It was primarily oriented toward single-processor high-performance computers with vector processors (i.e., Cray, Convex) or with hierarchical memory (i.e., SGI Origin, HP Exemplar, DEC Alpha workstations, etc.). The LAPACK was also designed to work well with shared memory parallel computers, providing parallelization inside the level 3 BLAS routines [26]. Unfortunately, while their performance was very good for the solution of linear systems (this was also the data used at many conferences to illustrate the success of the approach), the performance of eigenproblem solvers (for both single-processor and parallel machines) was highly dependent on the quality of the underlying BLAS implementation and very unsatisfactory in many cases [34, 35].

In algebraic notation, the BLAS operations have the following form (detailed description of the BLAS routines can be found in Refs. [25, 66]):

Level 1: vector–vector operations:

- $y \leftarrow \alpha x + y$, $x \leftarrow \alpha x$, $y \leftarrow x$, $y \leftrightarrow x$, $dot \leftarrow x^{\mathrm{T}} y$, nrm2 $\leftarrow \|x\|_2$, asum $\leftarrow \|re(x)\|_1 + \|im(x)\|_1$.

Level 2: matrix–vector operations:

- matrix–vector products: $y \leftarrow \alpha A x + \beta y$, $y \leftarrow \alpha A^{\mathrm{T}} x + \beta y$
- rank-1 update of a general matrix: $A \leftarrow \alpha x y^{\mathrm{T}} + A$
- rank-1 and rank-2 update of a symmetric matrix: $A \leftarrow \alpha x x^{\mathrm{T}} + A$, $A \leftarrow \alpha x y^{\mathrm{T}} + \alpha y x^{\mathrm{T}} + A$
- multiplication by a triangular matrix: $x \leftarrow T x, x \leftarrow T^{\mathrm{T}} x$
- solving a triangular system of equations: $x \leftarrow T^{-1} x, x \leftarrow T^{-\mathrm{T}} x$

Level 3: matrix–matrix operations:

- matrix–matrix products: $C \leftarrow \alpha A B + \beta C, C \leftarrow \alpha A^{\mathrm{T}} B + \beta C, C \leftarrow \alpha A B^{\mathrm{T}} + \beta C,$ $C \leftarrow \alpha A^{\mathrm{T}} B^{\mathrm{T}} + \beta C$
- rank-k and rank-$2k$ update of a symmetric matrix: $C \leftarrow \alpha A A^{\mathrm{T}} + \beta C, C \leftarrow \alpha A^{\mathrm{T}} A + \beta C,$ $C \leftarrow \alpha A^{\mathrm{T}} B + \alpha B^{\mathrm{T}} A + \beta C, C \leftarrow \alpha A B^{\mathrm{T}} + \alpha B A^{\mathrm{T}} + \beta C$
- multiplication by a triangular matrix: $B \leftarrow \alpha T B, B \leftarrow \alpha T^{\mathrm{T}} B, B \leftarrow \alpha B T, B \leftarrow \alpha B T^{\mathrm{T}}$
- solving a triangular system of equations: $B \leftarrow \alpha T^{-1} B, B \leftarrow \alpha T^{-\mathrm{T}} B, B \leftarrow \alpha B T^{-1},$ $B \leftarrow \alpha B T^{-\mathrm{T}}$

Observe that each operation from the BLAS 2 library can be expressed in terms of BLAS 1 operations. Consider, for instance, a matrix–vector multiplication

$$y \leftarrow \alpha A x + \beta y \tag{1.16}$$

can be conceptualized in terms of a sequence of dot-products (routine _DOT), vector scalings (_SCAL) or vector updates (_AXPY).

$$
\begin{aligned}
z_k &\leftarrow A_{k*} x, \text{ for } k = 1, \ldots, m \quad &\text{(\_DOT)} \\
y &\leftarrow \beta y \quad &\text{(\_SCAL)} \\
y &\leftarrow y + \alpha z \quad &\text{(\_AXPY)}
\end{aligned}
\tag{1.17}
$$

Observe also that regardless of the fact that the above described algorithm and procedure (1.16) are equivalent, the application of the BLAS 2-based approach can substantially reduce the amount of processor–memory communication and thus reduce the overall execution time. Similarly, operations represented by the BLAS 3 routines can be expressed in terms of lower-level BLAS. For instance, operation

$$C \leftarrow \alpha A B + \beta C \tag{1.18}$$

can be expressed as

$$C_{*k} \leftarrow \alpha A B_{*k} + \beta C_{*k}, \qquad \text{for } k = 1, ..., n, \tag{1.19}$$

which is a sequence of operations denoted by Equation (1.16). It should be also noted that, as in the case of replacing level 1 BLAS operations by level 2 BLAS, the application of Equation (1.18) instead of Equation (1.19) reduces the total amount of processor–memory communication. More precisely, to illustrate the advantages of the application of higher level BLAS, consider the total number of arithmetical operations and the amount of data exchanged between the processor and memory. Table 1.1 [25] depicts the ratio of the number of processor–memory communications to the number of arithmetical operations for $m = n = k$.

The higher the level of BLAS, the more favorable the ratio becomes. (The number of operations performed on data increases relative to the total amount of data movement.) This has a particularly positive effect in the case of hierarchical memory computers (see also Section 1.3.2).

To illustrate that this course of action plays an important role not only for "supercomputers" but also for more "ordinary" architectures, Table 1.2 presents processing speed (in Mflops) achieved during the completion of the task $C \leftarrow \alpha A B + \beta C$ using BLAS 1, 2, and 3 kernels (utilizing algorithms 1.17 and 1.19) for $m = n = k = 1000$ on a single-processor PC with Intel Pentium III 866 MHz processor with 512 MB RAM.

There are two ways of using BLAS routines in parallel computing. First, very often, BLAS routines are parallelized by the computer hardware vendors. For instance, a call to the level 3 BLAS routine _GEMM may result in parallel execution of matrix–matrix multiplication. Any code that utilizes _GEMM will automatically perform this operation in parallel. While some computer vendors spend considerable amount of time and resources to deliver highly optimized BLAS kernels (routines accumulated in Cray's *scilib* library were one of the best in delivered performance, while currently IBM's *ESSL* library is also very well optimized), this does not have to be the case. In addition, only some of BLAS kernels are parallelized (one of the typical and very important exceptions are routines for symmetric matrices stored in a compact form) and

TABLE 1.1
BLAS: Memory References and Arithmetic Operations

BLAS	Loads and stores	Flops	Ratio
$y \leftarrow y + \alpha x$	$3n$	$2n$	$3 : 2$
$y \leftarrow \alpha A x + \beta y$	$mn + n + 2m$	$2m + 2mn$	$1 : 2$
$C \leftarrow \alpha A B + \beta C$	$2mn + mk + kn$	$2mkn + 2mn$	$2 : n$

TABLE 1.2
Matrix Multiplication on the Pentium III 866 MHz

	Mflops	sec.
BLAS 1	93.81	21.32
BLAS 2	355.87	5.62
BLAS 3	1418.44	1.41

they are typically parallelized for shared memory environments only (for an example of problems encountered in parallelization of BLAS kernels, see Ref. [67]). In short, parallel performance of BLAS routines cannot be taken for granted (especially since they are primarily optimized for single-processor performance in the hierarchical memory environment). Taking this into account BLAS kernels should be rather utilized to develop parallel programs where the BLAS routines will run on separate processors. To illustrate the main idea behind such an approach, consider matrix update procedure based on the formula $C \leftarrow \alpha AB + \beta C$, which can be rewritten as:

$$
\begin{pmatrix} C_{11} & C_{12} \\ C_{21} & C_{22} \end{pmatrix} = \alpha \begin{pmatrix} A_{11} & A_{12} \\ A_{21} & A_{22} \end{pmatrix} \begin{pmatrix} B_{11} & B_{12} \\ B_{21} & B_{22} \end{pmatrix} + \beta \begin{pmatrix} C_{11} & C_{12} \\ C_{21} & C_{22} \end{pmatrix} \tag{1.20}
$$

Applying the definition of matrix multiplication, the following block algorithm to calculate matrix C is obtained.

$$
\begin{array}{ll}
C_{11} \leftarrow \alpha A_{11} B_{11} + \beta C_{11} & /1/ \\
C_{11} \leftarrow \alpha A_{12} B_{21} + C_{11} & /2/ \\
C_{12} \leftarrow \alpha A_{11} B_{12} + \beta C_{12} & /3/ \\
C_{12} \leftarrow \alpha A_{12} B_{22} + C_{11} & /4/ \\
C_{21} \leftarrow \alpha A_{22} B_{21} + \beta C_{21} & /5/ \\
C_{21} \leftarrow \alpha A_{21} B_{11} + C_{21} & /6/ \\
C_{22} \leftarrow \alpha A_{22} B_{22} + \beta C_{22} & /7/ \\
C_{22} \leftarrow \alpha A_{21} B_{12} + C_{22} & /8/
\end{array} \tag{1.21}
$$

Observe that this algorithm allows for parallel execution of operations /1/, /3/, /5/, /7/ and in the next phase of operations /2/, /4/, /6/, /8/. Obviously, it is desirable to divide large matrices into a larger number of blocks (i.e., to match the number of available processors). The order of operations may also need to be adjusted to reduce the memory access conflicts. The application of this approach can be illustrated by the block-Cholesky method for solving systems of linear equations for symmetric positive definite matrices.

It is well known that there exists a unique decomposition for such matrices

$$
A = LL^{\mathrm{T}}, \tag{1.22}
$$

where L is a lower triangular matrix. There exists also a simple algorithm for determining the matrix L with an arithmetical complexity of $O(n^3)$. Its analysis allows one to see immediately that it can be expressed in terms of calls to the level 1 BLAS. Consider how it can be translated into block operations expressed in terms of level 3 BLAS. Formula (1.22) can be rewritten in the following way:

$$
\begin{pmatrix} A_{11} & A_{12} & A_{13} \\ A_{21} & A_{22} & A_{23} \\ A_{31} & A_{32} & A_{33} \end{pmatrix} = \begin{pmatrix} L_{11} & & \\ L_{21} & L_{22} & \\ L_{31} & L_{32} & L_{33} \end{pmatrix} \begin{pmatrix} L_{11}^{\mathrm{T}} & L_{21}^{\mathrm{T}} & L_{31}^{\mathrm{T}} \\ & L_{22}^{\mathrm{T}} & L_{32}^{\mathrm{T}} \\ & & L_{33}^{\mathrm{T}} \end{pmatrix} \tag{1.23}
$$

Thus

$$
A = \begin{pmatrix} L_{11}L_{11}^{\mathrm{T}} & L_{11}L_{21}^{\mathrm{T}} & L_{11}L_{31}^{\mathrm{T}} \\ L_{21}L_{11}^{\mathrm{T}} & L_{21}L_{21}^{\mathrm{T}} + L_{22}L_{22}^{\mathrm{T}} & L_{21}L_{31}^{\mathrm{T}} + L_{22}L_{32}^{\mathrm{T}} \\ L_{31}L_{11}^{\mathrm{T}} & L_{31}L_{21}^{\mathrm{T}} + L_{32}L_{22}^{\mathrm{T}} & L_{31}L_{31}^{\mathrm{T}} + L_{32}L_{32}^{\mathrm{T}} + L_{33}L_{33}^{\mathrm{T}} \end{pmatrix}
$$

After the decomposition $A_{11} = L_{11}L_{11}^{\mathrm{T}}$, which is the same decomposition as the original one but of smaller size, appropriate BLAS 3 kernels can be applied in parallel to calculate matrices L_{21} and L_{31} by applying in parallel equalities $A_{21} = L_{21}L_{11}^{\mathrm{T}}$ and $A_{31} = L_{31}L_{11}^{\mathrm{T}}$. In the next step, equation

$$A_{22} = L_{21}L_{21}^{\mathrm{T}} + L_{22}L_{22}^{\mathrm{T}},$$

can be used. Thus the decomposition $L_{22}L_{22}^{\mathrm{T}}$ for the matrix $A_{22} - L_{21}L_{21}^{\mathrm{T}}$ is used to calculate the matrix L_{22}. Finally, L_{32} can be found from

$$L_{32} = (A_{32} - L_{31}L_{21}^{\mathrm{T}})(L_{22}^{\mathrm{T}})^{-1}.$$

In a similar way, subsequent block columns of the decomposition can be calculated. Finally, it should be noted that the parallelization of the matrix multiplication presented here as well as the Cholesky decomposition are examples of the divide-and-conquer method, which is one of the popular approaches to algorithm parallelization.

1.6.7.2 BLACS, PBLAS, and ScaLAPACK

At the time when the LAPACK project was completed, it became clear that there is a need to develop similar software to solve linear algebraic problems on distributed memory architectures. Obviously, this could have been done "by hand" using level 3 BLAS kernels and a software environment like PVM or MPI. However, this would have made such an approach dependent on their existence and backward compatibility. Since PVM is already slowly disappearing, while imposing strict backward compatibility on MPI may be holding it to too high a standard, the decision not to follow this path seems to be very good indeed. It has led in the first place to the development of the basic linear algebra communication subroutines (BLACS), a package that defines portable and machine-independent collection of communication subroutines for distributed memory linear algebra operations [68, 69]. The essential goals of BLACS are:

- Simplifying message passing in order to reduce programming errors
- Providing data structures to simplify at the level of matrices and their subblocks
- Portability across a wide range of parallel computers, including all distributed memory parallel machines and heterogenous clusters

In the BLACS, each process is treated as if it were a processor — it must exists for the lifetime of the BLACS run and its execution can affect other processes only through the use of message passing. Processes involved in the BLACS execution are organized in two-dimensional grids and each process is identified by its coordinates in a grid. For example, if a group of consists N_p processes then the grid will have P rows and Q columns, where $P \cdot Q = N_g \leq N_p$. A process can be referenced by its coordinates (p, q), where $0 \leq p < P$ and $0 \leq q < Q$.

The BLACS provides structured communication in a grid. Processes can communicate using the point-to-point paradigm or it is possible to organize communication (broadcasts) within a "scope" which can be a row or a column of a grid, or even the whole grid. Moreover, the performance of communication can be improved by indicating a particular hardware topology [68].

```
      integer iam, nprocs, cntx, nrow, ncol, myrow, mycol, i, j

      integer lrows, lcols

      real*8 a(3), h

*

      call blacs_pinfo(iam,nprocs)
```

```
      if (nprocs.eq.-1) then
        if (iam.eq.0) then
          print *,'How many processes ?'
          read *, nprocs, a(1), a(2)
        end if
        call blacs_setup(iam,nprocs)
      end if
*  determine grid size
      lrows=int(sqrt(real(nprocs)))
      lcols=lrows
*  init the grid
      call blacs_get(0,0,cntx)
      call blacs_gridinit(cntx,'C',lrows,lcols)
      call blacs_gridinfo(cntx,nrow,ncol,myrow,mycol)
*  broadcast or receive
      if ((myrow.eq.0).and.(mycol.eq.0)) then
        call dgebs2d(cntx,'A',' ',2,1,a,4)
      else
        call dgebr2d(cntx,'A',' ',2,1,a,4,0,0)
      end if
*

      h=(a(2)-a(1))/real(nprocs)
      a(1)=a(1)+real(iam)*h
      a(2)=a(1)+h
      a(3)=0.5*h*(f(a(1))+f(a(2)))
      call dgsum2d(cntx,'A',' ',1,1,a(3),4,0,0)
      if ((myrow.eq.0).and.(mycol.eq.0)) then
        print *,'result is ',a(3)
      end if
      call blacs_barrier(cntx,'A')
      call blacs_exit(0)
```

This program is analogous to the program presented in Section 1.6.4.1. This time, however, the communication infrastructure is expressed in terms of calls to BLACS routines. This illustrates the fact that while geared toward linear algebra, the BLACS can be also utilized as a general set of communication routines. It should be also noted that several version of BLACS were implemented based on PVM, MPI, and vendor-provided message passing routines.

The parallel basic linear algebra communication subprograms (PBLAS) [70] is a set of distributed vector–vector, matrix–vector, and matrix–matrix operations (analogous to the sequential BLAS) with the aim of simplifying the parallelization of linear algebra programs. The basic idea of PBLAS is to distribute matrices among distributed processors (i.e., BLACS processes) and utilize BLACS as the communication infrastructure. The general class of such distributions can be obtained by matrix partitioning like

$$
A = \begin{pmatrix} A_{11} & \ldots & A_{1m} \\ \vdots & & \vdots \\ A_{m1} & \ldots & A_{mm} \end{pmatrix},
$$

where each subblock A_{ij} is $n_b \times n_b$. These blocks are mapped to processes by assigning A_{ij} to the process whose coordinates in a grid are

$$
((i-1)\bmod P, (j-1)\bmod Q).
$$

Finally, ScaLAPACK is a library of high-performance linear algebra routines for distributed-memory message-passing MIMD computers and networks of heterogeneous computers [71]. It provides the same functionality as LAPACK for workstations, vector supercomputers, and shared-memory parallel computers. As LAPACK was developed by utilizing calls to the BLAS routines, ScaLAPACK is based on calls to the BLACS and PBLAS kernels.

Summarizing, the current state of the art of both compiler-based and language-based parallelization is such that neither can be relied on when considering efficient implementation of parallel algorithms. In the best case, the optimizing compiler should be able to fine-tune the hand-parallelized code to match the low level parallelism available in the hardware and to match various detailed hardware parameters of a given high-performance computer (i.e., processor characteristics, structure, sizes, and latencies of various levels of cache memories, etc.). When considering parallelization of an existing code to be executed on one of the available parallel computers, the availability and popularity of environments supporting such a process should be selected. At present, the best choices seem to be OpenMP, MPI, or a combination of them. Attention needs to be paid to the large and constantly growing number of existing libraries of modules out of which parallel programs can be assembled as well as to complete problem-solving environments that can be applied to efficiently find the solution.

1.7 CONCLUDING REMARKS

A few predictions for the future of parallel scientific computing can be risked on the basis of the above summary of the state of the art in parallel computing. The development of computer architectures is clearly pointing out to the increasing importance of parallel computing. The newest processors, which are about to become the standard for workstations, i.e., the Itanium architecture [72] from Intel or the Opteron and Athlon 64 architectures from AMD [73], involve a continuous increase of internal complexity (i.e., ever more sophisticated branch prediction), increased word size to 64 bits and introduce various forms of threading [72]. Each of these factors increases the total number of instructions that will be executed inside of the processor at any given time. While the available software tools (i.e., the optimizing compilers) are lagging behind the advances in computer hardware, their abilities are steadily improving and they should be able to support microparallelization successfully as well as to handle hierarchical memory latencies for a given architecture.

For smaller and medium size problems (where these notions are dynamic and their extensions change together with hardware capabilities), workstations with multiple processors and global shared memory become very popular. At the present time, dual-processor desktop computers have become so popular that their applicability to video processing and multimedia production has been thoroughly analyzed in the *AV Video Multimedia Producer* magazine [74]. This indicates clearly that parallel computing "has reached the masses." At the same time, on the high end of parallel computing, further substantial increase in computational power is about to take place. The three leading projects are: a 40 Tflop computer from Cray installed at Sandia National Laboratory in 2004 (Red Storm Project), the 100 Tflop ASCI Purple (consisting of 12,544 processors), and the 13,1072 processor BlueGene/L computer capable of peak performance of 360 Tflops. The latter machines will be built by IBM and installed in 2005 in Lawrence Livermore National Laboratory [75].

In the context of scientific computing, parallelization can be viewed on multiple levels that are nicely illustrated by the dense linear algebra software discussed above. On the low level, highly optimized building blocks will continue to be developed (i.e., BLAS kernels) with optimization coming from the hardware vendors or from research projects such as the ATLAS project. These building blocks will be combined into software libraries (i.e., ScaLAPACK). They will be also utilized in the development of environments designed for the solution of a class of problems (i.e., eigensolvers for complex symmetric non-Hermitian matrices, parallel solutions to various classes of constraint optimization problems, large-scale data mining problems, etc.).

For shared-memory parallel computers, it can be expected that tools similar to OpenMP will remain a standard for software writing, while the language extensions and parallelism supporting languages like Java will take some time to reach the required level of efficiency for the solution of larger problems. Both these approaches will be used in building software libraries, or will be combined with the use of software modules stored in various libraries to solve real-life applications. Due to the relative simplicity of the underlying architecture and a substantial body of knowledge about writing software for shared-memory computers that has been accumulated over more than 20 years of their existence, it will be possible to achieve a high level of efficiency relatively easily.

The situation will be slightly more complicated in the case of distributed memory environments (clusters and top-of-the-line supercomputers). Here, the distributed computing model based on message passing is the most likely to remain the standard for software writing and tools like the MPI (which has evolved into MPI 2 [76]) are most likely to be used to support it. Outside of the simpler well-structured problems that are easily amenable to parallelization, and afford high levels of efficiency, a lot of work in performance tuning will be required. In some cases, this work may be sometimes avoided due to the fact that the wall-clock problem-solving time is the most important for the user. The users may accordingly be willing to forgo the extra effort in code tuning and instead add more or faster hardware thus achieving the required or acceptable solution time. However, it is very likely that the solution of extremely large problems on computers with multiple layers of latencies (in current supercomputers of the ASCI project at least seven levels of latency can be accounted for) will still be mostly done "by hand" with the solution being individually developed for each particular problem to match the underlying hardware architecture.

Another approach, which also follows the general reasoning that more hardware can be substituted for fine tuning and "local" efficiency, is the grid. While it is unclear whether this is the computing paradigm for the future (many think so and are likely to be correct, but the real applications have not materialized yet), clearly there exist classes of computational problems that even today can benefit from the grid-like architecture and the availability of unused computational power. Any problem that can be divided into a number of relatively small (computational time between a few hours and a day on the weakest machines that are a part of the grid) and completely independent tasks that also has a favorable ratio of computation to communication (a small amount of data is transferred across the grid while a large amount of computation is then applied to it) is a definite candidate for a grid-based solution. Due to the steadily increasing network bandwidth and growing amount of available unused computational power, a very large amount of research is devoted to this area (with large companies such as IBM becoming seriously financially involved in such efforts). It can be therefore expected that while the fruits of the research will be available some time in the future, this is and will remain for some time to come one of the hottest research areas in distributed computing. In this context, it is worth mentioning that in addition to the focus on the plain computational power and efficiency, the grid also involves attempts of adding intelligence to problem solving. Here it is assumed that large problems will have separate parts that may run best on different computers. For instance, a part of the problem may work well on a shared memory vector machine, while another part may be matched with a cluster computer. The environment is to recognize this and try to use this information to optimize the solution process. One of the more interesting projects in this group is the NetSOLVE environment [77–79].

While some of predictions presented here may not materialize, one thing is for certain, parallel and distributed computing is taking over the world of computing — it is here to stay and to grow.

REFERENCES

[1] http://www.netlib.org/benchmark/performance.ps (accessed May 5, 2003).

[2] Paprzycki, M. (1992). Parallel matrix multiplication — can we learn anything new? *CHPC Newslett.* 7: 55–59.

[3] Paprzycki, M. and Cyphers, C. (1991). Multiplying matrices on the cray — practical considerations. *CHPC Newslett.* 6: 43–47.

[4] Paprzycki, M. and Cyphers, C. (1991). Gaussian elimination on Cray Y-MP. *CHPC Newslett.* 6: 77–82.

[5] Dongarra, J., Gustavson, F., and Karp, A. (1984). Implementing linear algebra algorithms for dense matrices on a vector pipeline machine. *SIAM Rev.* 26: 91–112.

[6] Whaley, R.C., Petitet, A., and Dongarra, J.J. (2001). Automated empirical optimizations of software and the ATLAS project. *Parallel Comput.* 27: 3–35.

[7] Flynn, M. (1972). Some computer organizations and their effectiveness. *IEEE Trans. Comput.* C–21: 94.

[8] Reddaway, S.F., Bowgen, G., and Berghe, S.V.D. (1989). High performance linear algebra on the AMT DAP 510. In Rodrigue, G., (ed.) *Proceedings of the 3rd Conference on Parallel Processing for Scientific Computing*, SIAM Publishers, Philadelphia, PA, pp. 45–49.

[9] http://www.beowulf.org (accessed May 5, 2003).

[10] Hochmuth, P. (2003). Gushing over linux. *NetworkWorld* 4: 46.

[11] Foster, I. and Kesselman, C. (1997). Globus: A metacomputing infrastructure toolkit. *Int. J. Supercomputer Applic. High Performance Comput.* 11: 115–128.

[12] Brodsky, I. (2003). Scale grid computing down to size. *NetworkWorld* January 27: 47.

[13] Burt, J. (2003). Grid puts supercomputing at enterprises' fingertips. *eWeek* January 20: 32.

[14] Franklin, C., Jr. (2003). Grid-dy determination. *NetworkWorld* June 1: 43, 46.

[15] Marsan, C.D. (2003). Grid vendors target corporate applications. *NetworkWorld* January 27: 26.

[16] Taft, D.K. (2003). IBM bolsters grid computing line. *eWeek* February 3: 24.

[17] Regan, K. Unsold gateway PCs to serve as on-demand grid network. http://www.techextreme.com/perl/story/20230.html (accessed May 5, 2003).

[18] Musgrove, M. (2002). Computers' shelf life gets livelier. *Washington Post* December, 10.

[19] *Intel Architecture Optimization.* Reference Manual. Intel Corp. (1999).

[20] Paprzycki, M., Lirkov, I., Margenov, S., and Owens, R. (1998). A shared memory parallel implementation of block-circulant preconditioners. In: Griebel, M. et al., (eds.), *Large Scale Scientific Computations of Engineering and Environmental Problems*, pp 319–327.

[21] Paprzycki, M., Lirkov, I., and Margenov, S. (1998). Parallel solution of 2d elliptic PDE's on silicon calgary, graphics supercomputers. In Pan, Y. et al. (eds.), *Proceedings of the 10th International Conference on Parallel and Distributed Computing and Systems*, IASTED/ACTA Press, Canada, pp. 575–580.

[22] Ekersand, R., Cullers, K., Billingham, J., and Scheffer, L. (2002). *A Roadmap for the Search for Extraterrestrial Intelligence*. SETI Press, Berkeley, CA.

[23] Hockney, R., and Jesshope, C. (1981). *Parallel Computers: Architecture, Programming and Algorithms*. Adam Hilger Ltd., Bristol.

[24] Stpiczyński, P., and Paprzycki, M. (2000). Fully vectorized solver for linear recurrence systems with constant coefficients. In: *Proceedings of VECPAR 2000 — 4th International Meeting on Vector and Parallel Processing*, Porto, June 2000, Facultade de Engerharia do Universidade do Porto, pp. 541–551.

[25] Dongarra, J., Duff, I., Sorensen, D., and Van der Vorst, H. (1991). *Solving Linear Systems on Vector and Shared Memory Computers*. SIAM Publishers, Philadelphia, PA.

[26] Dongarra, J., Duff, I., Sorensen, D., and Van der Vorst, H. (1998). *Numerical Linear Algebra for High Performance Computers*. SIAM Publishers, Philadelphia, PA.

[27] Paprzycki, M., and Stpiczyński, P. (1996). Parallel solution of linear recurrence systems. *Z. Angew. Math. Mech.* 76: 5–8.

[28] Stpiczyński, P. (2002). A new message passing algorithm for solving linear recurrence systems. *Lecture Notes Comput. Sci.* 2328: 466–473.

[29] Gustafson, J. (1988). Reevaluating Amdahl's law. *Comm. ACM* 31: 532–533.

[30] Gustafson, J., Montry, G., and Benner, R. (1988). Development of parallel methods for a 1024-processor hypercube. *SIAM J. Sci. Stat. Comput.* 9: 609–638.

[31] Wolfe, M. (1996). *High Performance Compilers for Parallel Computing*. Addison–Wesley, Boston, MA.

[32] Zima, H. (1990). *Supercompilers for Parallel and Vector Computers*. ACM Press, Washington, DC.

[33] Supercomputer Prospectives — 4th Jerusalem Conference on Information Technology, Jerusalem. IEEE Computer Society Press (1984), Los Alamitos, CA.

[34] Bar-On, I., and Paprzycki, M. (1998). High performance solution of complex symmetric eigenproblem. *Numerical Algorithms* 18: 195–208.

[35] Bar-On, I., and Paprzycki, M. (1998). A fast solver for the complex symmetric eigenproblem. *Computer Assisted Mechanics Eng. Sci.* 5: 85–92.

[36] Adams, J., Brainerd, W., Martin, J., Smith, B., and Wagner, J. (1997). *Fortran 95 Handbook*. MIT Press.

[37] Metcalf, M., and Reid, J. (1999). *Fortran 90/95 Explained*. Oxford University Press, Oxford.

[38] Koelbel, C., Loveman, D., Schreiber, R., Jr, Steele, G., and Zosel, M. (1994). *The High Perhormance Fortran Handbook*. MIT Press, Cambridge, MA.

[39] Allan, S.J., and Oldehoeft, R.R. (1985). HEP SISAL: Parallel functional programming. In Kowalik, J.S. (ed.), *Parallel MIMD Computation: HEP Supercomputer and Its Applications. Scientific Computation Series*. MIT Press, Cambridge, MA, pp. 123–150.

[40] Gurd, J.R. (1985). The manchester dataflow machine. In Duff, I.S. and Reid, J.K. (eds.), *Vector and Parallel Processors in Computational Science*. North-Holland, Amsterdam, pp. 49–62.

[41] Cann, D.C., Feo, J., and DeBoni, T. (1990). SISAL 1.2: High-performance applicative computing. In: *Proceedings of the 2nd IEEE Symposium on Parallel and Distributed Processing* (2nd SPDP'90), Dallas, TX.

[42] Bollman, D., Sanmiguel, F., and Seguel, J. (1992). Implementing FFTs in SISAL. In Feo, J.T., Frerking, C., and Miller, P.J. (eds.), *Proceedings of the Second SISAL User's Conference*, Livermore, pp. 59–65.

[43] Paprzycki, M. and Zalewski, J. (1997). Parallel computing in Ada: An overview and critique. *Ada Lett.* 17: 62–67.

[44] Carpenter, B. (1998). Toward a Java environment for SPMD programming. *Lecture Notes Comput. Sci.* 1470: 659–668.

[45] Ferrari, A. (1998). JPVM: Network parallel computing in Java. *Concurrency: Practice and Experience* 10.

[46] Chandra, R., Dagum, L., Kohr, D., Maydan, D., McDonald, J., and Menon, R. (2001). *Parallel Programming in OpenMP*. Morgan Kaufmann Publishers, San Francisco, CA.

[47] OpenMP Fortran application program interface. http://www.openmp.org (accessed May 5, 2003).

[48] OpenMP C and C++ application program interface. http://www.openmp.org (accessed May 5, 2003).

[49] Dongarra, J. et al. (1994). *PVM: A User's Guide and Tutorial for Networked Parallel Computing*. MIT Press, Cambridge, MA.

[50] Paprzycki, M., Hope, H., and Petrova, S. (1998). Parallel performance of a direct elliptic solver. In Griebel, M. et al., (eds.), *Large Scale Scientific Computations of Engineering and Environmental Problems*, VIEWEG, Wisbaden, pp. 310–318.

[51] Pacheco, P. (1996). *Parallel Programming with MPI*. Morgan Kaufmann, San Francisco, CA.

[52] Gropp, W. and Lusk, E. *A User's Guide for mpich, a Portable Implementation of MPI*, http://www-unix.mcs.anl.gov/mpi (accessed May 5, 2003).

[53] http://www.netlib.org (accessed May 5, 2003).

[54] ACM Trasaction on Mathematical Software. http://www.acm.org/toms (accessed May 5, 2003).

[55] Houstis, E.N., Rice, J.R., Chrisochoides, N.P., Karathanasis, H.C., Papachiou, P.N., Samartizs, M.K., Vavalis, E.A., Wang, K.Y., and Weerawarana, S. (1990). ELLPACK: A numerical simulation programming environment for parallel MIMD machines. In: *Proceedings 1990 International Conference on Supercomputing*, ACM SIGARCH Computer Architecture News, pp. 96–107.

[56] Rice, J.R. (1978). Ellpack 77 user's guide. Technical Report CSD–TR 226, Purdue University, West Lafayette.

[57] Eisenstat, S.C., Gursky, M.C., Schultz, M.H., and Sherman, A.H. (1982). Yale Sparse Matrix Package (YSMP) — I : The symmetric codes. *Int. J. Numer. Meth. in Eng.* 18: 1145–1151.

[58] George, A. and Ng, E. (1984). SPARSPAK : Waterloo sparse matrix package "User's Guide" for SPARSPAK-B. Research Report CS-84-37, Deptartment of Computer Science, University of Waterloo.

[59] Zlatev, Z., Wasniewski, J., and Schaumburg, K. (1981). Y12M solution of large and sparse systems of linear algebraic equations: documentation of subroutines. *Lecture Notes in Computer Science*, Vol. 121, Springer-Verlag, New York.

[60] Gustavson, F.G. (2002). New generalized data structures for matrices lead to a variety of high performance algorithms. *Lecture Notes Comput. Sci.* 2328: 418–436.

[61] Lawson, C., Hanson, R., Kincaid, D., and Krogh, F. (1979). Basic linear algebra subprograms for fortran usage. *ACM Trans. Math. Soft.* 5: 308–329.

[62] Dongarra, J., DuCroz, J., Hammarling, S., and Hanson, R. (1988). An extended set of fortran basic linear algebra subprograms. *ACM Trans. Math. Soft.* 14: 1–17.

[63] Dongarra, J., DuCroz, J., Duff, I., and Hammarling, S. (1990). A set of level 3 basic linear algebra subprograms. *ACM Trans. Math. Soft.* 16: 1–17.

[64] Dongarra, J., Bunsch, J., Moler, C., and Steward, G. (1979). *LINPACK User's Guide*. SIAM Publishers, Philadelphia, PA.

[65] Garbow, B., Boyle, J., Dongarra, J., and Moler, C. (1977). Matrix eiigensystems routines — EISPACK guide extension. *Lecture Notes in Computer Science*. Springer-Verlag, New York.

[66] Anderson, E., Bai, Z., Bischof, C., Demmel, J., Dongarra, J., Du Croz, J., Greenbaum, A., Hammarling, S., McKenney, A., Ostruchov, S., and Sorensen, D. (1992). *LAPACK User's Guide*. SIAM Publishers, Philadelphia, PA.

[67] Bar-On, I. and Paprzycki, M. (1997). A parallel algorithm for solving the complex symmetric eigenproblem. In Heath, M. et al. (eds.), *Proceedings of the SIAM Conference on Parallel Processing for Scientific Computing*, SIAM Publishers, Philadelphia, PA.

[68] Dongarra, J.J., and Whaley, R.C. (1997). LAPACK working note 94: A user's guide to the BLACS v1.1. http://www.netlib.org/blacs (accessed May 5, 2003).

[69] Whaley, R.C. (1994). LAPACK working note 73: Basic linear communication algebra subprograms: Analysis and implementation across multiple parallel architectures. http://www.netlib.org/blacs (accessed May 5, 2003).

[70] Choi, J., Dongarra, J., Ostrouchov, S., Petitet, A., Walker, D., and Whaley, R. (1995). LAPACK working note 100: A proposal for a set of parallel basic linear algebra subprograms. http://www.netlib.org/lapack/lawns (accessed May 5, 2003).

[71] Blackford, L. et al. (1997). *ScaLAPACK User's Guide*. SIAM Publishers, Philadelphia, PA.

[72] *Intel Itanium Architecture Software Developer's Manual*. Intel Press, Santa clara, CA.

[73] http://www.amd.com/us-en/Weblets/0,,7832_8366,00.html (accessed May 5, 2003).

[74] Gustavson, F.G. (2003). Waiting for your chip to come in. *AV Video Multimedia Producer* February 14:
      16.

[75] Cipra, B.A. (2003). Sc2002: A terable time for supercomputing. *SIAM News* 36.

[76] http://www.mpi-forum.org/docs/docs.html (accessed May 5, 2003).

[77] Casanova, H., Dongarra, J. (1997). NetSolve: A network-enabled server for solving computational
      science problems. *Int. J. Supercomput. Applic. High Performance Computing*, 11: 212–223.

[78] Casanova, H. and Dongarra, J. (1998). Applying NetSolve's network-enabled server. *IEEE Computa-
      tional Sc. Eng.* 5: 57–67.

[79] Plank, J.S., Casanova, H., Beck, M., and Dongarra, J.J. (1999). Deploying fault-tolerance and task
      migration with NetSolve. *Future Generation Comput. Sys.*, 15: 745–755.

FURTHER READING

[80] Dongarra, J. and Johnsson, L. (1987). Solving banded systems on parallel processor. *Parallel Computing*
      5: 219–246.

[81] Gallivan, K., Heath, M., Ng, E., Ortega, J., Peyton, B., Plemmons, R., Romine, C., Sameh, A., and
      Voight, R. (1991). *Parallel Algorithms for Matrix Computations*. SIAM Publishers, Philadelphia, PA.

[82] Ortega, J. and Voight, R. (1985). *Solution of Partial Differential Equations on Vector and Parallel Com-
      puters*. SIAM Publishers, Philadelphia, PA.

[83] Van Loan, C. (1992). *Computational Frameworks for the Fast Fourier Transform*. SIAM Publishers,
      Philadelphia, PA.

[84] Demmel, J.W. (1997). *Applied Numerical Linear Algebra*. SIAM Publishers, Philadelphia, PA.

[85] Trefethen, L.N., Bau, D. (1997). *Numerical Linear Algebra*. SIAM Publishers, Philadelphia, PA.

[86] Higham, N.J. (1996). *Accuracy and Stability of Numerical Algorithms*. SIAM Publishers, Philadelphia,
      PA.

[87] Modi, J. (1988). *Parallel Algorithms and Matrix Computation*. Oxford University Press, Oxford.

[88] Heller, D. (1978). A survey of parallel algorithms in numerical linear algebra. *SIAM Rev.* 20: 740–777.

[89] Dongarra, J., and Walker, D. (1995). Software libraries for linear algebra computations on high perfor-
      mance computers. *SIAM Rev.* 37: 73–83.

[90] Brainerd, W., Goldbergs, C., and Adams, J. (1990). *Programmers Guide to Fortran 90*. McGraw-Hill,
      New York.

[91] Lakshmivarahan, S. and Dhall, S.K. (1990). *Analysis and Design of Parallel Algorithms: Algebra and
      Matrix Problems*. McGraw-Hill, New York.

[92] Smith, B.T., Boyle, J.M., Dongarra, J.J., Garbow, B.S., Ikebe, Y., Klema, V.C., and Moler, C.B. (1976).
      Matrix Eigensystem Routines: EISPACK Guide. *Lecture Notes in Computer Science*. Vol. 6, Springer-
      Verlag, New York.

[93] Duff, I.S., Heroux, M.A., and Pozo, R. (2002). An overview of the Sparse Basic Linear Algebra Sub-
      programs: The new standard from the BLAS Technical Forum. *ACM Trans. Mathematical Software* 28:
      239–267.

[94] Amestoy, P.R., Daydé, M., and Duff, I.S. (1989). Use of level 3 BLAS in the solution of full and sparse
      linear equations. In Delhaye, J.L., and Gelenbe, E., (eds.), *High Performance Computing: Proceedings
      of the International Symposium on High Performance Computing*, Montpellier, France, 22–24 March,
      1989, Amsterdam, The Netherlands, North-Holland, Amerstdam, pp. 19–31.

[95] Daydé, M.J. and Duff, I.S. (1999). The RISC BLAS: a blocked implementation of level 3 BLAS for
      RISC processors. *ACM Trans. Mathematical Software* 25: 316–340.

[96] Beebe, N.H.F. (2001). EISPACK: numerical library for eigenvalue and eigenvector solutions.

[97] Guyer, S.Z., Lin, C. (2000). An annotation language for optimizing software libraries. *ACM SIGPLAN
      Notices* 35: 39–52.

[98] van de Geijn, R.A., and Overfelt, J. (1997). Advanced linear algebra object manipulation. In van de Geijn,
      R.A., (ed.), *Using PLAPACK: Parallel Linear Algebra Package. Scientific and Engineering Computing*.
      MIT Press, Cambridge, MA, pp. 42–57.

[99] van de Geijn, R.A. (1997). *Using PLAPACK: Parallel Linear Algebra Package. Scientific and Engineer-
      ing Computing*. MIT Press, Cambridge, MA.

[100] Alpatov, P., Baker, G., Edwards, H.C., Gunnels, J., Morrow, G., Overfelt, and J., van de Geijn, R. (1997). PLAPACK: Parallel linear algebra libraries design overview. In: *Proceedings of Supercomputing'97* (CD-ROM), San Jose, CA, ACM SIGARCH and IEEE.

[101] Baker, G., Gunnels, J., Morrow, G., Riviere, B., and van de Geijn, R. (1998). PLAPACK: High performance through high-level abstraction. In: *Proceedings of the 1998 International Conference on Parallel Processing* (ICPP '98), Washington–Brussels–Tokyo, pp. 414–423.

[102] Kuck, D. (1978). *Structure of Computers and Computations*. Wiley, New York.

2 Parallel Computer Architecture

Pedro Trancoso and Paraskevas Evripidou

CONTENTS

ABSTRACT

The different components of the parallel computer system are presented. The topics covered are the parallelism within the modern microprocessors, multiprocessor architectures of different scales, interconnection networks, and parallel input/output systems. The objective is to describe the different options available in terms of parallel architectures.

Currently it is easy, and somehow affordable, to buy the building blocks for a supercomputer. The understanding of the different available components is not only necessary for the engineers who are in charge of building the machine but also to the users, programmers, and decision makers. Current desktop machines are more powerful than many of the old supercomputers. The correct understanding of the basic features of each architecture is therefore essential. It is only with a good knowledge that it is possible to correctly select the architecture that best matches the needs of the user and the program. In addition, the understanding of the underlying system makes it possible for code optimizations to fully exploit the target system and consequently achieve better performance for the programs.

2.1 INTRODUCTION

As mentioned earlier, parallel computing is an efficient way to achieve a substantial reduction of the execution time for long running applications. Nevertheless, the best performance is only achieved if the programmer is aware of the underlying parallel hardware. In addition, understanding the characteristics of the different parallel architectures helps the programmers in two ways. On one hand, it gives the opportunity for the programmer to find the architecture that will be the best match for the application. On the other hand, it helps the programmer in tuning or choosing the algorithms that better match a certain parallel architecture.

This chapter extends the topic discussed in Chapter 1 by presenting in more detail the hardware used to exploit parallelism at different levels. Section 2.2 presents techniques used within the processor to exploit parallelism of the application. Sections 2.3 and 2.4 discuss the two major multiple instruction multiple data (MIMD) parallel architectures: shared-memory and message-passing. In addition, Section 2.3 also presents related issues such as synchronization, cache coherence, and memory consistency. Section 2.5 discusses a major component of the parallel architectures, the interconnection network, whereas Section 2.6 presents different parallel storage techniques and systems. The chapter ends with Section 2.7, which includes a brief summary of the material presented.

2.2 PROCESSOR PARALLELISM

In a system, parallelism may be exploited at all levels, starting at the processor. Pipelining is a technique that processors use in order to execute different stages of different instructions at the same time. This technique does not improve instruction latency. In reality, with the addition of the extra hardware, each instruction takes longer to complete its execution. The benefit of pipelining is in increasing the instruction throughput, i.e., number of instructions completed per unit of time. This section will address additional techniques used in modern processors to further exploit parallelism.

2.2.1 SINGLE INSTRUCTION MULTIPLE DATA (SIMD) INSTRUCTIONS

The increasing use of microprocessors for multimedia applications has led the major developers to extend the original set of instructions by adding instructions that are targeted specifically to optimize the execution of the most basic multimedia operations. One relevant characteristic of the multimedia applications, such as image processing, is the fact that they require to perform many repetitive operations on a large amount of data in a small amount of time. In addition, the data may be expressed by small integer numbers (e.g., pixel color in an image). Consequently, processor manufacturers engineered a smart solution where the existing resources are used in a more efficient way

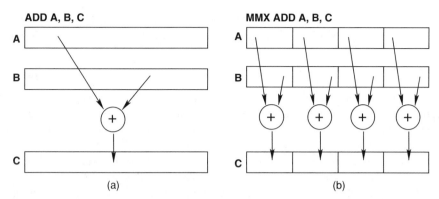

FIGURE 2.1 Hardware support for addition using: (a) regular instruction and (b) MMX instruction.

for multimedia operations. This solution consists of virtually splitting the general purpose registers into smaller blocks and executing the same operation simultaneously on each block independently. Using this technique, with the addition of extra Arithmetic and Logic Unit (ALU) components it is possible to significantly improve the performance. Using the example of the original Intel's multimedia instructions (MMX), the 128-bit floating point registers were divided into four blocks of 32-bit each and therefore, there is a potential for a fourfold improvement for some instructions.

In Figure 2.1 it is possible to observe that, assuming each addition executes in a single cycle, executing four 32-bit addition operations implemented using regular instructions require four cycles to complete. On the other hand, the same operations implemented using the MMX instruction require only a single cycle to complete resulting in a speedup of four.

Because a multimedia instruction performs the same operation simultaneously on different data items, these instructions fit in Flynn's SIMD category and consequently they are known as SIMD instructions. Examples of SIMD instruction sets found in the latest processors are the Streaming SIMD Extension (SSE2) included on Intel's Pentium 4 [1–3], *3D Now*! included on AMDs Athlon [4, 5], *AltiVec* included on Motorola's PowerPC [6, 7], and VIS included on Sun's Sparc [8–10]. Although introduced as instructions used to optimize multimedia applications, in reality they can be used for any application that applies the same operation to multiple data elements. The only issue is the fact that in order to benefit from this optimization, the applications have to be compiled to use the special SIMD instructions. Several successful examples exist of optimization of scientific as well as commercial workloads [11–14].

2.2.2 SUPERSCALAR

Pipelining exploits parallelism by partitioning the instruction execution into different stages and consequently allowing for different instructions to execute different stages of the execution simultaneously. One further step in exploiting parallelism is allowing different instructions to execute the same stage simultaneously. This can be achieved by providing multiple hardware units for the same stage of execution. For example, in a simple five-stage architecture, for the EXE state, which represents the stage executing arithmetic or logical functions, the processor may offer multiple ALU units that may be used simultaneously. A processor built this way, which supports the issue of multiple instructions in the same cycle, is known as *superscalar*. This technique has become very common. In fact, most modern processors, such as the Intel Pentium 4 [1], the AMD Athlon [4], the PowerPC [6], the MIPS R10K [15], and the Sun UltraSparc III [8, 9] are superscalar processors.

2.2.3 MULTITHREADING ARCHITECTURE

In a program there are usually circumstances where the processor is idle waiting for an input from the user, or waiting for an answer to a request from a resource such as the disk or memory. One approach to avoid the idle times and better utilize the available processing power is to submit several *jobs* or *processes* simultaneously and switch between them whenever one of them is blocked waiting. Although it increases the processor's utilization, this approach is costly due to the relatively large process switch time, known as *context switch*. To alleviate this problem researchers proposed the use of *threads*, which are lightweight units of execution, usually from the same program, i.e., belonging to the same process. The main advantage is that the switching among threads is very fast compared to the context switch among processes. The disadvantage is that the programmer has to be more careful as all threads share the same memory space of the process they belong to. Consequently there is no protection provided by the system against accesses to memory space of other threads.

To exploit the benefits mentioned above, the programmer has to divide the code into different threads and use the thread library supported by the target operating system. POSIX has proposed a standard for threads [16], which are known as *Pthreads*. Among others, one example of a threads library implementation is the LWP library [17] provided with Sun Solaris.

Threads may be supported by any regular processor architecture by executing at some point in time instructions from a single thread of execution. The superscalar architectures described in Section 2.2.2 would exploit parallelism by executing different instructions from the same thread. To execute multiple threads simultaneously, there is a need for additional hardware support. Such architectures that execute instructions from multiple threads at the same time are known as *multithreading* architectures. Although their potential is large, they are not very common. A few examples include the HEP [18], Tera [19], MASA [20], and Alewife [21].

2.2.4 SIMULTANEOUS MULTITHREADING

The two previously described techniques, superscalar and multithreading, focus on exploiting parallelism and using the architecture in a more efficient way. Nevertheless, both suffer from different limitations that do not allow for a full utilization of the resources. For the superscalar architecture, for example, the existence of dependencies between two instructions results in the fact that only one of them will be able to be issued and consequently the resources will be "wasted." This "waste" is known as *horizontal waste*. In addition, whenever an instruction issues a request from memory, its execution will block until the request is satisfied. This results in another type of "waste" known as *vertical waste*. The multithreading architectures focus on eliminating the *vertical waste* by switching the execution to another thread but as it can issue only instruction of a single thread at a time, this may still result in *horizontal waste*.

If both superscalar and multithreading techniques are combined, then it is possible to have an architecture that may issue, at the same time, instructions from different threads, resulting in a more efficient use of the resources. This technique is known as simultaneous multithreading (SMT) [22–24]. An example of the benefits of SMT for the execution of an application are presented in Figure 2.2.

The benefits of SMT architectures may be achieved by either executing multithreaded software, i.e., software that was written using threads, or alternatively executing different applications simultaneously.

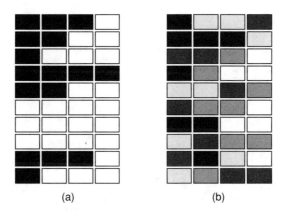

(a)   (b)

FIGURE 2.2 Comparison of the regular superscalar execution (a) with execution using SMT (b). The boxes represent processor units such as load/store or integer units. The colors represent the different threads occupying the respective unit. White represents unutilized unit.

Although SMT has been widely studied by the research community, a first example of an implementation of SMT in a large-scale product is Intel's Hyper-Threading (HT) introduced in the Xeon processor for server systems [25] and most recently the Pentium 4 3.06 GHz [26–28] for desktop systems. The current HT implementation offers in a single physical microprocessor the functionality of two logical processors. This seems to be an efficient optimization as the performance may improve up to 30%, with the cost of implementation with only a 5% increase in the chip size and maximum power dissipation [27].

2.2.5 VERY LONG INSTRUCTION WORD (VLIW) PROCESSORS

Contemporary microprocessors exploit Instruction Level Parallelism (ILP) by adding extra hardware and control logic to allow instructions to execute even out-of-order, as long as they are independent. As the demands for faster processors increase, the hardware becomes more complex, leading to difficulties in testing and debugging. An alternative solution is to transfer the responsibilities from the hardware to the software. This approach has the advantage that it allows the microprocessor to be simpler and consequently potentially faster. One such approach is known as *Very Long Instruction Word* where an instruction of the VLIW architecture is composed of several instructions of the regular architecture that are independent and may be executed simultaneously. The responsibility of finding these independent instructions lies on the compiler. The compiler performs a dependency analysis and finds independent instructions. All independent instructions that are found are then included in the same block as long as they do not violate the program correctness, i.e., some instruction reordering is allowed. The disadvantage of this approach is that the software is not always able to find many independent instructions leading to unoccupied slots in the VLIW instruction and consequently low utilization of the hardware resources. An example of execution of the process of arranging the instructions for a VLIW architecture is shown in Figure 2.3.

To increase the number of independent instructions, the compilers for the VLIW architectures use several different techniques such as *loop unrolling* [29] or *predicate execution* [30].

One recent example of an architecture that adopted the VLIW concept is the Explicitly Parallel Instruction Computing (EPIC) architecture by HP and Intel [31]. The Intel Itanium [32] was the first implementation of this architecture, also known as IA-64 [33]. Although conceptually the simpler hardware should be faster, this microprocessor was not very successful due to its poor performance.

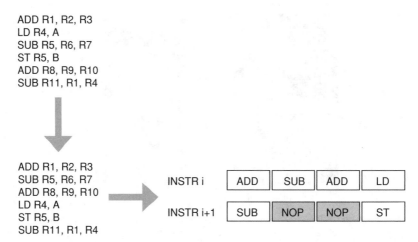

FIGURE 2.3 Example of transforming a program with regular instructions for execution on a VLIW architecture.

The solution of some of the initial problems and an increased hardware support for the execution are the determinant factors in the promising performance exhibited by Itanium's successor, the Intel Itanium2 [34].

2.2.6 CHIP MULTIPROCESSOR (CMP)

The increasing complexity in the microprocessor's hardware has led to the development of alternative architectures. One option is the one described in Section 2.2.5 where ILP is exploited by the compiler and not the hardware. A different approach is the one that considers exploiting parallelism by including several simpler microprocessor cores in the same die instead of a single complex one. The objective is to exploit parallelism and avoid high development and testing costs, which result from complex processors. This approach where multiple cores are included in the same die is known as Chip Multiprocessor (CMP). Initially Olukotun et al. [35] proposed the CMP work, which was then followed by several other attempts done in the research community, among which the proposed Piranha chip [36].

Commercial products that use this approach are not very common yet. Current examples of such products include the IBM Power4 [37] and the Sun Microprocessor Architecture for Java Computing (MAJC) [38]. The Power4 includes in its chip two identical microprocessors that have independent first-level caches but shared a common second-level cache. The first implementation of the MAJC technology, the MAJC 5200, is a CMP designed for real-time flow of multimedia data, which integrates two VLIW processors running at a speed of 500 MHz. One advantage of the CMP approach is that the multiprocessor resides on a single chip with high-speed data transfer between the processors. The sharing of the data, though, may create several problems which are described in detail in Section 2.3.1. An indication of the success of this approach is the fact that further CMP processors from Intel and Sun have been announced [39].

2.3 SHARED-MEMORY ARCHITECTURES

In this section the shared-memory parallel computer architecture is presented. This architecture may be considered as the natural extension to the single processor architecture as it provides a single system view to the user and programmer. Consequently, a program written for this architecture may access any memory location, independent of the processor the program is executing on. In addition, each memory location may be shared by the programs executing on different processors of the machine. These facts reveal one of the key advantages of this type of architecture, which is the easy programmability. In particular, a program that has been written for a uniprocessor environment may be easily ported to a shared-memory architecture as the application data may be automatically shared by all processors.

The following sections (see Sections 2.3.1 and 2.3.2) present the major issues that arise from supporting the single system view. The next sections (Sections 2.3.3–2.3.6) present the four most important implementations of this type of architecture.

2.3.1 CACHE COHERENCE

Although in the shared-memory architecture each memory address has a unique physical memory location, it is possible for two or more processors to obtain different values for the same memory location. This is because processors may have their own local copy of the data on their own cache. Upon modification of the data, the updates may not be visible to the rest of the processors on the system. An example of such misbehavior and consequent incorrect program execution is depicted in Figure 2.4. In this figure it is possible to observe that two CPUs execute a load of memory location A. They both read a value of 10, which was the original content of this memory location. This value is automatically placed on the cache of each processor. Later CPU_0 adds 10 to A and as it finds the memory location A in its local cache it modifies the cache contents. Later if CPU_1 wants to add 5 to A it will still get the initial value of 10 as it is the content of its local cache and update it to 15. At this point A has the value of 20 for CPU_0 and 15 for CPU_1. This problem, known as the *cache-coherence* problem, is solved using cache-coherence protocols [40].

For obvious reasons this problem does not present itself in the cases where the cache is shared by the different processors (which is not feasible for a large number of processors) and for systems which have no cache altogether. For the rest of the cases there are two major approaches to solve this problem known as the *snooping-* and *directory-based protocols*. The major difference between these two types of protocol is the amount of information that needs to be sent. In the first case, although it is more simple to implement, it requires the broadcasting of all memory operations. In the second case, the information is sent only to those processors that share the same data.

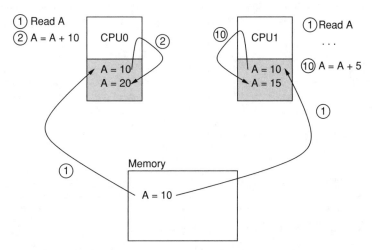

FIGURE 2.4 Example of a cache-coherence problem.

There are two types of *snooping-based protocols*: *write-invalidate* and *write-broadcast* or *write-update* protocols. Notice that an efficient implementation of the *snooping* protocols requires a shared media to broadcast its messages. It turns out that a *bus* is the most convenient communication media to support the *snooping-based protocols*. The bus offers not only the broadcasting capability but also it enforces automatically the serialization of messages sent by different processors. The latter is also determinant for the success of these protocols.

In the *write-invalidate* protocol, whenever a processor writes to a memory location, it sends an *invalidation message* to the shared communication media. At this time all the processors on the media listen to the message (they *snoop*) and if they hold a copy of that same data on their caches, they will invalidate their own copy. This invalidation will result in a cache miss for the next memory operation to that location. Whenever a processor reads a memory location, if it is a hit then the data may be read from cache, otherwise it may get the data from memory or it may have to force the last writer to write back the data to memory. The location of the latest valid copy of the data, in memory or cache, depends on whether the cache write policy is *write-through* or *write-back*, respectively.

The *write-update* protocol works in a similar way to the *write-invalidate*. The main difference is the fact that whenever a processor writes to memory, instead of sending an invalidation message, it broadcasts the new value for that memory location. All other processors that are *snooping* check the memory address and if they have a local copy of that data then they will update their own copy with the new value. Consequently, processors that share data in the same memory location will always have the latest value in their local cache. The update operation propagates not only to all processors holding a copy but also to the memory, which is also *snooping* on the bus. Therefore, as opposed to the *write-invalidate* protocol, with the *write-update* protocol the memory contains always the latest value for each data. This also simplifies the handling of the read misses as now the processor needs only to read the latest value from memory (Figure 2.5).

Notice that for a single-write operation, the traffic created by the write-invalidate protocol is smaller than the one created by the write-update protocol. This is because the former requires only to broadcast the memory address whereas the latter requires to broadcast the address and the content of the memory location (the new data value). Nevertheless, this does not imply that the total traffic for the complete execution of an application is smaller for the write-update protocol. The total traffic depends on the sharing pattern of the data. For example, if every time the data is updated its new value is used by the rest of the processors, the write-update protocol will be the most adequate. If, on the other hand, one processor updates the data several times before the rest of the processors use it, then the write-invalidate is the most appropriate protocol. Overall, the last type of data access pattern seems to be more common and consequently most commercial architectures implement the write-invalidate protocol to solve the cache-coherence problem.

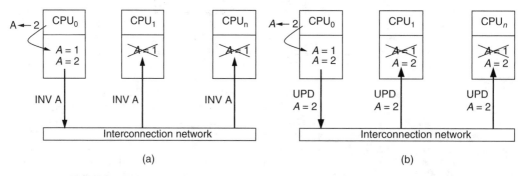

FIGURE 2.5 Cache-coherence protocols: (a) write-invalidate and (b) write-update.

In addition to the traffic generated as a result of *true-sharing*, as described above, these algorithms also generate coherence traffic in the cases where different processors access different data that belong to the same cache line. This effect is known as *false-sharing*. Therefore, while large cache lines are recommended to capture spatial locality, the coherence traffic increases as the cache line increases its size. Considering this effect, the programmers should be careful in order to avoid situations where two data variables, which are accessed independently by different processors, are placed in contiguous memory locations mapped to the same cache line.

An example of an implementation of the *snooping* protocol is the MESI protocol found on the Intel processors [41]. The name of the protocol, MESI, comes from the initials of the different possible states for a cache line: Modified, Exclusive, Shared, and Invalid. If the cache line is in the *Modified* state then the data is valid only on the local cache and its content is different from the one in memory. If the cache line is in the *Exclusive* state then the data is valid only on the local cache but its content is the same as the one in memory. The *Shared* state indicates that the cache line may be valid in caches of different processors. Finally the *Invalid* state indicates that the local copy is not valid. An extension of this protocol is the implementation found in the AMD processors that adds one more state (Owner) and is called MOESI [42, 43]. This extra state, *Owner*, indicates that there are different copies of the cache line in different processors but the local copy has been modified.

The fact that the *snooping* protocols are based on broadcasting the coherence messages makes them unacceptable for systems with a large number of processors. In addition, snooping protocols are unacceptable for situations where only a small subset of the processors actually share the same data. Therefore, one solution is to augment the cache line with information on the processors sharing the particular data line. With the aid of this extra information it is possible to multicast the coherence messages instead of broadcasting them. This scheme is known as the *directory-based* protocol. As previously stated, in this protocol each cache line will have a state and in addition a bit vector indicating which processors are currently sharing the data. The states of this protocol are similar to the ones of the *snooping* protocols. In case of a write operation, the processor will use the bit vector to send the coherence messages to the other sharing processors. Notice that this protocol works better with network topologies other than the bus, but requires the hardware to provide some sort of serialization for the messages.

2.3.2 MEMORY CONSISTENCY

In addition to keeping the cached data coherent, another issue is when the new values are made visible to the rest of the processors. The rules by which the memory updates are made visible are determined by the *memory consistency model*. The code in Figure 2.6 represents an example of a parallel code which may produce different results depending on the *memory consistency model*.

```
P1:    A = 0;              P2:    B = 0;

       ...                        ...

       A = 1;                     B = 1;

L1:    if (B == 0)...      L2:    if (A == 0)...

         (a)                        (b)
```

FIGURE 2.6 Example of code that produces different results depending on the memory consistency model.

Basically, the code in Figure 2.6(a) executes on a different processor than the one in Figure 2.6(b). Certain conditions, such as the delay of sending invalidation messages, may result in statements L1 and L2 returning *true*. This is obviously not the result that was expected.

Sequential consistency [44] is the most strict memory consistency model. In this model the memory accesses are kept in order within the same processor and are interleaved across processors. This model allows for simple development of parallel programs as it assures that if a process accesses a memory location, its contents will be the most up-to-date. The drawback of such a model is that its implementation results in a significant performance overhead. Therefore, other more relaxed models have been proposed such as the *processor consistency* [45], *weak consistency* [46], and *release consistency* [47].

Although a strict memory consistency model may enforce the memory accesses to perform as expected from a serial execution, a way to achieve better performance is to shift the responsibility of maintaining the consistency to the programmer. In such a scenario, the system itself maintains a relaxed or no consistency on the memory accesses and the programmer may choose at which occasions of the program the accesses must be committed to memory. To support this approach, which protects the accesses to shared-memory variables, the programmer should use a technique known as *synchronization* [48, 49]. The most common synchronization primitives are the *lock*, the *unlock*, and the *barrier*. The *lock* and *unlock* primitives are usually used to enforce exclusive access to part of the execution. As all processors may execute the same program at their own pace, the segments of the code that include updates to shared variables should be executed in mutual exclusion. These segments of the code are known as *critical sections*. The *lock* primitive guards the entry of a critical section, while the *unlock* primitive releases the critical section allowing another processor (or the same) to enter it next. The *barrier*, on the other hand, is used to synchronize all the processors in a certain line of code. With this primitive, the execution will only continue whenever all the processors have completed the execution of all tasks before the barrier.

Figure 2.7(a) shows a segment of program where the programmer uses locks to enforce mutual exclusion to the update of shared-variable A. In this way the programmer guarantees the memory consistency between memory accesses from different processors. Figure 2.7(b) shows the use of the barrier in order to synchronize the execution of all processors after each task.

Notice that a careless use of synchronization primitives may result in severe problems for the program such as *starvation* or *deadlock*. Starvation is the condition where a process execution waits indefinitely for a certain resource. A starvation situation may arise if, for example, in the code

```
        . . .
    LOCK(X)
                        . . .
                    WHILE (i=getTaskId()){
    A = A-1
                        processTask(i);
    UNLOCK(X)
                        BARRIER(Y);
        . . .
    (a)             }

                        . . .

                    (b)
```

FIGURE 2.7 Segments of programs showing: (a) an update to a shared variable that is protected by lock primitives; and (b) the synchronization of all processor's execution using the barrier primitive.

of Figure 2.7(a) the user had forgotten to include the UNLOCK primitive at the end of the critical section.

Deadlock is the condition where one process holds a certain resource which another process requests and requests a second resource which is currently held by the other process. In this situation none of the processes can advance their execution because they are on a *circular wait*: the first waits for the second whereas the second waits for the first. For example, to print a document, a process requires to reserve a disk for spooling and the printer. Suppose that there is only one disk and one printer resource and two processes trying to print a document. Process A decides to request the disk first. Then process B decides to request the printer first. Now process A tries to complete the printing by requesting the remaining resource, the printer. But it will have to wait for process B which currently holds the printer. Process B, on the other hand, requests the disk to complete the requests. But this process can also not proceed as the disk is currently held by process A. At this point the two processes have reached a deadlock situation.

In addition to careful coding, the operating systems may include algorithms to control these situations at different levels: prevention, avoidance, and recovery. A description of these algorithms may be found in Ref. [50].

The synchronization primitives described here may be implemented using either software algorithms such as Lamport's Bakery algorithm [51] or the spin locks [52]. Alternatively the primitives may be implemented using hardware instructions such as the *test-and-set* or the *fetch-and-increment* [53].

2.3.3 CENTRALIZED SHARED-MEMORY

The first type of shared-memory architecture presented here is the one where all the memory of the system is found in a single location of the system. Therefore, this type of architecture is known as the centralized shared-memory architecture.

As it can be observed in Figure 2.8 this architecture has multiple processors connected to some interconnection network, through which they can access a common set of memory banks. The distinct characteristic of this architecture is the fact that each processor may access any memory location with the same exact access time. Consequently, this type of architecture is also known as Uniform Memory Access (UMA) architecture.

As mentioned earlier, in this architecture all memory is located in a common set of memory banks. This fact turns out to be both an advantage and a disadvantage. On one hand it is an advantage as it makes the system easy to implement due to the simple design of the memory controller.

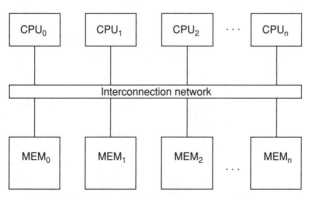

FIGURE 2.8 Centralized shared-memory architecture.

On the other hand, the system does not scale well. As it was shown in Figure 2.8, in this architecture all the processing elements are on one side of the interconnection network and all the memory elements on the other side. Therefore, it is obvious to conclude that the interconnection network becomes a bottleneck for a large number of processing elements as they may issue a large number of memory requests at the same time. Obviously, this disadvantage is more serious for simple interconnection networks such as the bus. More complex networks, as the ones described in Section 2.5, may overcome this problem at a higher cost and complexity.

Overall, implementations of such architecture which are more cost-effective use a bus-based interconnection network and are usually limited to 8–16 processors. Systems that use more complex networks such as a crossbar are able to scale better. In the market today it is possible to find small-scale two- or four-way bus-based multiprocessors such as Dell PowerEdge 2600 with up to 2 Intel Pentium Xeon at 2.4 GHz [54]. At the other end of the range it is possible to find high-end systems such as the Sun Fire 15 K, which is able to support up to 106 UltraSPARC III processors at 1.05 GHz and use a more complex crossbar interconnection network [55].

2.3.4 DISTRIBUTED SHARED-MEMORY

Another type of shared-memory multiprocessor is the distributed shared-memory. The main difference between the distributed and centralized types is that in the former, the memory banks are not all clustered together in a single location of the system but instead each processor has its own set of memory banks. Although physically the memory blocks are separated, logically the hardware offers support for the programmer to be able to see all the memory banks as one single large shared-memory space.

Figure 2.9 depicts the architecture of a distributed shared-memory system. One advantage of this type of architecture is that the memory access bottleneck is removed as the memory is now distributed among the different processors. Consequently, machines of this type of architecture are able to scale to a larger number of processors. One potential problem is that the memory accesses are not uniform in this architecture, i.e., if the memory location to be accessed belongs to the local physical memory then the access time will be small while if it belongs to a remote physical memory its access time will be large. As a consequence of this fact, this architecture is also known as non-uniform memory access (NUMA). Although this does not create any logical problems as the memory accesses are correctly handled independent of their physical location, it may create performance problems. For example, consider the case where two processors read, in parallel, a number n of entries in a data array. In the worst case, for one of the processors all the n entries are in its local memory whereas for the other processor all the n entries are in a remote memory. Although

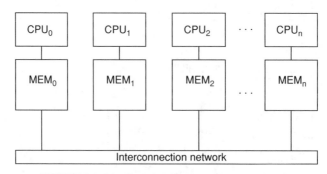

FIGURE 2.9 Distributed shared-memory architecture.

from the analysis of the program code it would be expected that this simple operation would result in a speedup of two, in practice the speedup may be much lower than that. From this small example it is possible to conclude that the performance of an application may depend significantly on the memory mapping scheme for the shared variables. A simple, static, page placement algorithm that tries to avoid the problem described above is the one which places the memory page in the processor that issues the first memory request to that page. For obvious reasons this scheme is called the *first-touch* page placement algorithm [56]. This placement scheme works well for applications that work mostly on independent parts of the shared-memory data. Notice that for this scheme to succeed it is required that each processor initializes the shared data that it will be working on. Allowing the initialization code to be executed by a single processor at the beginning of the execution would result in all data being placed on that processor.

A key component in this architecture is a table that includes, for each local memory block, an entry containing the state of the data block (e.g., one of the MESI states) and a bit vector indicating which are the processors that hold a copy of the block. This table is known as the *directory* and is very similar to the one described for the directory-based cache-consistency protocol. The main difference is that in this case there is one entry for each memory block as these blocks are *owned* by the local processors. Consequently it is the processor's responsibility to store the corresponding relevant information. A diagram representing a node in a distributed shared-memory architecture is shown in Figure 2.10.

Another key component is the memory controller, which must contain information indicating where each memory location is mapped. This is essential to determine which processor to send the memory request for any data that is mapped to a remote memory location.

Examples of machines that implement this type of architecture are the Stanford DASH [57] with 64 MIPS R3000/3010 (33 MHz) processors and the SGI Origin 3000 [58] with up to 512 MIPS R12000 (400 MHz) processors.

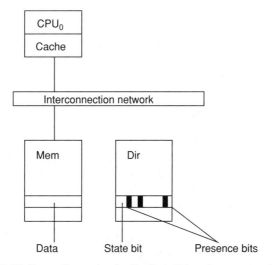

FIGURE 2.10 Computing node of a distributed shared-memory architecture.

2.3.5 Cache-Only Memory Access

One of the major problems with the NUMA architecture is that memory accesses to remote memory locations are more costly than to local memory locations. The overall performance of a program might degrade severely if the program keeps accessing remote memory locations. One way to avoid this problem is by having a location-conscious memory mapping to colocate the accesses with the local memory pages. One such scheme is the first-touch page placement algorithm described in Section 2.3.4. Unfortunately colocating the page and its use is not always possible or may be too difficult to do at run time. Additional difficulties arise if memory blocks accessed by different processors belong to the same memory page.

One solution to the problem is to consider the whole memory as a large cache. This type of architecture is known as the cache-only memory architecture (COMA). In this architecture, when a processor issues a request to a memory location, if this memory location is not local it tries to get its ownership and map it locally. Following accesses to that same location will then be local. With this strategy, complex memory mapping schemes are avoided as the memory will be moving to its real owner automatically. If the processors access mostly disjoint data sets then this scheme will result in optimal performance as on the steady state all processors would be issuing only local memory requests. On the other hand, if the processors access the same data in turns then the memory block will be changing owner for each access and consequently the performance of the system will degrade.

Examples of such architectures are the SICS Simple COMA [59], Illinois I-ACOMA [60], and the commercial product KSR-1 [61].

2.3.6 Software Distributed Shared-Memory

An alternative to the architectures described previously is to support shared-memory through the use of software instead of hardware. This solution is known as software distributed shared-memory.

In this approach, as well as in the previous ones, the shared-memory is transparent to the programmer. In this case the compiler analyzes and instruments the code with extra instructions to handle the sharing of the data.

The advantage of this approach is that the hardware is affordable, as it may be done out of off-the-shelf components. In addition, because there is no hardware support for the memory sharing, there are no overheads. The disadvantage of this software approach, compared to the hardware counterparts, is that the handling of the sharing is done by software and therefore it will never be as fast as the hardware. Nevertheless, efficient software implementations have been demonstrated.

Examples of this type of approach are Shasta [62] and Cashmere [63].

2.4 MESSAGE-PASSING ARCHITECTURES

Although many of the computational-intensive problems can currently be solved by single top-of-the-line microprocessor systems or small-scale shared-memory multiprocessors, there is still a vast number of problems which requires much larger computational power. Examples of applications that have such high demands are bioinformatics, environment, meteorological, engineering, physics, and defense applications. The type of architecture analyzed in this section, *message-passing*, is able to satisfy these requirements as they may scale to a very large degree of parallelism. The material presented starts with a general description of the concepts of this architecture, in particular the techniques used to achieve the desired scalability. This is followed by the presentation of the four types of systems that belong to this category of architecture: massively parallel processors, clusters, constellations, and the Grid.

2.4.1 SCALABILITY

As mentioned in the previous section, although shared-memory architectures offer the programmer the ability to transparently access any memory location of the system, the hardware support for maintaining the shared-memory is a limiting factor for a high degree of parallelism. In practice, shared-memory multiprocessors may be configured with up to 128 processors [58]. Nevertheless, there are applications that have demand for very large computing power. For example, the Earth Simulation Project [64] has a target requirement of 40 Tflops (40×10^{12} floating point operations per second). To have an idea of the scale of this problem, the latest model of the Intel Pentium 4, which runs at a speed of 3.06 GHz, has been rated with a peak performance of 6.12 Gflops (6.12×10^9 floating point operations per second). Consequently, to satisfy the 40 Tflops target we would need more than 5000 Pentium 4 processors. Such a large scale is out of reach for a regular shared-memory multiprocessor.

The key factor in creating machines that are able to reach a very large scale is to remove the bottleneck of the shared-memory architectures, i.e., the hardware support for shared-memory. Architectures that do not include any hardware component used to coordinate the operations on the different computing nodes are, theoretically, able to scale without limitations. Nevertheless, a potential limitation may come from the interconnection network used to connect the computing nodes.

Whereas at the hardware level there is no support for the execution of a parallel application, the required functionalities have to be offered at the software level. In such an architecture, the programmer has the responsibility of the distribution and exchange of data among the different computing nodes. This is achieved using explicit communication primitives to transport the data, also known as *messages*. As a consequence of this fact, this type of architecture is known as *message-passing* architecture.

The fact that the programmer is the one responsible for the data exchanges in the system leads to greater difficulties in developing new parallel applications and also in porting legacy codes to the parallel architecture. Helping the programmers in their task is the objective of different communication libraries such as parallel virtual machine (PVM) [65] and message passing interface (MPI) [66]. More details on these libraries and their use can be found in Chapter 1 (Section 1.6.4.1).

The performance achieved by a parallel application on a message-passing architecture may be very high depending on the data exchanges the application is required to perform during its execution. For example, if the programming model of the application is of the type *distribute–compute–collect*, the speedup achieved should be close to linear with the number of processing elements in the system. This leads to the conclusion that this type of architecture is a very good fit for many large-scale scientific applications. In fact, in a list compiled with the best performing computer systems known as the TOP500 Supercomputer Sites [67, 68], 94 out of the first 100 systems are message-passing.

2.4.2 MASSIVELY PARALLEL PROCESSORS

This section analyzes the category of parallel architectures that have as target the highest computing power. Such architectures include a large number of processing elements and therefore are named massively parallel processors (MPP). These processors are connected with a high performance, custom-designed interconnection network. According to the November 2002 TOP500 list [67], 39% of the machines on the list belong to this type of architecture. Also from the list it is possible to conclude that the MPP systems dominate in terms of maximum performance achieved. In particular, the sum of the maximum performance achieved by each machine belonging to this category adds up to 43% of the sum of the maximum performance achieved for all machines on the list. The next category accounts for only 27% of total maximum performance.

Depending on the type of processing nodes it is possible to classify the MPPs into two distinct categories: custom-designed or commodity processing nodes.

The first category includes the tightly coupled multiprocessors which used to be popular as the base for supercomputers. These are machines of very high cost, custom-designed, usually unique, and were by far the ones achieving the best performance. An impressive example of an MPP machine is the NEC Earth Simulator [64] which was already mentioned earlier. This machine occupies the first place in the November 2002 TOP500 list with a peak performance of 40 Tflops. This machine is composed of 640 nodes each with an eight-way shared-memory multiprocessor. The nodes are connected with a *crossbar* interconnection network (see Section 2.5.1). Another example of a machine belonging to this first category is the Cray T3E [69].

Although the machines belonging to this category are the prime choice for computational-intensive applications, due to the fast development of the microprocessor these machines are currently abandoned in favor of systems that use *commodities off-the-shelf* (COTS) components.

The second category of MPP architectures includes machines with COTS nodes, i.e., the nodes are regular single (or dual) processor nodes, but the interconnection network used to connect them together is proprietary. This is currently the most common type of machine used today for supercomputing. An example of such a machine is the IBM SP2 [70], which is the base machine for the ASCI White Supercomputer [71, 72]. The ASCI White Supercomputer is configured with 8192 IBM RS6000 processors running at 375 MHz, has 6-Terabyte of RAM memory, and 160-Terabyte of disk space. Its peak performance is rated to be 12.3-teraflops and it occupies the fourth place in the November 2002 TOP500 list [67]. The IBM SP2 was also the base machine for the Chess Computer known as *Deep Blue* which won against the world chess champion Gary Kasparov in 1997 [73].

2.4.3 CLUSTERS

With the increasing improvement of the commodity processor and network technology, a new interest arises on parallel systems using both processor and network COTS. The motivation for this type of parallel architecture was to produce an easy to scale, low-cost system that delivered good performance. Because of the fact that these systems are built either from readymade boxes or boards put up together on a rack, they are known as *clusters*.

The first proposed research systems were built using the existing workstations and connecting them altogether with regular 10 Mbps Ethernet. One of these pioneer systems was the *Network of Workstations* (NOW) project [74]. After some developing period this system led to a commercial product, the Inktomi server, one of the first large-scale systems used as an Internet search engine (HotBot).

A second wave of cluster projects started after the widespread of the Linux operating system as it broadened the selection of computing nodes to simple PC "boxes." One of the first such cluster systems was proposed by engineers at NASA which required the computing power but had no budget for MPP machines. The solution proposed was to assemble together several Linux desktop systems and connect them using regular 10 Mbps Ethernet. This type of system configuration was named BEOWULF [75]. Today there are several BEOWULF machines installed in different corporations and research centers around the world.

Although the initial motivation was focused on low cost, the dramatic improvement on microprocessor speeds and new networking technologies have led to cluster systems that surpass the performance of many MPP systems. In particular, a Linux cluster achieved the fifth place in the November 2002 TOP500 list [67]. That particular machine is configured with 2304 Intel Xeon 2.4 GHz processors, has 4.6-Terabyte of RAM memory and 138-Terabyte of disk space. The system is rated with a peak performance of 11.2 Tflops. Besides this machine, it is also relevant to mention that in that same TOP500 list, almost 20% of the machines reported are clusters. The increasing popularity of clusters is proven by the fact that all major computer companies offer cluster models such as the Apple XServer, the Sun LX50, and the IBM eServer xSeries 335.

In terms of applications, in addition to the traditional science, research, and business applications, one important use of clusters is as server for an Internet search engine. As it was mentioned earlier, one of the first clusters, the NOW project, led to a search engine server, the Inktomi server used at HotBot. Another example is Altavista which was powered by a Compaq cluster. The latest and most impressive example is the Google server [76, 77]. The cluster at Google is significantly large and by 2002 it included more than 10,000 processors, stored more than 3 billion web pages and was accessed at a rate of more than 150 million searches per day [78]. In this cluster the nodes are composed of a variety of processors among which are the Intel Celeron and Pentium III processors, which are connected using 100 Mbps and 1 Gbps Ethernet.

Finally it is relevant to notice two documents, the first by Baker et al. [79] and the second edited by Baker [80], which may serve as guideline on the design of a cluster system.

2.4.4 Constellations

In a cluster system, the computing nodes are usually off-the-shelf, single- or dual-processor based systems. It is possible to think of an extension of the cluster architecture to include large systems as the computing nodes. Because of their configuration of medium to large systems interconnected together, these machines are known as *constellations*.

An interesting constellation configuration is the one where the computing nodes are shared-memory multiprocessors. In this *hybrid* configuration of shared-memory within the node and message-passing across nodes, theoretically it is possible to obtain the best trade-off between performance and programmability. An application can be designed to spread the data among the system using the message-passing techniques but within the node the application may execute as multiple threads accessing the common memory address space. This results in two obvious benefits. The first is *scalability*, as without the global shared-memory support the machine may easily scale to a large number of nodes. The second is *load balancing*, as within the node, if one thread of execution finishes its work it may easily "help" in the processing of the data from other threads, because there is no extra data exchange overhead.

An example of such a machine is the ASCI Blue Mountain [81], which is composed of 48 nodes, each with an SGI Origin 3000 [58] shared-memory system configured with 128 processors. In this machine the nodes are connected using a *high-performance parallel interface* (HiPPI) network [82]. This machine achieved the 28th place in the November 2002 TOP500 list and was rated with a peak performance of 3.1 Tflops.

2.4.5 Grid

Large-scale computer applications have certain characteristics that are difficult to fulfill with regular computer systems. First, as already mentioned in earlier sections, they may require a large amount of computing power and data. Second, the computing elements required may be heterogeneous. Third, the data required may not be available from a single place, i.e., it may be physically distributed. Fourth, the requirements may change during the execution, i.e., the applications have phases such as serial execution, data collection, or highly parallel computational-intensive phase. Fifth, its success may require the cooperation between users, which may also be physically distributed. To satisfy these requirements, the Grid [83, 84], a new type of parallel and distributed system, has been proposed. At the beginning, the Grid started as a collection of networked supercomputers. As such, those systems differed from a constellation only by the fact that the network between the nodes was a wide-area network such as the Internet. Nevertheless, the Grid has evolved and currently may be classified as a system that enables the sharing of several distinct distributed resources such as computer systems, software, data and databases, special instruments and data-collecting devices, and people. Its name, Grid, was suggested as an analogy to the electrical distribution Grid. In a similar way, the goal of the Grid is to offer to the user its services and resources in a transparent way, i.e., independent of its location.

The different components of the Grid are: the *Grid fabric*, which includes all the resources; the *core Grid middleware*, which offers the main services such as security, authentication, naming, storage, and process management; the *user-level Grid middleware*, which includes developing environments and resource brokers; and the *Grid applications and portals*, which includes applications written using for example MPI [66] and their Web interface. An example of a core Grid middleware is the *Globus Toolkit* [85], a reference implementation which provides services enabling an application to see the distributed heterogeneous resources as a single virtual machine. Among the services offered by Globus are the *grid security infrastructure* (GSI), which is based on the X.509 certificates and the Public Key Infrastructure (PKI), the *grid resource allocation manager* (GRAM), and the *grid access to secondary storage* (GASS). An example of a user-level Grid middleware is *Nimrod-G* [86, 87], which is a resource broker that performs the scheduling and dispatching of *task-farms* on the Grid, on behalf of the user.

Currently, some of the existing Grid projects are the DataGrid [88], CrossGRID [89], DataTAG [90], GriPhyN [91], and ApGrid [92].

Some of today's typical Grid applications include bioinformatics applications such as applications to find new medicines, civil engineering applications to test structures under stress, climate and environment applications for emergency response analysis after natural disasters, aeronautics and aerospace applications, fraud detection of insurance claims, large-scale data mining, and business applications.

Finally, regarding the future of the Grid, it seems that it will also embrace another type of application called *Web Services*. To cover this new direction for the Grid, the Global Grid Forum has proposed a standard for this new Grid feature. This standard is known as the Open Grid Services Architecture (OGSA) [93] and its objective is to provide a standard interface for Web Services. OGSA is expected to be implemented in the next generation of the Globus Toolkit.

2.5 INTERCONNECTION NETWORK

One of the key components in the parallel system is the interconnection network, which connects together the processing and memory elements. Depending on the system requirements, networks can be very simple such as the *bus* or very complex such as the *Omega* network. To support high performance, the larger the system the more complex the network should be. This section examines both the topologies available and the technology for the interconnection network.

2.5.1 NETWORK TOPOLOGIES

The simplest way to connect multiple computing nodes is to use a *bus*. The bus is composed of several wires that are shared among all the nodes. This means that when it is inexpensive, it does not scale well. An increase in the number of nodes results in severe contention in accessing the media. Machines equipped with this type of interconnection usually do not scale beyond 16 or 32 processors [77].

The next topology is the *ring* interconnect, where each node has its own switch connecting to the switches of its two neighbors. This type of interconnect resolves the contention problem by introducing a token. The token gives the right to use the ring for communication. Although simple and inexpensive, this interconnect suffers from the problem of long communication latency for nodes that are "distant" in terms of their position on the ring.

Both of these topologies suffer from a low *bisection bandwidth*, which is defined as the sum of the bandwidth of lines that cross a cut dividing the network into two halves. In particular, for the bus the bisection bandwidth is the bandwidth of one line whereas for the ring it is the bandwidth of two lines. The bus and ring topologies are depicted in Figure 2.11.

Other interconnects with one switch per node but larger degree of connectivity are the *2D grid* or *mesh*, the *2D torus*, and the *hypercube*. In the 2D grid or mesh all nodes except the ones on the periphery communicate with their four neighbors. This is known as the North–East–West–South (NEWS) communication. By adding extra wiring to mesh, connecting the peripheral nodes, it is possible to obtain the torus. In the torus interconnect all the nodes have four immediate neighbors. The network with higher degree of connectivity is the hypercube. This network topology achieves higher bandwidth than the previously described ones. In particular, the bisection bandwidth for a $n \times n$ mesh is proportional to the bandwidth of n lines, whereas for the $n \times n$ torus it is proportional to the bandwidth of $2n$ lines. For a n binary hypercube which has 2^n nodes, the bisection bandwidth is proportional to the bandwidth of 2^{n-1} lines. Examples of mesh, torus, and hypercube topologies are depicted in Figure 2.12.

Examples of supercomputer machines that use these network topologies are the following: the Intel Delta (1991), Intel Paragon (1992), and Intel ASCI Red (1996) for the mesh network; the Cray T3E (1997) for the torus network; and the Thinking Machines CM-2 (1987) and SGI ASCI Blue Mountain (1998) for the hypercube.

Finally, there are the more complex networks which include a larger amount of switches than the number of nodes. This category of topologies include the *Fat Tree*, the *Omega* network, and the *Crossbar*. The Fat Tree includes more than one switch per internal tree node to improve the network's bandwidth. The Omega network arranges the switches in such a way that all nodes are able to directly communicate with each other but not without possible contention. In contrast, the *Crossbar* setup includes n^2 switches and consequently is able to deliver node-to-node communication without contention. The Fat Tree, Omega, and Crossbar networks are depicted in Figure 2.13.

Examples of supercomputers that use these topologies are: Thinking Machines CM-5 (1991) and IBM SP-2 (1993) using the Fat Tree topology; IBM ASCI Blue Horizon (1999), IBM SP (2000), and IBM ASCI White (2001) using the Omega network; and NEC Earth Simulator (2002) using the Crossbar.

2.5.2 NETWORK TECHNOLOGY

On one hand, for the custom-design machines, either shared-memory or message-passing, the network technology is proprietary and fixed for a certain system. Therefore, because there are no options, custom-design networks are of little interest within this context and consequently are not discussed in this section. On the other hand, for the machines that use commodity networks, there are several choices available in the market. The factors that are most relevant and that distinguish

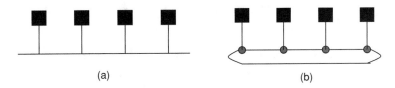

(a)   (b)

FIGURE 2.11 Bus and ring network topologies.

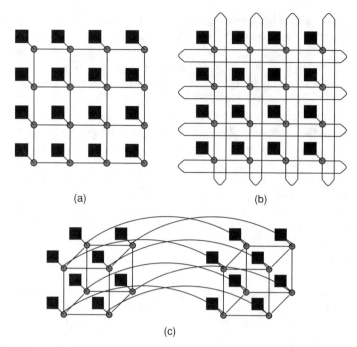

FIGURE 2.12 (a) Mesh, (b) torus, and (c) hypercube network topologies.

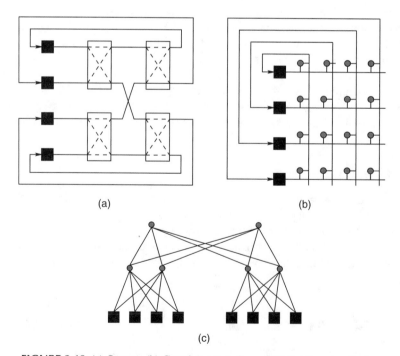

FIGURE 2.13 (a) Omega, (b) Crossbar network , and (c) Fat Tree topologies.

the different network technologies available are the communication latency, the maximum bandwidth, and the cost. The selection of a particular technology will not only depend on the budget but also on the characteristics of the application: if the application exchanges many small messages then it is most important to have low latency, but if the application exchanges large messages then it is most important to have large bandwidth.

One factor that affects the latency and bandwidth is the underlying communication protocol. The two most common protocols are the Transmission Control Protocol/Internet Protocol (TCP/IP) and the User Datagram Protocol (UDP). The former is the protocol used for communication on the Internet. It was the chosen one for that purpose as it was a standard at that time and therefore made it easy for different computers with different operating systems to communicate with each other. The TCP/IP protocol is a reliable, connection-oriented service whereas UDP is an unreliable, connection-less service. Although standard and with well-defined interfaces, the implementation of the TCP/IP protocol is layered and consequently there are significant overheads involved in the communication. These overheads are a result of numerous buffer copy operations between the different layers of the protocol. As a consequence, several research groups have proposed lower-latency protocols such as active messages [94], fast messages [95], VMMC [96], U-net [97], and BIP [98]. The common factor between all these protocols is that all of them reduce the latency using a technique known as *zero-copy*, which means that they eliminate the need for the copy operations of the buffers. In terms of standards for low-latency protocols, two relevant contributions have been presented recently: the *virtual interface architecture* (VIA) [99, 100], which is sponsored by major industry partners such as Intel, HP, and Microsoft; and Infiniband [101], which is also sponsored by major industry partners such as HP, Dell, IBM, Intel, Microsoft, and Sun Microsystems. VIA implementation, such as M-VIA [102], may be native or emulated. In a native implementation of VIA, the network hardware is responsible to process the messages whereas in the emulated VIA the host CPU will handle the protocol operations. Notice though that an emulated implementation of VIA still results in lower overhead than the traditional TCP/IP implementation.

Although the products described here are mostly focused for the cluster architectures, there are no specific restrictions that limit them to those systems. Traditionally, cluster nodes were connected using regular 10 Mbps *Ethernet*. This was the case for the NOW [74] and the BEOWULF [103] systems. Although the Ethernet technology has advanced since then, resulting in the 100 Mbps *Fast Ethernet* and the 1 Gbps *Gigabit Ethernet*, their latency is still large and their bandwidth is small compared to other products. In reality, the measured latency, after the protocol implementation, is between 100 and 200 μs and its bandwidth does not pass 50 Mbytes/s. This low performance is due to the fact that the native protocol supported is the TCP/IP, whereas VIA is emulated. The advantages of these products, though, are that they follow the same Ethernet standard and they are of lower cost. The first product to provide hardware implementation of VIA was the *Giganet cLAN* [104]. This product, which has a maximum bandwidth of 1.25 Gbps, was measured to deliver a bandwidth of 105 Mbyte/s and have a latency of 20 to 40 μs. *Myrinet* [105] is one of the most widespread network products for high-performance cluster interconnect. It is rated with a maximum 2 Gbps full-duplex data rate. It was measured to deliver over 220 Mbytes/s and having a latency of 11 μs. In addition to the high performance, this product offers some degree of fault tolerance as it allows for multiple paths between nodes and also because it includes a "heartbeat" monitor, i.e., it regularly sends a signal to assure the quality of the links. The efficient protocol implemented in Myrinet is called GM [106]. There are implementations of TCP/IP, VIA, and MPI running on top of GM. A cluster system equipped with a Myrinet network occupies the 17th place of the November 2002 TOP500 list [67]. The product which currently offers the lowest latency and highest bandwidth is the *Quadrics network* [107]. Quadrics implementation uses a global virtual memory concept to apply the zero-copy technique by copying the buffer from the origin to the target virtual memory. This product delivers approximately 210 Mbyte/s communication bandwidth and a latency of only 5 µs. Unfortunately, this is also the most expensive product of the ones described here. Nevertheless, if performance is the main goal, this is the product that may deliver it. To prove this statement, a cluster equipped with a Quadrics network occupies an impressive fifth place in the November 2002 TOP500 list. Other network products include *ServerNet* [108], which was proprietary of Tandem, now part of HP; the scalable coherent interface (SCI), which was designed for NUMA systems but now is also used for clusters; and the *atomic low latency* network (Atoll), which has comparable latency and bandwidth to Quadrics.

Overall, for the network technology it is possible to consider three major products: Fast Ethernet or Gigabit Ethernet, in the cases where standard and price are the major concerns; Myrinet when fault tolerance and better performance are the goals; and Quadrics whenever the best performance is required at any cost.

2.6 PARALLEL INPUT/OUTPUT

A result of the increasing scale of the problem size in parallel applications is the large amount of input data handled by those applications. In many cases the applications actually become I/O bound. Therefore, the storage operations can no longer be the simple serial input/output operations that exist in regular systems. To deliver large amounts of data in an acceptable amount of time there is a need to use parallel storage systems. The next sections will overview some techniques to achieve this goal.

2.6.1 FILE DISTRIBUTION

One opportunity for a file system of a parallel machine lies in exploiting the fact that whenever a node wishes to write a file to disk it should be able to do it in any disk of the whole machine and not only in its own local disk. The name *file distribution* was given to this property as it allows for the file to be distributed on the nodes of the system, independent of the node which is the owner of the file. One way to achieve this goal is to add a layer to the storage system, to make it node-independent. From that layer, an I/O request may be seen as a regular client–server operation where the client requests the file operation and either a local or remote server will execute the operation and return the result to the client.

2.6.2 FILE STRIPPING

After the machine offers the support for a file to be distributed on any node, it is possible to apply optimizations to increase the performance for file access. In particular, to increase the file access bandwidth, it is possible to partition the file into different portions called *stripes* and read/write from or to these stripes simultaneously. As long as there is available bandwidth in the interconnection network, this technique reduces the file access time by increasing the number of simultaneous disk accesses. The location of a stripe may be determined using a very complex algorithm to reduce the data exchanges or a simple *round-robin* algorithm. Figure 2.14 depicts an example of writing and reading a large file using stripping.

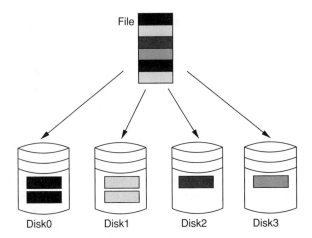

FIGURE 2.14 File access using stripping.

2.6.3 COLLECTIVE I/O

To improve the performance it was suggested to partition the data into small strips as was presented in Section 2.6.2. The result is that now the data is spread over a significant number of noncontiguous blocks on the disks. This in turn creates a severe overhead in retrieving the data from the disk. One technique used to alleviate this problem is called *collective I/O*, which basically combines different I/O requests. This technique though applies only to situations where all nodes are executing the same operation, at the same time. In addition, there is a need for each node to know what data the other nodes are requesting. The result is the following: instead of each node issuing requests for its own data blocks, which are noncontiguous, it requests a large contiguous block, it keeps its own blocks and sends the other blocks to the corresponding nodes. With this technique the messages exchanged are larger (blocks) to reduce the traffic and the message header overhead.

2.6.4 RAID

Although the disk drive capacity is increasing at a significant rate, this is still not enough for very large applications as they require, very large throughput. Consequently, Patterson et al. [109] suggested that instead of using large capacity disks, it is preferable to use an array of small capacity disks. This configuration is known as Redundant Array of Inexpensive Disks (RAID). Although there is only a single physical RAID configuration where all disks are connected through an interconnection network, logically RAIDs may be configured in different ways. Each different logical RAID configuration offers different characteristics. These logical configurations are also known as RAID levels. According to Chen et al. [110], there are seven RAID levels ranging from RAID Level 0 to RAID Level 6.

RAID Level 0, also known as nonredundant, is the lowest RAID level which supports only data stripping and no data redundancy. This configuration should be used if performance is the main concern.

RAID Level 1, also known as mirrored, is a RAID configuration where each data element is stored into two distinct disks simultaneously. This means that, in practice, we are using half of the disks as a mirror image of the other half. This configuration supports both stripping and redundancy. In case of a failure in a certain disk, with this configuration the system may immediately recover from the mirror disks. The high read/write performance of this configuration is achieved at a high hardware cost as only 50% of the disks contain useful data, whereas the rest include redundant data.

RAID Level 2, also known as error-correcting code memory (ECC), is a configuration that maintains the redundancy of Level 1 at a lower cost. This is achieved by using Hamming codes [111]. This technique requires a number of redundant disks, which is proportional to the logarithm of the number of data disks. In the case of having four data disks, using this configuration requires to have three redundancy disks. The reason to have multiple redundancy disks is to determine which data disk has suffered the failure. To recover the lost data, only one disk is needed.

RAID Level 3, also known as bit-interleaved parity, assumes that there is a knowledge of which is the disk that suffered the failure. Consequently it requires only one extra redundancy disk where the parity bits are stored. In this configuration, for each set of bits in the data disks there is a corresponding bit in the parity disk. Whenever there is a failure, all the data bits and corresponding parity bit are read and a parity algorithm is used to reconstruct the lost bit. Notice that write operations of large amounts of data will suffer from this configuration, as for each bit update there is a need to read all the other data bits and generate the new parity bit. The bandwidth required is therefore very large.

In RAID Level 4, also known as block-interleaved parity, the data granularity is changed from bit to block. This configuration improves the performance of writes for larger data sizes. In this case, whenever a block needs to be modified, the parity algorithm is called to calculate the "difference" between the old and the updated data, which together with the old parity information results in the calculation of the new parity block. The major disadvantage of this and other previous RAID levels is that although the system may be receiving write requests to data blocks on different disks, these requests will be serialized at the write on the parity disk. As a consequence the access to the parity disk is a major bottleneck.

To solve the parity disk bottleneck, in RAID Level 5, also known as block-interleaved distributed-parity, the parity blocks are distributed among the system's disks. This configuration reduces the potential conflicts between different requests to the storage system.

All previous RAID levels use redundancy to cover single disk errors. In larger systems where more disks are used, a larger number of errors becomes more probable. RAID Level 6, also known as $p+q$ redundancy, is a configuration that supports up to two disk failures using a minimum of two redundant disks. The $p+q$ redundant disks operate similarly to the redundant disks of the previous levels. The major difference is that in this configuration each write operation requires more disk accesses to complete to update the parity information on both disks.

To summarize, Table 2.1 includes the advantages and disadvantages for each of the most common RAID levels. The applications that best fit each different RAID level are the following: for RAID levels 0, 5, and 6 they are video production and editing, image editing, prepress applications, and any other application that requires high bandwidth; for RAID levels 1 and 6 the typical applications are accounting, payroll, financial, and any application requiring high availability.

Finally, it is relevant to notice that RAID technology is becoming more accessible as, for example, Intel server motherboards are equipped with RAID controllers as standard.

2.6.5 PARALLEL SYSTEM EXAMPLES

Among others, the techniques that were described in the previous sections have been successfully implemented in different parallel storage systems. In most cases these systems have been developed by the manufacturers of the parallel architectures to offer storage support to their machines. Examples of such systems are the Intel PFS [112] developed for the Intel Paragon multiprocessor systems, the IBM PIOFS [113] used on the IBM SP2 multiprocessor, and the SGI XFS [114] for

TABLE 2.1
Advantages and Disadvantages for Most Common RAID Levels

System	Advantages	Disadvantages
RAID 0	Simple, good performance	Not fault-tolerant
RAID 1	Simple, no rebuild necessary upon failure	Large disk overhead
RAID 4	Low ratio of parity disks	Bottleneck in parity disk access
RAID 5	Low ratio of parity disks	Parity calculation on regular and rebuild operations
RAID 6	Handles multiple disk failures	Larger number of parity disks

the SGI Challenge servers. Other systems developed with the same objectives, but for the cluster architecture are the SGI CXFS [115] and the Linux PVFS [116].

Although all the systems described above offer the necessary functionalities required for the storage support of a parallel application, they suffer from the problem of each one having its own different application interface (API). As part of their API, all parallel systems support the standard UNIX storage operations, but those do not include the efficient parallel file operations. Consequently, an application that uses the efficient parallel operations and is written for one of the systems cannot be immediately ported to another system. To address this problem, I/O was one of the topics in study for the preparation of the MPI-2 specification [117]. As a result MPI-IO [118] was proposed as a part of MPI-2. An example of an implementation of MPI-IO is ROMIO [119] which runs on several different machines such as IBM SP, Intel Paragon, HP Exemplar, and SGI Origin2000.

2.7 CONCLUSION

Although it is not necessary for a user or programmer to fully understand the architecture to execute a parallel application, the material presented in this chapter should help the user to decide which architecture is better for a certain program and how to tune a program for a certain architecture.

Regarding the type of architecture to choose, on one hand, if the program is a legacy code or its algorithm was originally done for serial execution then, the shared-memory architecture will be the best fit. This is justified by the fact that the programming effort to port a serial code to a shared-memory parallel system is minimal. On the other hand, if the program is explicitly parallel or belongs to the class also known as embarrassingly parallel, then the best match will be the message-passing architecture as it will be able to scale to larger systems. Another factor determining the decision of the system may be the required scale of the machine. If the program is to be parallelized on a small scale (up to 32 processors) then the best match will be a shared-memory system as it is easier to program. Otherwise, if the program is to be parallelized on a large scale then it will be required to use message-passing machines as they scale better.

In the cases that the target architecture is configured with a commodity network, then if cost or the use of a standard protocol is the major concern Ethernet should be chosen. If, on the other hand, performance and fault tolerance are the main goals, Myrinet should be the choice. Finally, if performance at any cost is the target, then the best choice would be the Quadrics network.

As far as the storage system is concerned, the machine should be equipped with a parallel system software layer allowing files to be distributed among the system. Regarding the hardware, if the application requires high bandwidth and does not require fault tolerance, then the best match would be the simple RAID Level 0. Otherwise, if fault tolerance is an issue for the application, the choice should be between RAID Levels 5 and 6, depending on the number of failures required to be supported.

Regarding the tuning of the application, it is important to mention that although the shared-memory architecture transparently handles the exchanges of data between the processors, it is important for the programmer to be aware of techniques to improve the performance. The first technique is to avoid placing close together variables that are going to be used by different processors. This is to avoid mapping these variables to the same cache line, creating unnecessary coherence traffic due to what is known as false-sharing. Also, the variables that are going to be mostly used by a certain processor should be initialized by that processor. This is to take benefit of the first-touch page placement algorithm, which places a memory page into the processor that issues the first reference to that page. Lastly, the programmer should choose systems with more relaxed memory consistency models and guarantee the consistency of the data using synchronization primitives, like the lock and unlock, to protect the updates of shared variables.

Concluding, this chapter has presented in more detail the characteristics of the different parallel architecture systems. This is helpful in order for the user or programmer to decide which architecture to choose for the execution of the target code. In addition, this description is also helpful in order for the programmer to tune the application resulting in a faster execution.

REFERENCES

[1] G. Hinton, D. Sager, M. Upton, D. Boggs, D. Carmean, A. Kyker, and P. Roussel. The microarchitecture of the Pentium 4 processor. *Intel Technology Journal*, 5(1):13–24, 2001.

[2] A. Peleg and U. Weiser. MMX technology extension to the Intel architecture. *IEEE Micro*, 16(4):42–50, 1996.

[3] P. Glaskowsky. Pentium 4 (partially) previewed. *Microprocessor Report*, 14(8): 10–13, 2000.

[4] K. Diefendorff. Athlon Outruns Pentium III. *Microprocessor Report*, 13(11):6–11, 1999.

[5] S. Oberman, G. Favor, and F. Weber. AMD 3DNow! technology: architecture and implementations. *IEEE Micro*, 19(2):37–48, 2000.

[6] C. May, E. Silha, R. Simpson, and H. Warren. *The PowerPC Architecture*. Morgan Kaufmann, San Franscisco, CA, 1994.

[7] K. Diefendroff, P.K. Dubey, R. Hochsprung, and H. Scales. Altivec extension to PowerPC accelarates media processing. *IEEE Micro*, 20(2): 85–95, 2000.

[8] T. Horel and G. Lauterbach. UltraSPARC-III: designing third-generation 64-bit performance. *IEEE Micro*, 19(3):73–85, 1999.

[9] Sun Microsystems. An Overview of the UltraSPARC III Cu processor. http://www.sun.com/processors/UltraSPARC-III/USIIICuoverview.pdf, 2002.

[10] M. Tremblay, J.M. O'Connor, V. Narayan, and L. He. VIS speeds new media processing. *IEEE Micro*, 16(4):10–20, 1996.

[11] C.J. Thompson, S. Hahn, and M. Oskin. Using modern graphics architectures for general-purpose computing: a framework and analysis. In *Proceedings of the 35th Annual IEEE/ACM International Symposium on Microarchitecture*, 2002, pp. 306–320.

[12] J. Zhou and K.A. Ross. Implementing database operations using SIMD instructions. In *Proceedings of the 2002 ACM SIGMOD International Conference on Management of Data*, 2002, pp. 145–156.

[13] J. Kruger and R. Westermann. Linear algebra oporators for GPU implementation of numerical algorithms. *ACM Transactions on Graphics*, 22(3):908–916, 2003.

[14] J. Bolz, I. Farmer, E. Grinspun, and P. Schroder. Sparse matrix solvers on the GPU: conjugate gradients and multigrid. *ACM Transactions on Graphics*, 22(3):917–924, 2003.

[15] K. Yeager. The MIPS R10000 superscalar microprocessor. *IEEE Micro*, 16(2):28–40, 1996.

[16] IEEE. *POSIX P1003.4a: Threads Extension for Portable Operating Systems*. IEEE Press, 1994.

[17] M.L. Powell, S.R. Kleiman, S. Barton, D. Shah, D. Stein, and M. Weeks. SunOS multi-thread architecture. In *Proceedings of the Winter 1991 USENIX Technical Conference and Exhibition*, 1991, pp. 65–80.

[18] B. Smith. Architecture and applications of the HEP multiprocessor computer system. In *Proceedings of the Fourth Symposium on Real Time Signal Processing IV (SPIE)*, 1981, pp. 241–248,

[19] R. Alverson, D. Callahan, D. Cummings, B. Koblenz, A. Porterfield, and B. Smith. The Tera Computer System. In *Proceedings of the 1990 International Conference on Supercomputing*, 1990, pp. 1–6,

[20] R.H. Halstead and T. Fujita. MASA: a multithreaded processor architecture for parallel symbolic computing. In *Proceedings of the 15th Annual International Symposium an Computer Architecture*, 1988, pp. 443–451.

[21] A. Agarwal, B.H. Lim, D. Kranz, and J. Kubiatowicz. APRIL: a processor architecture for multiprocessing. Technical Report MIT/LCS/TM-450, MIT, 1991.

[22] D. Tullsen, S. Eggers, J. Emer, H. Levy, J. Lo, and R. Stamm. Exploiting choice: instruction fetch and issue on an implementable simultaneous multithreading processor. In *Proceedings of the 23rd Annual International Symposium on Computer Architecture*, 1996, pp. 191–202.

[23] D. Tullsen, S. Eggers, and H. Levy. Simultaneous multithreading: maximizing on-chip parallelism. In *Proceedings of the 22nd Annual International Symposium on Computer Architecture*, 1995, pp. 392–403.

[24] J.L. Lo, J.S. Emer, H.M. Levy, R.L. Stamm, and D.M. Tullsen. Converting thread-level parallelism to instruction-level parallelism via simultaneous multithreading. *ACM Transactions on Computer Systems*, 15(3):322–354, 1997.

[25] Intel Corporation. Building cutting-edge server applications: Intel Xeon processor family features the Intel netburst microarchitecture with hyper-threading technology. http://www.intel.com/design/Xeon/whitepaper/WhitepaperNBHT.pdf, 2002.

[26] Intel Corporation. The Intel Pentium 4 processors: product overview. http://www.intel.com/design/Pentium4/prodbref/index.htm?iid=ipp_dlc_procp4p+prod_brief&, 2003.

[27] D. Marr, F. Binns, D. Hill, G. Hinton, D. Koufaty, J. Miller, and M. Upton. Hyper-threading technology architecture and microarchitecture. *Intel Technology Journal*, 6(1):4–15, 2002.

[28] Intel Corporation. Hyper-Threading Technology. http://developer.intel.com/technology/hyperthread/, 2003.

[29] D.F. Bacon, S.L. Graham, and O.J. Sharp. Compiler transformations for high-performance computing. *ACM Computing Surveys*, 26(4):345–420, 1994.

[30] B.R. Rau, D. Yen, W. Yen, and R. Towle. The Cydra 5 departmental supercomputer: design philosophies, decisions, and trade-offs. *IEEE Computer*, 22(1):12–35, 1989.

[31] M.S. Schlansker and B.R. Rau. EPIC: explicitly parallel instruction computing. *Computer*, 33(2): 37–45, 2000.

[32] H. Sharangpani. Intel Itanium processor microarchitecture overview. *Presentation at the Microprocessor Forum*, 1999.

[33] C. Dulong. The IA-64 Architecture at work. *IEEE Computer*, 31(7):24–32, 1998.

[34] S. Naffziger and G. Hammond. The Implementation of the next generation 64b Itanium microprocessor. Presentation at ISSCC 2002, also available at http://www.intel.com/design/itanium2/download/isscc_2002_1spdf, 2002.

[35] K. Olukotun, B.A. Nayfeh, L. Hammond, K. Wilson, and K. Chang. The case for a single chip multiprocessor. In *Proceedings of the Seventh International Conference on Architectural Support for Programming Languages and Operating Systems*, 1996, pp. 2–11.

[36] L. Barroso, K. Gharachorloo, R. McNamara, A. Nowatzyk, S. Qadeer, B. Sano, S. Smith, R. Stets, and B. Verghese. Piranha: a scalable architecture based on single-chip multiprocessing. In *Proceedings of the 27th Annual International Symposium on Computer Architecture*, 2000, pp. 282–293.

[37] K. Diefendorff. Power4 focuses on memory bandwidth: IBM Confronts IA-64, says ISA not important. *Microprocessor Report*, 13(16):1–6, 1999.

[38] M. Tremblay, J. Chan, S. Chaudhry, A.W. Conigliaro, and S.S. Tse. The MAJC architecture: a synthesis of parallelism and scalability. *IEEE Micro*, 20(6), 12–25, 2000.

[39] R. Merritt. Intel, sun sketch multiprocessor chip plans. EE Times, also available at http://www.eetimes.com/story/OEG20011210S0069, 2001.

[40] D.J. Lilja. Cache coherence in large-scale shared-memory multiprocessors: issues and comparisons. *ACM Computing Surveys*, 25(3):303–338, 1993.

[41] Intel Corporation. Ia-32 intel architecture software developers manual, volume 1 basic architecture. http://www.intel.com/design/Pentium4/manuals/24547010.pdf, 2003.

[42] J. Huynh. The Athlon MP processor. http://www.amd.com/us-en/assets/content_type/white_papers_and_tech_docs/26790A_Athlon_MP_white_paper_final.pdf, 2002.

[43] P. Sweazy and A.J. Smith. A class of compatible cache consistency protocols and their support by the IEEE futurebus. In *Proceedings of the 13th Annual International Symposium on Computer Architecture*, 1986, pp. 414–423.

[44] L. Lamport. How to make a multiprocessor computer that correctly executes multiprocess programs. *IEEE Transactions on Computers*, 28(9):690–691, 1979.

[45] J.R. Goodman. Cache consistency and sequential consistency. Technical Report 61, SCI Committee, 1989.

[46] M. Dubois, C. Scheurich, and F. Briggs. Memory access buffering in multiprocessors. In *Proceedings of the 13th Annual International Symposium on Computer Architecture*, 1986, pp. 434–442.

[47] K. Gharachorloo, D. Lenoski, J. Laudon, P.B. Gibbons, A. Gupta, and J.L. Hennessy. Memory consistency and event ordering in scalable shared-memory multiprocessors. In *25 Years ISCA: Retrospectives and Reprints*, pp. 376–387, 1998.

[48] J.R. Goodman, M.K. Vernon, and P. Woest. Efficient synchronization primitives for large-scale cache-coherent multiprocessors. In *Proceedings of the Third International Conference on Architectural support for Programming Languages and Operating Systems*, 1989, pp. 64–75.

[49] J. Mellor-Crummey and M. Scott. Algorithms for scalable synchronization on shared-memory multiprocessors. *ACM Transactions on Computer Systems*, 9(1):21–65, 1991.

[50] A. Silberschatz, P.B. Galvin, and G. Gagne. *Operating System Concepts*. John Wiley & Sons, New York, 2001.

[51] L. Lamport. A New Solution of Dijkstra concurrent programming problem. *Communications of the ACM*, 17(8): 453–455, 1974.

[52] T.E. Anderson. The performance of spin lock alternatives for shared-memory multiprocessors. *IEEE Transactions on Parallel and Distributed Systems*, 1(1):6–16, 1990.

[53] E. Freudenthal and A. Gottlieb. Process coordination with fetch-and-increment. In *Proceedings of the Fourth International Conference on Architectural Support for Programming Languages and Operating Systems*, 1991, pp. 260–268.

[54] Dell Computer Corporation. Dell poweredge 2600 server. http://www.dell.com/downloads/us/pedge/pedge_2600.pdf, 2002.

[55] Sun Microsystems. Sun fire 15k server. http://www.sun.com/servers/highend/whitepapers/SUN_SF15K_DS_01-03_V3.pdf, 2003.

[56] M. Marchetti, L. Kontothanassis, R. Bianchini, and M. Scott. Using simple page placement schemes to reduce the cost of cache fills in coherent shared-memory systems. In *Proceedings of the Ninthth IEEE International Parallel Processing Symposium*, 1995, pp. 380–385.

[57] D. Lenoski, J. Laudon, K. Gharachorloo, W-D. Weber, A. Gupta, J. Hennessy, M. Horowitz, and M. S. Lam. The Stanford dash multiprocessor. *IEEE Computer*, 25(3):63–79, 1992.

[58] Silicon Graphics Inc. Sgi origin 3000. http://www.sgi.com/pdfs/3399.pdf, 2002.

[59] A. Saulsbury, T. Wilkinson, J.B. Carter, and A. Landin. An argument for simple COMA. In *Proceedings of the First IEEE Symp. on High-Performance Computer Architecture*, 1995, pp. 276–285.

[60] J. Torrellas and D. Padua. The Illinois Aggressive Coma Multiprocessor Project (I-ACOMA). In *Proceedings of the Sixth Symposium on the Frontiers of Massively Parallel Computing*, 1996, pp. 106–117.

[61] D. Windheiser, E.L. Boyd, E. Hao, S.G. Abraham, and E.S. Davidson. KSR1 Multiprocessor: analysis of latency hiding techniques in a sparse solver. In *Proceedings of the Seventh International Parallel Processing Symposium*, 1993, pp. 454–461.

[62] D.J. Scales, K. Gharachorloo, and C.A. Thekkath. Shasta: a low overhead, software-only approach for supporting fine-grain shared memory. In *Proceedings of the Seventh International Conference on Architectural Support for Programming Languages and Operating Systems*, 1996, pp. 174–185.

[63] R. Stets, S. Dwarkadas, N. Hardavellas, G. Hunt, L. Kontothanassis, S. Parthasarathy, and M. L. Scott. CSM-2L: software coherent shared memory on a clustered remote-write network. In *Proceedings of the 16th ACM Symposium on Operating Systems Principles*, 1997, pp. 170–183.

[64] ESC. The Earth Simulation Center. http://www.es.jamstec.go.jp/esc/eng/index.html, 2003.

[65] V.S. Sunderam. PVM: A framework for parallel distributed computing. *Concurrency, Practice and Experience*, 2(4):315–340, 1990.

[66] Message Passing Interface Forum. MPI: a message-passing interface standard. Technical Report UT-CS-94-230, University of Tennessee, 1994.

[67] Top500. TOP500 list for November 2002. http://www.top500.org/list/2002/11, 2002.

[68] J.J. Dongarra, H.W. Meuer, and E. Strohmaier. TOP500 Supercomputer Sites, 11th ed. Technical Report UT-CS-98-391, University of Tennessee, 1998.

[69] S.L.Scott and G. Thorson. The Cray T3E network: adaptive routing in a high performance 3-D Torus. In *Proceedings of the Hot Interconnects Symposium IV*, 1996, pp. 147–155.

[70] T. Agerwala, J.L. Martin, J.H. Mirza, D.C. Sadler, D.M. Dias, and M. Snir. SP2 system architecture. *IBM Systems Journal*, 38(2-3):414–446, 1999.

[71] Lawrence Livermore National Lab. Hardware environment for ASCI white. http://www.llnl.gov/asci/platforms/white, 2002.

[72] IBM. ASCI white. http://www-1.ibm.com/servers/eserver/pseries/hardware/largescale/supercomputers/asciwhite/, 2002.

[73] M. Campbell. Knowledge discovery in deep blue. *Communications of the ACM*, 42(11):65–67, 1999.

[74] T. Anderson, D. Culler, and D. Patterson. A case for network of workstations. *IEEE Micro*, 15(1):54–64, 1995.

[75] T. Sterling, D. Savarese, D.J. Becker, J.E. Dorband, U.A. Ranawake, and C.V. Packer. BEOWULF: a parallel workstation for scientific computation. In *Proceedings of the 24th International Conference on Parallel Processing*, Vol 1, 1995, pp. 11–14.

[76] L.A. Barroso, J. Dean, and U. Holzle. Web Search for a planet: the Google cluster architecture. *IEEE Micro*, 23(2): 22–28.

[77] J.L. Hennessy and D.A. Patterson. *Computer Architecture: A Quantitative Approach*. 3rd ed. Morgan Kaufmann Publishers, San Francisco, CA, 2003.

[78] Google. Google's technical highlights. http://www.google.com/press/highlights.html, 2001.

[79] M. Baker, A. Apon, R. Buyya, and H. Jin. Cluster Computing and Applications, 2000. Cited in *Encyclopedia of Computer Science and Technology*, Vol. 45, A. Kent and J. Williams, Eds., Marcel Dekker, New York, 2002.

[80] M. Baker (Ed.) Cluster Computing White Paper, 2000.

[81] P. Lysne, G. Lee, L. Jones, and M. Roschke. HPSS at Los Alamos: experiences and analysis. In *Proceedings of the IEEE Symposium on Mass Storage Systems*, 1999, pp. 150–157.

[82] ANSI. High-performance Parallel Interface Mechanical, Electrical and Signalling Protocol Specification (HIPPI-PH). ANSI X3.183-1991, 1991.

[83] I. Foster and C. Kesselman. *The Grid: Blueprint for a Future Computing Infrastructure*. Morgan Kaufmann Publishers, San Francisco, 1999.

[84] M. Baker, R. Buyya, and D. Laforenza. Grids and grid technologies for wide-area distributed computing. *Software Practice and Experience*, 32(15):1437–1466, 2002.

[85] I. Foster and C. Kesselman. Globus: a metacomputing infrastructure toolkit. *International Journal of Supercomputer Applications*, 11(1):115–128, 1997.

[86] D. Abramson, J. Giddy, and L. Kotler. High performance parametric modeling with Nimrod/G: Killer applications for the global Grid? In *Proceedings of the International Parallel and Distributed Processing Symposium*, 2000, pp. 520–528.

[87] R. Buyya, D. Abramson, and J. Giddy. Nimrod/G: An architecture for a resource management and scheduling system in a global computational grid. In *Proceedings of the Fourth International Conference on High Performance Computing in Asia-Pacific Region (HPC Asia'2000)*, 2000, pp. 283–289.

[88] The DataGrid Project. http://eu-datagrid.web.cern.ch/eu-datagrid/.

[89] The CrossGrid Project. http://www.crossgrid.org/.

[90] DataTAG. http://datatag.web.cern.ch/datatag/.

[91] GriPhyN Grid Physics Network. http://www.griphyn.org/index.php.

[92] ApGrid. http://www.apgrid.org/.

[93] I. Foster, C. Kesselman, J. Nick, and S. Tuecke. The physiology of the grid: an open grid services architecture for distributed systems integration. http://www.globus.org/research/papers.html#OGSA, 2002.

[94] T. von Eicken, D. Culler, S. Goldstein, and K. Shauser. Active messages: a mechanism for integrated communications and computation. In *Proceedings of the International Symposium on Computer Architectures*, 1992, pp. 256–266.

[95] S. Pakin, M. Lauria, and A. Chien. High performance messaging on workstations: Illinois fast messages (FM) for Myrinet. In *Proceedings of the Supercomputing 95*, 1995, p. 55.

[96] M.A. Blumrich, C. Dubnicki, E.W. Felten, K. Li, and M.R. Mesarin. Virtual memory mapped network interfaces. *IEEE Micro*, 15(1):21–28, 1995.

[97] A. Basu, M. Welsh, and T. von Eicken. Incorporating memory management into user-level network interfaces. In *Proceedings of Hot Interconnects V*, 1997, pp. 27–36.

[98] L. Prylli and B. Tourancheau. BIP: a new protocol designed for high performance networking on Myrinet. In *Proceedings of PC-NOW workshop, IPPS/SPDP 1998*, 1998, pp. 472–485.

[99] Virtual Interface Architecture. http://www.viarch.org/.

[100] P. Buonadonna, A. Geweke, and D. Culler. An implementation and Analysis of the virtual interface architecture. In *Proceedings of the supercomputing 98*, 1998, pp. 1–15.

[101] InfiniBand Trade Association. Infiniband specification 1.0a, 2001.

[102] M-VIA: A high performance modular VIA for Linux. http://www.nersc.gov/research /FTG/via/.

[103] T.L. Sterling, J. Salmon, D.J. Becker, and D.F. Savarese. *How to Build a Beowulf. A Guide to the Implementation and Application of PC Clusters*. MIT Press, Cambridge, MA, 1999.

[104] Giganet. http://www.giganet.com.

[105] N.J. Boden, D. Cohen, R.E. Felderman, A.E. Kulawik, C.L. Seitz, J.N. Seizovic, and W.-K. Su. Myrinet: A Gigabit-per-Second Local Area Network. *IEEE Micro*, 15(1):29–36, 1995.

[106] Myricom. GM a message-passing system for Myrinet networks. http://www.myri.com/scs/GM/doc/html/, 2003.

[107] F. Petrini, W.-C. Feng, A. Hoisie, S. Coll, and E. Frachtenberg. The quadrics network: high-performance clustering technology. *IEEE Micro*, 22(1):46–57, 2002.

[108] ServerNet II. http://www.servernet.com.

[109] D.A. Patterson, G. Gibson, and R.H. Katz. A case for redundant arrays of inexpensive disks (RAID). In *Proceedings of the ACM SIGMOD International Conference on Management of Data*, 1988, pp. 109–116.

[110] P.M. Chen, E.K. Lee, G.A. Gibson, R.H. Katz, and D.A. Patterson. RAID: high-performance, reliable secondary storage. *ACM Computing Surveys*, 26(2):145–185, 1994.

[111] J.H. Conway and N.J.A. Sloane. Lexicographic codes: error-correcting codes from game theory. *IEEE Transactions on Information Theory*, 32(3):337–348, 1986.

[112] Intel Supercomputer Systems Division. *Using Parallel File I/O*, chap. 5. Paragon User's Guide. Intel Corporation, 1994.

[113] IBM Corporation. IBM AIX Parallel I/O File System: Installation, Administration, and Use. Document Number SH34-6065-01, 1995.

[114] M. Holton and R. Das. XFS: A Next Generation Journalled 64-Bit Filesystem with Guarateed Rate I/O. Silicon Graphics Inc.

[115] Silicon Graphics Inc. SGI CXFS: A High-Performance, Multi-OS SAN Filesystem from SGI. White Paper 2691, 2002.

[116] P.H. Carns, W.B. Ligon III, R.B. Ross, and R. Thakur. PVFS: a parallel file system for linux clusters. In *Proceedings of the Fourth Annual Linux Showcase and Conference*, 2000, pp. 317–327.

[117] Message Passing Interface Forum MPIF. MPI-2: Extensions to the Message-passing Interface. Technical Report, University of Tennessee, Knoxville, 1996.

[118] P. Corbett, D. Feitelson, S. Fineberg, Y. Hsu, B. Nitzberg, J.-P. Prost, M. Snir, B. Traversat, and P. Wong. Overview of the MPI-IO parallel I/O interface. In *High Performance Mass Storage and Parallel I/O: Technologies and Applications*, H. Jin, T. Cortes, and R. Buyya, Eds., IEEE Computer Society Press and Wiley, New York, 2001, pp. 477–487.

[119] R. Thakur, W. Gropp, and E. Lusk. On implementing MPI-IO portably and with high performance. In *Proceedings of the Sixth Workshop on Input/Output in Parallel and Distributed Systems*, 1999, pp. 23–32.

3 Fortran and Java for High-Performance Computing

R. Perrott, C. Phillips, and T. Stitt

CONTENTS

ABSTRACT

Sequential computing has benefited from the fact that there is a single model of computation on which language designers, compiler writers, computer architects etc. can base their activities. This has led to the success and widespread use of sequential computers. It has proved difficult to find a single model of parallel computation which can be applied to all aspects of parallel computing. As a result, parallel language design has been diverse and this has led to many different approaches to the provision of a language for high-performance computing.

One popular approach to providing a language for programming high-performance computers is to take an existing sequential language and extend it with facilities which permit the expression of parallelism. FORTRAN has been in the vanguard of such an approach and represents a viable approach which has manifested itself in a widely agreed version of FORTRAN known a high-performance FORTRAN. This has led to the wider use of and uptake of high-performance computing in science and engineering applications. The main features and attributes of this language are examined in this chapter. More recently Java has been used for such a task. Java has a much wider user base and can perhaps involve more users in high-performance computing. Again the main attributes of this approach are examined in order to give the reader a comparison basis on which to assess the two language extensions for the programming of high-performance computers.

3.1   INTRODUCTION

Over the last decade, substantial developments in the design and implementation of the hardware for parallel or high-performance computers have taken place. In the U.S., the Accelerated Strategic Computing Initiative was introduced to simulate nuclear testing and stockpiling and has accelerated the development of many aspects of high-performance computing. This Initiative has identified new hardware, software, and algorithms required for computational science and has provided substantial funding over a 10-year period to achieve the necessary advances. As a result, high-performance computers have been implemented, which have broken the teraflops/s barrier and are extending peak performance to the 100 teraflops/s range. At the same time, the requirements for programming languages to handle the increased parallelism in these machines are being better understood and realized but not at such a spectacular development rate.

The problems facing the development of reliable and efficient programming languages are more demanding than the development of new hardware. For example, in many cases, there is a large amount of legacy software available; there is a need to train programmers in new techniques to exploit parallelism; and a requirement for new compiler techniques to realize fast and efficient generated code.

In general, one of the major impeding factors to the development of parallel or high-performance computing is the lack of a single model of computation. Sequential computing has always had a single model, the von Neumann Model, on which to base the design and implementation of applications. High-performance computing has not been so fortunate as a single model of computation has proved difficult to identify. Currently there is no single model which incorporates all the different architectures that have been produced for high-performance computing. Such computers have been based on several different types of architectural models, the main types being the shared memory model and the distributed memory model. Both have been exploited to represent increasingly parallel machines and have been promoted by vendors and research groups as a viable model.

As the previous history of the area has shown, the development of effective software methods for programming such machines has not matched the pace of the hardware development, resulting in the inefficient use of these powerful machines in many instances.

One of the features that influence the acceptance of high-performance computing technology among potential users is the programming language. This is the first encounter that most users have when it comes to using high-performance computers. Essentially, there have been three main approaches which have been suggested for designing programming languages to promote the wider use of high-performance computers. These are as follows:

Invent a completely new language:
This approach ignores all existing languages and applications and proposes a new notation with features for the specification of parallel activity. The rationale is that it will facilitate the coherent

treatment of all aspects of programming high-performance machines, allowing the user to develop and to express a solution explicitly. The drawback, however, is that existing software will have to be reprogrammed in the new language, which is a labor-intensive and error-prone procedure.

Enhance an existing sequential language compiler to detect parallelism:
In this approach, the responsibility is placed on the compiler to detect which parts of the sequentially constructed program can be executed in parallel. The principal advantage of this approach is that existing sequential programs can be moved to the target parallel machine, exploiting its parallel facilities relatively inexpensively and quickly. However, not all inherent parallelism may be detected.

Introduce features to an existing language that deal explicitly with parallelism:
This approach should enable existing software to be adapted and transferred to parallel machines where appropriate by existing programmers. Many of the suggested extensions have been developed by different groups using the same language base, leading to definitions of nonstandard variants. This makes the production of a standard for such a language difficult.

This chapter concentrates on two approaches that have been used based on the languages Fortran and the relatively newcomer Java. The next section deals with Fortran and treats its development from a sequential language through its various standards, resulting in the language High-Performance Fortran. High-Performance Fortran is based on a data parallelism approach where most of the potential parallelism is expressed in the data with additional facilities given to the user to arrange the data in different ways to ensure that the overhead in the communication time to move data between the different processors is minimized. However, the counter to this criterion is trying to balance the load across the processors to ensure effective use of all the processors in the system. This has proved to be a difficult task on most existing machines. High-Performance Fortran does provide the facilities to help achieve these goals, however, the drawback has been that the compilers implementing these facilities have proved extremely difficult and demanding. This section gives a detailed example of the facilities available and a simple case study of matrix multiplication to demonstrate what new techniques are needed by the user and how to utilize them. The second language considered is Java, which originates in the object-orientated programming language group. Here, both the data and the methods of its manipulation are linked together and provided to the system by the user. This means that these components, data plus actions, can be reused in many similar situations. Although Java has been developed by a commercial company and has seen tremendous success because of the spread and development of the Internet, it does represent a serious base on which to introduce facilities for high-performance computing. There have been many attempts to do this and the one considered in this section is from the Java Grande Forum which produced High-Performance Java — HPJava. As in the Fortran language section, the features are examined before using the same case study, namely, matrix multiplication in the Java high-performance computing notation. Hence the reader has a means of comparing and evaluating the two approaches.

3.2 FORTRAN-BASED LANGUAGES

3.2.1 LANGUAGE PHILOSOPHY

Prior to the late 1950s, programmers of computers had to issue instructions to computers in machine/assembly code, which was a difficult and time-consuming task. The programmer had to analyze the application and break it down into simple operations for which equivalent machine code instructions were available. In 1954, John Backus [1] proposed the development of a programming language that would allow programmers to specify their requirements in a form, which was much closer to that of standard mathematical notation. An automatic translator would then be used to convert these high-level instructions identified by the programmer to the low-level instructions understood by the machine. A team of IBM developers led by Backus initiated a project to design the FORmula TRANslator System, which became known as FORTRAN. FORTRAN consisted of a set of standard higher level instructions and a compiler which performed the translation of these instructions into the equivalent machine code.

The first version of FORTRAN was released in 1957 and quickly became popular mainly because of its ease of use. Developing programs in FORTRAN proved much quicker and required less expertise than developing the equivalent programs in machine code. Several years after its release many variations of the original FORTRAN language had been developed by language designers wishing to adapt the language to suit their own requirements. An immediate consequence of this meant that it was difficult to transfer FORTRAN programs from one machine architecture to another.

In 1966, the American National Standards Association released the definitive FORTRAN specification, which became known as FORTRAN 66 [2]. This was the first time the international standardization process had been applied to a programming language. Since then, the language has been enhanced to provide additional features and to adopt more modern programming and data structuring abstractions. To date, three more standards have been issued, FORTRAN 77, Fortran 90, and Fortran 95 [3–5]. With the introduction of the Fortran 90, standard the name of the language was changed from upper to lower case letters.

3.2.2 EXTENSIONS FOR PARALLELISM

With the introduction and development of specifically parallel machines, new languages were required to enable a programmer to handle this parallelism and construct applications. The first parallel machines developed in the 1970s and 1980s, namely, the ILLIAC IV, the Distributed Array processor (DAP) and the Connection Machine (CM-2), [6–8] were single instruction, multiple data (SIMD) [9] architectures. In a SIMD computer, there are a large number of processors, one of which acts as a single controlling processor. This processor reads the program code and associated data and broadcasts the instruction to other processors in the system. Each processor then executes the broadcast instruction at the same time on its local data. It is not necessary for all the processors in the system to execute the broadcast instruction, giving flexibility to the introduction of parallelism. This became known as data parallelism.

In response to the demand for data parallelism, Fortran 90, a superset of FORTRAN 77, provided a range of data parallel operations based primarily on the use of array notation and corresponding intrinsic routines. The idea of an array "section" superseded the array "element" as operands in calculations, thereby providing implicit data parallelism. Shown below is a simple array section assignment operation in Fortran 90. In this example, all elements of the 4 by 4 matrix AA are assigned the value 3.0.

```
REAL, DIMENSION (4,4):: AA

AA(1:4,1:4) = 3.0
```

The colon symbol within the assignment specification indicates a range of indices i.e., 1:4 indicates the range 1 to 4 inclusive. A colon on its own indicates that the assignment is to range over all elements of the array dimension. The above statements could therefore be written as

```
AA(:,:) = 3.0
```

Data parallel aware compilers could schedule the operations so that the array section data could be operated on in parallel if an appropriate architecture was detected. For full data parallelism to be achieved, it was necessary that the array sections conform in shape and display no data dependence between them.

Below is a simple data parallel assignment instruction in Fortran 90. The scalar value 2 is broadcast to all array elements of A and used as the multiplicand.

```
REAL, DIMENSION(10) :: A

A = 2 * A
```

This instruction is equivalent to the following FORTRAN 77 elemental instructions:

```
A(1)  = 2 * A(1)

A(2)  = 2 * A(2)

....

A(10) = 2 * A(10)
```

The following example expresses an implicit data parallel operation in Fortran 90 using array sections:

```
REAL, DIMENSION(20) :: A, B, C

A(1:8) = B(10:17) + C(2:16:2)
```

In this example, the array sections conform in shape, each reducing to a vector of rank 8 with no data dependence. The equivalent elemental statements in FORTRAN 77 are

```
A(1) = B(10) + C(2)

A(2) = B(11) + C(4)

....

A(8) = B(17) + C(16)
```

Fortran 90 also provides a host of intrinsic routines that accept array sections as arguments. The following statements show the use of the data parallel SIN() function in Fortran 90:

```
REAL, DIMENSION(10) :: A, B

A = SIN( B )
```

The equivalent statements in FORTRAN 77 are

```
A(1)  = SIN( B(1) )

A(2)  = SIN( B(2) )

....

A(10) = SIN( B(10) )
```

The Fortran 90 WHERE statement is also ideally suited to processing on SIMD architectures. The WHERE statement provides a means to operate on only certain elements of an array determined by a masking condition, e.g.,

```
WHERE ( A < 0.0 ) A = A + 10.0
```

In this example, only negative values of the array A are operated on. Such an operation can be computed efficiently on data parallel machines.

Hence as a result of the emergence of SIMD machines FORTRAN 77 evolved into Fortran 90 to meet the demand from the scientific community for data parallel language support. As Fortran 90 compilers began to appear so did the shift in favor of large MIMD [9] architectures for scientific computing.

MIMD architectures consist of a number of processors linked by a communications network. Each processor of the machine is capable of executing its own program code and of communicating data to other processors. A program can therefore be split into a number of independent tasks (sequential subprograms) and each task allocated to a different processor. Common data required by a number of tasks can be passed via the communication network. These capabilities supported a new type of parallelism, task parallelism.

Task parallelism can be thought of as a group of concurrent operations. Each processor performs a particular operation on all the data and as the modified data becomes available it is passed to the next processor, which performs a different operation and so on until the end of the program is reached. In data parallelism, each processor would perform all the required operations but would only process a portion of the data. To give a simple example of task parallelism — if it was required to add two matrices A and B, divide the results by the matrix C and then add to matrix D:

```
E  =  (A  +  B)  /  C  +  D
```

then, in task parallelism, processor 1 would perform the addition of elements of matrices A and B. Processor 2 would divide the data received from processor 1 by C and processor 3 would take the output from processor 2, add it to D and store the result in E.

In data parallelism, the matrices A, B, C, D, and E would be divided equally among the three processors and each would perform the necessary addition, division, addition, and assignment on their third of the data. The output from each processor would then be collected together to form the complete matrix E.

Both task parallel and data parallel programs can be developed for MIMD machines using sequential programming languages supported by communication libraries. Task parallelism superficially appears more akin to sequential processing than data parallelism. Each task can be thought of as a subroutine with the data input and output requirements being similar to subroutine arguments. However, whereas in sequential processing, the subroutines are executed one after the other, in parallel processing, the subprograms are run concurrently and great care must be taken to ensure the integrity of the data shared by different tasks. The tasks must be synchronized to prevent deadlock or starvation of shared resources. This type of parallelism provides more flexibility than data parallelism, but is more difficult to use and debug.

Data parallelism allows a more sequential like processing of the program code but the insertion of the statements required to determine the sections of the data arrays on which each processor will work and the calls to the communication library can make the resulting program code cumbersome and difficult to maintain.

Based on the experience gained during the development of languages for SIMD machines, a quasistandard method of programming data parallel applications for MIMD machines emerged. This was defined as the single program, multiple data (SPMD) method. As the name implies, the same program is executed on all processors but the data used in the execution differs. The data, usually a large array, is distributed among the processors prior to program execution. Each processor then performs the program operations on its locally held data elements.

Many of the early data parallel languages were based on FORTRAN primarily because it was widely used in the scientific community where most data parallel applications were generated. Using FORTRAN as a basis meant that the transition from sequential to data parallel programming could be accomplished more easily. The languages, such as CFD and IVTRAN for the ILLIAC IV, and DAP FORTRAN for the ICL DAP [10, 11] included special FORTRAN-like statements, which gave the user some control over the distribution and manipulation of array elements. A major drawback of these languages was that the extent of the distributed dimensions of an array had to be equal to the number of processing elements available, 64 in the case of the ILLIAC IV and 4096 for the DAP. Therefore the addition of two matrices each containing 64 rows and 64 columns performed very efficiently on the DAP as each processor would receive one element of each matrix and add them together simultaneously. If, however, each of the two matrices contained 65 rows and 65 columns, the addition would have to be split into two stages, one, for say, the first 64 rows and columns and one for the last row and column. This "double" processing meant that much of the performance gained by the parallelization of the addition was lost.

In 1990, Thinking Machines Corporation made available a Fortran compiler for their Connection Machine series, which did not restrict the extent of a distributed array dimension. CM FORTRAN, as it was called, included all FORTRAN 77 syntax plus the array syntax introduced in Fortran 90. Unlike previous languages, CM FORTRAN enabled details of the data distribution to be coded in the form of directives to the compiler rather than in special Fortran instructions. This approach to data parallel programming was also being researched by other organizations. For example, Rice University in Houston, Texas, was developing the data parallel Fortran language, Fortran D, and the University of Vienna was working on a similar language, Vienna, Fortran [12, 13] . At Supercomputing '91, the Digital Equipment Corporation proposed that the work being carried out in this area should be co-ordinated and a standard set of Fortran extensions for data parallel programming drawn up. This resulted in the language High-Performance Fortran (HPF) and this name was adopted for the proposed standard. It is this language which has had the biggest influence on the field of Fortran parallel programming and is used as the basis for the next section.

3.2.3 HIGH-PERFORMANCE FORTRAN

In 1992, the first meeting of the High-Performance Fortran Forum (HPFF) [14] took place; this involved both academia and the commercial sector. The main institutional contributors to the formation of the HPF standard were Thinking Machines (Connection Machine Fortran), Rice and Syracuse Universities (Fortran77D and Fortran 90D), the University of Vienna (Vienna Fortran), and COMPASS, Inc. (various compiler projects).

The HPFF working group set itself the goal of designing and specifying extensions to the Fortran language to support parallel programming on a wide range of architectures and in particular to obtain high-performance on MIMD and SIMD computers with the ability to tune code across different architectures. Some secondary goals were also defined, for example:

- Simple conversion of existing sequential codes to parallel codes.
- Portability between parallel machines. Code that performs efficiently on one machine should also be efficient on other machines.
- Compatibility with existing Fortran standards.
- Simplicity. The resulting language should be easy to understand and use.
- Interoperability. Interfaces to other languages and programming styles should be provided.
- Availability and promptness. The language should be designed in such a way that compilers could be developed and available in a reasonable period of time.

The HPFF working group, which consisted of representatives from industry, academia, and government, met and corresponded regularly during 1992, and in May 1993 the first version of the High-Performance Fortran Language specification was produced.

The language consists of directive statements which, when inserted into standard Fortran code, instruct the compiler how data and computations are to be spread over a number of processors.

The directive statements are distinguished by the prefix !HPF$. There are two categories of directives, data mapping directives and data parallel directives. The first category indicates relationships between a program's arrays and dictates the method by which these arrays are to be distributed amongst the available processors. The second category instructs the compiler as to the sections of the code that can be executed in parallel. The HPFF working group decided to concentrate on the data parallel model, as this is well suited to problems involving manipulation of large arrays such as those found in many scientific and engineering applications. In practice, this choice was found to be somewhat limiting and extensions to the language to introduce task parallelism were approved in March 1997 (HPF version 2) [15].

In order to ensure that HPF compilers became available as soon as possible the Forum recommended a subset of the version 1 specification for initial development. The remainder of version 1 and, subsequently, the version 2 extensions could then be incorporated into the compilers in due course. The directives described below are those for which compilers are currently available.

HPF adopts the owner computes rule to handle array manipulation, that is, the processor owning the element appearing on the left-hand side of an assignment statement performs the assignment to that element in its local memory. The values of all the elements appearing on the right-hand side of the statement that do not reside in that processor's memory must be fetched before the commencement of the assignment. For example, if a vector A of 100 elements is divided evenly over two processors, then processor 1 will own elements 1–50 and processor 2 will own elements 51–100. When the following statements are executed,

```
DO I = 1, 100

   A(I) = A(I) / A(100)

ENDDO
```

Processor 1 will assign values to the elements A(1) to A(50) in its local memory after first fetching the value of element A(100) from processor 2. Processor 2 will assign elements A(51) to A(100) in its local memory and will require no communication with processor 1. The HPF compiler analyzes the assignment statements and where necessary inserts the communication primitives needed to fetch the required elements. In this way, the program computations are divided in the same manner as the data.

3.2.4 DATA MAPPING DIRECTIVES

HPF enables the user to distribute data according to the requirement of the application. To achieve this, a number of directives have been introduced which enable the user to map data onto processor configurations. Data arrays may be mapped directly onto processors by means of DISTRIBUTE/REDISTRIBUTE directives or may be aligned with other arrays which will result in them adopting the same distribution as the array to which they are aligned. The first directive that must be specified in the program is the PROCESSORS directive. This advises the compiler how many processors are to be used and how they are to be configured. For example, the directive

```
!HPF$ PROCESSORS(4:4)::Square
```

indicates that 16 processors are to be used and they will be configured in a 4×4 grid and referred to by the name Square. An illustration of this processor arrangement is shown in Figure 3.1.

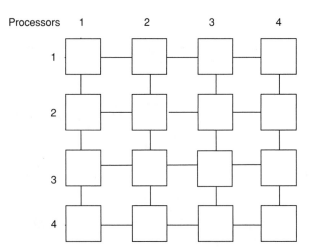

FIGURE 3.1 A 4 × 4 processor arrangement.

The shape of the processor arrangement is governed by the type of application but typically the processor arrangement will reflect the dimensionality of the data. That is, if the program deals mostly with one-dimensional arrays then a linear arrangement of processors will be specified. If two-dimensional arrays, then a two-dimensional arrangement, such as Square, will be used. Each processor will receive and execute its own copy of the program, communicating with other processors where necessary. HPF provides a data parallel directive, ON, which allows the user to select a subset of processors on which to perform certain sections of code thereby suppressing execution on the remaining processors. More details of this directive are given in the section dealing with the HPF data parallel directives.

The DISTRIBUTE and REDISTRIBUTE directives specify how the data is to be distributed to the processors. HPF gives the programmer a choice of four methods of distribution, BLOCK, GEN_BLOCK, CYCLIC, and Replicate (represented by an asterisk), which provide an efficient means of distributing the data for all standard scientific and engineering applications. A description of, and sample directives for, each of the methods are given below and illustrated in Figure 3.2.

In BLOCK distribution, the extent (total number of elements) of an array dimension is divided by the number of processors and each processor is allocated a contiguous block of array elements.

```
INTEGER, DIMENSION (16)::A

!HPF$ PROCESSORS :: Linear(4)

!HPF$ DISTRIBUTE A(BLOCK) ONTO Linear
```

The GEN_BLOCK distribution is similar to the BLOCK, in that contiguous blocks of elements are distributed, however, the number of elements assigned to each processor is specified by the user.

```
INTEGER, DIMENSION (16)::A

PARAMETER (N = /4, 2, 6, 4/)

!HPF$ PROCESSORS :: Linear(4)

!HPF$ DISTRIBUTE A(GEN_BLOCK(N)) ONTO Linear
```

In CYCLIC distribution, the elements of the array are distributed to the processors in a round robin fashion, either singly or in a group size defined by the user. Where the group size is greater than one the distribution is often referred to as block-cyclic.

	P1	P2	P2	P4
BLOCK	A(1), A(2), A(3), A(4)	A(5), A(6), A(7), A(8)	A(9), A(10), A(11), A(12)	A(13), A(14), A(15), A(16)
GEN_BLOCK	A(1), A(2), A(3), A(4)	A(5), A(6),	A(7), A(8) A(9), A(10) A(11), A(12)	A(13), A(14), A(15), A(16)
CYCLIC(1)	A(1), A(5), A(9), A(13)	A(2), A(6), A(10), A(14)	A(3), A(7), A(11), A(15)	A(4), A(8), A(12), A(16)
CYCLIC(2)	A(1), A(2), A(9), A(10)	A(3), A(4), A(11), A(12)	A(5), A(6), A(13), A(14)	A(7), A(8), A(15), A(16)
REPLICATE	A(1)............ ....A(16)	A(1)............ ....A(16)	A(1)............ ....A(16)	A(1)............ ....A(16)

FIGURE 3.2 Methods of distribution in HPF.

```
INTEGER, DIMENSION (16)::A

!HPF$ PROCESSORS :: Linear(4)

!HPF$ DISTRIBUTE A(CYCLIC(1)) ONTO Linear

or

!HPF$ DISTRIBUTE A(CYCLIC(2)) ONTO Linear
```

Replicated distribution results in each processor receiving a copy of an entire array dimension.

```
INTEGER, DIMENSION (16)::A

!HPF$ PROCESSORS :: Linear(4)

!HPF$ DISTRIBUTE A(*) ONTO Linear
```

The DISTRIBUTE directive is declarative, that is, it can only appear in the declaration section of the program and will therefore specify how the data is to be distributed at the start of the program. For certain applications, a static distribution scheme may prove inefficient as arrays are referenced in different ways in different sections of the program. For example, the ADI method of solving linear equations performs operations on the rows of an array and then on the columns. A static distribution, either by row or column, will be efficient for one set of operations but not for the other. HPF provides the executable directive, REDISTRIBUTE, to allow for a more flexible distribution strategy. The REDISTRIBUTE directive is of the same format, and performs the same function, as the DISTRIBUTE directive but can appear within the body of the program. During program execution, the data is re-distributed at the point indicated in the code by the insertion of the directive. The following directives would provide an efficient implementation of the ADI application.

```
REAL , DIMENSION ( 1000 , 1000 ) :: A , B , C
!HPF$ PROCESSORS(4,4):: Square
!HPF$ DYNAMIC :: A, B, C
!HPF$ DISTRIBUTE A (*, BLOCK ) ONTO Square
!HPF$ DISTRIBUTE B (*, BLOCK ) ONTO Square
!HPF$ DISTRIBUTE C (*, BLOCK ) ONTO Square
. . . .
. . . .
. . . .
!HPF$ REDISTRIBUTE A (BLOCK, * ) ONTO Square
!HPF$ REDISTRIBUTE B (BLOCK, * ) ONTO Square
!HPF$ DREISTRIBUTE C (BLOCK, * ) ONTO Square
. . . .
```

The above example includes the DYNAMIC directive. Any array that is re-mapped during program execution must be declared as DYNAMIC. The directive has no effect on the outcome of the program and is used solely to inform the compiler which arrays are to be re-mapped.

Relationships between items of data are specified using the ALIGN directive. This identifies corresponding elements of arrays and ensures that these elements will be distributed to the same processor. This in turn should reduce the amount of communication time required between processors when processing array elements. For example, given the following assignment statement, which copies array B to array A,

```
A(:) = B(:)
```

then the following ALIGN directive

```
!HPF$ ALIGN A(:) WITH B(:)
```

which aligns A(1) to B(1), and A(2) to B(2) etc. will ensure that A and B will be distributed in the same manner and so no interprocessor communication will be required to carry out the assignment. Arrays can be aligned with each other, as shown in the above example, or can be aligned with "templates." A template is an abstract indexed space that can be aligned and distributed. It enables the elements of an array to be repositioned along the axes, for alignment purposes, without the need to allocate additional memory space. For example, if the above assignment statement had been,

```
A(I) = B(I+6)
```

then aligning A(1) to B(7) and A(2) to B(8) etc. would link the arrays in the following way

```
index  1 2 3 4 5 6 7 8 9 10 11 12 13 14 15 16 17 18 19 20 21 22
-----------------------------------------------------------------
A                  1 2 3 4  5  6  7  8  9  10 11 12 13 14 15 16
B          1 2 3 4 5 6 7 8 9 10 11 12 13 14 15 16
```

This would result in the elements of array A extending across an index space of 22 elements whereas the array is only 16 elements in length. To obviate the need to declare an array of 22 elements with which to align A and B, a template can be used.

```
!HPF$ TEMPLATE C(22)

!HPF$ ALIGN A(I) WITH C(I+6)

!HPF$ ALIGN B(I) WITH C(I)
```

If the template C was then distributed cyclically, i.e.,

```
!HPF$ PROCESSORS (4):: Linear

!HPF$ DISTRIBUTE C(CYCLIC(1)) ONTO Linear
```

arrays A and B would be mapped as shown in Figure 3.3.

A template can also be used to expand a one-dimensional array to a two-dimensional array for distribution to a two-dimensional processor arrangement.

```
!HPF$ PROCESSORS(4, 4):: Square

!HPF$ TEMPLATE D(16,16)

!HPF$ ALIGN A(:) WITH D(*,:)

!HPF$ DISTRIBUTE D(BLOCK, BLOCK) ONTO Square
```

The asterisk in the alignment specification denotes replication and therefore this would have the effect of replicating vector A across all rows of D. The data mapping of vector A after alignment and distribution is illustrated in Figure 3.4.

HPF also includes a REALIGN directive, which allows arrays to be realigned during program execution. It has the same format as the ALIGN directive but can appear in the body of the program. An array that is to be re-aligned during program execution must be designated as aligned in the specification part of the program. For example, if the two assignment statements shown above

```
A(:) = B(:) and A(I) = B(I+6)
```

appeared in the same program, then, if A and B were initially distributed in the same manner, their referenced indices would reside in the same processors and therefore no alignment would be required. To avoid the need for interprocessor communication during the execution of the second assignment, statement alignment would be required. HPF will not accept a REALIGN directive before the second statement if the arrays were not originally aligned and therefore "dummy" alignment statements must be declared at the start of the program.

```
!HPF$ PROCESSORS (4):: Linear
```

Processor 1	Processor 2	Processor 3	Processor 4
A(3), A(7) A(11), A(15)	A(4), A(8) A(12), A(16)	A(1), A(5) A(9), A(13)	A(2), A(6) A(10), A(14)
B(1), B(5) B(9), B(13)	B(2), B(6) B(10), B(14)	B(3), B(7) B(11), B(15)	B(4), B(8) B(12), B(16)

FIGURE 3.3 An ALIGNED, CYCLIC(1) distribution.

Processor

Column Row	1	2	3	4
1	A(1), A(2), A(3), A(4)	A(5), A(6), A(7), A(8)	A(9), A(10), A(11), A(12)	A(13), A(14), A(15), A(16)
2	A(1), A(2), A(3), A(4)	A(5), A(6), A(7), A(8)	A(9), A(10), A(11), A(12)	A(13), A(14), A(15), A(16)
3	A(1), A(2), A(3), A(4)	A(5), A(6), A(7), A(8)	A(9), A(10), A(11), A(12)	A(13), A(14), A(15), A(16)
4	A(1), A(2), A(3), A(4)	A(5), A(6), A(7), A(8)	A(9), A(10), A(11), A(12)	A(13), A(14), A(15), A(16)

FIGURE 3.4 A replicated distribution.

```
!HPF$ TEMPLATE C(22)

!HPF$ ALIGN A(:) WITH C(:)

!HPF$ ALIGN B(:) WITH C(:)

!HPF$ DISTRIBUTE C(CYCLIC(1)) ONTO Linear

. . . .

A(:) = B(:)

. . . .

!HPF$ REALIGN A(I) WITH C(I+6)

!HPF$ REALIGN B(I) WITH C(I)

. . . .

A(I) = B(I+6)

. . . .
```

In addition to the above directives, HPF also includes the INHERIT directive, which is used at subroutine boundaries to provide link information between dummy and actual arguments. When it appears within a subroutine, the dummy argument to which it refers will inherit the same alignment and distribution characteristics as the actual argument. For example, the following code will specify that the dummy argument, array B, is mapped to the processors in the same manner as the actual argument, array A.

```
CALL Sub1(A)

. . . .
```

```
SUBROUTINE Sub1(B)

!HPF$ INHERIT(B)
```

3.2.5 DATA PARALLEL DIRECTIVES

Data parallelism in HPF is expressed primarily using the directives described above, however, there is an extended DO statement which can be used to introduce parallelism into a program. Before distributing the work to the processors, the compiler must ensure that data dependencies within the statements do not inhibit parallelization. For example, the iterations of the following DO loop could not be efficiently executed concurrently on different processors as each iteration requires the value of the element C, calculated in the previous iteration. Distribution of the iterations would therefore necessitate communication of previously calculated values between processors.

```
DO I = 2, 1000

    C(I) = C(I-1) + B(I)

ENDDO
```

Data dependency can usually be determined by examining the indices of the various elements in the statement, however in certain cases (such as indirect indexing) this is not possible and the compiler will err on the side of caution and assume that the statements must be executed sequentially. For example in the following DO loop, the compiler will not know the values contained in the vector, Vec, and will therefore assume that data dependency may be present between the values of C in each iteration and that the iterations must be executed sequentially.

```
DO I = 1,20

    C(Vec(I)) = C(I) + B(I)

ENDDO
```

However, if Vec contained values from 21 to 40, no such dependency would exist and the loop could be parallelised.

The data parallelism features of HPF allow the user to override the compiler's caution by explicitly stating that the code can be executed in parallel. In order to effect this, HPF provides three features, namely, the FORALL statement, the INDEPENDENT directive, and the PURE attribute. A brief description of each of these is given below.

The FORALL statement extends the Fortran 90 array assignment operations and as such is similar to a DO construct. However, it differs in two respects.

The elements of arrays to which data is assigned can be selected, thereby allowing a wider range of array shapes to be processed. No iteration order is performed over the indices of the statement. Each assignment to a left-hand side array element is considered atomic so the result of the FORALL will be the same regardless of the order in which the assignments are processed.

The following is an example of a FORALL statement, which will assign corresponding elements of array B to those elements of array A that initially contained values less than zero.

```
FORALL ( I = 1 : 1000,  J = 1 : 1000, A( I,J ) < 0)

    A( I,J ) = B( I,J )
```

The HPF specification also includes a FORALL construct, which allows multiple assignment statements to be included within the FORALL.

```
FORALL ( I = 1 : 1000,   J = 1 : 1000, A( I,J ) < 0)

   A( I,J ) = B( I,J )

   C( I,J ) = D( I,J )

ENDFORALL
```

FORALL constructs can be nested and can contain references to functions as long as the functions are PURE as defined below.

The PURE attribute applies to functions and subroutines. It is used to inform the compiler that the execution of a procedure will cause no side effects except to return a value or to modify the output arguments of the subroutine, that is, those designated as INTENT(OUT) or INTENT(INOUT).

The following subroutine, which calculates the total distance a traveling salesman must cover to visit eight towns is declared PURE as it will only modify the INTENT(OUT) parameter, Cost.

```
PURE   SUBROUTINE CalcCost(CostTable, Path, Cost)

     INTEGER,INTENT(IN)  ::  CostTable(8,8), Path(1:8)

     INTEGER,INTENT(OUT)::  Cost

     INTEGER            ::  I

     Cost = 0

     DO I = 1, 7

          Cost = CostTable(Path(I),Path(I+1)) + Cost

     ENDDO

END SUBROUTINE CalcCost
```

This subroutine can then be called from within a FORALL statement since it is guaranteed to change no other program variables apart from the subroutine's output argument, Cost. Each of the calls to the subroutine CalcCost, as specified in the FORALL statement below, can be performed in any order without affecting the outcome of the program and can therefore be distributed over the available processors.

```
INTEGER, PARAMETER:: N = 40320

INTEGER, PARAMETER:: NoOfTowns = 8

INTEGER, DIMENSION(N, NoOfTowns):: Routes

INTEGER, DIMENSION(NoOfTowns, NoOfTowns):: CostTable

INTEGER, DIMENSION(N):: Cost

FORALL   (I = 1, N) CALL CalcCost(CostTable,

          Routes(I, 1:8),Cost(I))
```

The attribute PURE is used to assert that functions and subroutines incur no side effects, the INDEPENDENT directive performs a similar function for DO loops and FORALL statements. It asserts to the compiler that the iterations of the DO loop or the operations in the FORALL statement can be executed independently in any order without changing the semantics of the program. Most compilers will try to extract as much parallelism as possible from a program but in certain cases, i.e., indirect addressing, the compiler does not have enough information to determine if the code can be parallelized without causing side effects. The INDEPENDENT directive, though it does not effect the outcome of the program, can improve the performance by allowing otherwise undetectable parallel loops to be executed concurrently.

For example, the FORALL statement below requires indirect addressing of the array X. The programmer may know that the vector, Vec, will have no two values the same and therefore that the order in which the assignments to X are performed is immaterial. The inclusion of the INDEPEN-DENT directive will pass this information to the compiler.

```
!HPF$   INDEPENDENT

FORALL (I = 1:100) X(Vec(I)) = A(I) * B(I)
```

An INDEPENDENT DO loop may have two associated clauses, NEW and REDUCTION. The NEW clause allows temporary variables to be written to within the body of the loop. The variables named in the NEW clause are considered to lose their values at the end of each iteration and therefore carry no interiteration dependency. An example of this is given below.

To conform to the rules of the INDEPENDENT directive, no data item can be read and written within the same iteration. The REDUCTION clause allows this rule to be broken if the data item is declared within the clause and is only used in a reduction operation, e.g., running sum, running product etc.

In the example below the variables, T1 and T2, are used only to build up the components of the final assignment and their values need not be retained once the assignment has been made. They are therefore declared as NEW. The variable Tot is used to accumulate all values of the assignment and is not referenced in any other way until the completion of the loop. Tot is therefore a reduction variable.

```
!HPF$ INDEPENDENT NEW(T1, T2), REDUCTION(Tot)

DO I = 1, N-2

    T1 = X(I) + X(I+1) + X(I+2)

    T2 = Y(I) + Y(I+1) + Y(I+2)

    Result(I) = T1 / T2

    Tot = Tot + Result(I)

ENDDO
```

Each iteration of the DO loop can be processed independently by any processor without affecting the values in the final Result array. T1 and T2 are no longer required once the Result element has been calculated and therefore carry no dependencies. The designation of the variable Tot as a REDUCTION variable instructs the compiler that the values calculated by each processor must be accumulated on conclusion of the loop.

HPF has one other data parallel directive, the ON directive, which allows the programmer to select the processors on which computations are to be carried out. It can refer to a set of processors or an array element. For example

```
!HPF$ PROCESSORS :: Linear(4)

!HPF$ ON HOME (Linear (1:3))

    ....
```

will execute the statement following the directive, on processors 1, 2, and 3.

```
INTEGER, DIMENSION(4):: X

!HPF$ ON HOME (X(1:3))

. . . .
```

will execute the statement following the directive, on the processors owning (the HOME processors of) elements X(1), X(2), and X(3).

Although only the processors specified, or implied, execute the statement other processors may be involved in communicating data to these processors. When it is known that the data needed to carry out a computation is resident on the set of processors involved in the execution of the statement (the active processor set), the information can be passed to the compiler by means of the RESIDENT clause. The compiler can then use the information to avoid unnecessary communication or simplify array address calculations. The directive

```
!HPF$ ON HOME (A(I)), RESIDENT(B, C, D)
```

asserts that B(I), C(I), and D(I) will be resident on the same processor as A(I). This may be because A, B, C, and D are all mapped in the same way or because arrays B, C, and D are replicated on the processors that contain A(I). Note: the RESIDENT clause does not change the location of the data in any way but merely confirms that, owing to previous alignment and distribution, the data will be resident on the specified processors.

HPF version 2.0 contains other directives such as the SHADOW directive to specify data buffering of boundary values for simplified processing of nearest neighbor operations and the TASK-REGION directive to implement task parallelism. Developing a compiler for a parallel language such as HPF is a long and complex process and as yet the complete range of directives have not been widely implemented.

3.2.6   EXISTING IMPLEMENTATIONS

Originally many commercial companies expressed an interest in developing compilers for HPF and many early implementations were initiated. Since then, however, several of the companies have gone out of business and many more have stopped supporting or developing HPF software. The commitment of time and expertise required to develop such a complex compiler was found to be commercially nonviable. Of the remaining companies, the Portland group has developed the most comprehensive compiler, PGHPF [16], which implements the full HPF version 1.0 plus some of the approved extensions from the version 2.0 specification and can be executed on a variety of platforms. Other commercially available compilers are Compaq FORTRAN [17] for the Alpha-based systems and xlhpf [18] for the IBM systems. There are also two public domain compilation systems Adaptor [19] from GMD-SCAI and SHPF [20] from the University of Southampton and VCPC, University of Vienna.

3.2.7   CASE STUDY

The following example illustrates how a matrix multiplication operation could be implemented in High-Performance Fortran. Matrix multiplication is chosen as it is frequently used in scientific computing and therefore will be familiar to many and also its implementation is concise and can be shown in full.

It is required to multiply matrix A by matrix B giving matrix C. A Fortran 90 version of this operation is shown below for three, 1000×1000 element arrays.

```
PROGRAM MULTAB

    IMPLICIT NONE

    REAL A(1000,1000), B(1000,1000), C(1000,1000)

    INTEGER I,J,K


    C = 0.0

    DO J=1,1000

        DO I=1,1000

            C(I,J) = SUM(A(I,:) * B(:,J))

        END DO

    END DO


    !!!!!!process results


END PROGRAM
```

To calculate element $C(i, j)$ it is necessary to multiply the ith row of A by the jth column of B. Therefore, when considering the parallelization of the program, the goal is to allocate the corresponding rows of A and columns of B on the same processor as result element C.

This can be achieved by inserting the following HPF directives at the start of the program. The resulting partitioning of arrays A, B, and C is shown in Figure 3.5. Figure 3.6 illustrates the contents of each processor after the distribution. The number of processors allocated to the program will depend on the number available. A 2×2 arrangement was chosen for this example so that the results of the distribution could be easily illustrated on a single page.

```
!HPF$ PROCESSORS(2,2)::Grid

!HPF$ DISTRIBUTE A(BLOCK, *) ONTO Grid

!HPF$ DISTRIBUTE B(*, BLOCK) ONTO Grid

!HPF$ DISTRIBUTE C(BLOCK,BLOCK) ONTO Grid
```

To calculate the value of an element of C, say C(100, 700), then row 100 of A must be multiplied by column 700 of B. As can be seen from Figure 3.6, these components reside on processor (1.2) as does the element C(100,700). The elements required to compute any element of C will always reside on the same processor as that element. Therefore this distribution will ensure that no interprocessor communication is required during the calculation of the elements of C. As no two elements of C are replicated, only one processor will own an individual element and only this processor will carry out the assignment. The computations of the elements of C will therefore be divided equally among the processors achieving an even load balance.

Although the distribution results in perfect parallelism, it does not bind together the corresponding rows and columns of A and B and therefore if the distribution of one was inadvertently changed, the efficiency in terms of communication and load balance would be lost. To guard against this, it is

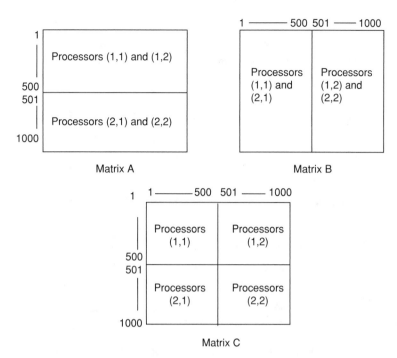

FIGURE 3.5 Partitioning of arrays for matrix multiplication.

best to align all the arrays before distribution thereby ensuring that their relationship remains fixed. This can be implemented using the following directives:

```
!HPF$ PROCESSORS(2,2)::Grid

!HPF$ ALIGN A(I,*) WITH C(I,*)

!HPF$ ALIGN B(*,J) WITH C(*,J)

!HPF$ DISTRIBUTE C(BLOCK, BLOCK) ONTO Grid
```

This will distribute copies of row i of A to all processors owning row i of C and copies of column j of B to all processors owning column j of C. Figure 3.7 shows this alignment for four elements of a set of matrices each containing four rows and four columns and named, for convenience, A, B, and C. If these arrays were distributed by BLOCK over four processors then processor (1,1) would

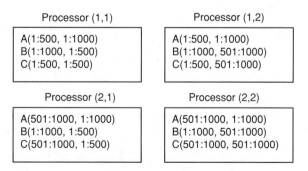

FIGURE 3.6 Data distribution for matrix multiplication.

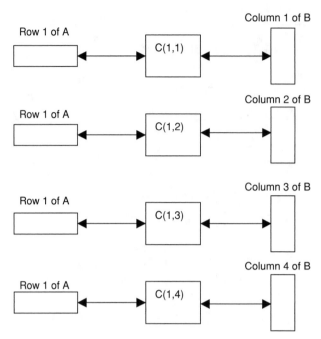

FIGURE 3.7 A detail of the alignment for matrix multiplication

hold the four elements of C shown in Figure 3.7, plus a copy of columns 1, 2, 3, and 4 of *B* plus four copies of row 1 of *A*. A good compiler should be able to detect the duplication of row 1 of *A* and only retain one copy.

The DO loops used to perform the matrix multiplication may be replaced by a FORALL statement or may remain but be declared INDEPENDENT to ensure that the iterations will be run concurrently.

```
!HPF$ INDEPENDENT

FORALL (J = 1,1000, I = 1,1000) C(I,J) = SUM(A(I,:) * B(:,J))


!HPF$ INDEPENDENT, NEW(I)

DO J=1,1000

    !HPF$ INDEPENDENT

    DO I=1,1000

        C(I,J) = SUM(A(I,:) * B(:,J))

    END DO

END DO
```

Incorporating both the data mapping and data parallel directives into the original Fortran 90 program results in the following fully parallelized version.

```
PROGRAM MULTAB
```

```
IMPLICIT NONE
REAL A(1000,1000), B(1000,1000),C(1000,1000)
INTEGER I,J,K
!HPF$ PROCESSORS(2,2)::Grid
!HPF$ ALIGN A(I,*) WITH C(I,*)
!HPF$ ALIGN B(*,J) WITH C(*,J)
!HPF$ DISTRIBUTE C(BLOCK, BLOCK) ONTO Grid
C = 0.0
!HPF$ INDEPENDENT
FORALL (J =1,1000, I = 1,1000) C(I,J) = SUM(A(I,:) * B(:,J))


!!!!!!process results


END PROGRAM
```

3.2.8 ADVANTAGES AND DISADVANTAGES

The production of the HPF specification was remarkable, not only in its original conception but also by the amount of cooperation, time, and expertise given by those involved in drawing it up. The basic aim of the project was to provide a standard method of developing data parallel programs that could be easily and efficiently ported to a wide variety of parallel machines. The language is relatively simple to learn and simple to program and meshes well with the sequential Fortran language so that existing programs can be easily converted. The HPF Forum fulfilled its objectives and the language designed, and since enhanced, does provide the user with a straightforward way of programming parallel applications.

The major disadvantages of HPF are:

- Compilers are difficult and expensive to write.
- The decision as to how the data is to be mapped is still the responsibility of the user.

The development of a compiler for HPF has turned out to be an extremely complex task exceeding the estimated effort required. The compiler must be able to analyze the program statements in great detail in order to recognize data dependencies, communication requirements, and opportunities for greater parallelism by simple program transformations. Although a lot of time and money has been spent on developing compilers for HPF because of the complexities involved the results are disappointing. Programs written in HPF and then compiled and executed can take much more time to run than their sequential equivalent [21]. It is necessary, therefore, to have a detailed knowledge of the language and of the particular compiler in order to produce an acceptably efficient program.

The adoption of the owner computes rule means that the data distribution strategy chosen has the greatest influence on the efficiency of the resulting parallel program; the data distribution dictates the amount of work each processor will do and the interprocessor communication required. Transferring data between processors is much slower than transfers from the processor's local memory and is a major factor in the degradation of performance. Likewise a poor distribution of work amongst the processors (load balance) will also degrade performance. For these reasons, a good distribution is considered to be one where the interprocessor communication is kept to a minimum and the load balance is even. If a bad distribution is chosen, then even the best compiler cannot produce efficient parallel code. It must also be considered that the benefit of the best choice of distribution can be entirely lost by a bad compiler. By leaving the choice of distribution to the user, HPF in effect leaves the major efficiency-influencing task of introducing parallelism into a sequential program to the user.

Once the programmer has determined the best distribution, coding in HPF is much more straightforward than a sequential language/communication library approach. As the compilers improve the performance gap between the two approaches should become narrower making HPF an easy to use and efficient development tool. In many ways, the concepts enshrined in HPF are ahead of their time; they address the main issues in expressing parallelism within an application using a sequentially constructed program. The features are comprehensive in their ability to express parallelism; however, the major problem has been the difficulty in constructing compilers across a range of machine architectures and to provide efficient execution. The commercial arguments have also mitigated against the promotion of HPF, however, it may be that the concepts will reappear in future languages when compiler technology is better understood and perfected.

3.3 JAVA-BASED LANGUAGES

3.3.1 OBJECT-ORIENTED LANGUAGES

Object-oriented programming (commonly referred to as OOP) is a programming methodology that allows the programmer to express programs as a collection of interacting objects, which have both state (i.e., the data) and behavior (i.e., the procedures). This approach differs from the more traditional structured programming methodology in which the programmer is primarily concerned with writing procedures and then defining the data, which the procedures manipulate. This traditional approach forces a separation between the functionality of the program and the data, which reduces the flexibility in modelling real-life entities through abstraction.

The emergence of OOP languages and the increasing impact of the software crisis created from the limitations of procedure-based methodologies [22] has meant that the approach encapsulated in OOP is well suited to tackling the complex problems of software engineering.

Although the rise to maturity of the OOP methodology has been relatively recent, its roots can be traced back to 1967 and the simulation language Simula designed by Ole-Johan Dahl and Kristen Nygaard of the Norwegian Computing Centre in Oslo [23]. Further research led to a popular successor Smalltalk (developed at Xerox PARC), which included classes, inheritance, and powerful dynamic typing mechanisms, all key components of modern object-orientated languages. The following decade saw the dominance of C++ with its object-orientated extensions to the C language.

More recently the Java language is competing for the title of most popular object-orientated programming language among software engineers. Java has been designed around object-orientated technologies unlike C++, which is an extension to the structured language C. With its suitability to Internet-related programming tasks and applicability to new grid-based applications, Java has the potential to dominate the software engineering industry for the foreseeable future.

There is still much debate about what defines a true object-based language. It is generally accepted that the ideas of abstraction, encapsulation, inheritance, and polymorphism are fundamental to the OOP paradigm. Abstraction can be loosely defined as the process of simplifying a complicated operation or idea and is a powerful tool in modern software engineering. In OOP languages, objects allow programmers to extend the traditional definition of an abstract data type (ADT). Not only can the state of the ADT be recorded (i.e., its data) but methods can be associated with the object defining its behavior. The ability to combine behavior with state is an important step forward in representing real-life entities and relationships. These extended data types are called classes and objects of this type are referred to as instances of that class. The following simple example in Java highlights these concepts:

```
class Simple_Clock
{
    int hours;  int minutes;

    int Get_Hours() { return hours; }
    int Get_Minutes() { return minutes; }
    void Set_Hours( int value ) {hours = value; }
    void Set_Minutes( int value ) {minutes = value; }
}
```

In the example above, the class *Simple_Clock* is defined whose state is represented by the variables *hours* and *minutes*. The state can be modified using the methods *Set_Hours* and *Set_Minutes* or accessed using the methods *Get_Hours* and *Get_Minutes*. In Java, one can specify the type of any values returned by a method by defining the return type before the method name. In this example, the methods *Get_Hours* and *Get_Minutes* return the current integer values of *hours* and *minutes*, respectively. The Java keyword *void* indicates no value is to be returned. Arguments passed to methods are defined inside round brackets after the method name as in most traditional languages.

A new instance (*My_Clock*) of the class *Simple_Clock* can be declared as follows:

```
Simple_Clock My_Clock = new Simple_Clock
```

When an instance of the class *Simple_Clock* has been created, its methods can be invoked in Java by

```
My_Clock.Set_Hours(7);
My_Clock.Set_Minutes(43);
int current_minutes = My_Clock.Get_Minutes();
int current_hours   = My_Clock.Get_Hours();
```

The first two statements initialize the state variables while the final two statements return the value of the state variables into the local variables *current_minutes* and *current_hours*.

Encapsulation or "information hiding" refers to the ability of an object to prevent unauthorized access to its internal state by another object; the object's own internal methods are the only interface to its state. Most OOP languages relax this constraint to varying degrees by allowing objects to access the internal state of other objects in a controlled way. In the next example, the class *Simple_Clock* has been modified to reflect the idea of encapsulation;

```
class Simple_Clock

    {


        public int hours;

        private int minutes;


        int Get_Hours() { return hours; }
        int Get_Minutes() { return minutes; }
        void Set_Hours( int value ) {hours = value; }
        void Set_Minutes( int value ) {minutes = value; }

    }
```

Here the Java keyword *private* restricts access to *minutes* to only the interface methods *Get_Minutes()* and *Set_Minutes()*. The keyword *public* allows all objects to access the state *hours* directly and by-pass the interface methods. Direct access to public *hours* data of *Simple_Clock* is obtained by the following;

```
My_Clock.hours
```

The property of inheritance allows classes to be defined that are extensions of previously created classes. The new class (referred to as a *subclass*) "inherits" the properties of the existing "parent" class or *superclass* and specialization can be achieved by adding or modifying only those properties that need to be changed. This facility encourages code reuse, minimizes the effort in duplication, and promotes consistency among similar classes and codes. To inherit a previously defined class, the programmer specifies the parent class to be inherited in the new class definition. The following simple example shows how inheritance is represented in Java.

```
class Better_Clock extends Simple_Clock

{

    private int seconds;


    int Get_Seconds() { return seconds; }
    void Set_Seconds( int value ) {seconds = value; }
    void Set_Hours( int value ) {hours = value;
    System.out.println("Hours set to" + hours); }

}
```

The new class (subclass) *Better_Clock* has inherited all the properties of *Simple_Class* (the parent or superclass) such as the state variables *hours* and *minutes* and their interface methods. It is also more specialized with the addition of the *seconds* state variable and corresponding interface methods. A further consequence of inheritance is method overloading. In the *Better_Clock* class, we have overwritten or overloaded the *Set_Hours* method to provide some extra functionality. A call to the *Set_Hours* method in the new class will now initialize the *hours* state and print a message to the terminal displaying its new value. An instance of *Better_Clock* is declared as follows:

```
Better_Clock New_Clock = new Better_Clock
```

The additional *seconds* state of *New_Clock* and the inherited *hours* state can be initialized using the following consistent statements in Java:

```
New_Clock.Set_Hours(5)

New_Clock.Set_Seconds(34)
```

Polymorphism is a powerful feature of OOP languages, which refers to their ability to assign a different meaning to the same entity in different contexts. Objects of different types can be collectively referenced by the same variable name; methods invoked on these references can be resolved depending on the type-specific behavior at run-time. Such resolution at run-time is called *dynamic binding*. The output to the terminal from the following statements on the new subclass defined above

```
Better_Clock Clock = new Better_Clock

Clock.Set_Hours(5)
```

will be

```
Hours set to 5
```

The following statement is equally valid in OOP languages such as Java

```
Simple_Clock Clock = new Better_Clock
```

As *Better_Clock* is a subclass of *Simple_Clock* it is legal to have a reference type which is a superclass of the actual object type. In general, if the reference type is a class, the object it refers to MUST be either that same type or a subclass of that type. Since *Better_Clock* is a subclass inherited from *Simple_Clock*, it can call any of the methods of *Simple_Clock* as if it were solely an object of that type.

A problem arises though when the following statements are called

```
Simple_Clock Clock = new Better_Clock

Clock.Set_Hours(7)
```

Which *Set_Hours* method is called?

The method that belongs to *Simple_Clock* class or the new overloaded method of the *Better_Clock* Class?

The answer lies in the "late binding" or "dynamic-binding" of the Java system. At run-time, Java knows what the "actual" object is rather than using the reference type. In this case, the overloaded method of the subclass *Better_Clock* will always be called. This is polymorphism — referring to an object in more than one way — which is a very powerful and flexible feature of many OOP languages, e.g., an array of reference type *Simple_Clock* may be set up, which contains many subclasses of this class as elements. By looping through the elements, the correct method will be called at run-time, depending on the object type. This method is much simpler than having many arrays each of a specific subclass type.

3.3.2 JAVA

The rapid rise in popularity of Internet applications over the past decade and recent interest in its e-commerce potential, has meant that Java has emerged as the most popular programming language for such applications. Not only is it a powerful, object-orientated language but its underlying interpreted nature is well suited to diverse heterogeneous environments found on the Internet.

Java's beginnings can be traced to the development of the language "Oak" [24] at Sun Microsystems in the early 1990s. A research team at Sun was interested in the idea of getting consumer products and devices to communicate through embedded computers. To achieve this goal, they needed to design a language that was hardware independent, robust, and enabled the quick development of small applications. It was decided that the new language would be based on C++ and would be completely object-based without the constructs of C++ that led to many programmer-induced errors. Over recent years, Java technology has been refined to produce an object-oriented language with powerful features [25].

It is an object-orientated language without the features of operator overloading and multiple inheritance. Such features are initially difficult to understand by novice OOP programmers, which leads to coding errors when applied incorrectly. In addition, garbage collection, namely, reclaiming storage locations was implemented automatically within the system to simplify its management. Various techniques are implemented in Java to determine automatically when memory associated with objects of dynamic extent (allocated with a new statement) are no longer referenced and can be reclaimed. In Java, programmers do not have the ability to deallocate objects explicitly and responsibility is passed to the system.

Java is also network-aware. Each Java system provides a library of routines, which allows programmers to handle TCP/IP protocols like hypertext transfer protocol (HTTP) and file transfer protocol (FTP); the former is the means to access information via world wide web pages written in platform-independent hypertext, while the latter is a protocol to send and receive files across a network both locally or remotely. As a result, network connections can be easily created and objects can be opened and accessed across the internet via web addresses or uniform resource locators (URLs).

As a language targeted towards heterogeneous distributed environments, a major requirement of Java is architecture independence and portability. As an interpreted language, it is compiled into a machine-independent intermediate language called "bytecode," which can then be interpreted into native machine code instructions with the use of a Java run-time system. The size of primitive data types are universally specified along with the arithmetic operations that can be applied to them, e.g., all integers are 32-bit integers and real numbers are represented by 32-bit IEEE 754 floating point numbers [26].

Security is also a motivating issue in the development of Java. As a language for distributed and networked environments, security is paramount to its success. Security provides the user with the confidence, that data and information will not be accessed illegally. To provide user confidence, authentication techniques based on public-key encryption are employed and secure controls prevent applications from unauthorized access to data structures and private data.

Another important feature of Java is its robustness. Java supports type checking at all levels including subscript-bound checks on array references. A modified pointer model also prevents the overwriting or corruption of data by controlling unauthorized access to memory locations.

Java also provides more advanced features such as client/server communication and concurrency at the language level. Java has a set of synchronization primitives based on the monitor and condition variable scheme proposed by Hoare [27]. Client/server distributed computing can be initiated through the use of Java sockets and remote method invocation (RMI).

3.3.3 PARALLELISM AND CONCURRENCY IN JAVA

A powerful feature of Java is its ability to support functional parallelism through its multithreaded language support. This support is based on coordinating access to data shared among multiple threads (which can be bound to separate processors). A thread, sometimes referred to as a *lightweight process*, is the sequential execution of instructions independent of other threads, which has its own stack but shares the same memory space (or heap).

The Java runtime system (or java virtual machine (JVM)) partitions data in a Java application into several data areas, namely, one or more Java stacks, a Java heap, and a method area. In a multithreaded application, each thread is allocated a Java stack, which contains data that no other thread can access, i.e., local variables, method parameters, and object references. In the example, classes discussed above the arguments passed to *Set.Hours* and *Set.Minutes* would be stored in the executing thread's stack. There is only one heap in the JVM which all threads share. The heap contains instances of objects dynamically allocated by the programmer. Primitive types can be stored on the heap only as components of classes. Arrays are also objects in Java and therefore are allocated on the heap. The method area contains method and constructor code and associated symbol tables. Class methods and constructors are compiled into Java bytecode and stored in the method area. Like the heap, the method area is shared by all threads.

As shared resources, the Java heap and method area, must be managed effectively when accessed concurrently to prevent undetermined behavior. To manage shared access to data, the JVM associates a unique lock with each object and class. A thread can only possess one lock at a time and can make a request to the JVM to lock a class or object. The JVM can decide to pass the lock to the thread sometime in the future or not at all. When a thread has finished with the lock it returns it to the JVM which in turn can allocate it to a new thread.

The JVM also uses monitors in conjunction with locks. A monitor guards certain sections of code ensuring that only one thread at a time can execute that code segment. Each monitor is associated with an object reference. When a thread wants to enter a section of code guarded by a monitor it must request the lock on that object before it can proceed. When the thread exits the code block, the lock is released on the referenced object.

In Java the process of sharing objects among threads is called synchronization. Two ways provided by Java to synchronize shared data are synchronized statements and synchronized methods.

To specify a synchronized statement a programmer uses the *"synchronized"* keyword with the object reference, e.g.,

```
class Example
{
    private int counter;
```

```
void Set_Value( int value)

{

    synchronized (this) {counter=value}

}

}
```

In the above example, the code within the synchronized block can only be executed by the thread if it acquires the lock on the current object (this). To synchronize a whole method a programmer must include the synchronize keyword as one of the method qualifiers, e.g.,

```
synchronize void Set_Value(int value)

{

    ....

}
```

In client–server applications, a server provides a service, which is used by a client. Over the Internet, a TCP/IP connection provides a reliable means for point-to-point communications for client–server applications. The client and the server establish a connection with each other over the network and bind a socket to each end of the communication. To communicate, the client and the server read from and write to the socket binding the connection. This low-level service is supported in Java using two socket classes, a class for the client and server sides, respectively.

Remote Method Invocation allows a client JVM to invoke a method on a server JVM. RMI is fully object-orientated and supports all Java data types with automatic garbage collection, this is also supported during RMI calls. The caller of a RMI can pass an object to the "callee" using Java's *object serialization*. Object serialization in Java is the process whereby an object is encoded in a machine-independent byte representation and transferred in such a way the "callee" receives the object (and not just a reference), which includes information on the class implementation. The class method can then invoked on the side of the "callee."

3.3.4 HIGH-PERFORMANCE JAVA

Recent studies in the use of Java for high-performance applications and numerical computing have raised some concerns over its suitability to handle such tasks. Recent research has highlighted various aspects of the Java language that contribute to poor performance in numerical and high-performance fields of calculation [28].

Java provides a host of exception handling routines that hinder the dynamic (re)arrangement of instructions undertaken by modern compilers to optimize code instructions and loop iterations. Such operations form the basis for highly efficient optimizations in numerical codes and provide a significant increase in performance in computationally intensive calculations. Loop unrolling is one such operation prevented by Java's exception model. Consider the following loop in Java which calls the function g() 100 times.

```
for (i = 0; i < 100; i++)

{

    g();

}
```

Using simple loop unrolling and replicating the loop body we can reduce the number of loop iterations from 100 to 50 as shown below.

```
for (i = 0; i < 100; i += 2)
{
    g ();
    g ();
}
```

Such dynamic reordering at compile time is not possible in Java.

Complex numbers are not supported as a primitive data type in Java. Applications involving complex numbers, interval mathematics, and multiprecision logic must make use of classes to model these abstract data types. Using classes affects the performance of numerical codes due to the time penalties acquired in the allocation and storage management of the class implementations.

Java also enforces strict semantics on floating point numbers. The use of the associative property for mathematical operators is forbidden in Java. For floating point numbers the associative law shown below does not strictly hold in most cases due to rounding.

```
( a + b ) + c /= a + ( b + c )
```

Languages such as Fortran use such a property in common optimizations.

Java does not support the fuse multiply add instruction $ax + y$; operations of this form are used extensively in matrix computations. Such an expression is typically computed in one floating point operation for languages that support it. As only a single rounding occurs for this instruction a more accurate result is produced. The absence of a fused multiply add instruction can be shown to have significant performance degradation effects on some platforms.

Java provides no support for multidimensional rectangular arrays, e.g., Figure 3.8 shows a 3×3 rectangular array A where all the rows and columns are of constant length. In Java, such two-dimensional arrays are represented as arrays of vectors with the extension of possibly different lengths producing jagged arrays such as shown in Figure 3.9. In Figure 3.9, each row of A can be a vector of varying length. To find the maximum row length in the array all rows need to be examined.

Languages such as Fortran that are used for their efficient array operations usually provide a contiguous memory space for rectangular arrays, providing fast efficient array access. In Java

A[1,1]	A[1,2]	A[1,3]
A[2,1]	A[2,2]	A[2,3]
A[3,1]	A[3,2]	A[3,3]

FIGURE 3.8 A 3×3 rectangular array.

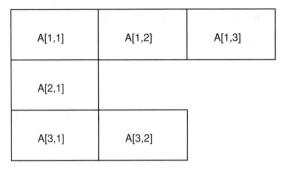

FIGURE 3.9 A standard multidimensional array in Java.

because of the array representation mechanism, access to an array element requires multiple pointer indirections which takes time. In addition, the enforced bound checks required by the Java security model also take time; all of which take place at run-time.

3.3.5 EXISTING IMPLEMENTATIONS

Currently there has been much research undertaken to solve the performance bottlenecks hindering Java's acceptance for high-performance computing applications. Several projects are underway to extend Java into a language capable of competing with Fortran and C in the world of scientific computing. These projects are showing potential in enhancing Java as a tool for high-performance numerical computing. The Hyperion project led by Phil Hatcher at the University of New Hampshire hopes to allow true parallel execution of multithreaded Java programs on distributed memory platforms [29]. Hyperion allows Java programmers to model a cluster of processors using a single Java Virtual Machine. Multithreaded Java programs can be directly mapped on to the cluster using Java's standard thread library. With the threads spread across the nodes, concurrent distributed execution can be achieved [30]. Hyperion also improves on performance by compiling Java "bytecodes" into native code.

The Manta project based at Vrije Universiteit, Amsterdam, uses highly efficient RMI to share objects among multiple threads. The shared memory model used by systems such as Hyperion is replaced by a model in which shared data is stored in remote objects which are accessed using Java's RMI. New implementations of Java and RMI were developed to enhance the efficiency of the Manta system. In addition, a native Java static compiler is used to compile Java source code into fast efficient executable binary code [30, 31].

The numerically intensive java (NINJA) programming environment has been designed to solve the performance problems encountered by Java in the field of intensive numerical computation. The NINJA project hopes to achieve this by supplying new class libraries for complex numbers and true multidimensional arrays while improving code optimizations using advanced compiler techniques. It is hoped that the Java compliant technology developed as part of the NINJA project can make Java programs perform as well on numerical codes as optimized Fortran and C codes [32, 33].

The Java Grande Forum is a collection of institutions, researchers, companies, and users who are working together to extend and promote Java as a programming environment for numerically intensive, high-performance applications or grande applications. The goals of the Java Grande Forum since its launch in 1998 are:

- Evaluating the suitability of the Java language and run-time environment for grande applications
- Initiating discussion among the Java Grande community
- Creating demonstrations, benchmarks, APIs, prototype codes
- Recommending improvements to the Java run time system applicable to grande applications.

The Java Grande Forum consists of two groups: the Numerics group and the Concurrency and Applications group. The Numerics group investigates issues of floating point representations and math libraries while the Concurrency group investigates message passing, benchmarking, and distributed computing. One development of this group has been message-passing java (MPJ) — a Java implementation of MPI [34].

Other projects including the Titanium project [35], Java-Party [36], and the development of just-in-time (JIT) [37] compilers show that there is a rich and diverse spectrum of research which holds much hope and promise for Java as a language for high-performance computing.

3.3.6 HIGH-PERFORMANCE JAVA (HPJAVA)

HPJava is "an environment for SPMD (single program, multiple data) parallel programming — especially, for SPMD programming with distributed arrays" [38]. Work on the design and development of HPJava was motivated by high-performance Fortran. The objective is to develop bindings to libraries for distributed arrays and general MIMD programming. The project has already released mpiJava — an object-orientated wrapper interface to the standard message-passing interface (MPI). The following is an overview of the main components of HPJava.

Two new classes have been used to extend the Java syntax in HPJava — the Group and Range classes. A group, or process group, defines a set of processes executing the SPMD program. A group is used to specify how data (e.g., arrays) are distributed across the processes as well as defining which processes execute a particular code fragment. Groups are represented in HPJava using subclasses (or inherited classes) of the Group class.

The class hierarchy is shown in Figure 3.10. Procs1 defines a "one-dimensional" process grid, Procs2 a "two-dimensional" process grid and so forth. Procs0 is a special case that defines a single process. The subclass constructors are called with a list of arguments defining the extent and shape of the grid. The required number of processes are requested from the total number of processes calling the constructor. For instance a 3×3 dimensional process grid is specified as:

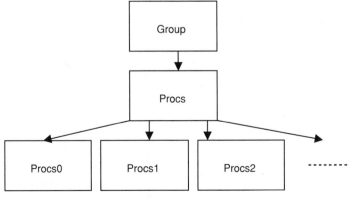

FIGURE 3.10 Hierarchy of the Group Class.

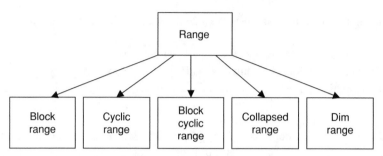

FIGURE 3.11 Hierarchy of the range class.

```
Procs2 grid1 = new Procs2(3,3) ;
```

while a one-dimensional array of four elements is specified as

```
Procs1 grid2 = new Procs1(4) ;
```

If the specified size of the process grid is larger than the set of processes, an exception occurs.

A range is a map from the integer interval $0, ..., N - 1$ to some dimension of a process grid. Ranges in Java are represented by objects of the Range class. To create a range object, the constructor associated with one of the subclasses representing a specific range distribution format is called. The hierarchy of the ranges class is outlined is Figure 3.11. The range subclasses are similar to their HPF counterparts. For example the CyclicRange class distributes the required data in a round-robin fashion. A grid containing four processes using CYCLIC distribution on a 16-element array A is shown in Figure 3.12. The CyclicRange class constructor is defined as follows

```
public CyclicRange(int N, DimRange D);
```

The DimRange subclass represents the range of co-ordinates of a process dimension. If a process grid p is defined in HPJava then the dimension range

```
p.dim(0)
```

specifies the first dimension of p while

```
p.dim(1)
```

the second dimension and so forth.

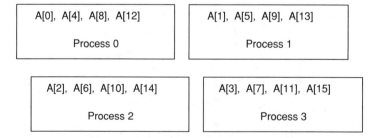

FIGURE 3.12 CYCLIC Distribution of 16 elements over a 2×2 processor grid.

The following HPJava statements construct the process grid shown in Figure 3.12

```
Procs2 p = new Procs2(2,2);

Range y  = new CyclicRange(16, p.dim(0))

float [[]] A = new float [[y]] on p;
```

The CollapsedRange class provides sequential undistributed dimensions of HPJava arrays. Distributed arrays in HPJava are represented as an in-built container class, similar to standard arrays in Java. They are true multidimensional arrays which support the array section notation available in Fortran 90. Elements of these arrays can be distributed over a process group. The notation used for arrays in HPJava are the square double-brackets [[and]]. A double-bracket pair contains one or more comma separate slots defining the rank of the array, e.g.,

```
Int [[,,]] A = new int [[2,4,5]]
```

This HPJava array statement creates a three dimensional rectangular integer array A with two rows, four columns, and a third dimension of length 5. The creation of distributed arrays in HPJava is shown below. Each section can contain a range object (if the dimension is distributed) or an integer expression (if the dimension is sequential):

```
Procs2 p = new Procs2(2,2);

Range x = new BlockRange(N, p.dim(0)) ;

Range y = new CyclicRange(N, p.dim(1)) ;

float [[,]] A = new float [[x, y]] on p;
```

In this example, a two-dimensional array A of type float is allocated with BLOCK distributed rows along the first dimension of p and CYCLIC distributed columns over the second dimension of p.

Accessing the element of a distributed array is not as trivial a matter as accessing an element of a standard array. Only the elements of the current process are local. The element of choice may be located on some other process. To overcome this problem, HPJava uses the idea of a *location entity*. A *location* is a particular element of a particular distributed range. The nth location (starting from zero) of range x is designated $x[n]$. Any subscript in a distributed dimension of an array must be a location in the range associated with that dimension of the array so that the location (array element) is local to the process. The location used as a subscript must be a named subscript scoped by an *at* or *overall* construct, e.g., the *at* construct in the example below scopes a named location. The body of the construct is only executed by processes that hold the specified location, e.g.,

```
Range x = new BlockRange(N, p.dim(0))

float [[]] A = new float [[x]] ;

at(i = x [0])

A[i] = 15
```

In this example, only the process containing the first element of the distributed range (pointed to by the location entity i, in the BLOCK distributed floating point array A), is active and initializes that element to the scalar value 15. Locations are comparable to FORALL indices or alignment dummies in HPF.

Subgroups or slices of the process grid can be constructed using HPJava's "/" operator. The "/" operator can be used to generate a slice of the distributed array to which the required location is

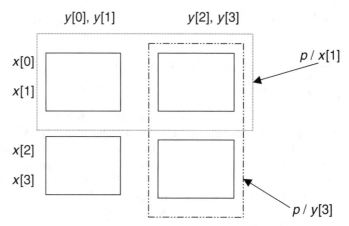

FIGURE 3.13 Array slices on the distributed ranges x and y on a 2×2 grid.

mapped. The following example identifies a 2×2 processor grid with BLOCK distributed ranges x and y distributed over its dimensions. The HPJava statement

```
p / i
```

generates a slice of the process grid dependent on the location "i" as shown in Figure 3.13. Additional operations are provided in HPJava such as sub-ranges and sub-arrays.

3.3.7   CASE STUDY: MATRIX MULTIPLICATION IN HPJAVA

To summarize the key features of HPJava, a simple case study is presented. The case study examines the multiplication of matrices from an HPJava viewpoint and provides an opportunity to compare HPJava with the high-performance Fortran (HPF) version described in an earlier section.

The case study focuses on the distributed multiplication of two matrices A, B with the result being stored in a further distributed matrix C. The matrices A, B, and C are represented globally using two-dimensional integer arrays of rank 1000×1000 (note that in the context of this example, the terms matrix and array are interchangeable).

The following Fortran 90 code segment can be used to compute the matrix multiplication on sequential and vector machines where A and B are suitably initialized arrays.

```
PROGRAM MULTAB

    IMPLICIT NONE

    REAL A(1000,1000), B(1000,1000),C(1000,1000)

    INTEGER I,J,K


    C = 0.0

    DO J=1,1000

        DO I=1,1000

            C(I,J) = SUM(A(I,:) * B(:,J))
```

```
        END  DO

    END  DO


    !!!!!!process  results


END  PROGRAM
```

HPJava provides a Java framework for constructing process grids and creating and manipulating distributed arrays whose elements are distributed over the extents of the process grid.

As discussed previously the efficiency of a parallel matrix multiplication lies in its ability to map the corresponding rows and columns of *A* and *B* onto the same process where the resultant *C* matrix element is located. It is this characteristic that determines the most efficient distribution of the arrays *A*, *B*, and *C*. With this in mind, the following HPJava program was developed to compute the multiplication of the distributed arrays *A* and *B*. For consistency with the HPF example, the arrays will be distributed over a 2×2 process grid.

```
Procs2 p = new Procs2(2, 2) ;

on(p)

{

    Range x = new BlockRange(1000, p.dim(0)) ;

    Range y = new BlockRange(1000, p.dim(1)) ;


    int [[,]]  C = new int [[x, y]] ;


    int [[,*]] A = new int [[x, 1000]] ;

    int [[*,]] B = new int [[1000, y]] ;


    ... initialize 'A', 'B'


    overall(i = x)

    overall(j = y)

    {

        int sum = 0 ;

        for(int k = 0 ; k < N ; k++) sum += A [i, k] * B [k, j];

        C [i, j] = sum ;

    }

}
```

Initially a 2×2 dimensional grid p is created which represents a four process subset of the total number of processes that started the program. The HPJava construct (*on p*) ensures that only these 4 processes will be active in the code block.

Arrays A and B are declared using HPJava's distributed array syntax. These declarations use the double-bracket notation, which differentiate them from Java's standard array declarations.

Ranges in HPJava specify the distribution format of the dimensions of distributed arrays. In HPJava, these ranges are represented as objects of the Range class. The range x uses BLOCK distribution to distribute 1000 elements over the first dimension of the process grid indicated by $p.dim(0)$. BLOCK distribution is specified by using the BlockRange subclass constructor of the Range class.

Similarly the range y distributes 1000 elements in BLOCK format over the second dimension of the process grid indicated by $p.dim(1)$. Additional subclasses of the Range class are provided in HPJava, which provide alternative distribution formats. By mixing distribution formats over different dimensions of the process grid, many complex and varied distribution schemes can be achieved.

Using the distribution formats x and y, the distributed array C is declared. Both dimensions of C are BLOCK distributed over the process grid. This distribution is shown in Figure 3.14. The distributed array A is BLOCK distributed by row. The asterisk in the array declaration indicates that the column elements are sequential as opposed to distributed and are of length 1000 (the CollapsedRange class could alternatively be used to define a sequential range on this dimension). Figure 3.15 shows the distribution of array A over the process grid p.

The first dimension of A is BLOCK distributed over the vertical extent of the process grid p. As the second dimension of A is sequential the whole column is duplicated on all processes.

A similar but opposite distribution is used for matrix B. The columns of B are BLOCK distributed over the horizontal extent of the process grid p while the rows are replicated on all processes as shown in Figure 3.16. By using these distributions for A, B and C the parallel computation of the matrix multiplication is trivial. As row i and column j are local to the process containing $C(i,j)$, the calculation is straightforward and no interprocess communication is required.

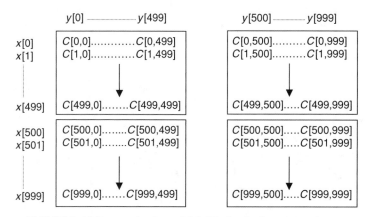

FIGURE 3.14 Row and column BLOCK distribution of the array C.

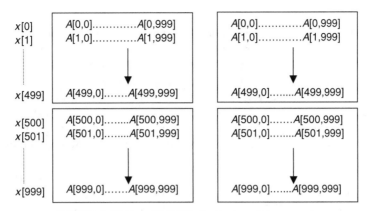

FIGURE 3.15 Row BLOCK distribution of the array A.

In HPJava parallel-distributed loops are provided using the overall construct. The overall construct takes a Range object as an argument. In our example the statement

```
overall(i = x)
```

. . . .

will be executed 500 times by the processes in the first row of the process grid and 500 times by the processes in the second row of the grid. The symbol i is known in HPJava as a location entity. In each iteration of the overall construct, i will point to the location of a particular locally held element in the range. These location entities are then used to subscript elements in the distributed array. As the columns of A and the rows of B are sequential, they can be subscripted using typical integer expressions.

Within the pair of distributed parallel loops a standard sequential Java *for* loop evaluates the sum of a row in A with its corresponding column in B. The variable *sum* stores this accumulation, which is finally assigned to its respective element in C outside the scope of the *for* loop.

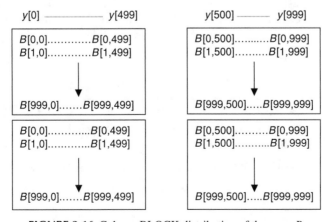

FIGURE 3.16 Column BLOCK distribution of the array B.

3.3.8 HPJAVA PERFORMANCE

The future of HPJava will depend on its ability to execute real applications efficiently and accurately. Currently there is real focus on benchmarking HPJava applications, which has highlighted some limitations of the HPJava system. Firstly, the HPJava base language is Java. Therefore commercial Java virtual machines (JVMs) might be required to perform the source translation. The question that concerns HPJava developers is " are these JVMs good enough ?" [39]. The second limitation concerns HPJava's evaluation of the *distribution format* of the distributed data at runtime. HPF uses static or compile-time information about the data distribution to optimize node code. It remains to be seen how HPJava's performance is affected by this lack of compile-time optimization.

The matrix multiplication operation described in the case study has been benchmarked by HPJava researchers on a single processor [39]. Table 3.1 shows the performance of a HPJava matrix multiplication operation on a 100×100 grid against other sequential languages. The table also shows the effect of various optimizations to the naive HPJava code on overall performance. The other sequential languages were compiled at their highest optimization levels.

The table shows that Java comes within 70–75% of the performance of C and F77 after the loop unrolling and strength reduction optimizations are applied. More importantly, HPJava with more complex array subscripting operations has slower but comparable performance with Java after similiar optimizations. Further benchmark results using HPJava are produced in Ref. [39].

Development of large applications in HPJava is still in its infancy but with the release of the *HPJava Development Kit v1.0* [40] Researchers worldwide can now evaluate Java as a new tool for high-performance computing. Already HPJava applications in various problem domains have started to appear such such as a HPJava implementation of a neural network to mine information on sunspot activity [41].

3.3.9 JAVA AND OPENMP

It is possible to implement shared memory parallel programs in Java using Java's native thread model [42]. Unfortunately, the use of threads can complicate program design and maximum efficiency on shared memory parallel architectures usually requires one thread per processor and all threads running during the whole lifetime of the parallel program [43].

An alternative and fast growing approach to shared memory parallel programming is OpenMP, which is a collection of compiler directives, library functions, and environment variables. Currently, OpenMP standards are defined for C/C++ [44] and Fortran [45].

TABLE 3.1
Matrix Multiplication Performance. All Timings in Mflops

HPJava	Naive	Strength reduction	Loop unrolling
HPJava	125.0	166.7	333.4
Java	251.3	403.2	388.3
C++	533.3	436.7	552.5
F77	536.2	531.9	327.9

In general, compiler directives allow the programmer to generate resultant code that is very close to the sequential version of the same program. Typically, an OpenMP program can be written, which will compile and execute even when the compiler is instructed to ignore the OpenMP directives. Therefore programmers can implement and maintain codes much more easily.

An important directive in OpenMP is the *parallel* construct. When a thread encounters this directive, a new thread team is created and executes the code in the immediately proceeding code block. The master thread awaits the completion of the code block for all other threads before continuing execution alone. The *for/do* directive is another common construct which can be used to divide loop iterations among a thread team which are then executed concurrently.

OpenMP directives are normally embedded in the source code as comments and allow programmers to apply basic but powerful parallelization to their code with little programming overhead.

Recent work has led to an OpenMP specification for Java [46]. This specification is very similar to the C/C++ OpenMP standard and describes the JOMP application programmer interface (API) and JOMP runtime library.

Results from early JOMP compilers look encouraging. Table 3.2 compares synchronization constructs in a JOMP code with a comparable code in Fortran using the equivalent OpenMP constructs for a two-dimensional fluid flow in a box application.

In general, all the JOMP directives using barrier synchronization outperform their Fortran equivalents and other analysis shows that the resulting code scales well. Further comparisons of the JOMP compiler with Java threads and MPI Java are available in Ref. [43].

With continuing improvements to the JOMP system an OpenMP-approach to shared-memory parallelization of Java programs looks very promising. Familiarity with OpenMP is increasing rapidly among both C/C++ and Fortran programmers and it is hoped that JOMP will make parallel programming in Java on shared memory architectures a more attractive alternative.

3.3.10 ADVANTAGES AND DISADVANTAGES

As explained in the previous sections, there is great potential for Java to become the language of choice for application developers in the area of high-performance and numerical computing. With the current formation of portals and the growing trend towards distributed applications, Java could

TABLE 3.2
Synchronization Overheads (in microseconds) on 16-Processor Sun HPC 6500

Directive	F90	JOMP
PARALLEL	78.4	34.3
PARALLEL+REDUCTION	166.5	58.4
DO/FOR	42.3	24.6
PARALLEL DO/FOR	87.2	42.9
BARRIER	41.7	11.0
SINGLE	83.0	1293
CRITICAL	11.2	19.1
LOCK/UNLOCK	12.0	20.9
ORDERED	12.4	47.0

TABLE 3.3
Advantages and Disadvantages of Java in HPC

Advantages	Disadvantages
Portability — Java code compiled to bytecode and interpreted	Interpreted languages are slow
Threads and concurrency part of language	Multithreading only for shared memory systems
Object orientated	Object-orientation not as rich as other OOP languages such as C++ - restricted pointer model, operator overloading, etc.
No pointers which reduces programmer errors and aids optimization	Array bound access checks on all array references reduces performance
Automatic garbage collection	Complex numbers only supported by classes. Reduces performance
Strict security model — not allowed to access objects or memory locations	No code or loop re-ordering
Similar to C and C++ for learning	Strict floating point semantics — no use of associative properties of mathematical operators and no fuse–multiply–add instruction
Good visualization and user interface facilities	No language level support for multidimensional arrays
Unified data format	RMI based on client/server model and not distributed memory model

become the dominant language if it can match its strengths in graphical visualization and Internet programming with improved performance and numerical accuracy. Table 3.3 summarizes some of the main reasons for and against the use of Java. It will remain to be seen if the reasons against can be overcome and finally signal Java as a practical and powerful language for high-performance computing.

3.4 SUMMARY

The topic of high-performance computing is receiving considerable attention in the US, Europe, and Japan. It is being hailed as a technology that can make contributions to the improvement of the quality of life in many sectors of human activity. It is now possible to construct parallel hardware that is both efficient and reliable and teraflop/s machines have become a reality. However, the software to utilize the tremendous power offered by these machines is proving difficult to master; the widespread use of parallel computing will not occur if this problem is not solved. The situation is further complicated in that there is no widely accepted single model of parallel computation on which software and algorithm designers can base their products.

It now appears that language designers can identify those features which are required, certainly in the area of data parallelism, to provide an effective notation for a user to construct scalable parallel programs in languages such as high-performance Fortran and HPJava. However, the difficulty has been that the compiler technology, to realize these features effectively and efficiently, is not yet mature enough to make these features attractive to a widespread audience of programmers. The previous sections have dealt in detail with two of the more promising approaches to providing a language for high-performance computing; it has illustrated the features and shown how to employ them to introduce a parallel solution. Although these features provide many of the abstractions needed in this area, it is the compiler implementation techniques which are now urgently required.

REFERENCES

[1] Jones Telecommunications and Multimedia Encyclopaedia http://www.digitalcentury.com/encyclo/update/backus.html

[2] American Standards Association, Fortran 66. ANSI X3.9-1966.

[3] American National Standards Institute, American National Standard Programming language FORTRAN 77. ANSI X3.9-1978, 1978.

[4] International Standards Organization, ISO/IEC 1539:1991, Information Technology — Programming Languages — Fortran, Geneva, 1991 (Fortran 90).

[5] International Standards Organization, ISO/IEC 1539:1997, Information Technology — Programming Languages — Fortran, Geneva, 1997 (Fortran 95).

[6] G. H. Barnes, R. M. Brown, M. Kato, D. J. Kuck, D. L. Slotnick, and R. A. Stokes. The ILLIAC IV Computer. *IEEE Transactions on Computers* C-17(8), 1968, 746–757.

[7] C. R. Jesshope and R. W. Hockney, editors. *The DAP Approach*, vol. 2, pp. 311-329. Infotech Intl. Ltd., Maidenhead, 1979. Infotech State of the Art Report: Supercomputers.

[8] T. Thiel. *The Design of the Connection Machine, Design Issues*, vol. 10, No. 1, Spring 1994, published by The MIT Press, Cambridge, MA. http://mission.base.com/tamiko/theory/cm_txts/di-frames.html

[9] M. Flynn. Some Computer organisations and their effectiveness, *IEEE Transactions on Computing*, C-21, 1972, 948–960.

[10] K. G. Stevens Jr., CFD; A FORTRAN-like language for the ILLIAC IV. Proc. Conference on Programming languages and compilers for parallel and vector machines, pp. 72–76, 1975.

[11] ICL. DAP: *Fortran Language Reference Manual*. ICL Technical Pub. 6918, ICL, London, 1979.

[12] S. Hiranandani, K. Kennedy, C. Koelbel, U. Kremer, and C. W. Tseng. An overview of the Fortran D programming system, languages and compilers for parallel computing, 4th Intl. Workshop, 1991.

[13] B. Chapman, P. Mehrotra, and H. Zima. Programming in Vienna Fortran, *Scientific Programming*, 1(1), 1992, 31–50.

[14] High-Performance Fortran Forum. http://www.crpc.rice.edu/HPFF/.

[15] J. H. Merlin and B. M. Chapman. High-Performance Fortran 2.0, Proc. Summer School on Modern Programming Languages and Models, Technical University of Hamburg, Hamburg, 1997.

[16] Z. Bozkus, L. Meadows, S. Nakamoto, V. Schuster, and M. Young. PGHPF — An Optimizing High-Performance Fortran Compiler for Distributed Memory Machines, *Scientific Programming*, 6(1), 1997, 29–40.

[17] Compaq Fortran. http://www.compaq.com/info/SP3754/SP3754PF.PDF

[18] IBM, *XL HPF Language Reference*, First edition, March 1995.

[19] C. A. Thole and T. Brandes. ADAPTOR. A Transformation Tool for HPF Programs, *Programming Environments for Massively Parallel Distributed Systems*, Birkhäuser Verlag, 1994, pp 91–96.

[20] J. H. Merlin, D. B. Carpenter, and A. J. G. Hey. SHPF: a Subset High-Performance Fortran compilation system. *Fortran Journal*, March/April 1996, 2–6.

[21] C. Phillips and R. Perrott. Problems with Data Parallelism, *Parallel Processing Letters*, 11(1), 2001, 77–94

[22] G. Booch. *Software Engineering with Ada*. The Benjamin/Cummings Publishing Company, Menlo Park, CA, 1987.

[23] J. R. Holmevik. Compiling SIMULA: a historical study of technological genesis. *Annals of the History of Computing*, 16(4), 1994, 25–37.

[24] OAK Language Specification. http://today.java.net/jag/old/green/OakSpec0.2.ps

[25] The Java^{TM} Language specification, Sun Microsystems. http://java.sun.com/docs/books/jls

[26] 754-1985 IEEE Standard for Binary Floating-Point Arithmetic, 1985.

[27] C. A. R. Hoare, Monitors: An Operating System Structuring Concept, *Comm. ACM* 17, 10:549–557. 1974.

[28] Java Numerics Issues: Preliminary Assessment. First meeting of the Numerics Working Group of the Java Grande Forum held in Palo Alto on March 1, 1998.

[29] G. Antoniu, L. Bougè, P. Hatcher, M. MacBeth, K. McGuigan, and R. Namyst. The Hyperion system: Compiling multithreaded Java bytecode for distributed execution. *Parallel Computing*, 27, 2001, 1279–1297.

[30] T. Kielmann, P. Hatcher, L. Bougs, and H. E. Bal. Enabling Java for high-performance computing. *Commun. ACM.* 44(10), 2001, 110–117.

[31] Manta fast parallel java. http://www.cs.vu.nl/manta/

[32] J. E. Moreira, S. P. Midkiff, M. Gupta, P. Wu, G. S. Almasi, P. V. Artigas: NINJA: Java for high-performance numerical computing. *Scientific Programming*, Vol. 10(1), 19–33, 2000.

[33] Ninja: Numerically intensive java. http://www.research.ibm.com/ninja/

[34] B. Carpenter, V. Getov, G. Judd, A. Skjellum, and G. Fox. MPJ: MPI-like Message Passing for Java. *Concurrency: Practice and Experience*, 12(11), 2000, 1019–1038.

[35] K. A. Yelick, L. Semenzato, G. Pike, C. Miyamoto, B. Liblit, A. Krishnamurthy, P. Hilfinger, S. L. Graham, D. Gay, P. Colella, A. Aiken, Titanium: A High-Performance Java Dialect. *Concurrency: Practice and Experience*, Vol. 10, pp. 825–836, 1998.

[36] M. Philippsen and M. Zenger. JavaParty — transparent remote objects in Java. *Concurrency: Practice and Experience* 9(11), 1997, 1225–1242.

[37] A. Adl-Tabatabai, M. Cierniak, G. Lueh, V. M. Parikh, and J. M. Stichnoth. Fast, Effective Code Generation in a Just-In-Time Java Compiler. Proceedings of ACM Programming Languages Design and Implementation, pp. 280–290, 1998.

[38] B. Carpenter, G. Zhang, G. Fox, X. Li, and Y. Wen, HPJava: Data parallel extensions to Java, ACM 1998 workshop on Java for high-performance network computing. Palo Alto, California. *Concurrency: Practice and Experience*, 10(11–13): 873–877, 1998.

[39] H. K. Lee, B. Carpenter, G. Fox, and S. B. Lim. Benchmarking HPJava: Prospects for Performance In 6th Workshop on Languages, Compilers and Run-time Systems for Scalable Computers, March 2002.

[40] The HPJava Project, http://www.hpjava.org

[41] D. Walker and O. Rana: The Use of Java in High Performance Computing: A Data Mining Example. *Proceedings of the Seventh International Conference on High Performance Computing and Networks.* Springer-Verlag LNS 1593, eds. P. Sloot, M. Bubak, A. Hoekstra, and B. Hertzberger, pp. 863–872, 1999.

[42] D. Lea. *Concurrent Programming in Java: Design Principles and Patterns.* Addison-Wesley, Boston, USA, 1996.

[43] J. M. Bull, M. D. Westhead, M. E. Kambites, and J. Obdrzalek. Towards OpenMP for Java. Proceedings of Second European Workshop on OpenMP, Edinburgh, UK, September 2000.

[44] OpenMP Architecture Review Board. OpenMP C and C++ Application Program Interface, October 1998.

[45] OpenMP Architecture Review Board. OpenMP Fortran Application Program Interface, October 1997.

[46] M. Kambites and Java OpenMP. Technical Report EPCC-SS99-05, Edinburgh Parallel Computer Centre, 1999.

4 Parallel Algorithms for the Singular Value Decomposition

Michael W. Berry, Dani Mezher, Bernard Philippe, and Ahmed Sameh

CONTENTS

ABSTRACT

The goal of the survey is to review the state-of-the-art of computing the singular value decomposition (SVD) of dense and sparse matrices, with some emphasis on those schemes that are suitable for parallel computing platforms. For dense matrices, we present those schemes that yield the complete decomposition, whereas for sparse matrices we describe schemes that yield only the extremal singular triplets. Special attention is devoted to the computation of the smallest singular values which are normally the most difficult to evaluate but which provide a measure of the distance to singularity of the matrix under consideration. Also, we conclude with the presentation of a parallel method for computing pseudospectra, which depends on computing the smallest singular values.

4.1 INTRODUCTION

4.1.1 BASICS

The SVD is a powerful computational tool. Modern algorithms for obtaining such a decomposition of general matrices have had a profound impact on numerous applications in science and engineering disciplines. The SVD is commonly used in the solution of unconstrained linear least squares problems, matrix rank estimation, and canonical correlation analysis. In computational science, it is commonly applied in domains such as information retrieval, seismic reflection tomography, and real-time signal processing [1].

In what follows, we will provide a brief survey of some parallel algorithms for obtaining the SVD for dense and large sparse matrices. For sparse matrices, however, we focus mainly on the problem of obtaining the smallest singular triplets.

To introduce the notations of the chapter, the basic facts related to SVD are presented without proof. Complete presentations are given in many text books, as for instance Refs. [2, 3].

Theorem 4.1 *[SVD]*

If $A \in \mathbb{R}^{m \times n}$ is a real matrix, then there exist orthogonal matrices

$$U = [u_1, \ldots, u_m] \in \mathbb{R}^{m \times m} \quad and \quad V = [v_1, \ldots, v_n] \in \mathbb{R}^{n \times n}$$

such that

$$\Sigma = U^{\mathrm{T}} A V = \mathrm{diag}(\sigma_1, \ldots, \sigma_p) \in \mathbb{R}^{m \times n}, \quad p = \min(m, n) \tag{4.1}$$

where $\sigma_1 \geq \sigma_2 \geq \cdots \geq \sigma_p \geq 0$.

Definition 4.1 *The singular values of A are the real numbers $\sigma_1 \geq \sigma_2 \geq \cdots \geq \sigma_p \geq 0$. They are uniquely defined. For every singular value σ_i $(i = 1, \ldots, p)$, the vectors u_i and v_i are respectively the left and right singular vectors of A associated with σ_i.*

More generally, the theorem holds in the complex field but with the inner products and orthogonal matrices replaced by the Hermitian products and unitary matrices, respectively. The singular values, however, remain real nonnegative values.

Theorem 4.2 *Let $A \in \mathbb{R}^{m \times n}$ $(m \geq n)$ have the singular value decomposition*

$$U^{\mathrm{T}} A V = \Sigma.$$

Then, the symmetric matrix

$$B = A^{\mathrm{T}} A \in \mathbb{R}^{n \times n}, \tag{4.2}$$

has eigenvalues $\sigma_1^2 \geq \cdots \geq \sigma_n^2 \geq 0$, *corresponding to the eigenvectors* (v_i), $(i = 1, \cdots, n)$.
  The symmetric matrix

$$A_{\text{aug}} = \begin{pmatrix} 0 & A \\ A^{\mathrm{T}} & 0 \end{pmatrix} \tag{4.3}$$

has eigenvalues $\pm\sigma_1, \ldots, \pm\sigma_n$, *corresponding to the eigenvectors*

$$\frac{1}{\sqrt{2}} \begin{pmatrix} u_i \\ \pm v_i \end{pmatrix}, \quad i = 1, \ldots, n.$$

The matrix A_{aug} *is called the augmented matrix.*

Every method for computing singular values is based on one of these two matrices.

The numerical accuracy of the ith approximate singular triplet $(\tilde{u}_i, \tilde{\sigma}_i, \tilde{v}_i)$ determined via the eigensystem of the *2-cyclic* matrix A_{aug} is then measured by the norm of the eigenpair residual vector r_i defined by

$$\| r_i \|_2 = [\| A_{\text{aug}}(\tilde{u}_i, \tilde{v}_i)^{\mathrm{T}} - \tilde{\sigma}_i(\tilde{u}_i, \tilde{v}_i)^{\mathrm{T}} \|_2] / [\| \tilde{u}_i \|_2^2 + \| \tilde{v}_i \|_2^2]^{1/2},$$

which can also be written as

$$\| r_i \|_2 = [(\| A\tilde{v}_i - \tilde{\sigma}_i\tilde{u}_i \|_2^2 + \| A^{\mathrm{T}}\tilde{u}_i - \tilde{\sigma}_i\tilde{v}_i \|_2^2)^{1/2}] / \| \tilde{u}_i \|_2^2 + \| \tilde{v}_i \|_2^2]^{1/2}. \tag{4.4}$$

The backward error [4]

$$\eta_i = \max\{\| A\tilde{v}_i - \tilde{\sigma}_i\tilde{u}_i \|_2, \quad \| A^{\mathrm{T}}\tilde{u}_i - \tilde{\sigma}_i\tilde{v}_i \|_2^2\}$$

may also be used as a measure of absolute accuracy. A normalizing factor can be introduced for assessing relative errors.

Alternatively, we may compute the SVD of A indirectly by the eigenpairs of either the $n \times n$ matrix $A^{\mathrm{T}}A$ or the $m \times m$ matrix AA^{T}. If $V = \{v_1, v_2, \ldots, v_n\}$ is the $n \times n$ orthogonal matrix representing the eigenvectors of $A^{\mathrm{T}}A$, then

$$V^{\mathrm{T}}(A^{\mathrm{T}}A)V = \text{diag}(\sigma_1^2, \sigma_2^2, \ldots, \sigma_r^2, \underbrace{0, \ldots, 0}_{n-r}),$$

where σ_i is the ith nonzero singular value of A corresponding to the right singular vector v_i. The corresponding left singular vector, u_i, is then obtained as $u_i = (1/\sigma_i)Av_i$. Similarly, if $U = \{u_1, u_2, \ldots, u_m\}$ is the $m \times m$ orthogonal matrix representing the eigenvectors of AA^{T}, then

$$U^{\mathrm{T}}(AA^{\mathrm{T}})U = \text{diag}(\sigma_1^2, \sigma_2^2, \ldots, \sigma_r^2, \underbrace{0, \ldots, 0}_{m-r}),$$

where σ_i is the ith nonzero singular value of A corresponding to the left singular vector u_i. The corresponding right singular vector, v_i, is then obtained as $v_i = (1/\sigma_i)A^{\mathrm{T}}u_i$.

Computing the SVD of A via the eigensystems of either $A^{\mathrm{T}}A$ or AA^{T} may be adequate for determining the largest singular triplets of A, but some loss of accuracy may be observed for the smallest singular triplets (see Ref. [5]). In fact, extremely small singular values of A (i.e., smaller than $\sqrt{\epsilon}\|A\|$, where ϵ is the machine precision parameter) may be computed as zero eigenvalues of $A^{\mathrm{T}}A$ (or AA^{T}). Whereas the smallest and largest singular values of A are the lower and upper bounds of the spectrum of $A^{\mathrm{T}}A$ or AA^{T}, the smallest singular values of A lie at the center of the spectrum of A_{aug} in (4.3). For computed eigenpairs of $A^T A$ and $A A^T$, the norms of the ith eigenpair residuals (corresponding to (4.4)) are given by

$$\| r_i \|_2 = \| A^T A \tilde{v}_i - \tilde{\sigma}_i^2 \tilde{v}_i \|_2 \,/\, \| \tilde{v}_i \|_2 \quad \text{and} \quad \| r_i \|_2 = \| A A^T \tilde{u}_i - \tilde{\sigma}_i^2 \tilde{u}_i \|_2 \,/\, \| \tilde{u}_i \|_2,$$

respectively. Thus, extremely high precision in computed eigenpairs may be necessary to compute the smallest singular triplets of A. This fact is analyzed in Section 4.1.2. Difficulties in approximating the smallest singular values by any of the three equivalent symmetric eigenvalue problems will be discussed in Section 4.4.

When A is a square nonsingular matrix, it may be advantageous in certain cases to compute the singular values of A^{-1} which are $(1/\sigma_n) \geq \cdots \geq (1/\sigma_1)$. This approach has the drawback of solving linear systems involving the matrix A, but when manageable, it provides a more robust algorithm. Such an alternative is of interest for some subspace methods (see Section 4.3). Actually, the method can be extended to rectangular matrices of full rank by considering a QR-factorization:

Proposition 4.1 *Let $A \in \mathbb{R}^{m \times n}$ ($m \geq n$) be of rank n. Let*

$$A = QR, \text{ where } Q \in \mathbb{R}^{m \times n} \text{ and } R \in \mathbb{R}^{n \times n},$$

such that $Q^T Q = I_n$ and R is upper triangular.
The singular values of R are the same as the singular values of A.

Therefore, the smallest singular value of A can be computed from the largest eigenvalue of $(R^{-1} R^{-T})$ or of $\begin{pmatrix} 0 & R^{-1} \\ R^{-T} & 0 \end{pmatrix}$.

4.1.2 SENSITIVITY OF THE SMALLEST SINGULAR VALUE

To compute the smallest singular value in a reliable way, one must investigate the sensitivity of the singular values with respect to perturbations of the matrix at hand.

Theorem 4.3 *Let $A \in \mathbb{R}^{n \times n}$ and $\Delta \in \mathbb{R}^{n \times n}$. The singular values of A and $A + \Delta$ are respectively denoted*

$$\sigma_1 \geq \sigma_2 \geq \cdots \geq \sigma_n$$
$$\tilde{\sigma}_1 \geq \tilde{\sigma}_2 \geq \cdots \geq \tilde{\sigma}_n.$$

They satisfy the following bounds:

$$|\sigma_i - \tilde{\sigma}_i| \leq \| \Delta \|_2, \quad \text{for } i = 1, \ldots, n.$$

Proof. See Ref. [3]. □

When applied to the smallest singular value, this result ends up with the following estimation.

Proposition 4.2 *The relative condition number of the smallest singular value of a nonsingular matrix A is equal to $\chi_2(A) = \| A \|_2 \| A^{-1} \|_2$.*

Proof. The result is obtained from

$$\frac{|\sigma_n - \tilde{\sigma}_n|}{\sigma_n} \leq \left(\frac{\| \Delta \|_2}{\| A \|_2} \right) \frac{\| A \|_2}{\sigma_n}.$$

□

This means that the smallest singular value of an ill-conditioned matrix cannot be computed with high accuracy even with an algorithm of perfect arithmetic behavior (i.e., backward stable).

Recently, some progress has been made [6]. It is shown that for some special class of matrices, an accurate computation of the smallest singular value may be obtained via a combination of some QR-factorization with column pivoting and a one-sided Jacobi algorithm (see Section 4.2.2).

Because the nonzero singular values are roots of a polynomial (e.g., roots of the characteristic polynomial of the augmented matrix), then when simple, they are differentiable with respect to the entries of the matrix. More precisely, one can states that:

Theorem 4.4 *Let σ be a nonzero simple singular value of the matrix $A = (a_{ij})$ with $u = (u_i)$ and $v = (v_i)$ being the corresponding normalized left and right singular vectors. Then the singular value is differentiable with respect to the matrix A, or*

$$\frac{\partial \sigma}{\partial a_{ij}} = u_i v_j, \quad \forall i, j = 1, \dots, n.$$

Proof. See Ref. [7]. □

The effect of a perturbation of the matrix on the singular vectors can be more significant than that on the singular values. The sensitivity of the singular vectors depend on the singular value distribution. When a simple singular value is not well separated from the rest, the corresponding left and right singular vectors are poorly determined. This is made precise by the following theorem, see Ref. [3], which we state here without proof. Let $A \in \mathbb{R}^{n \times m}$ $(n \geq m)$ have the SVD

$$U^T A V = \begin{pmatrix} \Sigma \\ 0 \end{pmatrix}.$$

Partition $U = (u_1 \ U_2 \ U_3)$ and $V = (v_1 \ V_2)$ where $u_1 \in \mathbb{R}^n$, $U_2 \in \mathbb{R}^{n \times (m-1)}$, $U_3 \in \mathbb{R}^{n \times (n-m)}$, $v_1 \in \mathbb{R}^m$, and $U_2 \in \mathbb{R}^{m \times (m-1)}$. Partition conformally

$$U^T A V = \begin{pmatrix} \sigma_1 & 0 \\ 0 & \Sigma_2 \\ 0 & 0 \end{pmatrix}.$$

Given a perturbation $\tilde{A} = A + E$ of A, let

$$U^T E V = \begin{pmatrix} \gamma_{11} & g_{12}^T \\ g_{21} & G_{22} \\ g_{31} & G_{32} \end{pmatrix}.$$

Theorem 4.5 *Let $h = \sigma_1 g_{12} + \Sigma_2 g_{21}$. If $(\sigma_1 I - \Sigma_2)$ is nonsingular (i.e., if σ_1 is a simple singular value of A), then the matrix*

$$U^T \tilde{A} V = \begin{pmatrix} \sigma_1 + \gamma_{11} & g_{12}^T \\ g_{21} & \Sigma_2 + G_{22} \\ g_{31} & G_{32} \end{pmatrix}$$

has a right singular vector of the form

$$\begin{pmatrix} 1 \\ (\sigma_1^2 I - \Sigma_2^2)^{-1} h \end{pmatrix} + O(\|E\|^2).$$

It was remarked in Theorem 4.1 that computing the SVD of A could be obtained from the eigendecomposition of the matrix $C = A^T A$ or of the augmented matrix $A_{\text{aug}} = \begin{pmatrix} 0 & A \\ A^T & 0 \end{pmatrix}$. It is clear, however, that using C to compute the smallest singular value of A is bound to yield poorer result as the condition number of C is the square of the condition number of A_{aug}. It can be shown, however, that even with an ill-conditioned matrix A, the matrix C can be used to compute accurate singular values.

4.1.3 DISTANCE TO SINGULARITY — PSEUDOSPECTRUM

Let us consider a linear system defined by the square matrix $A \in \mathbb{R}^{n \times n}$. However, one needs to quantify how far is the system under consideration from being singular. It turns out that the smallest singular value $\sigma_{\min}(A)$ is equal to that distance.

Let $\mathcal{S}$ be the set of all singular matrices in $\mathbb{R}^{n \times n}$ and the distance corresponding to the 2-norm: $d(A, B) = \|A - B\|_2$ for $A, B \in \mathbb{R}^{n \times n}$.

Theorem 4.6 $d(A, S) = \sigma_{\min}(A)$.

Proof. Let us denote $\sigma = \sigma_{\min}(A)$ and $d = d(A, \mathcal{S})$. There exist two unitary vectors u and v such that $Au = \sigma v$. Therefore $(A - \sigma v u^{\mathrm{T}})u = 0$. Since $\|\sigma v u^{\mathrm{T}}\|_2 = \sigma$, apparently $B = A - \sigma v u^{\mathrm{T}} \in \mathcal{S}$ and $d(A, B) = \sigma$ which proves that $d \leq \sigma$.

Conversely, let us consider any matrix $\varDelta \in \mathbb{R}^{n \times n}$ such that $(A + \varDelta) \in \mathcal{S}$. There exists a unitary vector u such that $(A + \varDelta)u = 0$. Therefore:

$$\sigma \leq \|Au\|_2 = \|\varDelta u\|_2 \leq \|\varDelta\|_2,$$

which concludes the proof. □

This result leads to a useful lower bound of the condition number of a matrix.

Proposition 4.3 *The condition number* $\chi_2(A) = \|A\|_2 \|A^{-1}\|_2$ *satisfies*

$$\chi_2(A) \geq \frac{\|A\|_2}{\|A - B\|_2},$$

for any singular matrix $B \in \mathcal{S}$.

Proof. The result follows from the fact that $B = A + (B - A)$ and $\|A^{-1}\|_2 = 1/\sigma_{\min}(A)$. □

For instance in Ref. [8], this property is used to illustrate that the condition number of the linear systems arising from the simulation of flow in porous media, using mixed finite element methods, is of the order of the ratio of the extreme values of conductivity.

Let us now consider the sensitivity of eigenvalues with respect to matrix perturbations. Towards this goal, the notion of pseudospectrum [9] or of ε-spectrum [10] was introduced:

Definition 4.2 For a matrix $A \in \mathbb{R}^{n \times n}$ (or $A \in \mathbb{C}^{n \times n}$) and a parameter $\epsilon > 0$, the pseudospectrum is the set:

$$\Lambda_\epsilon(A) = \{z \in \mathbb{C} \mid \exists \varDelta \in \mathbb{C}^{n \times n} \text{ such that } \|\varDelta\| \leq \epsilon \text{ and } z \text{ is an eigenvalue of } (A + \varDelta)\}. \quad (4.5)$$

This definition does not provide a constructive method for determining the pseudospectrum. Fortunately, a constructive method can be drawn from the following property.

Proposition 4.4 *The pseudospectrum is the set:*

$$\Lambda_\epsilon(A) = \{z \in \mathbb{C} \mid \sigma_{\min}(A - zI) \leq \epsilon\}, \quad (4.6)$$

where I is the identity matrix of order n.

Proof. For any $z \in \mathbb{C}$, z is an eigenvalue of $(A + \Delta)$ if and only if the matrix $(A - zI) + \Delta$ is singular. The proposition is therefore a straight application of Theorem 4.6. □

This proposition provides a criterion for deciding whether z belongs to $\Lambda_\epsilon(A)$. To represent the pseudospectrum graphically, one can define a grid in the complex region under consideration and compute $\sigma_{\min}(A - z_{ij}I)$, for all the z_{ij} determined by the grid. Although highly parallel, this approach involves a very high volume of operations. Presently, one prefers to use path-following techniques [11–13]. Section 4.5 describes one of these methods. For illustration, the pseudospectrum of a matrix from the Matrix Market [14] test suite is displayed in Figure 4.1, where several values of ϵ are shown.

In what follows, we present a selection of parallel algorithms for computing the SVD of dense and sparse matrices. For dense matrices, we restrict our survey to the Jacobi methods for obtaining all the singular triplets, a class of methods not contained in ScaLAPACK [15, 16]. The ScaLAPACK (or Scalable LAPACK) library includes a subset of LAPACK routines redesigned for distributed memory Multiple Instruction Multiple Data (MIMD) parallel computers. It is currently written in a Single-Program-Multiple-Data style using explicit message passing for interprocessor communication. It assumes matrices are laid out in a two-dimensional block cyclic fashion. The routines of the library achieve respectable efficiency on parallel computers and the software is considered to be robust. Some projects are under way, however, for making the use of ScaLAPACK in large-scale applications more user-friendly. Examples include, the PLAPACK project [17, 18], the OURAGAN project which is based on SCILAB [19], as well as projects based on MATLAB [20]. None of these projects, however, provide all the capabilities of ScaLAPACK.

Whereas for sparse matrices, we concentrate our exposition on those schemes that obtain the smallest singular triplets. We devote the last section to the vital primitives that help to assure the realization of high performance on parallel computing platforms.

4.2 JACOBI METHODS FOR DENSE MATRICES

4.2.1 TWO-SIDED JACOBI SCHEME [2JAC]

Here, we consider the standard eigenvalue problem

$$Bx = \lambda x \tag{4.7}$$

where B is a real $n \times n$-dense symmetric matrix. One of the best known methods for determining all the eigenpairs of (4.7) was developed by the 19th century mathematician, Jacobi. We recall that Jacobi's sequential method reduces the matrix B to the diagonal form by an infinite sequence of plane rotations

$$B_{k+1} = U_k B_k U_k^{\mathrm{T}}, \quad k = 1, 2, \ldots,$$

where $B_1 \equiv B$, and $U_k = U_k(i, j, \theta_{ij}^k)$ is a rotation of the (i, j)-plane where

$$u_{ii}^k = u_{jj}^k = c_k = \cos\theta_{ij}^k \quad \text{and} \quad u_{ij}^k = -u_{ji}^k = s_k = \sin\theta_{ij}^k.$$

The angle θ_{ij}^k is determined so that $b_{ij}^{k+1} = b_{ji}^{k+1} = 0$, or

$$\tan 2\theta_{ij}^k = \frac{2b_{ij}^k}{b_{ii}^k - b_{jj}^k},$$

where $|\theta_{ij}^k| \le \frac{1}{4}\pi$.

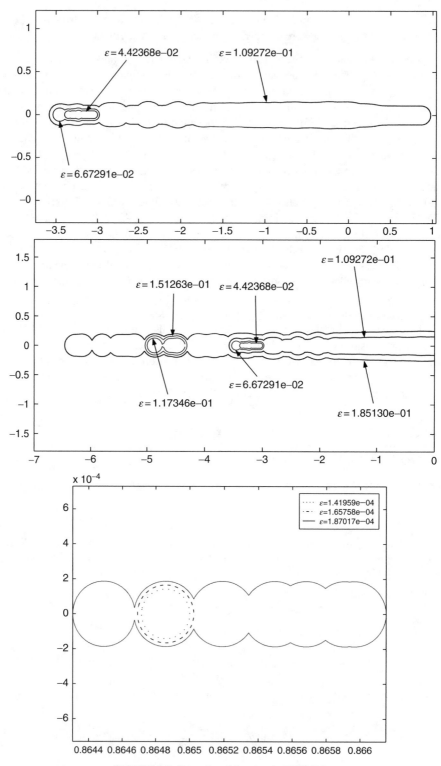

FIGURE 4.1 Portraits of the matrix DW8192.

For numerical stability, we determine the plane rotation by

$$c_k = \frac{1}{\sqrt{1 + t_k^2}} \quad \text{and} \quad s_k = c_k t_k,$$

where t_k is the smaller root (in magnitude) of the quadratic equation

$$t_k^2 + 2\alpha_k t_k - 1 = 0, \quad \alpha_k = \cot 2\theta_{ij}^k.$$

Hence, t_k may be written as

$$t_k = \frac{\text{sign } \alpha_k}{|\alpha_k| + \sqrt{1 + \alpha_k^2}}.$$

Each B_{k+1} remains symmetric and differs from B_k only in rows and columns i and j, where the modified elements are given by

$$b_{ii}^{k+1} = b_{ii}^k + t_k b_{ij}^k,$$
$$b_{jj}^{k+1} = b_{jj}^k - t_k b_{ij}^k,$$

and

$$b_{ir}^{k+1} = c_k b_{ir}^k + s_k b_{jr}^k, \tag{4.8}$$
$$b_{jr}^{k+1} = -s_k b_{ir}^k + c_k b_{jr}^k, \tag{4.9}$$

in which $r \neq i, j$. If we represent B_k by

$$B_k = D_k + E_k + E_k^{\mathrm{T}}, \tag{4.10}$$

where D_k is diagonal and E_k is strictly upper triangular, then as k increases $\| E_k \|_{\mathrm{F}}$ approaches zero, and B_k approaches the diagonal matrix $\Lambda = \text{diag}(\lambda_1, \lambda_2, \ldots, \lambda_n)$ ($\| \cdot \|_{\mathrm{F}}$ denotes the Frobenius norm). Similarly, the transpose of the product $(U_k \cdots U_2 U_1)$ approaches a matrix whose jth column is the eigenvector corresponding to λ_j.

Several schemes are possible for selecting the sequence of elements b_{ij}^k to be eliminated via the plane rotations U_k. Unfortunately, Jacobi's original scheme, which consists of sequentially searching for the largest off-diagonal element, is too time consuming for implementation on a multiprocessor. Instead, a simpler scheme in which the off-diagonal elements (i, j) are annihilated in the cycle fashion $(1, 2), (1, 3), \ldots, (1, n), (2, 3), \ldots, (2, n), \ldots, (n - 1, n)$ is usually adopted as its convergence is assured [21]. We refer to each sequence of n rotations as a sweep. Furthermore, quadratic convergence for this sequential cyclic Jacobi scheme has been well documented (see Refs. [22, 23]). Convergence usually occurs within 6 to 10 sweeps, i.e., from $3n^2$ to $5n^2$ Jacobi rotations.

A parallel version of this cyclic Jacobi algorithm is obtained by the simultaneous annihilation of several off-diagonal elements by a given U_k, rather than only one as is done in the serial version. For example, let B be of order 8 and consider the orthogonal matrix U_k as the direct sum of four independent plane rotations, where the c_i's and s_i's for $i = 1, 2, 3, 4$ are simultaneously determined. An example of such a matrix is

$$R_k(1, 3) \oplus R_k(2, 4) \oplus R_k(5, 7) \oplus R_k(6, 8),$$

where $R_k(i, j)$ is that rotation which annihilates the (i, j) off-diagonal element. If we consider one sweep to be a collection of orthogonal similarity transformations that annihilate the element in each of the $\frac{1}{2}n(n - 1)$ off-diagonal positions (above the main diagonal) only once, then for a matrix of order 8, the first sweep will consist of eight successive orthogonal transformations with each one annihilating distinct groups of four elements simultaneously. For the remaining sweeps, the structure of each subsequent transformation $U_k, k > 8$, is chosen to be the same as that of U_j where $j = 1 + (k - 1) \bmod 8$. In general, the most efficient annihilation scheme consists of $(2r - 1)$ similarity transformations per sweep, where $r = \left\lfloor \frac{1}{2}(n + 1) \right\rfloor$, in which each transformation annihilates different $\left\lfloor \frac{1}{2}n \right\rfloor$ off-diagonal elements (see Ref. [24]). Although several annihilation schemes are possible, the Jacobi algorithm we present below utilizes an annihilation scheme which requires a minimal amount of indexing for computer implementation. Moreover, Luk and Park [25, 26] have demonstrated that various parallel Jacobi rotation ordering schemes are equivalent to the sequential row ordering scheme, and hence share the same convergence properties.

Algorithm [2JAC]
Step 1: (Apply orthogonal similarity transformations via U_k for current sweep).

1. (a) For $k = 1, 2, 3, \ldots, n - 1$ (serial loop)

    simultaneously annihilate elements in position (i, j), where

    $$\begin{cases} i = 1, 2, 3, \ldots, \left\lceil \frac{1}{2}(n - k) \right\rceil, \\ j = (n - k + 2) - i. \end{cases}$$

    for $k > 2$,

    $$\begin{cases} i = (n - k + 2), (n - k + 3), \ldots, n - \left\lfloor \frac{1}{2}k \right\rfloor, \\ j = (2n - k + 2) - i. \end{cases}$$

   (b) For $k = n$ simultaneously annihilate elements in positions (i, j), where

    $$\begin{cases} i = 2, 3, \ldots, \left\lceil \frac{1}{2}n \right\rceil \\ j = (n + 2) - i \end{cases}$$

Step 2: (Convergence test).

1. (a) Compute $\| D_k \|_F$ and $\| E_k \|_F$ (see (4.10)).
   (b) If

    $$\frac{\| E_k \|_F}{\| D_k \|_F} < \text{tolerance}, \tag{4.11}$$

    then stop. Otherwise, go to Step 1 to begin next sweep.

We note that this algorithm requires n similarity transformations per sweep for a dense real symmetric matrix of order n (n may be even or odd). Each U_k is the direct sum of either $\left\lfloor \frac{1}{2}n \right\rfloor$ or $\left\lfloor \frac{1}{2}(n - 1) \right\rfloor$ plane rotations, depending on whether k is odd or even, respectively. The annihilation pattern for $n = 8$ is shown in Table 4.1, where the integer k denotes an element annihilation via U_k.

TABLE 4.1
Annihilation Scheme for [2JAC]

x	7	6	5	4	3	2	1
	x	5	4	3	2	1	8
		x	3	2	1	8	7
			x	1	8	7	6
				x	7	6	5
					x	5	4
						x	3
							x

In the annihilation of a particular (i, j)-element in Step 1 above, we update the off-diagonal elements in rows and columns i and j as specified by (4.8) and (4.9). With regard to storage requirements, it would be advantageous to modify only those row or column entries above the main diagonal and utilize the guaranteed symmetry of B_k. However, if one wishes to take advantage of the vectorization that may be supported by the parallel computing platform, we disregard the symmetry of B_k and operate with full vectors on the entirety of rows and columns i and j in (4.8) and (4.9), i.e., we are using a full matrix scheme. The product of the U_k's, which eventually yields the eigenvectors for B, is accumulated in a separate two-dimensional array by applying (4.8) and (4.9) to the $n \times n$-identity matrix.

In Step 2, we monitor the convergence of the algorithm by using the ratio of the computed norms to measure the systematic decrease in the relative magnitudes of the off-diagonal elements with respect to the relative magnitudes of the diagonal elements. For double precision accuracy in the eigenvalues and eigenvectors, a *tolerance* of order 10^{-16} will suffice for Step 2(b). If we assume convergence (see Ref. [25]), this multiprocessor algorithm can be shown to converge quadratically by following the work of Wilkinson [23] and Henrici [27].

4.2.2 ONE-SIDED JACOBI SCHEME [1JAC]

Suppose that A is a real $m \times n$-matrix with $m \gg n$ and rank $A = r$. The SVD of A can be defined as

$$A = U \Sigma V^{\mathrm{T}}, \tag{4.12}$$

where $U^{\mathrm{T}}U = V^{\mathrm{T}}V = I_n$ and $\Sigma = \mathrm{diag}(\sigma_1, \ldots, \sigma_n)$, $\sigma_i > 0$ for $1 \leq i \leq r$, $\sigma_j = 0$ for $j \geq r + 1$. The first r columns of the orthonormal matrix U and the orthogonal matrix V define the orthonormalized eigenvectors associated with the r nonzero eigenvalues of AA^{T} or $A^{\mathrm{T}}A$.

As indicated in Ref. [28] for a ring of processors, using a method based on the one-sided iterative orthogonalization method of Hestenes (see also Refs. [29, 30]) is an efficient way to compute the decomposition (4.12). Luk [31] recommended this singular value decomposition scheme on the Illiac IV, and corresponding systolic algorithms associated with *two-sided* schemes have been presented in Refs. [32, 33]. We now consider a few modifications to the scheme discussed in Refs. [28] for the determination of (4.12) on shared-memory multiprocessors, e.g., see Refs. [34].

Our main goal is to determine the orthogonal matrix $V = [\tilde{V}, \tilde{W}]$, where $\tilde{V}$ is $n \times r$, so that

$$A\tilde{V} = Q = (q_1, q_2, \dots, q_r), \tag{4.13}$$

and

$$q_i^\mathrm{T} q_j = \sigma_i^2 \delta_{ij},$$

where the columns of A are orthogonal and δ_{ij} is the Kronecker-delta. Writing Q as

$$Q = \tilde{U}\tilde{\Sigma} \quad \text{with} \quad \tilde{U}^\mathrm{T}\tilde{U} = I_r, \quad \text{and} \quad \tilde{\Sigma} = \mathrm{diag}(\sigma_1, \dots, \sigma_r),$$

then

$$A = \tilde{U}\tilde{\Sigma}\tilde{V}^\mathrm{T}.$$

We construct the matrix V via the plane rotations

$$(a_i, a_j) \begin{bmatrix} c & -s \\ s & c \end{bmatrix} = (\tilde{a}_i, \tilde{a}_j), \quad i < j,$$

so that

$$\tilde{a}_i^\mathrm{T}\tilde{a}_j = 0 \quad \text{and} \quad \parallel \tilde{a}_i \parallel_2 > \parallel \tilde{a}_j \parallel_2, \tag{4.14}$$

where a_i designates the ith column of matrix A. This is accomplished by choosing

$$c = \left[\frac{\beta + \gamma}{2\gamma} \right]^{1/2} \quad \text{and} \quad s = \left[\frac{\alpha}{2\gamma c} \right], \quad \text{if} \quad \beta > 0, \tag{4.15}$$

or

$$s = \left[\frac{\gamma - \beta}{2\gamma} \right]^{1/2} \quad \text{and} \quad c = \left[\frac{\alpha}{2\gamma s} \right], \quad \text{if} \quad \beta < 0, \tag{4.16}$$

where $\alpha = 2a_i^\mathrm{T} a_j$, $\beta = \parallel a_j \parallel_2^2$, and $\gamma = (\alpha^2 + \beta^2)^{1/2}$. Note that (4.14) requires the columns of Q to decrease in norm from left to right, and hence the resulting σ_i to be in monotonic nonincreasing order. Several schemes can be used to select the order of the (i, j)-plane rotations. Following the annihilation pattern of the off-diagonal elements in the sequential Jacobi algorithm mentioned in Section 4.1, we could certainly orthogonalize the columns in the same cyclic fashion and thus perform the one-sided orthogonalization serially. This process is iterative with each sweep consisting of $\frac{1}{2}n(n - 1)$ plane rotations selected in cyclic fashion.

By adopting the ordering of the annihilation scheme in [2JAC], we obtain a parallel version of the one-sided Jacobi method for computing the singular value decomposition on a multiprocessor. For example, let $n = 8$ and $m \gg n$ so that in each sweep of our one-sided Jacobi algorithm we simultaneously orthogonalize pairs of columns of A (see Table 4.1). For example, for $n = 8$ we can orthogonalized the pairs (1,8), (2,7), (3,6), (4,5) simultaneously via postmultiplication by a matrix V_i which consists of the direct sum of four plane rotations. In general, each V_k will have the same form as U_k so that at the end of any particular sweep s_i we have

$$V_{s_1} = V_1 V_2 \cdots V_n,$$

and hence

$$V = V_{s_1} V_{s_2} \cdots V_{s_t}, \tag{4.17}$$

where t is the number of sweeps required for convergence.

Algorithm [1JAC]

Step 1: (Postmultiply matrix A by orthogonal matrix V_k for current sweep).

1. (a) Initialize the convergence counter, *istop*, to zero.
   (b) For $k = 1, 2, 3, \ldots, n - 1$ (serial loop)
   simultaneously orthogonalize the column pairs (i, j), where i and j are given by 1(a) in Step 1 of [2JAC], provided that for each (i, j) we have

   $$\frac{(a_i^T a_j)^2}{(a_i^T a_i)(a_j^T a_j)} > \text{tolerance}, \tag{4.18}$$

   and $i, j \in \{k | k < k_{\min}\}$, where $k_{\min}$ is the minimal column index k such that $\| a_k \|_2^2 < \text{tolerance}$. Upon the termination of [1JAC], $r = \text{rank } A = k_{\min}$. *Note:* if (4.18) is not satisfied for any particular pair (i, j), *istop* is incremented by 1 and that rotation is not performed.
   (c) For $k = n$
   simultaneously orthogonalized the column pairs (i, j), where i and j are given by 1(b) in Step 1 of [2JAC].

Step 2: (Convergence test).

If *istop* $= \frac{1}{2} n(n - 1)$, then compute $\sigma_i = \sqrt{(A^T A)_{ii}}$, $i = 1, 2, \ldots, k_{\min} = r$, and stop. Otherwise, go to beginning of Step 1 to start next sweep.

In the orthogonalization of columns in Step 1, we are implementing the plane rotations specified by (4.15) and (4.16), and hence guaranteeing the ordering of column norms and singular values upon termination. Whereas [2JAC] must update rows and columns following each similarity transformation, [1JAC] performs only postmultiplication of A by each V_k and hence the plane rotation (i, j) changes only the elements in columns i and j of matrix A. The changed elements can be represented by

$$a_i^{k+1} = ca_i^k + sa_j^k, \tag{4.19}$$

$$a_j^{k+1} = -sa_i^k + ca_j^k, \tag{4.20}$$

where a_i denotes the ith column of matrix A, and c, s are determined by either (4.15) or (4.16). Since no row accesses are required and no columns are interchanged, one would expect good performance for this method on a machine which can apply vector operations to compute (4.19) and (4.20). Each processor is assigned one rotation and hence orthogonalizes one pair of the n columns of matrix A.

Following the convergence test used in Ref. [30], in Step 2, we count the number of times the quantity

$$\frac{a_i^T a_j}{(a_i^T a_i)(a_j^T a_j)} \tag{4.21}$$

falls, in any sweep, below a given *tolerance*. The algorithm terminates when the counter *istop* reaches $\frac{1}{2} n(n - 1)$, the total number of column pairs, after any sweep. Upon termination, the first r columns of the matrix A are overwritten by the matrix Q from (4.13) and hence the nonzero singular values σ_i can be obtained via the r square roots of the first r diagonal entries of $A^T A$. The matrix $\tilde{U}$ in (4.12), which contains the leading r, left singular vectors of the original matrix A, is readily obtained by column scaling of the resulting matrix A (now overwritten by $Q = \tilde{U} \tilde{\Sigma}$) by the nonzero singular values σ_i. Similarly, the matrix V, which contains the right singular vectors of the original matrix A, is obtained as in (4.17) as the product of the orthogonal V_k's. This product is accumulated in a separate two-dimensional array by applying the rotations specified by (4.19) and (4.20) to the $n \times n$-identity matrix. It is important to note that the use of the ratio in (4.21) is preferable over the use of $a_i^T a_j$, since this dot-product is necessarily small for relatively small singular values.

Although our derivation of [1JAC] is concerned with the singular value decomposition of rectangular matrices, it is most effective for solving the eigenvalue problem in (4.7) for symmetric positive definite matrices. If $m = n = r$, B is a positive definite matrix, and Q in (4.13) is an orthogonal matrix. Consequently, it is not difficult to show that

$$\begin{cases} \sigma_i = \lambda_i \\ x_i = \frac{q_i}{\lambda_i}, \end{cases} \quad i = 1, 2, \ldots, n,$$

where λ_i denotes the ith eigenvalue of B, x_i the corresponding normalized eigenvector, and q_i the ith column of matrix Q. If B is symmetric, but perhaps not positive definite, we can obtain its eigenvectors by considering instead $B + \alpha I$, where α is the smallest quantity that ensures definiteness of $B + \alpha I$, and retrieve the eigenvalues of B via Rayleigh quotients.

[1JAC] has two advantages over [2JAC]: (i) no row accesses are needed and (ii) the matrix Q need not be accumulated.

4.2.3 Algorithm [QJAC]

As discussed above, [1JAC] is certainly a viable candidate for computing the SVD (4.12) on multiprocessors. However, for $m \times n$-matrices A in which $m \gg n$, the problem complexity can be reduced if an initial orthogonal factorization of A is performed. One can then apply the *one-sided* Jacobi method, [1JAC], to the resulting upper-triangular matrix R (which may be singular) and obtain the decomposition (4.12). In this section, we present a multiprocessor method, QJAC, which can be quite effective for computing (4.12) on parallel machines.

Given the $m \times n$-matrix A, where $m \gg n$, we perform a block generalization of Householder's reduction for the orthogonal factorization

$$A = QR, \tag{4.22}$$

where Q is $m \times n$-orthonormal matrix, and R is an $n \times n$-upper-triangular matrix. The block schemes of LAPACK are used for computing (4.22) to make use of vector–matrix, matrix–vector (BLAS2), and matrix–matrix (BLAS3) multiplication modules. The [1JAC] algorithm can then be used to obtain the SVD of the upper-triangular matrix R.

Hence, the SVD of an $m \times n$-matrix ($m \gg n$) A (having rank r) defined by

$$A = U \Sigma V^T,$$

where $U^T U = V^T V = I_r$, and $\Sigma = \text{diag}(\sigma_1, \ldots, \sigma_r)$, $\sigma_i > 0$ for $1 \le i \le r$, can be efficiently determined as follows:

Block Householder-Jacobi [QJAC]
Step 1: Apply block Householder reduction via (3.6) to the matrix A to obtain the factorization

$$A = QR, \tag{4.23}$$

where Q is $m \times n$ with orthonormal columns, and R is upper triangular of order n.

Step 2: Determine the SVD of the upper-triangular matrix via [1JAC],

$$R = \tilde{U} \begin{bmatrix} \tilde{\Sigma} \\ 0 \end{bmatrix} V^{\mathrm{T}}, \tag{4.24}$$

where $\tilde{U}$ and V are $n \times r$-matrices having orthogonal columns ($r \equiv$ rank A) and $\tilde{\Sigma} = $ diag σ_i contains the r nonzero singular values of A.

Step 3: Recover the left singular vectors u_i of A by back-transforming the columns of $\tilde{U}$:

$$U = Q\tilde{U}, \tag{4.25}$$

where Q is the product of the Householder transformations applied in Step 1 and u_i is the ith column of U.

Note that in using [1JAC] for computing the SVD of R, we must iterate on a full $n \times n$-matrix which is initially upper-triangular. This sacrifice in storage must be made to capitalize upon the potential vectorization and parallelism inherent in [1JAC] on parallel machines with vector processors.

Charlier et al. [35] demonstrate that an implementation of Kogbetliantz's algorithm for computing the SVD of upper-triangular matrices is quite effective on a systolic array of processors. We recall that Kogbetliantz's method for computing the SVD of a real square matrix A mirrors the [2JAC] method for symmetric matrices, in that the matrix A is reduced to the diagonal form by an infinite sequence of plane rotations:

$$A_{k+1} = U_k A_k V_k^{\mathrm{T}}, \quad k = 1, 2, \ldots, \tag{4.26}$$

where $A_1 \equiv A$, and $V_k = V_k(i, j, \phi_{ij}^k)$, $U_k = U_k(i, j, \theta_{ij}^k)$ are orthogonal plane rotation matrices which deviate from I_n and I_m, respectively, in the (i, i)-, (j, j)-, (i, j)-, and (j, i)-entries. It follows that A_k approaches the diagonal matrix $\Sigma = $ diag$(\sigma_1, \sigma_2, \ldots, \sigma_n)$, where σ_i is the ith singular value of A, and the products $(U_k \cdots U_2 U_1)$, $(V_k \cdots V_2 V_1)$ approach matrices whose ith column is the respective left and right singular vector corresponding to σ_i. For the case when the σ_i's are not pathologically close, Paige and Van Dooren [36] have shown that the row (or column) cyclic Kogbetliantz's method ultimately converges quadratically. For triangular matrices, Charlier and Van Dooren [37] have demonstrated that Kogbetliantz's algorithm converges quadratically for those matrices having multiple or clustered singular values provided that singular values of the same cluster occupy adjacent diagonal position of A_v, where v is the number of sweeps required for convergence. Even if we were to assume that R in (4.22) satisfies this condition for quadratic convergence of the parallel Kogbetliantz's method in [36], the ordering of the rotations and subsequent row (or column) permutations needed to maintain the upper-triangular form is more efficient for systolic architectures than for shared-memory parallel machines. One clear advantage of using [1JAC] is to determine the SVD of R lies in that the rotations defined by (4.15) or (4.16), as applied via the parallel ordering illustrated in Table 4.1, require no processor synchronization among any set of the $\lfloor \frac{1}{2}n \rfloor$ or $\lfloor \frac{1}{2}(n-1) \rfloor$ simultaneous plane rotations. The convergence rate of [1JAC], however, does not necessarily match that of Kogbetliantz's algorithm.

Let

$$R_k = D_k + E_k + E_k^{\mathrm{T}},$$

and

$$S_k = R_k^T R_k = \tilde{D}_k + \tilde{E}_k + \tilde{E}_k^T, \tag{4.27}$$

where D_k, $\tilde{D}_k$ are diagonal matrices and E_k, $\tilde{E}_k$ are strictly upper-triangular.

Although we cannot guarantee quadratic convergence for [1JAC], we can always produce clustered singular values on adjacent positions of $\tilde{D}_k$ for any matrix A. If we monitor the magnitudes of the elements of $\tilde{D}_k$ and $\tilde{E}_k$ in (4.28) for successive values of k in [1JAC] (for clustered singular values), S_k will converge to a diagonal form through an intermediate *block* diagonal form, where each of the principal submatrices (positioned along $\tilde{D}_k$) has diagonal elements which comprise one cluster of singular values of A (see Refs. [34, 35]). Thus, after a particular number of critical sweeps k_{cr}, we obtain

$$S_{k_{cr}} = \begin{array}{c} \boxed{T_1} \\ \quad \boxed{T_2} \\ \qquad \boxed{T_3} \\ \qquad\qquad \ddots \\ \qquad\qquad\qquad \ddots \\ \qquad\qquad\qquad\qquad \boxed{T_{n_c}} \end{array} \tag{4.28}$$

so that the SVD of each T_i, $i = 1, 2, \ldots, n_c$, can be computed in parallel by either a Jacobi or Kogbetliantz's method. Each symmetric matrix T_i will, in general, be dense of order q_i, representing the number of singular values of A contained in the ith cluster. Since the quadratic convergence of Kogbetliantz's method for upper-triangular matrices [37] mirrors the quadratic convergence of the two-sided Jacobi method, [2JAC], for symmetric matrices having clustered spectra [38], we would obtain a faster global convergence for $k > k_{cr}$ if [2JAC], rather than [1JAC], were applied to each of the T_i. Thus, a *hybrid* method consisting of an initial phase of several [1JAC] iterations followed by [2JAC] on the resulting subproblems would combine the optimal parallelism of [1JAC] and the fast convergence of the [2JAC] method. Of course, the difficulty in implementing such a method lies in the determination of the critical number of sweeps k_{cr}. We note that such a hybrid SVD method would be quite suitable for implementation on multiprocessors with hierarchical memory structure and/or vector processors.

4.2.4 BLOCK-JACOBI ALGORITHMS

The above algorithms are well-suited for shared-memory computers. Although they can also be implemented on distributed memory systems, their efficiency on such systems may suffer due to communication costs. To increase the granularity of the computation (i.e., to increase the number of floating point operations between two communications), block algorithms are considered. For one-sided algorithms, each processor is allocated a block of columns instead of a single column. The computation remains the same as discussed above with the ordering of the rotations within a sweep is as given in Ref. [26].

For the two-sided version, the allocation manipulates two-dimensional blocks instead of single entries of the matrix. Some authors [39–41] propose a modification of the basic algorithm in which one annihilates, in each step, two off-diagonal blocks by performing a full SVD on a small-sized matrix. Good efficiencies are realized on distributed memory systems but this block strategy increases the number of sweeps needed to reach convergence.

We note that parallel Jacobi algorithms can only surpass the speed of the bidiagonalization schemes of ScaLAPACK when the number of processors available are much larger than the size of the matrix under consideration.

4.3 METHODS FOR LARGE AND SPARSE MATRICES

4.3.1 SPARSE STORAGES AND LINEAR SYSTEMS

When the matrix is large and sparse, a compact storage scheme must be considered. The principle behind such storage schemes is to store only the nonzero entries and sometimes more data as in band or profile storage schemes. For a more detailed presentation of various compact storage schemes, we refer for instance to Refs. [42, 43]. Here, we consider only the compressed sparse row (CSR) format storage scheme.

Let us consider a sparse matrix A of order n with n_z (denoted nz in algorithms) nonzero entries. The CSR format is organized into three one-dimensional arrays:

array a(1:nz): contains all the nonzero entries of the matrix sorted by rows; within a row no special ordering is assumed although it is often preferable to sort the entries by increasing column indices.

array ja(1:nz): contains all the column indices of the nonzero entries in the same order as the order of the entries in array a.

array ia(1:n+1): , ia(i) $(i = 1, \ldots, n)$ is the index of the first nonzero entry of the ith row which is stored in array a and ia(n+1) is set to $n_z + 1$.

The main procedures which use a matrix stored in that way are the multiplications of a matrix, A or A^T, by a vector $x \in \mathbb{R}^n$. The corresponding algorithms are:

ALGORITHM : y := y + A*x	ALGORITHM : y := y + A^T*x
for i = 1:n,	for i = 1:n,
for l = ia(i) : ia(i+1)-1,	for l = ia(i) : ia(i+1)-1,
y(i) = y(i) + a(l)*x(ja(l)) ;	y(ja(l)) = y(ja(l)) + a(l)*x(i) ;
end ;	end ;
end ;	end

Solving linear systems which are defined by sparse matrices is not an easy task. One may consider direct methods, invariably consisting of matrix factorization, or consider iterative schemes. In direct solvers, care is needed to minimize the fill-in in the triangular factors, whereas in iterative methods, adopting effective preconditioning techniques is vital for fast convergence.

It is usually admitted that direct methods are more robust but are economical only when the triangular factors are not too dense, and when the size of the linear system is not too large. Reordering schemes are almost necessary to keep the level of fill-in as low as possible. Also, while pivoting strategies for dense linear systems are relaxed to minimize fill-in, the most effective sparse factorization schemes forbid the use of pivots below a given threshold. It is well known that the QR-factorization schemes, one of the most robust, is not used often in direct solvers as it suffers from a high level of fill-in. Such a high level of fill-in occurs because the upper-triangular factor, R, is the transpose of the Cholesky factor of matrix $A^T A$, which is much more dense than the original matrix A. Nevertheless, we shall see that this orthogonal factorization is a viable tool for computing the smallest singular value of sparse matrices. A survey of the state-of-the-art of sparse matrix factorization may be found in Refs. [42, 44–51]. Current efficient packages for LU-factorization include UMFPACK [52], SuperLU [53], and MUMPS [54].

When the order of the matrix is so large as to make the use of direct methods prohibitively expensive in storage and time, one resorts to iterative solvers. Classic iterative solvers such as the relaxation schemes methods of Jacobi, Gauss Seidel, SOR, or SSOR, are easy to use but not as effective as Krylov subspace methods. The latter class uses the matrix only through the procedures of matrix–vector multiplications as defined above. Moreover, under certain conditions, Krylov subspace schemes exhibit superlinear convergence. Often, however, Krylov subspace schemes are successful only in conjunction with a preconditioning strategy. This is especially true for nonsymmetric ill-conditioned linear systems. Such preconditioners may be one of the above relaxation methods, approximate factorizations or approximate inverses of the matrix of coefficients. A state-of-the-art survey of iterative solvers may be found in Refs. [43, 48, 55–58] or other references therein. When the matrix is symmetric positive definite, a preconditioned conjugate gradient scheme (PCG) may be an optimal choice as an iterative solver [2]. For symmetric indefinite systems, methods like SYMMLQ and MINRES [59] are adequate, but surprisingly PCG is often used with great success even though it may fail in theory. For nonsymmetric systems, the situation is less clear because available iterative schemes cannot combine minimization of the residual, at any given step, for a given norm within the Krylov subspace, and orthogonalizing it with respect to the same subspace for some scalar product. Therefore, two classes of methods arise. The most popular methods include GMRES [60], Bi-CGSTAB [61], QMR [62], and TFQMR [63].

Before presenting methods for computing the sparse SVD, we note that classical methods for determining the SVD of dense matrices: the Golub–Kahan–Reinsch method [64, 65] and Jacobi-like SVD methods [34, 66] are not viable for large sparse matrices. Because these methods apply orthogonal transformations (Householder or Givens) directly to the sparse matrix A, they incur excessive fill-ins and thereby require tremendous amounts of storage. Another drawback to these methods is that they will compute all the singular triplets of A, and hence may be computationally wasteful when only a few of the smallest, or largest, singular triplets are desired. We demonstrate how canonical sparse symmetric eigenvalue problems can be used to (indirectly) compute the sparse SVD.

Since the computation of the smallest singular value is equivalent to computing an eigenvalue of a symmetric matrix, which is the augmented matrix or the matrix of the normal equations, in what follows we present methods that are specifically designed for such problems.

4.3.2 SUBSPACE ITERATION [SISVD]

Subspace iteration is perhaps one of the simplest algorithms used to solve large sparse eigenvalue problems. As discussed in Ref. [67], subspace iteration may be viewed as a block generalization of the classical power method. The basic version of subspace iteration was first introduced by Bauer [68] and if adapted to the matrix

$$\hat{B} = \gamma^2 I_n - A^{\mathrm{T}} A, \tag{4.29}$$

would involve forming the sequence

$$Z_k = \hat{B}^k Z_0,$$

where γ^2 is chosen so that $\hat{B}$ is (symmetric) positive definite and $Z_0 = [z_1, z_2, \ldots, z_s]$ is an $n \times s$ matrix. If the column vectors, z_i are normalized separately (as done in the power method), then these vectors will converge to the dominant eigenvectors of $\hat{B}$, which are also the right singular vectors corresponding to the smallest singular values of A. Thus, the columns of the matrix Z_k will progressively lose linear independence. To approximate the p-largest eigenpairs of $\hat{B}$, Bauer demonstrated that linear independence among the z_i's could be maintained via reorthogonalization at each step, by a modified Gram–Schmidt procedure, for example. However, the convergence rate of the z_i's to eigenvectors of $\hat{B}$ will only be linear.

The most sophisticated implementation of subspace iteration is that of Rutishauser's RITZIT (see Ref. [69]). This particular algorithm incorporates both a Rayleigh–Ritz procedure and acceleration via Chebyshev polynomials. The iteration which embodies the RITZIT program is given in Table 4.2. The Rayleigh quotient matrix, H_k, in step (3) is essentially the projection of $\hat{B}^2$ onto the span Z_{k-1}. The three-term recurrence in step (6) follows from the adaptation of the Chebyshev polynomial of degree q, say $T_q(x)$, to the interval $[-e, e]$, where e is chosen to be the smallest eigenvalue of H_k. This use of Chebyshev polynomials has the desired effects of damping unwanted eigenvalues of $\hat{B}$ and producing an improved rate of convergence which is considerably higher than the original rate of convergence, governed by θ_s/θ_1 ($\theta_1 \geq \theta_2 \geq \cdots \geq \theta_s$), where $\theta_i's$ are the eigenvalues of H_k, given by the square roots of the diagonal matrix Δ_k^2 in step (4) of Table 4.2.

We note that one could alternatively compute the eigenpairs of the positive definite two-cyclic matrix:

$$\tilde{A}_{\text{aug}} = \begin{pmatrix} \gamma I & A \\ A^{\mathrm{T}} & \gamma I \end{pmatrix}, \tag{4.30}$$

where γ is an estimate of the largest singular value of A. The smallest singular values of A, in this case, will lie in the center of the spectrum of $\tilde{A}_{\text{aug}}$ (see Section 4.1.1), and thus prohibit suppression of the unwanted (largest) singular values by the use of Chebyshev polynomials defined on the symmetric interval $[-e, e]$. Thus, it is preferable to approximate eigenpairs of $\hat{B} = \gamma^2 I_n - A^{\mathrm{T}} A$ instead. In practice, γ can be chosen as either $\| A \|_1$ or $\| A \|_\infty$, depending upon the sparse data structure used to store the nonzero elements (row or columnwise). Once the eigenvectors of $\hat{B}$ (right singular vectors of A) have been determined, one can recover the left singular vectors (u_i) of A via $u_i = (1/\sigma_i) A v_i$ (see Section 4.1.1).

The orthogonal factorization in step (2) of Table 4.2 may be computed by a modified Gram–Schmidt procedure or by Householder transformations provided that the orthogonal matrix Q_k is explicitly available for the computation of Z_k in step (5). On multiprocessor architectures, especially those having hierarchical memories, one may achieve high performance by using either a block Gram–Schmidt [70] or the block Householder orthogonalization method in step (2). To improve upon the two-sided Jacobi algorithm, originally suggested by Rutishauser [71] for the spectral decomposition in step (4) of ritzit, one may employ a parallel two- or one-sided Jacobi method on a multiprocessor. In fact, the one-sided Jacobi scheme, when appropriately adapted for symmetric positive definite matrices (see Ref. [34]), is quite efficient for step (4) provided the dimension

---

TABLE 4.2

Subspace Iteration as Implemented in Rutishauser's ritzit [SISVD]

(1)	Compute $C_k = \hat{B} Z_{k-1}$
(2)	Factor $C_k = Q_k R_k$
(3)	Form $H_k = R_k R_k^{\mathrm{T}}$
(4)	Factor $H_k = P_k \Delta_k^2 P_k^{\mathrm{T}}$
(5)	Form $Z_k = Q_k P_k$
(6)	Iterate $Z_{k+j} = \frac{2}{e} \hat{B} Z_{k+j-1} - Z_{k+j-2}$ ($j = 2, \ldots, q$)

of the current subspace, s, is not too large. For larger subspaces, an optimized implementation of the classical EISPACK [72] pair, TRED2 and TQL2, or Cuppen's algorithm as parallelized by Dongarra and Sorensen [73] may be used in step (4).

The success of Rutishauser's subspace iteration method using Chebyshev acceleration relies upon the following strategy for delimiting the degree of the Chebyshev polynomial, $T_q(x/e)$, on the interval $[-e, e]$, where $e = \theta_s$ (assuming s vectors carried and $k = 1$ initially), $\xi_1 = 0.04$ and $\xi_2 = 4$:

$$q_{\text{new}} = \min\{2q_{\text{old}}, \hat{q}\},$$

where

$$\hat{q} = \begin{cases} 1, & \text{if } \theta_1 < \xi_1 \theta_s \\[2ex] 2 \times \max\left[\dfrac{\xi_2}{\text{arccosh}\left(\frac{\theta_s}{\theta_1}\right)}, 1 \right] & \text{otherwise.} \end{cases} \tag{4.31}$$

The polynomial degree of the current iteration is then taken to be $q = q_{\text{new}}$. It can easily be shown that the strategy in (4.31) insures that

$$\left\| T_q\left[\frac{\theta_1}{\theta_s}\right] \right\|_2 = \cosh\left[q \, \text{arccosh}\left(\frac{\theta_1}{\theta_s}\right) \right] \leq \cosh(8) < 1500.$$

Although this bound has been quite successful for `ritzit`, we can easily generate several variations of polynomial-accelerated subspace iteration schemes (`SISVD`) using a more flexible bound. Specifically, we consider an *adaptive* strategy for selecting the degree q in which ξ_1 and ξ_2 are treated as control parameters for determining the *frequency* and the *degree* of polynomial acceleration, respectively. In other words, large (small) values of ξ_1 inhibit (invoke) polynomial acceleration and large (small) values of ξ_2 yield larger (smaller) polynomial degrees when acceleration is selected. Correspondingly, the number of matrix–vector multiplications will increase with ξ_2 and the total number of iterations may well increase with ξ_1. Controlling the parameters, ξ_1 and ξ_2 allows us to monitor the method's complexity so as to maintain an optimal balance between dominating kernels (e.g., sparse matrix multiplication, orthogonalization, and spectral decomposition). We will demonstrate these controls in the polynomial acceleration-based trace minimization SVD method discussed in Section 4.3.4.

4.3.3 LANCZOS METHODS

4.3.3.1 The Single-Vector Lanczos Method [LASVD]

Other popular methods for solving large, sparse, symmetric eigenproblems originated from a method attributed to Lanczos (1950). This method generates a sequence of tridiagonal matrices T_j with the property that the extremal eigenvalues of the $j \times j$ matrix T_j are progressively better estimates of the extremal eigenvalues of the original matrix. Let us consider the $(m + n) \times (m + n)$ two-cyclic matrix A_{aug} given in (4.3), where A is the $m \times n$ matrix whose singular triplets are sought. Also, let w_1 be a randomly generated starting vector such that $\|w_1\|_2 = 1$. For $j = 1, 2, \ldots, l$ define the corresponding Lanczos matrices T_j using the following recursion [74]. Define $\beta_1 \equiv 0$ and $v_0 \equiv 0$, then for $i = 1, 2, \ldots, l$ define Lanczos vectors w_i and scalars α_i and β_{i+1} where

$$\beta_{i+1} w_{i+1} = A_{\text{aug}} w_i - \alpha_i w_i - \beta_i w_{i-1}, \text{ and } \alpha_i = w_i^{\text{T}}(A_{\text{aug}} w_i - \beta_i w_{i-1}),$$

$$|\beta_{i+1}| = \|A_{\text{aug}} w_i - \alpha_i w_i - \beta_i w_{i-1}\|_2. \tag{4.32}$$

For each j, the corresponding Lanczos matrix T_j is defined as a real symmetric, tridiagonal matrix having diagonal entries $\alpha_i (1 \leq i \leq j)$, and subdiagonal (superdiagonal) entries β_{i+1} ($1 \leq i \leq (j-1)$), i.e.,

$$T_j \equiv \begin{pmatrix} \alpha_1 & \beta_2 & & & & \\ \beta_2 & \alpha_2 & \beta_3 & & & \\ & \beta_3 & \cdot & \cdot & & \\ & & \cdot & \cdot & \cdot & \\ & & & \cdot & \cdot & \beta_j \\ & & & & \beta_j & \alpha_j \end{pmatrix}. \tag{4.33}$$

By definition, the vectors $\alpha_i w_i$ and $\beta_i w_{i-1}$ in (4.32) are, respectively, the orthogonal projections of $A_{\text{aug}} w_i$ onto the most recent w_i and w_{i-1}. Hence for each i, the next Lanczos vector w_{i+1} is obtained by orthogonalizing $A_{\text{aug}} w_i$ with respect to w_i and w_{i-1}. The resulting α_i, β_{i+1} obtained in these orthogonalizations define the corresponding Lanczos matrices. If we rewrite (4.32) in matrix form, then for each j we have

$$A_{\text{aug}} W_j = W_j T_j + \beta_{j+1} w_{j+1} e_j, \tag{4.34}$$

where $W_j \equiv [w_1, w_2, \ldots, w_j]$ is an $n \times j$ matrix whose kth column is the kth Lanczos vector, and e_j is the jth column of the $(m+n) \times (m+n)$ identity matrix. Thus, the Lanczos recursion (4.34) generates a family of real symmetric tridiagonal matrices related to both A_{aug} and w_1. Table 4.3 outlines the basic Lanczos procedure for computing the eigenvalues and eigenvectors of the symmetric two-cyclic matrix A_{aug}.

As in subspace iteration, the matrix A_{aug} is only referenced through matrix–vector multiplication. At each iteration, the basic Lanczos recursion requires only the two most recently generated vectors, although for finite-precision arithmetic modifications suggested by Grcar [75], Parlett and Scott [76], and Simon [77] require additional Lanczos vectors to be readily accessible via secondary storage. We note that on a multiprocessor architecture, step (2) in Table 4.3 may benefit from any available optimized library routine that solves the symmetric tridiagonal eigenvalue problem (e.g., multisectioning in Ref. [78] or divide and conquer in Ref. [73]).

In using finite-precision arithmetic, any practical Lanczos procedure must address problems created by losses in the orthogonality of the Lanczos vectors, w_i. Such problems include the occurrence of numerically multiple eigenvalues of T_j (for large j) for simple eigenvalues of A_{aug} and the appearance of spurious eigenvalues among the computed eigenvalues for some T_j. Approaches to

---

TABLE 4.3
Single-Vector Lanczos Recursion [LASVD]

(1) Use any variant of the Lanczos recursion (4.32) to generate a family of real symmetric tridiagonal matrices, T_j ($j = 1, 2, \ldots, q$)

(2) For some $k \leq q$, compute relevant eigenvalues of T_k

(3) Select some or all of these eigenvalues as approximations to the eigenvalues of the matrix A_{aug}, and hence singular values of A

(4) For each eigenvalue λ compute a corresponding unit eigenvector z such that $T_k z = \lambda z$. Map such vectors onto the corresponding Ritz vectors $y \equiv W_q z$, which are then used as approximations to the desired eigenvectors of the matrix A_{aug} or the singular vectors of A

deal with these problems range between two different extremes. The first involves total reorthogonalization of every Lanczos vector with respect to every previously generated vector [79]. The other approach accepts the loss of orthogonality and deals with these problems directly. Total reorthogonalization is certainly one way of maintaining orthogonality, however, it will require additional storage and additional arithmetic operations. As a result, the number of eigenvalues which can be computed is limited by the amount of available secondary storage. On the other hand, a Lanczos procedure with no reorthogonalization needs only the two most recently generated Lanczos vectors at each stage, and hence has minimal storage requirements. Such a procedure requires, however, the tracking [5] of the resulting spurious eigenvalues of A_{aug} (singular values of A) associated with the loss of orthogonality in the Lanczos vectors, w_i.

We employ a version of a single-vector Lanczos algorithm (4.32) equipped with a selective reorthogonalization strategy, LANSO, designed by Parlett and Scott [76] and Simon [77]. This particular method (LASVD) is primarily designed for the standard and generalized symmetric eigenvalue problem. We simply apply it to either $B = A^{T}A$ or the two-cyclic matrix A_{aug} defined in (4.3).

4.3.3.2 The Block Lanczos Method [BLSVD]

Here, we consider a block analog of the single-vector Lanczos method. Exploiting the structure of the matrix A_{aug} in (4.3), we can obtain an alternative form for the Lanczos recursion (4.32). If we apply the Lanczos recursion specified by (4.32) to A_{aug} with a starting vector $\tilde{u} = (u, 0)^{T}$ such that $\|\tilde{u}\|_2 = 1$, then the diagonal entries of the real symmetric tridiagonal Lanczos matrices generated are all identically zero. The Lanczos recursion in (4.32) reduces to the following: define $u_1 \equiv u$, $v_0 \equiv 0$, and $\beta_1 \equiv 0$, then for $i = 1, 2, \ldots, k$:

$$
\begin{aligned}
\beta_{2i}v_i &= A^{T}u_i - \beta_{2i-1}v_{i-1}, \\
\beta_{2i+1}u_{i+1} &= Av_i - \beta_{2i}u_i.
\end{aligned}
\tag{4.35}
$$

The Lanczos recursion (4.35), however, can only compute the distinct singular values of an $m \times n$ matrix A and not their multiplicities. Following the block Lanczos recursion for the sparse symmetric eigenvalue problem [80, 81], (4.35) can be represented in matrix form as

$$
\begin{aligned}
A^{T}\hat{U}_k &= \hat{V}_k J_k^{T} + Z_k, \\
A\hat{V}_k &= \hat{U}_k J_k + \hat{Z}_k,
\end{aligned}
\tag{4.36}
$$

where $\hat{U}_k = [u_1, \ldots, u_k]$, $\hat{V}_k = [v_1, \ldots, v_k]$, J_k is a $k \times k$ bidiagonal matrix with $J_k[j, j] = \beta_{2j}$ and $J_k[j, j+1] = \beta_{2j+1}$, and Z_k, $\tilde{Z}_k$ contain remainder terms. It is easy to show that the nonzero singular values of J_k are the same as the positive eigenvalues of

$$
K_k \equiv \begin{pmatrix} O & J_k \\ J_k^{T} & O \end{pmatrix}.
\tag{4.37}
$$

For the block analog of (4.36), we make the simple substitutions

$$
u_i \leftrightarrow U_i, \quad v_i \leftrightarrow V_i,
$$

where U_i is $m \times b$, V_i is $n \times b$, and b is the current block size. The matrix J_k is now a upper block bidiagonal matrix of order bk

$$
J_k \equiv \begin{pmatrix}
S_1 & R_1^{T} & & & \\
& S_2 & R_2^{T} & & \\
& & \cdot & \cdot & \\
& & & \cdot & \cdot \\
& & & & \cdot & R_{k-1}^{T} \\
& & & & & S_k
\end{pmatrix},
\tag{4.38}
$$

where the S_i's and R_i's are $b \times b$ upper-triangular matrices. If U_i's and V_i's form mutually orthogonal sets of bk vectors so that $\hat{U}_k$ and $\hat{V}_k$ are orthonormal matrices, then the singular values of the matrix J_k will be identical to those of the original $m \times n$ matrix A. Given the upper block bidiagonal matrix J_k, we approximate the singular triplets of A by first computing the singular triplets of J_k. To determine the left and right singular vectors of A from those of J_k, we must retain the Lanczos vectors of $\hat{U}_k$ and $\hat{V}_k$. Specifically, if $\{\sigma_i^{(k)}, y_i^{(k)}, z_i^{(k)}\}$ is the ith singular triplet of J_k, then the approximation to the ith singular triplet of A is given by $\{\sigma_i^{(k)}, \hat{U}_k y_i^{(k)}, \hat{V}_k z_i^{(k)}\}$, where $\hat{U}_k y_i^{(k)}$, $\hat{V}_k z_i^{(k)}$ are the left and right approximate singular vectors, respectively. The computation of singular triplets for J_k requires two phases. The first phase reduces J_k to a bidiagonal matrix C_k having diagonal elements $\{\alpha_1, \alpha_2, \ldots, \alpha_{bk}\}$ and superdiagonal elements $\{\beta_1, \beta_2, \ldots, \beta_{bk-1}\}$ via a finite sequence of orthogonal transformations (thus preserving the singular values of J_k). The second phase reduces C_k to diagonal form by a modified QR-algorithm. This diagonalization procedure is discussed in detail in Ref. [65]. The resulting diagonalized C_k will yield the approximate singular values of A, whereas the corresponding left and right singular vectors are determined through multiplications by all the left and right transformations used in both phases of the SVD of J_k.

There are a few options for the reduction of J_k to the bidiagonal matrix, C_k. Golub et al. [82] advocated the use of either band Householder or band Givens methods which in effect *chase off* (or zero) elements on the diagonals above the first superdiagonal of J_k. In either reduction (bidiagonalization or diagonalization), the computations are primarily sequential and offer limited data locality or parallelism for possible exploitation on a multiprocessor. For this reason, we adopt the single-vector Lanczos bidiagonalization recursion defined by (4.35) and (4.36) as our strategy for reducing the upper block bidiagonal matrix J_k to the bidiagonal form (C_k), i.e.,

$$\begin{aligned} J_k^{\mathrm{T}} \hat{Q} &= \hat{P} C_k^{\mathrm{T}}, \\ J_k \hat{P} &= \hat{Q} C_k, \end{aligned} \tag{4.39}$$

or

$$\begin{aligned} J_k p_j &= \alpha_j q_j + \beta_{j-1} q_{j-1}, \\ J_k^{\mathrm{T}} q_j &= \alpha_j p_j + \beta_j p_{j+1}, \end{aligned} \tag{4.40}$$

where $\hat{P} \equiv \{p_1, p_2, \ldots, p_{bk}\}$ and $\hat{Q} \equiv \{q_1, q_2, \ldots, q_{bk}\}$ are orthonormal matrices of order $bk \times bk$. The recursions in (4.40) require band matrix–vector multiplications which can be easily exploited by optimized level-2 BLAS routines [83] now resident in optimized mathematical libraries on most high-performance computers. For orthogonalization of the outermost Lanczos vectors, $\{U_i\}$ and

---

TABLE 4.4

Hybrid Lanczos Outer Iteration [BLSVD]

(1)	[Formation of symmetric block tridiagonal matrix H_k]
	Choose V_1 ($n \times b$ and orthonormal) and $c = \max\{bk\}$
	Compute $S_1 = V_1^{\mathrm{T}} A^{\mathrm{T}} A V_1$. ($V_0$, $R_0^{\mathrm{T}} = 0$ initially)
	For $i = 2, 3, \ldots, k$ do: ($k = \lfloor c/b \rfloor$)
(2)	Compute $Y_{i-1} = A^{\mathrm{T}} A V_{i-1} - V_{i-1} S_{i-1} - V_{i-1} R_{i-2}^{\mathrm{T}}$
(3)	Orthogonalize Y_{i-1} against $\{V_\ell\}_{\ell=0}^{i-1}$
(4)	Factor $Y_{i-1} = V_i R_{i-1}$
(5)	Compute $S_i = V_i^{\mathrm{T}} A^{\mathrm{T}} A V_i$

$\{V_i\}$, as well as the innermost Lanczos vectors, $\{p_i\}$ and $\{q_i\}$, we have chosen to apply a complete or total reorthogonalization [79] strategy to insure robustness in our triplet approximations for the matrix A. This hybrid Lanczos approach which incorporates inner iterations of single-vector Lanczos bidiagonalization within the outer iterations of a block Lanczos SVD recursion is also discussed in Ref. [1].

As an alternative to the outer recursion defined by (4.36), which is derived from the equivalent two-cyclic matrix A_{aug}, Table 4.4 depicts the simplified outer block Lanczos recursion for approximating the eigensystem of $A^T A$. Combining the equations in (4.36), we obtain

$$A^T A \hat{V}_k = \hat{V}_k H_k,$$

where $H_k = J_k^T J_k$ is the $k \times k$ symmetric block tridiagonal matrix

$$
H_k \equiv \begin{pmatrix}
S_1 & R_1^T & & & & \\
R_1 & S_2 & R_2^T & & & \\
& R_2 & \cdot & \cdot & & \\
& & \cdot & \cdot & \cdot & \\
& & & \cdot & \cdot & R_{k-1}^T \\
& & & & R_{k-1} & S_k
\end{pmatrix}, \tag{4.41}
$$

having block size b. We then apply the block Lanczos recursion [79] in Table 4.4 for computing the eigenpairs of the $n \times n$ symmetric positive definite matrix $A^T A$. The tridiagonalization of H_k via an inner Lanczos recursion follows from simple modifications of (4.34). Analogous to the reduction of J_k in (4.38), the computation of eigenpairs of the resulting tridiagonal matrix can be performed via a Jacobi or QR-based symmetric eigensolver.

As with the previous iterative SVD methods, we access the sparse matrices A and A^T for this hybrid Lanczos method only through sparse matrix–vector multiplications. Some efficiency, however, is gained in the outer (block) Lanczos iterations by the multiplication of b vectors rather than by a single vector. These dense vectors may be stored in a fast local memory (cache) of a hierarchical memory-based architecture, and thus yield more effective data reuse. A stable variant of Gram–Schmidt orthogonalization [69], which requires efficient dense matrix–vector multiplication (level-2 BLAS) routines [83] or efficient dense matrix–matrix multiplication (level-3 BLAS) routines [84], is used to produce the orthogonal projections of Y_i (i.e., R_{i-1}) and W_i (i.e., S_i) onto $\tilde{V}^\perp$ and $\tilde{U}^\perp$, respectively, where U_0 and V_0 contain converged left and right singular vectors, respectively, and

$$\tilde{V} = (V_0, V_1, \ldots, V_{i-1}) \quad \text{and} \quad \tilde{U} = (U_0, U_1, \ldots, U_{i-1}).$$

4.3.4 THE TRACE MINIMIZATION METHOD [TRSVD]

Another candidate subspace method for the SVD of sparse matrices is based upon the trace minimization algorithm discussed in Ref. [85] for the generalized eigenvalue problem

$$Hx = \lambda Gx, \tag{4.42}$$

where H and G are symmetric and G is also positive definite. To compute the SVD of an $m \times n$ matrix A, we initially replace H with $\tilde{A}_{\text{aug}}$ as defined in (4.30) or set $H = A^T A$. Since we need to only consider equivalent standard symmetric eigenvalue problems, we simply define $G = I_{m+n}$ (or I_n if $H = A^T A$). Without loss of generality, let us assume that $H = A^T A$, $G = I_n$ and consider the associated symmetric eigensystem of order n. If $\mathcal{Y}$ is defined as the set of all $n \times p$ matrices Y for which $Y^T Y = I_p$, then using the Courant–Fischer theorem (see Ref. [86]) we obtain

$$\min_{Y \in \mathcal{Y}} \text{trace}(Y^T H Y) = \sum_{i=1}^{p} \tilde{\sigma}_{n-i+1}, \tag{4.43}$$

where $\sqrt{\tilde{\sigma}_i}$ is a singular value of A, $\lambda_i = \tilde{\sigma}_i$ is an eigenvalue of H, and $\tilde{\sigma}_1 \geq \tilde{\sigma}_2 \geq \cdots \geq \tilde{\sigma}_n$. In other words, given an $n \times p$ matrix Y which forms a *section* of the eigenvalue problem

$$H z = \lambda z, \tag{4.44}$$

i.e.,

$$Y^T H Y = \tilde{\Sigma}, \quad Y^T Y = I_p, \tag{4.45}$$

$$\tilde{\Sigma} = \text{diag}(\tilde{\sigma}_n, \tilde{\sigma}_{n-1}, \dots, \tilde{\sigma}_{n-p+1}),$$

our trace minimization scheme [TRSVD] Ref. [85], see also [87], finds a sequence of iterates $Y_{k+1} = F(Y_k)$, where both Y_k and Y_{k+1} form a section of (4.44), and have the property trace $(Y_{k+1}^T H Y_{k+1}) < \text{trace}(Y_k^T H Y_k)$. From (4.43), the matrix Y in (4.45) which minimizes $\text{trace}(Y^T H Y)$ is the matrix of H-eigenvectors associated with the p-smallest eigenvalues of the problem (4.44). As discussed in Ref. [85], $F(Y)$ can be chosen so that global convergence is assured. Moreover, (4.45) can be regarded as the quadratic minimization problem

$$\text{minimize } \text{trace}(Y^T H Y) \tag{4.46}$$

subject to the constraints

$$Y^T Y = I_p. \tag{4.47}$$

Using Lagrange multipliers, this quadratic minimization problem leads to solving the $(n + p) \times (n + p)$ system of linear equations

$$\begin{pmatrix} H & Y_k \\ Y_k^T & 0 \end{pmatrix} \begin{pmatrix} \Delta_k \\ L \end{pmatrix} = \begin{pmatrix} H Y_k \\ 0 \end{pmatrix}, \tag{4.48}$$

so that $Y_{k+1} \equiv Y_k - \Delta_k$ will be an optimal subspace iterate.

Since the matrix H is positive definite, one can alternatively consider the p-independent (parallel) subproblems

$$\text{minimize } ((y_j^{(k)} - d_j^{(k)})^T H (y_j^{(k)} - d_j^{(k)})) \tag{4.49}$$

subject to the constraints

$$Y^T d_j^{(k)} = 0, \quad j = 1, 2, \dots, p,$$

where $d_j^{(k)} = \Delta_k e_j$, e_j the jth column of the identity, and $Y_k = [y_1^{(k)}, y_2^{(k)}, \dots, y_p^{(k)}]$. The corrections Δ_k in this case are selected to be orthogonal to the previous estimates Y_k, i.e., so that (see Ref. [88])

$$\Delta_k^T Y_k = 0.$$

We then recast (4.48) as

$$\begin{pmatrix} H & Y_k \\ Y_k^T & 0 \end{pmatrix} \begin{pmatrix} d_j^{(k)} \\ l \end{pmatrix} = \begin{pmatrix} Hy_j^{(k)} \\ 0 \end{pmatrix}, \quad j = 1, 2, \ldots, p, \tag{4.50}$$

where l is a vector of order p reflecting the Lagrange multipliers.

The solution of the p-systems of linear equations in (4.50) can be done in parallel by either a direct or iterative solver. Since the original matrix A is assumed to be large, sparse and without any particular sparsity structure (pattern of nonzeros) we have chosen an iterative method (conjugate gradient) for the systems in (4.50). As discussed in Refs. [1, 85], a major reason for using the conjugate gradient (CG) method for the solution of (4.50) stems from the ability to terminate CG iterations early without obtaining fully accurate corrections $d_j^{(k)}$ that are more accurate than warranted. In later stages, however, as Y_k converges to the desired set of eigenvectors of B, one needs full accuracy in computing the correction matrix Δ_k.

4.3.4.1 Polynomial Acceleration Techniques for [TRSVD]

The Chebyshev acceleration strategy used within subspace iteration (see Section 4.3.2), can also be applied to [TRSVD]. However, to dampen unwanted (largest) singular values of A in this context, we must solve the generalized eigenvalue problem

$$x = \frac{1}{P_q(\lambda)} P_q(H)x, \tag{4.51}$$

where $P_q(x) = T_q(x) + \epsilon I_n$, $T_q(x)$ is the Chebyshev polynomial of degree q and ϵ is chosen so that $P_q(H)$ is (symmetric) positive definite. The appropriate quadratic minimization problem similar to (4.49) here can be expressed as

$$\text{minimize } ((y_j^{(k)} - d_j^{(k)})^T (y_j^{(k)} - d_j^{(k)})) \tag{4.52}$$

subject to the constraints

$$Y^T P_q(H) d_j^{(k)} = 0, \quad j = 1, 2, \ldots, p.$$

In effect, we approximate the smallest eigenvalues of H as the *largest* eigenvalues of the matrix $P_q(H)$ whose gaps are considerably larger than those of the eigenvalues of H.

Although the additional number of sparse matrix–vector multiplications associated with the multiplication by $P_q(H)$ will be significant for high degrees q, the system of equations via Lagrange multipliers in (4.50) becomes much easier to solve, i.e.,

$$\begin{pmatrix} I & P_q(H)Y_k \\ Y_k^T P_q(H) & 0 \end{pmatrix} \begin{pmatrix} d_j^{(k)} \\ l \end{pmatrix} = \begin{pmatrix} y_j^{(k)} \\ 0 \end{pmatrix}, \quad j = 1, 2, \ldots, p. \tag{4.53}$$

It is easy to show that the updated eigenvector approximation, $y_j^{(k+1)}$, is determined by

$$y_j^{(k+1)} = y_j^{(k)} - d_j^{(k)} = P_q(H)Y_k[Y_k^T P_q^2(H)Y_k]^{-1} Y_k^T P_q(H)y_j^{(k)}.$$

Thus, we may not need to use an iterative solver for determining Y_{k+1} since the matrix $[Y_k^T P_q^2(H)Y_k]^{-1}$ is of relatively small order p. Using the orthogonal factorization

$$P_q(H)Y_k = \hat{Q}\hat{R},$$

we have

$$[Y_k^T P_q^2(H)Y_k]^{-1} = \hat{R}^{-T} \hat{R}^{-1},$$

where the polynomial degree, q, is determined by the strategy defined in Section 4.3.2.

4.3.4.2 Shifting Strategy for [TRSVD]

As discussed in Ref. [85], we can also accelerate the convergence of the Y_k's to eigenvectors of H by incorporating Ritz shifts (see Ref. [67]) into TRSVD. Specifically, we modify the symmetric eigenvalue problem in (4.44) as follows,

$$(H - \nu_j^{(k)} I)z_j = (\lambda_j - \nu_j^{(k)})z_j, \quad j = 1, 2, \dots, s, \tag{4.54}$$

where $\nu_j^{(k)} = \tilde{\sigma}_{n-j+1}^{(k)}$ is the jth approximate eigenvalue at the kth iteration of [TRSVD], with λ_j, z_j an exact eigenpair of H. In other words, we simply use our most recent approximations to the eigenvalues of H from our kth section within [TRSVD] as Ritz shifts. As was shown by Wilkinson [89], the Rayleigh quotient iteration associated with (4.54) will ultimately achieve cubic convergence to the square of an exact singular value A, σ_{n-j+1}^2, provided $\nu_j^{(k)}$ is sufficiently close to σ_{n-j+1}^2. However, we have $\nu_j^{(k+1)} < \nu_j^{(k)}$ for all k, i.e., we approximate eigenvalues of H from above, $H - \nu_j^{(k)} I$ will not be positive definite and thus we cannot guarantee the convergence of this shifted [TRSVD] method for any particular singular triplet of A. However, the strategy outlined in Ref. [85] has been quite successful in maintaining global convergence with shifting.

Table 4.5 outlines the basic steps of [TRSVD]. The scheme appropriately utilizes polynomial (Chebyshev) acceleration before the Ritz shifts. It is important to note that once shifting has been invoked (Step (4)) we abandon the use of Chebyshev polynomials $P_q(H)$ and solve shifted systems (H replaced by $H - \nu_j^{(k)} I$) of the form (4.48) via (4.50) and the CG-algorithm. The *context switch* from either nonaccelerated (or polynomial-accelerated trace minimization iterations) to trace minimization iterations with Ritz shifting, is accomplished by monitoring the reduction of the residuals in Step (4) for isolated eigenvalues ($r_j^{(k)}$) or clusters of eigenvalues ($R_j^{(k)}$).

The Chebyshev acceleration and Ritz shifting within [TRSVD] for approximating smallest singular value of a 374×82 matrix with only 1343 nonzero elements has cut down the number

---

TABLE 4.5

[TRSVD] Algorithm with Chebyshev Acceleration and Ritz Shifts

(Step 0)	Set $k = 0$, choose an initial $n \times s$ subspace iterate $Y_0 = [y_1^{(0)}, y_2^{(0)}, \dots, y_s^{(0)}]$
(Step 1)	Form a section as in (4.45), or determine Y_k such that $Y_k^T P_q(H)Y_k = I_p$, $Y_k^T Y_k = \tilde{\Sigma}$
(Step 2)	Compute residuals: $r_j^{(k)} = Hy_j^{(k)} - \tilde{\sigma}_{n-j+1}^{(k)} y_j^{(k)}$, and access accuracy
(Step 3)	Analyze the current approximate spectrum (Gershgorin disks)
(Step 4)	Invoke Ritz shifting strategy (see Ref. [85]):
	For isolated eigenvalues:
	$\quad$ if $\|r_j^{(k)}\|_2 \leq \eta \|r_j^{(k_0)}\|_2$, where $\eta \in [10^{-3}, 10^0]$ and $k_0 < k$ for some j
	For a cluster of eigenvalues (size c):
	$\quad$ if $\|R_j^{(k)}\|_F \leq \eta \|R_j^{(k_0)}\|_F$, where $R_j^{(k)} \equiv \{r_j^{(k)}, \dots, r_{j+c}^{(k)}\}$ and $k_0 < k$ for some j
	(Disable polynomial acceleration if shifting is selected)
(Step 5)	Deflation: reduce subspace dimension, s, by number of H-eigenpairs accepted.
(Step 6)	Adjust polynomial degree q via (4.31) for $P_q(H)$ in iteration $k + 1$ (if needed)
(Step 7)	Update subspace iterate $Y_{k+1} \equiv Y_k - \Delta_k$ via (4.48) or (4.50)
(Step 8)	Set $k = k + 1$ and go to (Step 1)

of iterations needed by [TRSVD] with no acceleration by a factor of 3, using the same stopping criteria. The convergence rate of [TRSVD] with Chebyshev acceleration plus Ritz shifts (CA + RS) is three times higher than [TRSVD] with no acceleration.

4.3.5 REFINEMENT OF LEFT SINGULAR VECTORS

As discussed in Section 4.1.1, the smallest singular values of the matrix A lie in the interior of the spectrum of either the two-cyclic matrix A_{aug} (4.3) or the shifted matrix $\tilde{A}_{\text{aug}}$ (4.30), all four candidate methods will have difficulties in approximating these singular values. The Lanczos-based methods, [LASVD] and [BLSVD] cannot be expected to effectively approximate the interior eigenvalues of either A_{aug} or $\tilde{A}_{\text{aug}}$. Similarly, subspace iteration [SISVD] would not be able to suppress the unwanted (largest) singular values if Chebyshev polynomials are defined on symmetric intervals of the form $[-e, e]$. [TRSVD], however, via direct or iterative schemes for solving the systems in (4.50), could indeed obtain accurate approximation of the smallest singular triplets. Recently, some progress has been realized by considering an implicitly restarted Lanczos bidiagonalization [90] that avoids direct inversion.

To target the p-smallest singular triplets of A, we compute the p-smallest eigenvalues and eigenvectors of the (operator) matrices listed in Table 4.6 for LASVD, BLSVD, and TRSVD. With SISVD, we determine the p-largest eigenpairs of $\gamma^2 I_n - A^T A$. As discussed in Section 4.1.1, we may determine the singular values, σ_i, of A and their corresponding right singular vectors, v_i, as eigenpairs of either $A^T A$ or $\gamma^2 I_n - A^T A$, where A is $m \times n$ and $\gamma = \|A\|_{1,\infty}$. The corresponding left singular vector, u_i, must then be determined by

$$u_i = \frac{1}{\sigma_i} A v_i, \qquad (4.55)$$

which may not be of sufficient accuracy (see Section 4.1.1) for a given precision in the residual (4.4). We point out, however, that alternative operators of the form (*shift and invert*)

$$(A^T A - \tilde{\sigma}^2 I)^{-1},$$

where $\tilde{\sigma}$ is a good approximation to an exact singular value of A, can also be used to determine the p-smallest singular values of A (see Refs. [67, 69]). The effectiveness of this approach depends on an *accurate* approximation of the Cholesky factorization of $A^T A$, or preferably of the QR-factorization of A, when A is large and sparse rectangular matrix.

TABLE 4.6

Operators of Equivalent Symmetric Eigenvalue Problems for Computing the p-Smallest Singular Triplets of the $m \times n$ Matrix A

Method	Label	Operator
Single-vector Lanczos	LASVD	$A^T A$
Block Lanczos	BLSVD	$A^T A$
Subspace iteration	SISVD	$\gamma^2 I_n - A^T A$
Trace minimization	TRSVD	$A^T A$

For [LASVD], we simply define $\tilde{A}_{\text{aug}} = A^{T}A$ in (4.32) and apply the strategy outlined in Table 4.3. For [SISVD] we make the substitution $\tilde{A}_{\text{aug}} = \gamma^{2}I_{n} - A^{T}A$ for (4.30) and apply the method in Table 4.2. Similarly, we solve the generalized eigenvalue problem in (4.44) with

$$H = A^{T}A, \quad G = I_{n},$$

and apply the strategy in Table 4.5 for the appropriate [TRSVD] method. Here, we do not require the shift γ^{2} because the trace minimization will converge to the smallest eigenvalues of $A^{T}A$ (and hence squares of singular values of A) by default. For [BLSVD], we combine the equations in (4.36) and obtain

$$A^{T}A\hat{V}_{k} = \hat{V}_{k}H_{k} + \hat{Z}_{k},$$

where $H_{k} = J_{k}^{T}J_{k}$ is the $k \times k$ symmetric block tridiagonal matrix in (4.41), with the $n \times k$ matrix $\hat{Z}_{k}$ containing the remainder terms. The appropriate block Lanczos (outer) recursion is then given by Table 4.4.

Having determined approximate singular values, $\tilde{\sigma}_{i}$, and corresponding right singular vectors, $\tilde{v}_{i}$, to a user-specified tolerance for the residual

$$\hat{r}_{i} = A^{T}A\tilde{v}_{i} - \tilde{\sigma}_{i}^{2}\tilde{v}_{i}, \tag{4.56}$$

we must then obtain an approximation to the corresponding left singular vector, u_{i}, via (4.55). As mentioned in Section 4.1.1, it is quite possible that square roots of the approximate eigenvalues of either $A^{T}A$ or $\gamma^{2}I_{n} - A^{T}A$ will be poor approximations to exact singular values of A which are extremely small. This phenomenon, of course, will lead to poor approximations to the left singular vectors. Even if $\tilde{\sigma}_{i}$ (computed by any of the four methods) is an acceptable singular value approximation, the residual corresponding to the singular triplet $\{\tilde{\sigma}_{i}, \tilde{u}_{i}, \tilde{v}_{i}\}$, defined by (4.4), will be bounded by

$$\|r_{i}\|_{2} \leq \|\hat{r}_{i}\|_{2}/[\tilde{\sigma}_{i}(\|\tilde{u}_{i}\|_{2}^{2} + \|\tilde{v}_{i}\|_{2}^{2})^{\frac{1}{2}}], \tag{4.57}$$

where $\hat{r}_{i}$ is the residual given in (4.56) for the symmetric eigenvalue problem for $A^{T}A$ or $(\gamma^{2}I_{n} - A^{T}A)$. Scaling by $\tilde{\sigma}_{i}$ can easily lead to significant loss of accuracy in estimating the triplet residual norm, $\|r_{i}\|_{2}$, especially when $\tilde{\sigma}_{i}$ approaches the machine unit roundoff error, μ.

One remedy is to refine the initial approximation of the left singular vector via (inverse iteration)

$$AA^{T}\tilde{u}_{i+1}^{(k)} = \tilde{\sigma}_{i}^{2}\tilde{u}_{i}^{(k)}, \tag{4.58}$$

where $\tilde{u}_{i}^{(0)} \equiv u_{i}$ from (4.55). Since AA^{T} is symmetric semidefinite, direct methods developed by Aasen [91], Bunch and Kaufman [92], or Parlett and Reid [93] could be used in each step of (4.58) if AA^{T} is explicitly formed. With regard to iterative methods, the SYMMLQ algorithm developed by Paige and Saunders [94] can be used to solve these symmetric indefinite systems of equations. As an alternative, we consider the following equivalent eigensystem for the SVD of an $m \times n$ matrix A.

$$\begin{pmatrix} \gamma I_{n} & A^{T} \\ A & \gamma I_{m} \end{pmatrix} \begin{pmatrix} v_{i} \\ u_{i} \end{pmatrix} = (\gamma + \sigma_{i}) \begin{pmatrix} v_{i} \\ u_{i} \end{pmatrix}, \tag{4.59}$$

where $\{\sigma_{i}, u_{i}, v_{i}\}$ is the ith singular triplet of A and

$$\gamma = \min[1, \max\{\sigma_{i}\}]. \tag{4.60}$$

One possible refinement recursion (inverse iteration) is thus given by

$$
\begin{pmatrix} \gamma I_n & A^{\mathrm{T}} \\ A & \gamma I_m \end{pmatrix} \begin{pmatrix} \tilde{v}_i^{(k+1)} \\ \tilde{u}_i^{(k+1)} \end{pmatrix} = (\gamma + \tilde{\sigma}_i) \begin{pmatrix} \tilde{v}_i^{(k)} \\ \tilde{u}_i^{(k)} \end{pmatrix}, \tag{4.61}
$$

By applying block Gaussian elimination to (4.61) we obtain a more optimal form (reduced system) of the recursion

$$
\begin{pmatrix} \gamma I_n & A^{\mathrm{T}} \\ 0 & \gamma I_m - \frac{1}{\gamma} A A^{\mathrm{T}} \end{pmatrix} \begin{pmatrix} \tilde{v}_i^{(k+1)} \\ \tilde{u}_i^{(k+1)} \end{pmatrix} = (\gamma + \tilde{\sigma}_i) \begin{pmatrix} \tilde{v}_i^{(k)} \\ \tilde{u}_i^{(k)} - \frac{1}{\gamma} A \tilde{v}_i^{(k)} \end{pmatrix}. \tag{4.62}
$$

Our iterative refinement strategy for an approximate singular triplet of A, $\{\tilde{\sigma}_i, \tilde{u}_i, \tilde{v}_i\}$, is then defined by the last m equations of (4.62), i.e.,

$$
\left(\gamma I_m - \frac{1}{\gamma} A A^{\mathrm{T}} \right) \tilde{u}_i^{(k+1)} = (\gamma + \tilde{\sigma}_i) \left(\tilde{u}_i^{(k)} - \frac{1}{\gamma} A \tilde{v}_i \right), \tag{4.63}
$$

where the superscript k is dropped from $\tilde{v}_i$, since we refine only our left singular vector approximation, $\tilde{u}_i$. If $\tilde{u}_i^{(0)} \equiv u_i$ from (4.59), then (4.63) can be rewritten as

$$
\left(\gamma I_m - \frac{1}{\gamma} A A^{\mathrm{T}} \right) \tilde{u}_i^{(k+1)} = (\gamma - \tilde{\sigma}_i^2/\gamma) \tilde{u}_i^{(k)}, \tag{4.64}
$$

with (normalization)

$$
\tilde{u}_i^{(k+1)} = \tilde{u}_i^{(k+1)} / \|\tilde{u}_i^{(k+1)}\|_2.
$$

It is easy to show that the left-hand-side matrix in (4.63) is symmetric positive definite provided (4.60) holds. Accordingly, we may use parallel conjugate gradient iterations to refine each singular triplet approximation. Hence, the refinement procedure outlined in Table 4.6 may be considered as a *black box* procedure to follow the eigensolution of $A^{\mathrm{T}} A$ or $\gamma^2 I_n - A^{\mathrm{T}} A$ by any one of our four candidate methods for the sparse SVD. The iterations in Step (3) of the refinement scheme in Table 4.7 terminate once the norms of the residuals of all p approximate singular triplets ($\|r_i\|_2$) fall below a user-specified tolerance or after $k_{\max}$ iterations:

$$
(\alpha I_n + \beta A^{\mathrm{T}} A) \tilde{u}_i = (\alpha + \beta \tilde{\sigma}_i^2), \tilde{v}_i
$$

TABLE 4.7
**Refinement Procedure for the Left Singular
Vector Approximations Obtained via Scaling**

(1) Solve eigensystems of the form $i = 1, 2, \ldots, p$, where
$\alpha = 0$ and $\beta = 1$ for LASVD, BLSVD, and TRSVD, or
$\alpha = \gamma^2 > \sigma_{max}^2$ and $\beta = -1$ for SISVD

(2) Define $\tilde{u}_i^{(0)} \equiv \dfrac{1}{\tilde{\sigma}_i} A \tilde{v}_i$, $\quad i = 1, 2, \ldots, p$

(3) For $k = 0, 1, 2, \ldots$ (until $\|r_i\|_2 \leq$ tolerance or $k \geq k_{\max}$)
Solve $\left(\gamma I_m - \frac{1}{\gamma} A A^{\mathrm{T}} \right) \tilde{u}_i^{(k+1)} = (\gamma - \tilde{\sigma}_i^2/\gamma) \tilde{u}_i^{(k)}$,
$i = 1, 2, \ldots, p$
Set $\tilde{u}_i^{(k+1)} = \tilde{u}_i^{(k+1)} / \|\tilde{u}_i^{(k+1)}\|_2$

Summary

In this work, we demonstrated that a trace minimization strategy [TRSVD] using Chebyshev acceleration, Ritz shifting, and an iterative refinement method for improving left singular vector approximations, can be quite economical and robust when computing several of the smallest singular triplets of sparse matrices arising from a variety of applications. A single-vector Lanczos method [LASVD] can be quite competitive in speed compared to [TRSVD] at the risk of missing some of the desired triplets due to parameter selection difficulties. Whereas [TRSVD] is effective in achieving high accuracy in approximating all the desired singular triplets. [LASVD] is acceptable only when moderate to low accurate triplets are needed. A subspace iteration-based method [SISVD] using Chebyshev acceleration performed poorly for clustered singular values when polynomials of relatively high degree are used. Typically, this scheme requires less memory than the other three methods. A block Lanczos SVD method [BLSVD], on the other hand, is quite robust for obtaining all the desired singular triplets at the expense of large memory requirements. The use of alternative reorthogonalization strategies could be a remedy for this difficulty. Finally, we have proposed a hybrid [TRSVD] method which circumvents the potential loss of accuracy when solving eigensystems of $A^{\mathrm{T}}A$.

4.3.6 The Davidson Methods

4.3.6.1 General Framework of the Methods

In 1975, Davidson [95] introduced a method for computing the smallest eigenvalues of the Schrödinger operator. Later, the method was generalized in two papers by Morgan and Scott [96] and Crouzeix et al. [97] in which the convergence of the method is proved. More recently, another version of the algorithm, under the name Jacobi–Davidson, was introduced in Ref. [98]. We present here a general framework for the class of Davidson methods, and point out how the various versions differ from one another. We should point out that for symmetric eigenvalue problems, the Davidson method version in Ref. [98] is strongly related to the trace minimization scheme discussed in Section 4.3.4, see Refs. [85, 99]. All versions of the Davidson method may be regarded as various forms of preconditioning the basic Lanczos method. To illustrate this point, let us consider a symmetric matrix $A \in \mathbb{R}^{n \times n}$. Both classes of algorithms generate, at some iteration k, an orthonormal basis $V_k = [v_1, ..., v_k]$ of a k-dimensional subspace $\mathcal{V}_k$ of $\mathbb{R}^{n \times n}$. In the Lanczos algorithm, $\mathcal{V}_k$ is a Krylov subspace, but for Davidson methods, this is not the case. In both classes, however, the interaction matrix is given by the symmetric matrix $H_k = V_k^{\mathrm{T}} A V_k \in \mathbb{R}^{k \times k}$. Likely with the Lanczos method, the goal is to obtain V_k such that some eigenvalues of H_k are good approximations of some eigenvalues of A: if (λ, y) is an eigenpair of H_k, then it is expected that the Ritz pair (λ, x), where $x = V_k y$, is a good approximation of an eigenpair of A. Note that this occurs only for some eigenpairs of H_k and at convergence.

Davidson methods differ from Lanczos in the definition of the new direction w which will be incorporated in the subspace $\mathcal{V}_k$ to obtain $\mathcal{V}_{k+1}$. For Lanczos schemes the vector w is given by $w = Av_k$, whereas for Davidson methods, a local improvement of the direction of the Ritz vector towards the sought after eigenvector is obtained by a quasi-Newton step (similar to the trace minimization scheme in Refs. [85, 99]). In Lanczos, the following vector v_{k+1} is computed by the three-term recursion, if reorthogonalization is not considered, whereas in Davidson methods, the next vector is obtained by reorthogonalizing w with respect to $\mathcal{V}_k$. Moreover, in this case, the matrix H_k is no longer tridiagonal. Therefore, one iteration of Davidson methods involves more arithmetic operations than the basic Lanczos scheme; it is at least as expensive as a Lanczos scheme with full reorthogonalization. Moreover, the basis V_k must be stored which implies the need for limiting the maximum value $k_{\max}$ to control storage requirements. Consequently an algorithm with periodic restarts must be implemented.

To compute the smallest eigenvalue of matrix A, the skeleton of the basic algorithm is:

ALGORITHM : Generic Davidson

Choose an initial normalized vector $V_1 = [v_1]$
Repeat
    for $k = 1 : k_{\max}$,
        compute $W_k = AV_k$;
        compute the interaction matrix $H_k = V_k^{\mathrm{T}} AV_k$;
        compute the smallest eigenpair (λ_k, y_k) of H_k ;
        compute the Ritz vector $x_k = V_k y_k$;
        compute the residual $r_k = W_k y_k - \lambda_k x_k$;
        if convergence then exit ;
        compute the new direction t_k **;** (Davidson correction)
        $V_{k+1} = \mathrm{MGS}(V_k, t_k)$;
    end ;
    until convergence ;
    $V_1 = \mathrm{MGS}(x_{k,1}, t_k)$;
end repeat ;

Algorithms of the Davidson family differ only in how the vector t_k is determined. Note that the above generic algorithm is expressed in a compact form, the steps which compute W_k or H_k, however, only determine updates to W_{k-1} or H_{k-1}. In this algorithm MGS denotes the modified Gram–Schmidt algorithm which is used to orthogonalize t_k with respect to V_k and then to normalize the resulting vector to obtain v_{k+1}.

Similar to the Lanczos algorithm, a block version of Davidson methods may be considered for approximating the s smallest eigenvalues of A. These are even more closely related to the trace minimization scheme [85] and [99].

ALGORITHM : Generic Block Davidson

Choose an initial orthonormal matrix $V_1 = [v_1, \cdots, v_s] \in \mathbb{R}^{n \times s}$
Repeat
    for $k = 1 : k_{\max}/s$,
        compute $W_k = AV_k$;
        compute the interaction matrix $H_k = V_k^{\mathrm{T}} AV_k$;
        compute the s smallest eigenpairs $(\lambda_{k,i}, y_{k,i})_{1 \leq i \leq s}$ of H_k ;
        compute the Ritz vectors $x_{k,i} = V_k y_{k,i}$ for $i = 1, \cdots, s$;
        compute the residuals $r_{k,i} = W_k y_{k,i} - \lambda_{k,i} x_{k,i}$ for $i = 1, \cdots, s$;
        if convergence then exit ;
        compute the new directions $(t_{k,i})_{1 \leq i \leq s}$ **;** (Davidson correction)
        $V_{k+1} = \mathrm{MGS}(V_k, t_{k,1}, \cdots, t_{k,s})$;
    end ;
    until convergence ;
    $V_1 = \mathrm{MGS}(x_{k,1}, \cdots, x_{k,s}, t_{k,1}, \cdots, t_{k,s})$;
end repeat ;

4.3.6.2 How Do the Davidson Methods Differ?

To introduce the correction vectors $(t_{k,i})_{1 \le i \le s}$, we assume that a known normalized vector x approximates an unknown eigenvector $x + y$ of A, where y is chosen orthogonal to x. The quantity $\lambda = \rho(x)$ (where $\rho(x) = x^T A x$ denotes the Rayleigh quotient of x) approximates the eigenvalue $\lambda + \delta$ which corresponds to the eigenvector $x + y$: $\lambda + \delta = \rho(x+y) = (x+y)^T A(x+y)/\|x+y\|^2$. The quality of the approximation is measured by the norm of the residual $r = Ax - \lambda x$. Since $r = (I - xx^T)Ax$, the residual is orthogonal to vector x. Let us denote θ the angle $\angle(x, x+y)$; let t be the orthogonal projection of x onto $x + y$ and $z = t - x$.

Lemma 4.1 *The norms of the involved vectors are:*

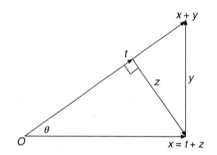

$$\|x + y\| = \frac{1}{\cos^2 \theta}, \qquad (4.65)$$
$$\|y\| = \tan \theta, \qquad (4.66)$$
$$\|z\| = \sin \theta, \qquad (4.67)$$
$$\|t\| = \cos \theta. \qquad (4.68)$$

Proof. Obvious. $\qquad\qquad\qquad\square$

Proposition 4.5 *With the previous notations, the correction δ to the approximation λ of an eigenvalue, and the orthogonal correction y of the corresponding approximate eigenvector x, satisfy:*

$$\begin{cases} (A - \lambda I)y = -r + \delta(x + y), \\ y \perp x, \end{cases} \qquad (4.69)$$

in which the following bounds hold:

$$|\delta| \le 2\|A\| \tan^2 \theta, \qquad (4.70)$$
$$\|r\| \le 2\|A\| \tan \theta \left(\frac{\sin \theta}{\cos^2 \theta} + 1 \right). \qquad (4.71)$$

Proof. Equation (4.69) is directly obtained from

$$A(x + y) = (\lambda + \delta)(x + y).$$

Moreover

$$\begin{aligned} \lambda + \delta &= \rho(t), \\ &= (1 + \tan^2 \theta)\, t^T A t, \\ &= (1 + \tan^2 \theta)(x - z)^T A(x - z), \\ &= (1 + \tan^2 \theta)(\lambda - 2x^T Az + z^T Az). \end{aligned}$$

Since z is orthogonal to the eigenvector t, we obtain $x^T Az = (t + z)^T Az = z^T Az$ and therefore

$$\begin{aligned} \delta &= -z^T Az + \tan^2 \theta(\lambda - z^T Az), \\ &= \tan^2 \theta(\lambda - \rho(z)), \text{ or} \\ |\delta| &\le 2\tan^2 \theta\, \|A\|, \end{aligned}$$

which proves (4.70). Bound (4.71) is a straight consequence of relations (4.69), (4.66), and (4.70). □

Thus the various version of the Davidson method are based on the following: at iteration k, the Ritz pair (λ, x) under consideration is computed and the determined correction y is added to the space $\mathcal{V}_k$ to obtain $\mathcal{V}_{k+1}$.

The system (4.69) is not easy to solve. Since the norm of the nonlinear term $\delta(x + y)$ is $O(\theta^2)$ whereas $\|r\| = O(\theta)$, one may instead consider the approximate problem

$$\begin{cases} (A - \lambda I)y = -r, \\ y \perp x, \end{cases} \tag{4.72}$$

which has no solution except the trivial solution when λ is an eigenvalue. Two alternatives may be considered to define another approximate problem.

The first one, which corresponds to the original Davidson's method, is to approximately solve the problem:

$$(A - \lambda I)y = -r, \tag{4.73}$$

and then orthogonalize y with respect to the subspace $\mathcal{V}_k$. The solution cannot be the exact one else it would provide $y = -x$ and no new direction would be incorporated to yield $\mathcal{V}_{k+1}$.

The second approach is that of trace minimization and the Jacobi–Davidson schemes in which one projects the linear system (4.72) onto the hyperplane $(x)^{\perp}$ orthogonal to x.

More precisely, the different approaches for the correction are:

Davidson with a preconditionner: solve

$$(M - \lambda I)y = -r, \tag{4.74}$$

where M is some preconditionner of A. In the original Davidson's method, M is taken as the diagonal of A.

Davidson and Inverse Iteration: when λ is close to an eigenvalue, solve approximately

$$(A - \lambda I)y = -r, \tag{4.75}$$

by an iterative method with a fixed number of iterations. In such situation, the method is similar to inverse iteration in which the ill-conditioning of the system provokes an error which is in the direction of the sought after eigenvector. The iterative solver, however, must be adapted to symmetric indefinite systems.

Trace Minimization and Jacobi–Davidson schemes: solve the problem:

$$\begin{cases} Q(x)(A - \lambda I)y = -r, \\ y \perp x, \end{cases} \tag{4.76}$$

where $Q(x) = I - xx^{\mathrm{T}}$ is the orthogonal projection with respect to x. The system is solved iteratively. As in the previous situation, the iterative solver must be adapted for symmetric indefinite systems (the system matrix can be expressed as $Q(x)(A - \lambda I)Q(x)$).

A recent study by Simoncini and Eldén [100] compares the Rayleigh quotient method, correction via "Davidson and Inverse Iteration," and the Newton–Grassmann method which corresponds to corrections via trace minimization or Jacobi–Davidson. The study concludes that the two correction schemes have comparable behavior. Simoncini and Eldén [100] also provides a stopping criterion for controlling the inner iterations of an iterative solver for the correction vectors.

4.3.6.3   Application to the Computation of the Smallest Singular Value

The smallest singular value of a matrix A may be obtained by applying one of the various versions of the Davidson methods to obtain the smallest eigenvalue of the matrix $C = A^T A$, or to obtain the innermost positive eigenvalue of the two-cyclic augmented matrix in (4.3). We assume that one has the basic kernels for matrix–vector multiplications, including the multiplication of the transpose of the original matrix A by a vector. Multiplying the transpose of a matrix by a vector is considered a drawback, and whenever possible, the so-called "transpose-free" methods should be used. Even though one can avoid such a drawback when dealing with the interaction matrix $H_k = V_k^T C V_k = A V_k^T A V_k$, we still have to compute the residuals corresponding to the Ritz pairs which do not involve multiplication of the transpose of a matrix by a vector.

For the regular single-vector Davidson method, the correction vector is obtained by approximately solving the system

$$A^T A t_k \;=\; r_k. \tag{4.77}$$

Obtaining an exact solution of (4.77) would yield the Lanczos algorithm applied to C^{-1}. Once the Ritz value approaches the square of the sought after smallest singular value, it is recommended that we solve (4.77) without any shifts; the benefit is that we deal with a fixed symmetric positive definite system matrix.

The approximate solution of (4.77) can be obtained by performing a fixed number of iterations of the conjugate gradient scheme, or by solving an approximate linear system $M t_k = r_k$ with a direct method. The latter has been theoretically studied in Ref. [101] in which M is obtained from approximate factorizations:

Incomplete LU-factorization of A: Here, $M = U^{-1} L^{-1} L^{-T} U^{-T}$, where L and U are the products of an incomplete LU-factorization of A obtained via the so-called ILUTH. In such a factorization, one drops entries of the reduced matrix A which are below a given threshold. This version of the Davidson method is called DAVIDLU.

Incomplete QR-factorization of A: Here, $M = R^{-1} R^{-T}$, where R is the upper-triangular factor of an incomplete QR-factorization of A. This version of the Davidson method is called DAVIDQR.

Incomplete Cholesky of $A^T A$: Here, $M = L^{-T} L^{-1}$, where L is the lower-triangular factor of an incomplete Cholesky factorization of the normal equations. This version of the Davidson method is called DAVIDIC.

Even though the construction of any of the above approximate factorizations may fail, experiments presented in Ref. [101] show the effectiveness of the above three preconditioners whenever they exist. It is also shown that DAVIDQR is slightly more effective than either DAVIDLU or DAVIDIC.

Similar to trace minimization, the Jacobi–Davidson method can be used directly on the matrix $A^T A$ to compute the smallest eigenvalue and the corresponding eigenvector. More recently, the Jacobi–Davidson method has been adapted in Ref. [102] for obtaining the singular values of A by considering the eigenvalue problem corresponding to the two-cyclic augmented matrix.

SUMMARY

Computing the smallest singular value often involves a shift and invert strategy. The advantage of trace minimization method and of Davidson methods over Lanczos or subspace iteration methods is to only use an approximation of the inverse through a preconditionner or through a partial resolution of an iterative procedure. Trace minimization and Davidson methods are all based on a Newton-like correction. With the former a fixed-sized basis is updated at each step whereas with the latter an extension of the subspace is invoked at each iteration except when restarting. One may assume that expanding the subspace usually speeds up the convergence. Often, however, using a subspace with fixed dimension results in a more robust scheme, especially if the size of the subspace is well adapted to the eigenvalue separation (i.e., singular value separation for our problem).

4.4 PARALLEL COMPUTATION FOR SPARSE MATRICES

In this section, we present approaches for the parallelization of the two most important kernels: matrix–vector products and basis orthogonalization. The experiments were done on a 56-processor distributed memory machine, an Intel Paragon XP/S i860, where communications between processors were handled by the MPI library. We conclude with the description of a parallel computation of the smallest singular value of a family of sparse matrices. In that case, the experiment was done on a cluster of workstation.

4.4.1 PARALLEL SPARSE MATRIX-VECTOR MULTIPLICATIONS

Matrix-vector multiplications are usually the most CPU-time-consuming operations in most iterative solvers. As mentioned earlier, the methods devoted to SVD computations involve multiplication by the original matrix and by its transpose as well. Obviously, only one copy of the matrix is stored. We present here the approach described in Ref. [103] which considers a matrix stored using the CSR format outlined in Section 4.3.1.

The main goal of data partitioning is to define large-grain local tasks and to overlap communications with computations. This is not always possible as efficiency depends on the matrix structure. Because of the row-oriented compact storage of the sparse matrix, data allocation is performed by rows:

Data allocation:

$$
A = \begin{bmatrix} A^0 \\ \hline A^1 \\ \hline \vdots \ \vdots \quad \vdots \quad \vdots \\ \hline A^{p-1} \end{bmatrix} \qquad v = \begin{bmatrix} v_0 \\ \hline v_1 \\ \hline \vdots \\ \hline v_{p-1} \end{bmatrix} \begin{array}{l} \longrightarrow \text{Node } 0 \\ \longrightarrow \text{Node } 1 \\ \longrightarrow \vdots \\ \longrightarrow \text{Node } (p-1) \end{array}
$$

On each processor, the matrix components are stored row-wise as well. These, in turn, are organized into two parts: *local* and *exterior*. On processor k, the local part, denoted $A_{\mathrm{loc},k}$ defines the component which do not require communication during the multiplication. On the same processor, the exterior part is split into the blocks $A_{\mathrm{ext},k}^j$ $(j \neq k)$ as shown below.

Data partition for the multiplications by A_z and A_z^H :

$$
A = \begin{bmatrix} A_{\mathrm{loc},0} & A_{\mathrm{ext},0}^1 & \cdots & \cdots & A_{\mathrm{ext},0}^{p-1} \\ \hline A_{\mathrm{ext},1}^0 & A_{\mathrm{loc},1} & \cdots & \cdots & A_{\mathrm{ext},1}^{p-1} \\ \hline \vdots & & & \vdots & \vdots \\ \vdots & & & \vdots & \vdots \\ \hline A_{\mathrm{ext},p-1}^0 & A_{\mathrm{ext},p-1}^1 & \cdots & \cdots & A_{\mathrm{loc},p-1} \end{bmatrix} \qquad v = \begin{bmatrix} v_{\mathrm{loc}}^0 \\ \hline v_{\mathrm{loc}}^1 \\ \hline \vdots \\ \vdots \\ \hline v_{\mathrm{loc}}^{p-1} \end{bmatrix} \begin{array}{l} \longrightarrow \text{Node } 0 \\ \longrightarrow \text{Node } 1 \\ \longrightarrow \vdots \\ \longrightarrow \vdots \\ \longrightarrow \text{Node } (p-1) \end{array}
$$

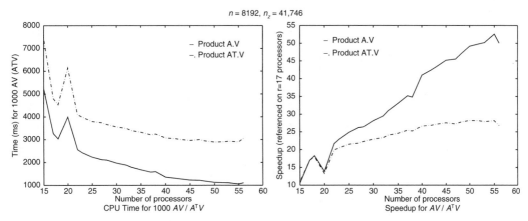

FIGURE 4.2 Timings for matrix–vector multiplications.

Algorithms for the matrix–vector multiplication

The algorithms for the two matrix–vector multiplications are expressed in the two follow-ing tables in which the communicating routine <u>Send</u> is preferably a nonblocking one.

k = processor number; p = number of processors.

Matrix–vector multiplication $v \rightarrow Av$:

Step 1: <u>Send</u> the needed components of v_{loc}^k to other processors.
Step 2: Compute $w^k = A_{\text{loc},k} v_{\text{loc}}^k$.
Step 3: <u>Receive</u> the needed components of v_{loc}^j for $j \neq k$ from other processors.
Step 4: Compute $w^k = w^k + \sum_{j \neq k} A_{\text{ext},k}^j v_{\text{loc}}^j$.

Matrix transpose–vector multiplication $v \rightarrow A^{\text{T}} v$:

Step 1: Compute $y^j = A_{\text{ext}}^{j\text{T}} v_{\text{loc},j}^k$ for $j \neq k$.
Step 2: <u>Send</u> y^j for $j \neq k$ to node number j.
Step 3: Compute $w^k = A_{loc,k}^{\text{T}} v_{\text{loc}}^k$.
Step 4: <u>Receive</u> $y^{j'}$ with $j' \neq k$ from other processors.
Step 5: Compute $w^k = w^k + \sum_{j' \neq k} y^{j'}$.

Matrix DW8192 of the Matrix Market [14] test suite was used to measure the performance of the above two matrix–vector multiplication schemes. Run times and speedups are displayed in Table 4.8, showing that one should avoid matrix-vector multiplications involving the transpose of the stored matrix, at least on architectures similar to the Intel Paragon.

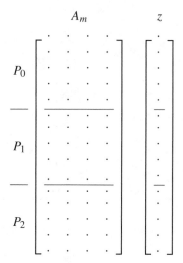

TABLE 4.8 Distribution by contiguous rows

4.4.2 A PARALLEL SCHEME FOR BASIS ORTHOGONALIZATION

One of the crucial procedures, which must be optimized for maximum performance is the construction of an orthogonal basis. Let us assume that at some stage, the factor Q of the QR-factorization of a matrix $A \in \mathbb{R}^{m \times n}$ ($m \gg n$) must be computed on a parallel computer with distributed memory.

The most common procedure, that of the MGS-algorithm, does not offer effective parallelism for the case at hand ($m \gg n$). The classical version is often considered due to its suitability on parallel architectures. In this case, however, to maintain numerical reliability, it is necessary to reapply the procedure a second time. We present here a better procedure that we outlined in a previous study, e.g., see Ref. [104], in which the QR-factorization of A is obtained into two steps: ROCDEC which computes the triangular factor R by combining Householder and Givens transformations, and ROCVEC which accumulates the orthogonal transformations to compute the vector Qy for any vector y. Assuming that the columns of A are distributed on a ring of r processors, see Table 4.8, the two procedures are given below.

ALGORITHM : ROCDEC

initialize $\left(N, m, N_{\text{loc}}, A_{\text{loc}}(1 : N_{\text{loc}}, :) \right)$;

for $j := 1 : m$ do

   create <u>my</u> reflector[j] and update my local columns

   if myid() = 0 then

      send row $A_{\text{loc}}(j, j : m)$ to myright()

   else

      receive row(j:m) from myleft()

      create my rotation[1,j] to annihilate $A_{\text{loc}}(1, j)$

      if myid() $\neq r - 1$ update and send row(j:m) to myright()

   endif

endfor

At completion the factor Q is explicitly available but is shared in the same fashion that the original matrix A_m at the beginning. Hence, the product $z = Qy$ is trivial.

```
ALGORITHM : ROCVEC
k := myid();
z_loc := 0;
{ apply orthogonal transformations in reverse order }
for j := m : 1 step −1 do
   if (k = 0) then
      receive z_loc(j) from myright()
      i := j;
   else
      if (k = r − 1) then
         apply my rotation[1, j] on [y(j), z_loc(1)]
         send the updated y(j) to myleft()
      else
         receive yj from myright()
         apply my rotation[1, j] on [yj, z_loc(1)]
         send the updated yj to myleft()
      endif
      i := 1;
   endif
   apply my reflector[j] on z_loc(i : N_loc)
endfor
```

Data distribution. The factor R is sent to processor P_{r-1}. To compute the product $z = Qy$ the vector $y \in \mathbb{R}^m$ must be at the beginning in processor P_{r-1}. The resulting vector $z \in \mathbb{R}^N$ is split at completion.

Parallelism. The application of Householder reflectors is a completely independent stage. However, to annihilate the remaining nonzero elements through Givens rotations, it is necessary to transport a row portion along r processors. But as soon as that portion is passed through a processor, the latter applies its next reflector so that there is an overlapping of its computations and the transfer of that row portion to the other processors. Since there will be a total of m transfers of this sort during the whole factorization, communication overheads will be masked if $m \gg r$. In summary, we expect a reasonable speedup when $(n/r) \gg m$ and $m \gg r$.

Experiment. The following experiment illustrates the superior scalability of the scheme ROC (ROC=ROCDEC+ROCVEC) with respect to the MGS procedure. The goal is to compute Qy where Q is the Q-factor in the QR-factorization of A. In Table 4.9, we give the running times (in seconds) of the two algorithms with a constant computing load per processor. The column length is $n = 10^4 r$ and the number of columns is $m = 40$. The algorithm ROC appears to be much more scalable than MGS.

TABLE 4.9
Scalability of the Orthogonalization Process
(As Illustrated by the Running Time in Seconds)

r	MGS	ROC
1	0.16E + 01	0.18E + 01
2	0.21E + 01	0.19E + 01
4	0.25E + 01	0.20E + 01
8	0.28E + 01	0.20E + 01
16	0.32E + 01	0.21E + 01
24	0.34E + 01	0.21E + 01
32	0.36E + 01	0.22E + 01
56	0.40E + 01	0.24E + 01

4.4.3 Computing the Smallest Singular Value on Several Processors

As mentioned earlier (see Proposition 4.1) the smallest singular value of A can be computed from the largest eigenvalue of the matrix

$$B = \begin{pmatrix} 0 & R^{-1} \\ R^{-T} & 0 \end{pmatrix}, \tag{4.78}$$

where $A = QR$ is a QR-decomposition of A. Typically, one would couple a Lanczos algorithm to a QR-decomposition to compute the largest eigenvalue of B and hence compute the smallest singular value of A.

For large matrix dimensions, the matrix R can be too large to fit in the fast memory. Furthermore, each iteration of the Lanczos algorithm requires the solutions of two linear systems based on R and R^T. Therefore, parallel QR-decomposition algorithms and parallel system solvers are required to efficiently compute the smallest singular value.

The multifrontal QR-decomposition (MFQRD), presented in Ref. [105], allows a large granularity for parallelism. The MFQRD starts by building a dependency graph that expresses the connections between successive steps involved in the Householder reflections. The tree nodes in the elimination tree correspond to matrix columns where column a is a parent of column b if the application of the Householder reflection corresponding to column b alters column a. The computation of the Householder reflections are totally independent for separated tree leafs. Therefore, the reduction of independent subtrees can be done in parallel. On a cluster of workstations, one processor computes the elimination tree, isolates a set of independent subtrees and scatters the subtrees among the different processors of the cluster. Each processor retrieves a subtree and operates the Householder reflections. This process leads to a scattered R matrix (rowwise) adequate for parallel solvers. For a detailed description of this procedure, the reader is refered to Ref. [13].

Table 4.10 shows the wall-clock time needed to compute σ_{min} for three test matrices $S_1 \in \mathbb{R}^{1890 \times 1890}$, $S_2 \in \mathbb{R}^{3906 \times 3906}$, and $S_3 \in \mathbb{R}^{32130 \times 32130}$. Although the test matrices are obtained from actual applications, we do not discuss here the physical interpretation of the results. It can be observed that parallelizing the computation of σ_{min} is not beneficial for S_1 and S_2, but for S_3

TABLE 4.10
Parallel Computation of σ_{min} on a Small Cluster of Machines

Matrix	1 Proc.	3 Procs.	Speedup	Efficiency
Multi-frontal QR-decomposition				
S_1	3.7	3.6	1.02	0.34
S_2	41.8	30.8	1.35	0.45
S_3	2353.0	1410.0	1.66	0.55
Lanczos algorithm				
S_1	0.7	5.2	0.13	0.04
S_2	2.1	15.6	0.14	0.05
S_3	1479.0	647.0	2.29	0.76
σ_{min}				
S_1	4.4	8.8	0.49	0.16
S_2	43.9	46.5	0.94	0.32
S_3	3822.0	2057.0	1.85	0.61

speedups of 1.66 and 2.29 are achieved for the QR-decomposition and the Lanczos algorithm, respectively. The total speedup for obtaining σ_{min} is 1.85 with a corresponding efficiency of 61%.

4.5   APPLICATION: PARALLEL COMPUTATION OF A PSEUDOSPECTRUM

As discussed in Section 4.1.3, the computation of the pseudospectrum of a matrix A involves a large volume of arithmetic operations. It is now commonly accepted that path-following algorithms which compute the level curve

$$\Gamma_\epsilon = \{z \in \mathbb{C} \mid \sigma_{min}(A - zI) = \epsilon\}, \tag{4.79}$$

are of much less complexity than methods based on grid discretization [106]. The first attempt in this direction was published by Brühl [11]. Based on a continuation with a predictor–corrector scheme, the process may fail for angular discontinuities along the level curve [12, 107]. Trefethen and Wright [108] use the upper Hessenberg matrix constructed after successive iterations of the implicitly restarted Arnoldi algorithm to cheaply compute an approximation of the pseudospectrum. However, they show that for highly nonnormal matrices the computed pseudospectrum is a low approximation of the exact one.

4.5.1   PARALLEL PATH-FOLLOWING ALGORITHM USING TRIANGLES

In this section, we present the parallel path-following algorithm using triangles (PPAT) for computing the level curve Γ_ϵ. For a detailed description of this algorithm, the reader is referred to Refs. [13, 109].

PPAT uses a numerically stable algorithm that offers a guarantee of termination even in the presence of round-off errors. Furthermore, the underlying algorithm can handle singular points along the level curve of interest without difficulty.

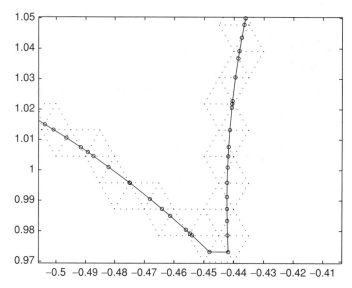

FIGURE 4.3 Computing a level curve using PPAT.

The main idea is to line up a set of equilateral triangles along the level curve as presented in Figure 4.3 and use a bisection algorithm to compute a numerical approximation of the pseudo-spectrum. More specifically, given a mesh of the complex plane with equilateral triangles, for any triangle T_i of the mesh which intersects the sought-level curve, an adjacent triangle T_{i+1} of the same type is defined as successor of T_i. Therefore, from a starting triangle T_0, a chain of triangles $T_1, \ldots, T_n$ such that $T_n = T_0$ is defined. In Ref. [109], it is proven that to compute a level curve of length l, the number of equilateral triangles of size τ satisfies

$$\frac{l}{\tau} \leq n \leq \frac{10l}{\sqrt{3}\tau}. \tag{4.80}$$

PPAT is based on master–slave model where a master node controls a set of slave nodes capable of computing $\sigma_{\min}(A - zI)$ for a given complex value z and of extracting $z \in \Gamma_\epsilon$ from a segment $[z_1, z_g]$, assuming $\sigma_{\min}(A - z_1I) \leq \epsilon < \sigma_{\min}(A - z_gI)$. For that purpose, A is broadcast to all workers. Tasks are queued in a task list managed by the master. Given a triangle T_i along the level curve, the master spawns two tasks; the first computes T_{i+1} whereas the second extracts a new point of the level curve. The dynamic task scheduling allows better load balancing among the different processors of a heterogeneous network of workstations.

PPAT exploits the fact that multiple level curve slices can be computed simultaneously. The main idea is to locate different starting triangles along the level curve and use each triangle to compute a level curve slice. To get through successfully, the triangles of the different computed slices should align perfectly. Therefore, a prefixed lattice is used for all level curve slices. Furthermore, PPAT proceeds in both directions on a single level curve slice to achieve higher speedups.

4.5.2 SPEEDUP AND EFFICIENCY

It is shown in Ref. [13] that if n is the number of equilateral triangles built to compute an approximation of a given level curve using a single slice, p the number of processors, and q the number of evaluations of $\sigma_{\min}$ for each bisection process, then the speedup is given by

$$S_p = \begin{cases} \dfrac{2n(q+1)}{n+2q} & \text{if } p \geq 2q+2, \\[3mm] \dfrac{2np(q+1)}{p^2 - 2p + 2n(q+1)} & \text{if } p < 2q+2. \end{cases} \tag{4.81}$$

For large values of n, the speedup and efficiency can be expressed as

$$S_p = \min(p, 2q+2) + O(1/n),$$

and

$$E_p = \min\left(1, \frac{2q+2}{p}\right) + O(1/n).$$

The upper bound of the speedup is given by $S_{\max} = 2q + 2$. This limit is theoretically achieved whenever $p \geq 2q + 2$.

4.5.3 TEST PROBLEMS

Three test matrices were selected from the Matrix Market suite [14]. Table 4.11 summarizes their characteristics (Nz is the number of nonzero entries and $t_{\sigma_{\min}}$ is the average computation time for $\sigma_{\min}(A - zI)$). The application uses up to 20 workers, where the master process shares the same physical processor as the first worker. The underlying hardware is a low-cost general purpose network of personal computers (Pentium III, 600 MHz, 128 MB RAM).

Figure 4.4 displays speedups and efficiencies for computing 100 points on a given level curve with PPAT. The best performance was observed for matrix Dw8192 which is the largest of the set; a speedup of 10.85 using 12 processors which corresponds to a 90% efficiency. The lowest efficiency obtained was 63% for Olm1000 on 13 processors. In a more realistic approach, we have used the 70 processors of the PARASKI cluster at IRISA to compute, in 41 s, the level curve $\epsilon = 0.05$ of Olm1000 split into 15 slices. A single processor required 4020 s to compute the same level curve. The corresponding speedup is 98 with an efficiency greater than 1, indicating higher data locality realized for each processor. This result indicates the favorable scalability of PPAT because PARASKI is a heterogeneous cluster of processors ranging from PII to PIV; the sequential time is measured on a randomly selected PIII of the cluster.

TABLE 4.11
Test Matrices from the Non-Hermitian Eigenvalue Problem (NEP) Collection

Matrix	Order	Nz	$\|A\|_F$	$t_{\sigma_{\min}}$ (s)
Olm1000	1,000	3,996	1.3×10^6	0.27
Rdb3200L	3,200	18,880	2.8×10^3	15.14
Dw8192	8,192	41,746	1.6×10^3	114.36

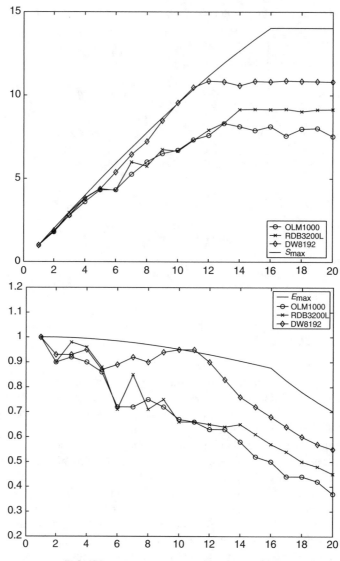

FIGURE 4.4 Speedup and efficiency of PPAT.

REFERENCES

[1] M.W. Berry. Large scale singular value computations. *Int. J. Supercomputer Appl.*, 6:13–49, 1992.

[2] G.H. Golub and C.F.Van Loan. *Matrix Computations*, 3rd ed. Johns Hopkins University Press, Baltimore, MD, 1996.

[3] G.W. Stewart and J.-G. Sun. *Matrix Perturbation Theory*. Academic Press, New York, 1990.

[4] J.-G. Sun. Condition number and backward error for the generalized singular value decomposition. *SIAM J. Matrix Anal. Appl.*, 22(2):323–341, 2000.

[5] J.K. Cullum and R.A. Willoughby. *Lanczos Algorithm for Large Symmetric Eigenvalue Computations*, Vol. 1. Birkhauser, Basel, 1985.

[6] J. Demmel, M. Gu, S. Eisenstat, Z. Drmač, I. Slapničar, and K. Veselić. Computing the singular value decomposition with high accuracy. *Linear Algebra Appl.*, 299(1–3):21–80, 1999.

[7] J.-G. Sun. A note on simple non-zero singular values. *J. Comput. Math.*, (6):258–266, 1988.

[8] H. Hoteit, J. Erhel, R. Mosé, B. Philippe, and P. Ackerer. Numerical reliability for mixed methods applied to flow problems in porous media. *J. Comput. Geosci.*, 6:161–194, 2002.

[9] L.N. Trefethen. Pseudospectra of matrices. In D.F. Griffiths and G.A. Watson, Eds., *Numerical Analysis*, Dundee 1991. Longman, Chicago, 1992, pp. 234–266.

[10] S.K. Godunov. Spectral portraits of matrices and criteria of spectrum dichotomy. In L. Atanassova and J. Hezberger, Eds., *3rd Int. IMACS-CAMM Symp. Comp. Arithmetic Enclosure Meth.*, Amsterdam, 1992.

[11] M. Brühl. A curve tracing algorithm for computing the pseudospectrum. *BIT*, 36(3): 441–454, 1996.

[12] C. Bekas and E. Gallopoulos. Cobra: parallel path following for computing the matrix pseudospectrum. *Parallel Comput.*, 27(14):1879–1896, 2001.

[13] D. Mezher and B. Philippe. Parallel computation of pseudospectra of large sparse matrices. *Parallel Comput.*, 28(2):199–221, 2002.

[14] Matrix Market. http://math.nist.gov/MatrixMarket/.

[15] L.S. Blackford, J. Choi, A. Cleary, E.D'Azevedo, J. Demmel, I. Dhillon, J. Dongarra, S. Hammarling, G. Henry, A. Petitet, K. Stanley, D. Walker, and R.C. Whaley. *ScaLAPACK Users' Guide*. SIAM, Philadelphia, 1997.

[16] ScaLAPACK. The scalapack project. http://www.netlib.org/scalapack/.

[17] P. Alpatov, G. Baker, C. Edwards, J. Gunnels, G. Morrow, J. Overfelt, Y.-J.J. Wu, and R. van de Geijn. PLAPACK: parallel linear algebra package. In *P. SIAM Parallel Process. Conf.*, 1997.

[18] PLAPACK. PLAPACK: parallel linear algebra package.

[19] E. Caron, S. Chaumette, S. Contassot-Vivier, F. Desprez, E. Fleury, C. Gomez, M. Goursat, E. Jeannot, D. Lazure, F. Lombard, J.-M. Nicod, L. Philippe, M. Quinson, P. Ramet, J. Roman, F. Rubi, S. Steer, F. Suter, and G. Utard. Scilab to scilab//, the ouragan project. *Parallel Comput.*, 11(27):1497–1519, 2001.

[20] Mathtools.net. Mathtools.net > matlab > parallel. http://www.mathtools.net/MATLAB/Parallel/.

[21] G. Forsythe and P. Henrici. The cyclic Jacobi method for computing the principal values of a complex matrix. *Trans. Am. Math. Soc.*, 94:1–23, 1960.

[22] A. Schonhage. Zur konvergenz des Jacobi-Verfahrens. *Numer. Math.*, 3:374–380, 1961.

[23] J.H. Wilkinson. Note on the quadratic convergence of the cyclic jacobi process. *Numer. Math.*, 5: 296–300, 1962.

[24] A. Sameh. On Jacobi and Jacobi-like algorithms for a parallel computer. *Math. Comput.*, 25:579–590, 1971.

[25] F. Luk and H. Park. A proof of convergence for two parallel Jacobi SVD algorithms. *IEEE Trans. Comput.*, 38(6):806–811, 1989.

[26] F. Luk and H. Park. On parallel Jacobi orderings. *SIAM J. Sci. Stat. Comput.*, 10(1):18–26, 1989.

[27] P. Henrici. On the speed of convergence of cyclic and quasicyclic Jacobi methods for computing eigenvalues of Hermitian matrices. *Soc. Ind. Appl. Math.*, 6:144–162, 1958.

[28] A. Sameh. *Solving the Linear Least Squares Problem on a Linear Array of Processors*. Academic Press, New York, 1985, pp. 191–200.

[29] H.F. Kaiser. The JK method: a procedure for finding the eigenvectors and eigenvalues of a real symmetric matrix. *Comput. J.*, 15(33):271–273, 1972.

[30] J.C. Nash. A one-sided transformation method for the singular value decomposition and algebraic eigenproblems. *Comput. J.*, 18(1):74–76, 1975.

[31] F.T. Luk. Computing the singular value decomposition on the Illiac IV. *ACM Trans. Math. Soft.*, 6(4):524–539, 1980.

[32] R.P. Brent and F.T. Luk. The solution of singular value and symmetric eigenproblems on multiprocessor arrays. *SIAM J. Sci. Stat.*, 6:69–84, 1985.

[33] R.P. Brent, F.T. Luk, and C. Van Loan. Computation of the singular value decomposition using mesh connected processors. *VLSI Comput. Syst.*, 1(3):242–270, 1985.

[34] M.W. Berry and A.H. Sameh. An overview of parallel algorithms for the singular value and symmetric eigenvalue problems. *Comput. Appl. Math.*, 27:191–213, 1989.

[35] J.-P. Charlier, M. Vanbegin, and P. Van Dooren. On efficient implementations of Kogbetliantz's algorithm for computing the singular value decomposition. *Numer. Math.*, 52:279–300, 1988.

[36] C. Paige and P. Van Dooren. On the quadratic convergence of Kogbetliantz's algorithm for computing the singular value decomposition. *Linear Algebra Appl.*, 77:301–313, 1986.

[37] J.-P. Charlier and P. Van Dooren. On Kogbetliantz's SVD algorithm in the presence of clusters. *Linear Algebra Appl.*, 95:135–160, 1987.

[38] J.H. Wilkinson. Almost diagonal matrices with multiple or close eigenvalues. *Linear Algebra Appl.*, 1:1–12, 1968.

[39] M. Bečka and M. Vajtešic. Block–Jacobi SVD algorithms for distributed memory systems I. *Parallel Algorithms Appl.*, 13:265–267, 1999.

[40] M. Bečka and M. Vajtešic. Block–Jacobi SVD algorithms for distributed memory systems II. *Parallel Algorithms Appl.*, 14:37–56, 1999.

[41] M. Bečka, G. Okša, and M. Vajtešic. Dynamic ordering for a Block–Jacobi SVD algorithm. *Parallel Comput.*, 28:243–262, 2002.

[42] I. Duff, A. Erisman, and J Reid. *Direct Methods for Sparse Matrices.* Oxford Science Publications, Oxford, 1992.

[43] Y. Saad. *Iterative Methods for Sparse Linear Systems.* PWS Publishing Company, Boston, MA 1996.

[44] P.R. Amestoy, I.S. Duff, and C. Puglisi. Multifrontal QR factorization in a multiprocessor environment. *Numer. Linear Algebra Appl.*, 3:275–300, 1996.

[45] P.R. Amestoy, I.S. Duff, and J.Y. L'Excellent. Multifrontal parallel distributed symmetric and unsymmetric solvers. *Comput. Meth. Appl. Mech. Eng.*, 184(2–4):501–520, 2000.

[46] I. Brainman and S. Toledo. Nested-dissection orderings for sparse LU with partial pivoting. *SIAM J. Matrix Anal. Appl.*, 23(4):998–1012, 2002.

[47] T.A. Davis and I.S. Duff. A combined unifrontal/multifrontal method for unsymmetric sparse matrices. *ACM Trans. Math. Software*, 25:1–19, 1999.

[48] J.J. Dongarra, I.S. Duff, D.S. Sorensen, and H.A. van der Vorst. *Solving Linear Systems on Vector and Shared Memory Computers.* SIAM, Philadelphia, 1991.

[49] J.W. Demmel, S.C. Eisenstat, J.R. Gilbert, X.S. Li, and J.W.H. Joseph. A supernodal approach to sparse partial pivoting. *SIAM J. Matrix Anal. Appl.*, 20(3):720–755, 1999.

[50] I.S. Duff. A review of frontal methods for solving linear systems. *Comput. Phys. Commun.*, 97(1–2):45–52, 1996.

[51] J.R. Gilbert, X.S. Li, E.G. Ng, and B.W. Peyton. Computing row and column counts for sparse QR and LU factorization. *BIT*, 41(4):693–710, 2001.

[52] T.A. Davis. Umfpack version 4.0. http://www.cise.ufl.edu/research/sparse/umfpack, April 2002.

[53] X.S. Li. SuperLU version 2. http://www.nersc.gov/ xiaoye/SuperLU/, September 1999.

[54] P.R. Amestoy, I.S. Duff, J.Y. L'Excellent, J. Koster, and M. Tuma. Mumps 4.1.6. http://www.enseeiht.fr/lima/apo/MUMPS/, March 2000.

[55] S.F. Ashby, T.A. Manteuffel, and P.E. Saylor. A taxonomy for conjugate gradient methods. *SIAM J. Numer. Anal.*, 26:1542–1568, 1990.

[56] R. Barret, M. Berry, T. Chan, J. Demmel, J. Donato, J. Dongarra, V. Eijkhout, R. Pozo, C. Romine, and H. van der Vorst. *Templates for the Solution of Linear Systems: Building Blocks for Iterative Methods*, 2nd ed. SIAM / netlib, Philadelphia, PA, 1994.

[57] W. Hackbusch. *Iterative Solution of Large Sparse Systems of Equations*, Vol. 95 of *Applied Mathematical Sciences.* Springer-Verlag, Heidelberg 1994.

[58] G. Meurant. *Computer Solution of Large Linear Systems.* North-Holland, Amsterdam, 1999.

[59] C. Paige and M. Saunders. Solution of sparse indefinite systems of linear equations. *SIAM J Numeri. Anal.*, 12:617–629, 1975.

[60] Y. Saad and H. Schultz. GMRES: a generalized minimal residual algorithm for solving nonsymmetric linear systems. *SIAM J. Sci. Stat. Comput.*, 7:856–869, 1986.

[61] H.A. van der Vorst. Bi-CGSTAB: a fast and smoothly converging variant of Bi-CG for the solution of nonsymmetric linear systems. *SIAM J. Sci. Statist. Comput.*, 13:631–644, 1992.

[62] R. Freund and N. Nachtigal. QMR: a quasi-minimal residual method for non-Hermitian linear systems. *Numer. Math.*, 60:315–339, 1991.

[63] R. Freund. A transpose-free quasi-minimal residual algorithm for non-Hermitian linear systems. *SIAM J. Sci. comput.*, 14:470–482, 1993.

[64] G.H. Golub and W. Kahan. Calculating the singular values and pseudo-inverse of a matrix. *SIAM J. Numer. Anal.*, 2(3):205–224, 1965.

[65] G.H. Golub and C. Reinsch. *Singular Value Decomposition and Least Squares Solutions*. Springer-Verlag, Heidelberg, 1971.

[66] M.R. Hestens. Inversion of matrices by biorthogonalization and related results. *Soc. Ind. Appl. Math.*, 6:51–90, 1958.

[67] B.N. Parlett. *The Symmetric Eigenvalue Problem*. Prentice Hall, New York, 1980.

[68] F.L. Bauer. Das verfahred der treppeniteration und verwandte verfahren zur losung algebraischer eigenwertprobleme. *ZAMP*, 8:214–235, 1957.

[69] H. Rutishauser. Simultaneous iteration method for symmetric matrices. *Numer. Math.*, 16:205–223, 1970.

[70] K. Gallivan, W. Jalby, and U. Meier. The use of BLAS3 in linear algebra on a parallel processor with a hierarchical memory. *SIAM J. Sci. Stat. Comput.*, 18(6):1079–1084, 1987.

[71] H. Rutishauser. Computational aspects of F.L. Bauer's simultaneous iteration method. *Numer. Math.*, 13:4–13, 1969.

[72] B. Smith, J. Boyce, J. Dongarra, B. Garbow, Y. Ikebe, V. Klema, and C. Moler. *Matrix Eigensystem Routines — EISPACK Guide*, 2nd ed. Springer-Verlag, Heidelberg, 1976.

[73] J. Dongarra and D. Sorensen. A fast algorithm for the symmetric eigenvalue problem. *SIAM J. Sci. Stat. Comput.*, 8(2):s139–s154, 1987.

[74] C.C. Paige. Error analysis of the Lanczos algorithms for tridiagonalizing a symmetric matrix. *Inst. Math. Appl.*, 18:341–349, 1976.

[75] J. Grcar. Analysis of the Lanczos Algorithm and of the Approximation Problem in Richardson's Method, Ph.D. Thesis. Technical report, The University of Illinois, Chicago, 1981.

[76] B.N. Parlett and D.S. Scott. The Lanczos algorithm with selective reorthogonalization. *Math. Comput.*, 33:217–238, 1979.

[77] H. Simon. Analysis of the symmetric Lanczos algorithm with reorthogonalization methods. *Linear Algebra Appl.*, 61:101–131, 1984.

[78] S. Lo, B. Philippe, and A.H. Sameh. A multiprocessor algorithm for the symmetric tridiagonal eigenvalue problem. *SIAM J. Sci. Stat. Comput.*, 8(2):s155–s165, 1987.

[79] G.H. Golub and C. Van Loan. *Matrix Computations*, 2nd ed. Johns Hopkins University Press, Battimore, MD, 1989.

[80] R.R. Underwood. An Iterative Block Lanczos Method for the Solution of Large Sparse Symmetric Eigenproblems, Ph.D. thesis. Technical report, Stanford University, Stanford, 1975.

[81] G.H. Golub and R.R. Underwood. *The Block Lanczos Method for Computing Eigenvalues*, Academic Press, New York, 1977, pp. 361–377.

[82] G.H. Golub, F.T. Luk, and M.L. Overton. A block Lanczos method for computing the singular values and corresponding singular vectors of a matrix. *ACM Trans. Math. Software*, 7(2):149–169, 1981.

[83] J. Dongarra, J. Du Croz, S. Hammarling, and R. Hanson. An extended set of FORTRAN basic linear algebra subprograms. *ACM Trans. Math. Software*, 14(1):1–17, 1988.

[84] W. Jalby and B. Philippe. Stability analysis and improvement of the block Gram–Schmidt algorithm. *SIAM J. Sci. Stat. Comput.*, 12(5):1058–1073, 1991.

[85] A.H. Sameh and J.A. Wisniewski. A trace minimization algorithm for the generalized eigenvalue problem. *SIAM J. Numer. Anal.*, 19(6):1243–1259, 1982.

[86] J.H. Wilkinson. *The Algebraic Eigenvalue Problem*. Clarendon Press, Oxford, 1965.

[87] M.W. Berry, B. Parlett, and A.H. Sameh. Computing extremal singular triplets of sparse matrices on a shared-memory multiprocessor. *Int. J. High Speed Comput.*, 2:239–275, 1994.

[88] D.G. Luenberger. *Introduction to Linear and Nonlinear Programming*. Addison-Wesley, Reading, Massachusetts, 1973.

[89] J.H. Wilkinson. *Inverse Iteration in Theory and in Practice*. Academic Press, New York, 1972.

[90] E. Kokiopoulou, C. Bekas, and E. Gallopoulos. Computing Smallest Singular Triplets with Implicitly Restarted Lanczos Bidiagonalization. *J. Appl. Numer.*, 49(1):39–61, 2004.

[91] J.O. Aasen. On the reduction of a symmetric matrix to tridiagonal form. *BIT*, 11:233–242, 1971.

[92] J.R. Bunch and L. Kaufman. Some stable methods for calculating inertia and solving symmetric linear systems. *Math. Comput.*, 31:162–179, 1977.

[93] B.N. Parlett and J.K. Reid. On the solution of a system of linear equations whose system is symmetric but not definite. *BIT*, 10:386–397, 1970.

[94] C.C. Paige and M.A. Saunders. Solution of sparse indefinite systems of linear equations. *SIAM J. Numer. Anal.*, 12(4):617–629, 1975.

[95] E.R. Davidson. The iterative calculation of a few of the lowest eigenvalues and corresponding eigenvectors of large real-symmetric matrices. *Comput. Phys.*, 17:87–94, 1975.

[96] R.B. Morgan and D.S. Scott. Generalizations of Davidson's method for computing eigenvalues of sparse symmetric matrices. *SIAM J. Sci. Stat. Comput.*, 7:817–825, 1986.

[97] M. Crouzeix, B. Philippe, and M. Sadkane. The Davidson method. *SIAM J. Sci. Stat. Comput.*, 15:1: 62–76, 1994.

[98] G.L.G. Sleipen and H.A. van der Vorst. A Jacobi–Davidson iteration method for linear eigenvalue problems. *SIAM J. Matrix Anal. Appl.*, 17(2):401–425, 1996. Also SIGEST in *SIAM Rev.* 42(2): 267–293.

[99] A. Sameh and Z. Tong. The trace minimization method for the symmetric generalized eigenvalue problem. *J. Comput. Appl. Math.*, 123:155–175, 2000.

[100] V. Simoncini and L. Eldén. Inexact Rayleigh quotient-type methods for eigenvalue computations. *BIT*, 42(1):159–182, 2002.

[101] B. Philippe and M. Sadkane. Computation of the fundamental singular subspace of a large matrix. *Linear Algebra Appl.*, 257:77–104, 1997.

[102] M.E. Hochstenbach. A Jacobi–Davidson type SVD method. *SIAM J. Sci. Comput.*, 23(2):606–628, 2001.

[103] V. Heuveline, B. Philippe, and M. Sadkane. Parallel computation of spectral portrait of large matrices by Davidson type methods. *Numer. Algorithms*, 16(1):55–75, 1997.

[104] R. Sidje and B. Philippe. Parallel Krylov subspace basis computation. In J. Tankoano, Ed., *CARI'94 Proc., Ouagadougou*. INRIA — ORSTOM, ORSTOM Editions, Paris, 1994, pp. 421–440.

[105] P.R. Amestoy, I.S. Duff, and C. Puglisi. Multifrontal QR factorization in a multiprocessor environment. *Numer. Linear Algebra Appl.*, 3:275–300, 1996.

[106] L.N. Trefethen. Computation of pseudospectra. *Acta Numer.*, 8:247–295, 1999.

[107] S.H. Lui. Computation of pseudospectra by continuation. *SIAM J. Sci. Comput.*, 28(2): 565–573, 1997.

[108] L.N. Trefethen and T. Wright. Large-scale Computation of Pseudospectra using ARPACK and Eigs. Technical report, Oxford University Computing Laboratory, Oxford, June 2000.

[109] D. Mezher and B. Philippe. PAT — a reliable path-following algorithm. *Numer. Algorithms*, 29: 131–152, 2002.

5 Iterative Methods for the Partial Eigensolution of Symmetric Matrices on Parallel Machines

Maurice Clint

CONTENTS

ABSTRACT

The problem of computing a subset of the eigenvalues and corresponding eigenvectors of large, often sparse, symmetric matrices is addressed. The only tractable methods for solving this problem are iterative in nature. In these the matrix of interest is used only as a multiplier, thereby allowing its sparsity to be preserved. Two main approaches are discussed — subspace iteration methods and Krylov space-based methods. The fundamental operations employed in both approaches — matrix–

vector products, vector orthogonalization, etc., are amenable to efficient parallel implementation on a wide range of machine architectures. Several variants of the iterative algorithms, designed to improve computational efficiency and reduce storage demands, are discussed. The parallel implementation of the basic operations is addressed. The performances of the methods on machines having different architectures is illustrated by their application to a number of matrices selected from standard library collections. The experiments show that the choice of the best parallel method to use is significantly influenced by the nature of the machine on which it is to be executed. The natural extension of the methods to the solution of the unsymmetric problem is briefly discussed. A short overview of the extensive literature devoted to computational aspects of solving eigenproblems is given.

5.1 INTRODUCTION

The computation of a few eigenvalues and their corresponding eigenvectors of (very) large matrices is an important requirement of many scientific and engineering applications. The eigenvalues to be computed may be extremal or may be located in some internal region of the eigenspectrum. The matrices involved are frequently large, with orders exceeding 10^6 are not uncommon; in addition, the matrices are often very sparse. It is infeasible to compute *full* eigensolutions using methods such as the QR and Householder algorithms because, in addition to the excessive computational load incurred, the sparsity of the target matrices is destroyed by the application in these methods of rotations or reflections, leading to unsatisfiable storage demands. Iterative methods, on the other hand, offer practicable means for computing *partial* solutions of such eigenproblems because they are characterized by the use of the target matrix solely as a multiplier: thus its sparsity is preserved. In addition, the most computationally expensive kernels from which they are built — vector innerproducts, vector updates, and matrix–vector products — may be efficiently implemented on a wide range of parallel architectures. The eigensolution of Hermitian matrices (which, if real, are symmetric) is much less problematic than that of unsymmetric matrices, although iterative methods for both have the same basic structure. Real symmetric eigenproblems frequently arise in computational statistics and, often, only partial eigensolutions are required.

5.2 THE REAL SYMMETRIC EIGENVALUE PROBLEM

The eigenproblem for a real symmetric matrix A $(A^T = A)$ may be stated as follows: Given $A \in \mathcal{R}^{n \times n}$, find an orthonormal matrix $X \in \mathcal{R}^{n \times n}$ $(X^T X = I)$ and a diagonal matrix $\Lambda \in \mathcal{R}^{n \times n}$ such that $AX = X\Lambda$. The diagonal elements of Λ, λ_j, $j = 1, \ldots, n$, are the eigenvalues of A. The columns of X are their associated eigenvectors. Thus $Ax_j = \lambda_j x_j$, $j = 1, \ldots, n$, where x_j is the jth column of X and λ_j is the jth diagonal element of Λ. All of the eigenvalues of A are real. Even where eigenvalues coincide n orthonormal eigenvectors may be found so that the columns of X span the n-space.

A *partial* eigensolution of A is a set of eigenpairs $\{(\lambda_k, x_k) | k \in S \subset \{1, \ldots, n\}\}$. It is usually the case that it is required to compute *adjacent* eigenpairs — for example, the set of $p(<n)$ dominant eigenpairs in which the p-eigenvalues are those having the greatest magnitude, or the set of $r(<n)$ eigenpairs with eigenvalues closest to a given value σ and so on.

5.3 ITERATIVE METHODS

Iterative methods for the partial eigensolution of an eigenproblem of order n are (in general) characterized by projection of the problem onto an m-dimensional subspace, $\mathcal{K}$, of $\mathcal{R}^n$. It is *usually* the case that $m << n$. In practice the projection is accomplished using an orthonormal basis for $\mathcal{K}$, the constituent vectors of this basis serving as approximations to the required dominant m eigenvectors of A. A symmetric eigenproblem for a matrix, $B \in \mathcal{R}^{m \times m}$, is generated whose m eigenvalues (usually) approximate the m-dominant eigenvalues of A. The m eigenvectors of B are used to construct a new orthonormal basis for $\mathcal{K}$ from the current one, yielding more accurate approximations to the m required eigenvectors. This process of iterative improvement is repeated until sufficiently accurate eigenpairs of A have been obtained.

Although, in general, the order of B is very much lower than that of A its eigensolution is computed using a direct method, such as the QR method, Householder's method, etc. [28, 73, 85]. These methods often comprise several phases: reduction of the target matrix to a simpler form, extraction of the eigenvalues of this simpler matrix, computation of its eigenvectors using inverse iteration, and recovery of the eigenvectors of the original matrix using the transformations employed in the initial reduction phase. The eigenvectors of B are used to improve approximate eigenvectors of A. The required eigenvalues are eventually approximated by the eigenvalues (Ritz values) of the projected problem.

Invariably, iterative methods of this kind require the computationally intensive orthogonalization of a collection of n-vectors. This operation, together with matrix–vector products involving A, comprise the main computational burden of the iterative methods. If interior eigenpairs are required then so-called shift-and-invert techniques may be employed to drive the required eigenvalues (in fact, simple functions of them) to one or other extreme of the eigenspectrum of a modified eigenproblem whose partial solution is then computed. The formal statement of the transformed problem includes matrix inversion, which, in computational practice, is circumvented by solving sets of linear equations. In the symmetric case, this might involve an LU decomposition or Cholesky factorization of a matrix $(A - \sigma I)$. However, when A is large and sparse an iterative method such as a conjugate gradient (CG) algorithm is used. In practice a *preconditioner* is first constructed which is used to improve the convergence rate of the CG algorithm. In effect, an *approximate LU* factorization of $(A - \sigma I)$ $(= LU + E)$ is constructed and then a conjugate gradient method is applied to a linear system with coefficient matrix $U^{-1}L^{-1}(A - \sigma I)$.

5.4   THE POWER METHOD

The power method [4, 61, 81] is the simplest of the iterative methods and the progenitor of the more powerful subspace iteration (sometimes called simultaneous iteration) methods. It is a projection method in which the dimension of the subspace is one and is used to compute a *single* eigenpair (λ_1, x_1) where λ_1 is the eigenvalue of A of largest modulus. The method proceeds by iteratively improving a randomly chosen initial approximation to x_1 until an acceptably accurate eigenvector has been obtained. The improvement is achieved simply by premultiplying the current eigenvector approximation by A and normalizing the resulting vector.

Let the eigenvalues of A be ordered according to: $|\lambda_1| > |\lambda_2| \geq |\lambda_3| \geq \cdots \geq |\lambda_n|$ and let $q_1 = \sum_{j=1}^{n} \gamma_j x_j$ be an initial approximation to x_1. Then,

$$q_k = (1/\lambda_1^k) A^k q_1 = (1/\lambda_1^k) \sum_{j=1}^{n} \gamma_j A^k x_j$$

$$= \sum_{j=1}^{n} \gamma_j (\lambda_j / \lambda_1)^k x_j$$

Since λ_1 is an eigenvalue of multiplicity one then, as $k \to \infty$,

$$(\lambda_j / \lambda_1) \to 0 \quad \text{for } 2 \leq j \leq n \quad \text{and } q_k \to x_1.$$

These observations lead to the implementation outlined in Algorithm 5.1. Here, rather than normalizing the eigenvector approximations using $\| \cdot \|_\infty$, the Euclidean norm, $\| \cdot \|_2$, is used. Eigenvalue approximations are generated using the Rayleigh quotient [81] values θ. The convergence rate of the method is determined by $|\lambda_2|/|\lambda_1|$. If $|\lambda_1| = |\lambda_2|$ then the power method will not converge whereas, if $|\lambda_1| \approx |\lambda_2|$, convergence will be very slow. The power method may be be used when A is unsymmetric.

> **Algorithm 5.1:** Power(A, q_1, tol)
>
> $\quad q \leftarrow q_1, converged \leftarrow$ false
> $\quad$**while not** $converged$ **do**
> $\quad\quad\quad v \leftarrow q/\|q\|_2$
> $\quad\quad\quad q \leftarrow Av$
> $\quad\quad\quad \theta \leftarrow v^{\mathrm{T}} Av$
> $\quad\quad\quad converged \leftarrow (\| Av - \theta v \|_2 < tol)$
> $\quad$**end_while**
> $\quad (\lambda_1, x_1) \leftarrow (\theta, v)$

where *tol* defines the accuracy required and is chosen to be proportional to the norm of A.

The main deficiency of the power method is that only one (dominant) eigenpair can be computed. However, when this eigenpair has been found an extension of the method allows subdominant eigenpairs to be computed. The extension consists in the incorporation of a deflation process which effectively replaces the dominant eigenvalue of A with a zero eigenvalue. To find the eigenpair with eigenvalue closest to a given value, σ, the power method is applied with the matrix $(A - \sigma I)^{-1}$. This is called a shift-and-invert technique. The eigenvalues of $(A - \sigma I)^{-1}$ are $\{(\lambda_k - \sigma)^{-1} | k \in \{1, \ldots, n\}\}$. The dominant eigenvalue of $(A - \sigma I)^{-1}$ is $(\lambda_c - \sigma)^{-1}$, where λ_c is the eigenvalue of A closest to σ. Its associated eigenvector is x_c. If σ is zero the value computed is λ_n^{-1}, the reciprocal of the smallest eigenvalue of A. In practice, the matrix inversion is not performed explicitly. Instead of computing the product $(A - \sigma I)^{-1} q_i = \tilde{q}_{i+1}$, the linear system $(A - \sigma I)\tilde{q}_{i+1} = q_i$ is solved. The next eigenvector approximation q_{i+1} is then generated by normalizing $\tilde{q}_{i+1}$. This variant of the power method is referred to as the inverse power method or inverse iteration. It is an integral part of many *direct* eigensolution algorithms where it is used to compute the eigenvectors of a matrix after highly accurate approximations to its eigenvalues have been calculated. The accurate eigenvalue approximations are used as shifts and the associated eigenvector for each is computed by the inverse power method in a few iterations (usually, two or three) [57].

Efficient parallel implementation of the power method depends on the efficiency with which matrix–vector products can be realized on the architecture of the machine. Such products are ideally suited to parallel implementation on both shared-memory and distributed-memory machines. If A is dense it is subdivided into panels of rows of (roughly) equal depth and distributed over the processors of the machine. Each processor then computes, independently, a subvector of the product of length equal to the depth of its panel. If A is sparse then, if good load balance is to be achieved, the numbers of rows allocated to processors may vary considerably. It is a rule of parallel computing that interprocessor communication should be kept to a minimum. Thus the way in which a sparse matrix is represented may strongly influence the implementation strategy. If, for example, A is represented by a graph then effective data distributions may be derived using graph partitioning techniques.

Implementation of shift-and-invert methods requires (in the dense case) the LU factorization of the matrix $(A - \sigma I)$ followed by the solution of triangular linear systems. These operations are fundamental building blocks for many methods in numerical linear algebra with the result that research into the development of parallel algorithms for their realization has been intense. A range of parallel algorithms for use with matrices having different characteristics are available for use on shared-memory and distributed-memory machines.

5.5 SUBSPACE ITERATION

The power method may be thought of as a subspace iteration method [4, 61, 81] of order one. Methods referred to as subspace (simultaneous) iteration methods may be regarded as block power methods in which a set of dominant eigenpairs $\{(\lambda_k, x_k)|k \in \{1, \ldots, m\}\}$ is computed, usually with $1 < m << n$. Instead of iterating with a *single* vector (with initial value q_1), m mutually orthonormal n-vectors $Q_m^{(1)} = [q_1^{(1)}, q_2^{(1)}, \ldots, q_m^{(1)}]$ are progressively improved until acceptable approximations are obtained for the required m dominant eigenvectors $X_m = [x_1, \ldots, x_m]$.

A straightforward extension of the power method entails iteratively premultiplying a set of mutually orthogonal vectors, $Q_m^{(i)} = [q_1^{(i)}, q_2^{(i)}, \ldots, q_m^{(i)}]$ until convergence is reached. After each premultiplication the vectors lose their orthogonality and before further iteration the m vectors have to be reorthogonalized. Eigenvalue approximations, $\tilde{\lambda}_j$, $j = 1, \ldots, m$ may be extracted as the m Rayleigh quotients $\tilde{\lambda}_j = q_j^{(i)^T} A q_j^{(i)}$, $j = 1, \ldots, m$.

Orthogonalization of the vectors is an expensive operation. To reduce the overall cost of orthogonalization more than one premultiplication by A may be used in each iteration. However, care must be exercised in order to avoid the premultiplied vectors approaching linear dependence thereby precipitating numerical difficulties in the orthogonalization process.

In subspace iteration with projection the original problem is projected onto an m-dimensional subspace of $\mathcal{R}^n$ to yield an eigenproblem of order m. The eigenvalues of this smaller problem, $\{\tilde{\lambda}_1, \tilde{\lambda}_2, \ldots, \tilde{\lambda}_m\}$, approximate the m dominant eigenvalues, $\{\lambda_1, \lambda_2, \ldots, \lambda_m\}$, of A. Its eigenvectors are used to reorient the current orthogonal eigenvector approximations, $Q_m^{(i)}$. The reoriented approximations are orthonormalized and the process is repeated until, for some k, the columns of $Q_m^{(k)}$ are acceptably accurate approximations to the m dominant eigenvectors of A. Identifying $Q_m^{(k)}$ with $X_m = [x_1, \ldots, x_m]$, the m dominant eigenvalues of A are recovered as the diagonal elements of the diagonal matrix, $\Lambda_m \in \mathcal{R}^{m \times m}$ given by: $\Lambda_m = X_m^T A X_m$.

The method proceeds as follows. Choose randomly a set of m orthonormal vectors $Q_m^{(0)}$ and iteratively generate sets of orthonormal vectors according to:

$$Q_m^{(i+1)} = \textbf{orthonormalize } (A Q_m^{(i)} S_m^{(i)})$$

where $S_m^{(i)} \in \mathcal{R}^{m \times m}$ is the matrix of eigenvectors of $B_m^{(i)} = Q_m^{(i)^T} A Q_m^{(i)}$ and **orthonormalize** orthonormalizes the m columns of its argument. As in the power method matrix–vector products constitute a significant part of the computational load. An implementation of subspace iteration with projection is sketched in Algorithm 5.2.

Algorithm 5.2: Subspace(A, Q, tol)
    $Q \leftarrow Q_1, converged \leftarrow$ false
    while not *converged* **do**
        $V \leftarrow AQ$
        $B \leftarrow Q^T V$
        $(\Theta, S) \leftarrow$ **eigenreduction**(B)
        $Q \leftarrow$ **orthonormalize**(VS)

$$converged \leftarrow (\| AQ - Q\Theta \|_2 < tol)$$

end_while

$$(\Lambda_m, X_m) \leftarrow (\Theta, Q)$$

where $Q_1 = [q_1^{(1)}, q_2^{(1)}, \ldots, q_m^{(1)}]$ and *tol* defines the accuracy required.

However, as for the simple extension of the power method, the main burden is associated with the orthonormalization of a set of vectors, $Q_m^{(i)}$. The order of the symmetric matrix eigenproblem (associated with $B_m^{(i)}$) is very small compared with that of A and the computational load associated with its solution is often negligible. In a parallel implementation of subspace iteration this phase of the algorithm is usually performed using a serial *direct* algorithm — such as the QR method executing on a single processor. In fact, it is unnecessary to compute the eigenvectors $V_m^{(i)}$ to high accuracy and the use of even a crude approximation to these impairs the convergence rate of the method only marginally [13].

In contrast to the power method where only a single vector has to be normalized on each iteration, subspace iteration requires, at each step, the orthonormalization of m vectors. This may be achieved using the modified Gram–Schmidt (MGS) algorithm [81] outlined in Algorithm 5.3.

Algorithm 5.3: Modified Gram–Schmidt (Q)

> **for** $k := 1$ **to** m **do**
>
> $q_k \leftarrow q_k / \| q_k \|_2$
>
> > **for** $l := k + 1$ **to** m **do**
> >
> > $q_l \leftarrow q_l - (q_k^T q_l) q_k$
> >
> > **end_for**
>
> **end_for**

where $Q = [q_1, q_2, \ldots, q_m]$.

The orthonormalization process is computationally intensive and, since it is applied to n-vectors, it is worthwhile to use a parallel implementation of this operation. This may be organized by splitting the m vectors in blocks of rows across the processors of the machine. The vector dot products, which constitute the kernel operations of the algorithm, are partially computed, independently, on each processor. These partial results are combined on each processor to yield the requisite normalizing and vector-update factors. Vector normalization and vector-updates (the corrections to the q_l in the inner loop of Algorithm 5.3 may be performed in parallel) are carried out independently on the processors. MGS can be implemented with very high parallel efficiency on a distributed memory machine especially when $n >> m$. Several different implementation strategies are available which exploit the particular topologies of different types of multiprocessor machine. In the parallel implementation of this operation it is important to minimize the number of interprocessor messages transmitted — perhaps by sending fewer, longer messages. Opportunities to overlap computation and message transmission should also be exploited.

If the eigenvalues of A satisfy: $|\lambda_1| \geq |\lambda_2| \geq \cdots \geq |\lambda_m| > |\lambda_{m+1}| \geq \cdots \geq |\lambda_n|$ then the convergence rate of subspace iteration is determined by the ratio, $|\lambda_{m+1}| / |\lambda_m|$. Coincident eigenvalues may be computed without difficulty. To improve the convergence rate a subspace larger than the smallest necessary to compute the required partial eigensolution may be used. Thus to compute a solution of order m, r ($> m$) vectors may be used, thereby improving the convergence rate to $|\lambda_{r+1}| / |\lambda_m|$. The performance of the method may also be enhanced through a process of deflation whereby, when an eigenvector has been accurately computed, it is *locked*. This means that it is no longer premultiplied by A and its role in the orthonormalization process is reduced in that the nonconverged eigenvector approximations need only to be orthogonalized with respect to it.

An important subspace iteration-like method, the trace minimization algorithm [6,64,65], may be used to compute simultaneously the *smallest* eigenpairs of a positive definite matrix A. The method is based on the following observation: let $\mathcal{Q} = \left\{ Q_m \in \mathcal{R}^{n \times m} | Q_m^T Q_m = I \right\}$ then (by the Courant–Fischer theorem, see Ref. [85])

$$\min_{Q_m \in \mathcal{Q}} \text{trace}(Q_m^T A Q_m) = \sum_{i=1}^{m} \lambda_{n-i+1}$$

(i.e., the sum of the m smallest eigenvalues of A). The Q_m which achieves the minimum comprises the eigenvectors corresponding to the m smallest eigenvalues of A. The method proceeds as follows. A sequence of approximations, $Q_m^{(i)}$, to the m required eigenvectors is generated, with the property that

$$\text{trace}(Q_m^{(i+1)^T} A Q_m^{(i+1)}) < \text{trace}(Q_m^{(i)^T} A Q_m^{(i)}).$$

On convergence, the columns of $Q_m^{(i)}$ are the required eigenvectors. The algorithm is built from the same operations as Algorithm 5.2 — viz. matrix products, eigensystem computation, and vector orthonormalization. In addition, the $Q_m^{(i)}$ are generated in exactly the same way but with the inclusion of an extra step. In each iteration, before the Ritz vector approximations — $A Q_m^{(i)} S_m^{(i)} (= V_m^{(i)} = [v_1^{(i)}, v_2^{(i)}, \ldots, v_m^{(i)}])$ — are orthonormalized, $v_k^{(i)}$, $k = 1, \ldots, m$, is *corrected* by a vector, $\delta_k^{(i)}$ computed by minimizing $(v_k^{(i)} - \delta_k^{(i)})^T A (v_k^{(i)} - \delta_k^{(i)})$ subject to the constraint $V_m^{(i)^T} \delta_k^{(i)} = 0$. The $\delta_k^{(i)}$ may be computed independently, in parallel, but require the solution of systems of linear equations of order $n + m$: when A is large and sparse iterative methods such as the CG should be used for this purpose to avoid the undesirable fill-in precipitated by the use of direct methods. In its structure the trace minimization method has many similarities to the Jacobi–Davidson method discussed later.

An important feature of subspace iteration methods is that their storage requirements are fixed so that, when executed in a parallel environment, data distributions can be determined at the outset. This desirable property is not universally enjoyed by another class of iterative methods — based on Krylov spaces — which are now discussed.

5.6 THE LANCZOS METHOD

Successive vectors q_j generated by the power method almost invariably contain information along eigenvector directions other than in the direction of x_1. Because only a single vector is retained in the power method this useful information is discarded. Where an eigensolution of order greater than one is to be computed it is advantageous to exploit this discarded information. Thus it is natural to consider the Krylov subspace:

$$\mathcal{K}(A, q_1, j) = \text{span}(q_1, Aq_1, A^2 Q_1, \ldots, A^{j-1} q_1)$$

which comprises the sequence of vectors generated in the first j steps of the power method.

A vector $y \in \mathcal{K}(A, q_1, j)$ is called a *Ritz vector* of A with corresponding *Ritz value*, θ, if $q^T(Ay - \theta y) = 0$ for all $q \in \mathcal{K}(A, q_1, j)$. Let $Q_j \in \mathcal{R}^{n \times j}$ be an orthonormal basis for $\mathcal{K}(A, q_1, j)$ and let $y = Q_j s$ where $s \in \mathcal{R}^j$. Then:

$$Q_j^T (A - \theta I) y = (Q_j^T A Q_j) s - \theta s = 0$$

Thus, $(Q_j^T A Q_j) s = \theta s$ and (θ, s) is an eigenpair of the symmetric matrix $Q_j^T A Q_j \in \mathcal{R}^{j \times j}$. The eigenvalues of $Q_j^T A Q_j$ serve as *approximations* to a subset of the eigenvalues of A.

The idea underpinning the Lanczos algorithm is to construct an orthonormal basis Q_k for a sufficiently large Krylov subspace $\mathcal{K}(A, q_1, k)$ so that the p dominant eigenvalues of $T_k = Q_k^T A Q_k$ approximate, to acceptable accuracy, the p dominant eigenvalues of A. The associated eigenvectors can be recovered from Q_k and the eigenvectors of T_k.

The structure of the basic Lanczos algorithm is derived from the following observation. Let $Q \in \mathcal{R}^{n \times n}$ be orthonormal ($Q^T Q = I$), $T = Q^T A Q$ be tridiagonal (thus $AQ = QT$) and $q_1 = Q e_1$ (e_i is the ith column of the unit matrix of order n). Then, since $AQ = QT$, $Q\{e_1, T e_1, T^2 e_1, \ldots, T^{(n-1)} e_1\} = \{q_1, A q_1, A^2 q_1, \ldots, A^{(n-1)} q_1\} = \mathcal{K}(A, q_1, n)$. Note that, since the matrix $R = [e_1, T e_1, T^2 e_1, \ldots, T^{(n-1)} e_1]$ is upper-triangular, QR is the QR factorization of $\mathcal{K}(A, q_1, n)$.

Let the dominant submatrix of T, of order j, be denoted by

$$
T_j = \begin{bmatrix}
\alpha_1 & \beta_1 & & \cdots & & 0 \\
\beta_1 & \alpha_2 & \ddots & & & \vdots \\
& \ddots & \ddots & \ddots & & \\
\vdots & & \ddots & \ddots & \beta_{j-1} \\
0 & \cdots & & & \beta_{j-1} & \alpha_j
\end{bmatrix}
$$

then, equating columns in $AQ = QT$, gives $A q_j = \beta_{j-1} q_{j-1} + \alpha_j q_j + \beta_j q_{j+1}$, with $\beta_0 q_0 = 0$ for $j = 1, \ldots, n-1$. Thus $\alpha_j = q_j^T A q_j$, $\beta_j = \|r_j\|_2$, and $q_{j+1} = r_j / \beta_j$, where $r_j = (A - \alpha_j I) q_j - \beta_{j-1} q_{j-1}$.

These results and observations provide the basis for a class of important iterative methods, Lanczos methods, for the partial eigensolution of symmetric matrices. (The method is named after its creator C. Lanczos whose seminal paper, Ref. [36], has spawned a rich variety of partial eigensolvers [4, 56, 61].)

Let $A \in \mathcal{R}^{n \times n}$. Then a relation of the form

$$
AQ_j = Q_j T_j + r_j e_j^T,
$$

where $Q_j \in \mathcal{R}^{n \times j}$ has orthonormal columns, $Q_j^T r_j = 0$ and $T_j \in \mathcal{R}^{j \times j}$ is tridiagonal, is called a j-step *Lanczos Factorization* of A. The columns of Q_j are referred to as Lanczos vectors.

Suppose that it is required to compute the p dominant eigenpairs of A. Then the classical Lanczos proceeds as follows. As in the power method, choose a random vector $q_1 \in \mathcal{R}^n$ and progressively generate a sequence of tridiagonal matrices $T_1, T_2, \ldots$, according to $T_j = Q_j^T A Q_j$ where the columns of Q_j, other than the first, are generated using the characteristic three-term recurrence relation given above. When j exceeds p, the p largest eigenvalues of T_j (Ritz values of A) are (usually) approximations to the required eigenvalues. It is a remarkable fact that, if n is very large, the order of the Lanczos factorization required for T_j to yield acceptably accurate approximations to the p dominant eigenvalues is, by comparison, very small.

Let $S_j \in \mathcal{R}^{j \times j}$ be orthogonal and satisfy $S_j^T T_j S_j = \Theta_j$ where $\Theta_j \in \mathcal{R}^{j \times j}$ is diagonal with (Ritz values of A) $\theta_1, \theta_2, \ldots, \theta_j$ as its diagonal elements. Then the columns of S_j are the eigenvectors of T_j and the required p Ritz vectors — approximations to the eigenvectors of A — may be extracted from the columns of $Y_j \in \mathcal{R}^{n \times j}$ where $Y_j = [y_1, y_2, \ldots, y_j] = Q_j S_j$. If the ratio $|\lambda_i| / \|A\|_2$ is close to unity it transpires that, for surprisingly small j, the ith Ritz pair $(\theta_i, y_i)_j$ generated in the jth step of the Lanczos process closely approximates the ith eigenpair of A. If the eigenvalue, λ_i, is sought then (for this value) the Lanczos process terminates when $|\hat{\beta}_j| |\hat{s}_{ji}| \leq \epsilon_1$, where $|\hat{\beta}_j| |\hat{s}_{ji}|$ is a computed error bound on λ_i and ϵ_1 is the required accuracy. This error bound requires the computation of the matrix of eigenvectors (S_j) of $T_j - \hat{s}_{ji}$ is the computed value of the last component of the eigenvector of T_j associated with the Ritz value θ_i. The eigenvectors of

T_j are usually computed using inverse iteration. As well as being relatively expensive, use of this stopping criterion may precipitate premature termination if λ_i is poorly separated from the rest of the spectrum of A. Alternatively, termination may be decided by comparing successive Ritz value approximations to λ_i [77]. Thus, $|\hat{\theta}_i(T_j) - \hat{\theta}_i(T_{j-1})| < \epsilon_2$, where ϵ_2 is the desired accuracy. The accuracy demanded for the eigenvalues of the T_j may be more stringent than that required of the Lanczos procedure itself.

By construction (using the three-term recurrence relation) the Lanczos vectors are mutually orthogonal in exact arithmetic. However, in limited precision arithmetic a loss of orthogonality occurs among the Lanczos vectors due to rounding errors with the result that, in computational practice, reorthogonalization of the Lanczos vectors is required. Unfortunately, the length of the Lanczos factorization needed to compute the required eigenpairs may be large. Consequently, orthonormalization of the Lanczos vectors may constitute a major part of the overall computational burden. Note that, by contrast, in the case of subspace iteration the number of vectors to be orthonormalized in each iteration is fixed at the outset. Two other problems associated with the Lanczos algorithm and from which subspace iteration methods are free are: (i) an inability to detect multiple (or indeed closely clustered) eigenvalues and (ii) misconvergence to unwanted eigenpairs [69].

An outline of the basic Lanczos algorithm is given in Algorithm 5.4, where p is the order of eigensolution sought and *tol* is the accuracy required. The most expensive operations in the implementation of the algorithm are: (i) matrix–vector products in the generation of the Lanczos vectors; (ii) orthonormalization of the Lanczos vectors; (iii) computation of the eigensolution of symmetric tridiagonal matrices; and (iv) monitoring convergence. The first two of these are common to all iterative methods for computing partial eigensolutions. In all cases their parallel implementation is realized as sketched above for subspace iteration. Note, however, that, in contrast to subspace iteration in which *all m* eigenvector approximations (except in implementations where they have converged and are locked) are orthonormalized, in each iteration the *new* Lanczos vector only is orthogonalized with respect to the previously generated Lanczos vectors (which remain orthonormal). The convergence test is the classical one discussed earlier.

Algorithm 5.4: Lanczos(A, q_1, p, tol)

    $j \leftarrow 0, \beta_0 \leftarrow 1, q_0 \leftarrow 0, converged \leftarrow$ false

    while not *converged* **do**)

         $j \leftarrow j + 1$

         $\alpha_j \leftarrow q_j^T A q_j$

         $q_{j+1} \leftarrow (A - \alpha_j I) q_j - \beta_{j-1} q_{j-1}$

         $\beta_j \leftarrow \| q_{j+1} \|_2$

         orthonormalize q_{j+1} **against** $q_1, \ldots, q_j$

         $(\Theta_j, S_j) \leftarrow$ **eigenreduction**(T_j)

         $converged \leftarrow$ **convergence_test**(S_j, β_j, p, tol)

    end_while

    compute $(\lambda_1, x_1), \ldots, (\lambda_p, x_p)$

When striving to reduce the amount of orthonormalization required the aim is to achieve accuracy close to that produced by full orthonormalization while incurring minimum cost — that associated with no orthonormalization. The basis for selective orthonormalization is Paige's work [53–55], which establishes that Lanczos vectors lose orthogonality through contamination by Ritz vectors corresponding to Ritz values which have already converged. The convergence of the Ritz values is monitored and only when a Ritz value satisfies its accuracy requirement are Lanczos vectors orthogonalized with respect to its associated Ritz vector. Thus, in effect, Lanczos vectors are orthonormalized against accepted eigenvectors. Among strategies to reduce the amount of orthogonalization are the selective orthogonalization of Parlett and Scott [58], the partial orthogonalization of Simon [67], and the reorthogonalization techniques of Grimes et al. [30]. However, in all of these, the same suboperations — vector dot products and vector updates — are required as in full reorthogonalization and the parallel implementation of this phase is similar to that when full orthonormalization is performed. When only eigenvalues are sought, the orthonormalization phase may be dispensed with completely [17] — the drawback is the possible appearance of spurious eigenvalues, though these may be discarded by the use of heuristic techniques.

When the classical Lanczos algorithm is used, the length of the factorization required to produce the solution may be significantly large — while remaining much smaller than n. In such cases there is an opportunity to exploit parallelism in the eigensolution of T_j when its order becomes sufficiently large. Here, the use of the Sturm sequence property for tridiagonal matrices allows intervals within which batches of eigenvalues fall to be isolated. These intervals can be distributed over the available processors and the eigenvalues computed accurately and independently without necessitating interprocessor communication. If the eigenvectors of T_j are computed in order to apply the classical convergence test then the availability of accurate eigenvalue approximations allows the eigenvectors to be computed, by inverse iteration, independently and in parallel on the available processors. The possible occurrence of clustered eigenvalues must, of course, be catered for. More efficient convergence monitoring techniques which do not require the *explicit* computation of the eigensolution of T_j on *every* iteration and which can exploit parallelism are also available [77].

Where multiple or closely clustered eigenvalues occur the performance of the (single-vector) Lanczos methods may be enhanced by using block versions of the algorithms [15, 29, 30]. (This approach is analogous to the use of subspace iteration as an extension of the power method.) Block versions of the characteristic three-term recurrence relation are employed and matrix–vector products are replaced with matrix-block multipliers and simple solvers with block solvers. An additional benefit of using block algorithms is that the use of the higher order BLAS is facilitated.

For many reasons it is now rare for the classical Lanczos discussed earlier to be used in practice. Firstly, the storage requirements for the Lanczos vectors can become prohibitive. To alleviate this problem *restarted* Lanczos methods (see below) are employed. In these the lengths of the Lanczos factorizations are reduced or are explicitly constrained. In the latter case the length of the factorizations are fixed with the result that the *maximum* order of T_j is bounded. Consequently, the eigensolution of T_j may constitute such a minor part of the overall computational load that it is not worthwhile to use parallel execution in the convergence monitoring phase. In constrained restarted methods it is usual to compute the eigensolution of T_j using a serial algorithm. Secondly, full reorthogonalization of the Lanczos vectors in the classical method imposes severe computational demands. This burden is automatically reduced when shorter factorizations are employed.

5.7   RESTARTED LANCZOS METHODS

A major problem associated with the classical Lanczos algorithm is that, j, the length of the Lanczos factorization must be sufficiently large to ensure the required accuracy in the p eigenpairs sought. This value of j is not known in advance. The length of the factorization required depends to some extent on the order of the partial eigensolution being computed and on the nature of the eigenspectrum of A. Since this factorization may be *very* long with high storage demands it is imperative to seek ways to alleviate this problem. This may be done by restarting the Lanczos periodically with a new initial vector, a suggestion put forward by Karush [35].

5.7.1 EXPLICITLY RESTARTED LANCZOS METHODS

One way to avoid very long factorizations is to compute the p required eigenpairs *one at a time* by applying the Lanczos algorithm p times [78]. The effect of this strategy is to reduce the length of the longest factorization required to complete the computation resulting in lower storage requirements and less orthogonalization. In this approach newly generated Lanczos vectors in each application of the process are orthogonalized, not only with respect to their predecessors in that application, but also with respect to already computed Ritz vectors which have been accepted as eigenvectors of A in the previous applications of the algorithm. This is analogous to locking converged eigenvectors in subspace iteration.

The question arises as to how the starting vector should be chosen for each of the p executions of the Lanczos algorithm. For the first a random vector, $q_1^1 \in \mathcal{R}^n$, is chosen, as before. After i eigenpairs have been computed the $(i+1)$th Lanczos execution may be started using a vector, $q_1^{i+1} \in \mathcal{X}_i^\perp$, where $\mathcal{X}_i^\perp$ is the orthogonal complement of $\mathcal{X}_i = \text{span}(\hat{x}_1, \ldots, \hat{x}_i)$ and $\hat{x}_1, \ldots, \hat{x}_i$ are accepted Ritz vector approximations to the eigenvectors $x_1, \ldots, x_i$ of A. Many possibilities suggest themselves. At the end of the isolation process for the ith eigenpair an approximation to the $(i+1)$th eigenpair (λ_{i+1}, x_{i+1}) will be available from the eigensolution of the final T_j. Suppose that this approximation is $(\hat{\lambda}_{i+1}, \hat{x}_{i+1})$. Then the starting vector for the $(i+1)$th application of the Lanczos algorithm, q_1^{i+1}, may be generated by orthogonalizing $\hat{x}_{i+1}$ with respect to $x_1, \ldots, x_i$. A variation on this approach is to compute an approximation to $\hat{x}_{i+1}$ at the end of *each* of the i previous applications of the Lanczos algorithm, select the most accurate of them, orthogonalize it with respect to the accepted Ritz vectors and use the resulting vector as q_1^{i+1}. In this approach to restarting the lengths of some of the factorizations required to compute the eigenpairs may still be large (with the attendant storage problems) and be sufficiently large to make it profitable to employ parallelism in the eigensolution of the T_j and/or in convergence monitoring. Another benefit of restarting in this way is that multiple and clustered eigenvalues can be isolated. The methods just described are called *explicitly* restarted methods. Such a method is sketched in Algorithm 5.5, where the call, Lanczos$(A, q_1, 1, tol)$ computes a *single* Ritz pair using the classical Lanczos method.

Algorithm 5.5 Restarted_Lanczos (A, q_1, p, tol)

$\quad \mathcal{X} \leftarrow 0$

$\quad$ **for** $i = 1$ **to** p **do**

$\quad\quad\quad (\lambda_i, x_i) \longleftarrow \text{Lanczos}(A, q_1, 1, tol, \mathcal{X})$

$\quad\quad\quad$ **if** $i < p$ **then**

$\quad\quad\quad\quad\quad \mathcal{X} \leftarrow \mathcal{X} \oplus \text{span}\{x_i\}$

$\quad\quad\quad\quad\quad \text{choose } q_1 \in \mathcal{X}^\perp$

$\quad\quad\quad$ **end_if**

$\quad$ **end_for**

where *tol* is the specified accuracy.
(Note: the parameter list in the internal invocation of Algorithm 5.4 has been extended to include $\mathcal{X}$ since the Lanczos vectors are to be orthonormalized with respect to the converged Ritz vectors in $\mathcal{X}$.)

In a more radical and commonly used explicit restarting technique the maximum number of Lanczos vectors to be generated during a *run* of the Lanczos process is fixed, thereby allowing the storage requirement to be determined at the outset. The idea is to compute the required eigensolution using a *sequence* of Lanczos factorizations each of length not greater than some small value, k.

At the end of each run a new Lanczos factorization of length k is initiated using an appropriately chosen starting vector. Such methods are called k-step methods.

Suppose that only the dominant eigenpair of A, (λ_1, x_1), is to be computed. Then the obvious choice for the restart vector to be used after completion of a k-step run is the most recently available Ritz vector approximation to x_1. This single eigenpair method is guaranteed to converge after a sufficiently long sequence of restarts.

If a solution of order $p(>1)$ is required then the length of a run, k, must exceed $(p-1)$ since T_k must be sufficiently large to allow the extraction of p Ritz values. At the end of each run a restart vector, which must carry significant information about the entire subspace spanned by the p Ritz vectors, has to be chosen. The obvious choice is some linear combination of these Ritz vectors. However, it is not clear what the optimal combination is and heuristic arguments are advanced in favor of particular choices of coefficients: proposals in this regard have been made by Cullum and coworkers [15–17]. Approaches, based on polynomial preconditioning, developed for use in the general case and designed to damp the contributions of unwanted Ritz vectors from the restarting vector have been suggested by, among others, Saad [60] and Chatelin and Ho [12].

The difficulty associated with the choice of a restarting vector after each run can be circumvented by using a k-step method [79] to compute a single eigenpair at a time as described earlier. In this case k may be chosen independently of p and convergence to each required eigenpair is guaranteed. The method uses the technique of deflation which is also an intrinsic part of the unbounded Lanczos factorization method discussed earlier.

Suppose that the eigenpairs (Ritz pairs) $(\hat{\lambda}_1, \hat{x}_1), \ldots, (\hat{\lambda}_i, \hat{x}_i)$ have already been computed. Then a new sequence of k-step runs is initiated with q_1 chosen to be orthogonal to the columns of $\widehat{X} = [\hat{x}_1, \ldots, \hat{x}_1]$ (thus, ignoring rounding errors, $q_1 \in \mathcal{X}_i^{\perp}$, the invariant subspace of the uncomputed $(n-i)$ eigenvectors of A). In effect, this process implicitly implements a deflation scheme to remove the already computed eigenvalues (Ritz values). Explicitly computing $(\widehat{X}_i^{\perp})^{\mathrm{T}} A \widehat{X}_i^{\perp}$ (where $X_i^{\perp}$ is a orthonormal basis for $\mathcal{X}_i^{\perp}$) would destroy the sparsity of A which, for practical reasons, it is imperative to preserve. At the end of each k-step run the most recent Ritz vector approximation (suitably orthonormalized) is used to restart the next run. In k-step methods the order of T_j never exceeds k. Thus, the computational effort involved in the eigensolution of these small matrices is negligible compared with that of performing matrix–vector operations involving A or vector orthogonalization. Consequently, there is no advantage to be gained by using parallelism in this phase of the algorithm. A serial implementation of the QR algorithm is usually employed to compute the Ritz values and the eigenvectors of the T_j. The short Lanczos factorizations associated with k-step methods often produce T_j whose eigensolutions are difficult to compute numerically and great care should be exercised in this phase.

5.7.2   IMPLICITLY RESTARTED LANCZOS METHODS

In the explicit methods just described the restarting vector for each unbounded Lanczos factorization or k-step run is explicitly computed. A powerful implicit restarting method proposed by Sorensen [71] and developed with Lehoucq [37], Bai et al. [4] obviates the necessity to compute new restart vectors. Sorensen based his implicit algorithm on the Arnoldi method [3], the unsymmetric equivalent of the Lanczos method, but the approach is the same for the symmetric case. A major problem associated with explicit restarting is that a single vector must carry useful information about the entire subspace of the p Ritz vectors sought.

The implicit approach avoids this drastic requirement by updating p Lanczos vectors after each k-step run. This compression of a factorization of length k to a factorization of length p is achieved by applying $(k-p)$ implicitly shifted QR steps, yielding a factorization rich in information relating to the required eigenspace of order p. At the end of a k-step run the Lanczos factorization is

$$A Q_k = Q_k T_k + r_k e_k^{\mathrm{T}}$$

The restriction of the factorization is accomplished in two steps. First $(k - p)$ implicit QR steps are applied to yield

$$A Q_k^{\odot} = Q_k^{\odot} T_k^{\odot} + r_k e_k^{\mathrm{T}} \Psi$$

where $Q_k^{\odot} = Q_k \Psi$, $T_k^{\odot} = \Psi^T T_k \Psi$, and $\Psi = \Psi_1 \Psi_2, \ldots, \Psi_{k-p}$ where Ψ_s is the orthogonal matrix generated from the QR factorization associated with the sth shift in the shifted QR algorithm applied to T_k. The Ψ_s are Hessenberg matrices and so the first $(p - 1)$ elements of $e^{\mathrm{T}} \Psi$ are zero. Thus, the Lanczos property is preserved among the first p columns of the modified Lanczos factorization. Equating the first p columns of this factorization, an *updated* p-step Lanczos factorization

$$A Q_p^{\odot} = Q_p^{\odot} T_p^{\odot} + r_p^{\odot} e_p^{\mathrm{T}}$$

with an updated residual $r_p^{\odot}$ is generated. Starting from this compressed factorization a further $(k - p)$ Lanczos steps may be applied to reestablish a k-step method.

The requisite $(k - p)$ shifts may be chosen in many ways. When *exact* shifts are used the eigenvalues of T_k are sorted into disjoint sets comprising the required p Ritz values and the $(k - p)$ unwanted Ritz values. The unwanted values are used as shifts with the result that, after application of all $(k - p)$ shifts, $T_p^{\odot}$ has approximations to the required Ritz values as its eigenvalues. Eventually, $T_p^{\odot}$ becomes diagonal and the required eigenvalues may be recovered from the $p \times p$ leading submatrix of T_k. In addition, the columns of Q_k converge to the associated orthogonal eigenvectors of A. There are several ways in which required eigenvalues may be defined — for example, the p eigenvalues with smallest magnitude, the p algebraically largest eigenvalues, etc.

In each k-step run the implicit method may be interpreted as the implicit application of a polynomial in A of degree $(k - p)$ to the starting vector q_1, the effect of which is to damp the influence in this vector of unwanted eigenvectors. Thus, q_1 is transformed to $\rho(A)q_1$ where

$$\rho(\lambda) = \prod_{j=1}^{(k-p)} (\lambda - \mu_j) \tag{5.1}$$

where the μ_j are the shifts used in the QR process and may be chosen so as to enrich q_1 with the required eigenvectors. The *implicit* application of polynomial filters of this kind can simulate any of the *explicit* restarting techniques discussed above. Note that if k is chosen to be n, the order of A, then the implicitly restarted Lanczos method becomes the implicitly shifted QR iteration method.

The implicit k-step method requires $(k - p)$ matrix–vector products per run compared with the k required by the explicitly restarted k-step methods. This may be significant on machines where the cost of this operation is relatively expensive. However, on parallel architectures where the cost of this operation is cheap it may pay to use a method such as one of the explicit methods discussed earlier which make profligate use of matrix–vector products. Thus, for very sparse matrices explicit restarting methods may significantly outperform the implicit method. As the sparsity declines the implicit method comes into its own.

5.8 THE JACOBI–DAVIDSON METHOD

Lanczos methods are useful when computing eigenpairs at either end of the eigenspectrum of A, particularly if the required eigenvalues are well separated. Lanczos methods can also be used to compute eigenpairs close to a specified value, σ. In this case, A is replaced by $(A - \sigma I)^{-1}$ so that, when generating the Krylov subspace, it is required to compute products of the form $(A - \sigma I)^{-1} q_k$. In computational practice these are realized by solving sets of linear equations. Solving these equations accurately may be prohibitively expensive: in such circumstances, the Davidson method [14,19,45,46] or Jacobi–Davidson methods [4,68], the structure of which closely resembles that of Lanczos methods, may be employed.

In the Jacobi–Davidson approach, an expanding collection of vectors $V_j = [v_1, v_2, \ldots, v_j]$ is iteratively maintained. The j eigenpairs, (θ_i, s_i), of $V_j^T A V_j$ yield, as usual, the Ritz pairs $(\theta_i, V s_i)$ of A with respect to the subspace spanned by the columns of V_j. The aim of the Jacobi–Davidson method is to improve these approximations to yield the required eigenpairs of A. Let $u_i = V_j s_i$ and suppose that (θ_i, u_i) approximates the eigenpair (λ_i, x_i). Then, following Jacobi's suggestion for the case when A is strongly diagonal, an orthogonal correction, t_i, for the approximate eigenvector, u_i, satisfies

$$A(u_i + t_i) = \lambda_i (u_i + t_i)$$

Since t_i is orthogonal to u_i attention may be restricted to the subspace orthogonal to u_i so that, after restriction of A to this subspace, t_i satisfies

$$(I - u_i u_i^T)(A - \lambda_i I)(I - u_i u_i^T) t_i = -(A - \lambda_i I) u_i$$

Since λ_i is not known it is replaced by θ_i yielding the Jacobi–Davidson correction equation for t_i:

$$(I - u_i u_i^T)(A - I)(I - u_i u_i^T) t_i = -r_i$$

where $r_i = (A - \theta_i I) u_i$.

The correction equation is solved *approximately* to yield $\tilde{t}_i$ and this vector is orthogonalized with respect to $v_1, v_2, \ldots, v_j$ and added to the subspace V_j to yield V_{j+1}. Note that this subspace is *not* a Krylov subspace. To improve efficiency an approximate solution of the correction equation may be computed using, for example, a suitable preconditioner, W_j, for $(A - \theta_i I)$. W_j is usually an approximate inverse of $(A - \theta_i I)$. Note that if $W_j = I$ for all j, then the Lanczos and Jacobi–Davidson methods coincide. In the Lanczos method $Q_j^T A Q_j$ $(= T_j)$ is symmetric and tridiagonal; in the Jacobi–Davidson method $V_j^T A V_j$ $(= M_j)$ is symmetric and full. However, just as T_{j+1} is a simple extension of T_j generated by adding three new elements, M_{j+1} is generated from M_j by adding a new row and column. The outline of a Jacobi–Davidson method for computing the dominant eigenpair of a matrix is given in Algorithm 5.6.

Algorithm 5.6: Jacobi–Davidson(A, v_1, tol)

$j \leftarrow 1, V \leftarrow v_1, converged \leftarrow$ false

while not $converged$ **do**

$\quad j \leftarrow j + 1$
$\quad M \leftarrow V^T A V$
$\quad (\theta_1, s_1) \leftarrow$ **max_eigenpair**(M)
$\quad u \leftarrow V s_1$
$\quad r \leftarrow A u - \theta u$
$\quad t \leftarrow$ **approx_ortho_t_solve** $((I - u u^T)(A - I)(I - u u^T) t = -r)$

$v_j \leftarrow$ **orthogonalize** t **against** $v_1, v_2, \ldots, v_{i-1}$

$V \leftarrow [V | v_j]$

$converged \leftarrow (\| r \|_2 < tol)$

end_while

$(\lambda_1, x_1) \leftarrow (\theta, u)$

The possible unacceptable growth in the number of vectors in the subspace V_j suggests the employment of restarting strategies like those used with the Lanczos methods. Restarting with a single vector means that potentially useful information is discarded. It is often better to begin again with a small set of vectors, including the most recent approximation to the eigenvector sought and Ritz vectors corresponding to a few Ritz values neighboring the eigenvalue of interest. When more than one eigenpair at either end of the eigenspectrum are required a deflation process, similar to that employed in explicitly restarted Lanczos methods, may be used in which the vectors of the Jacobi–Davidson subspace are kept orthogonal to already converged eigenvectors [25].

In all iteration methods the Ritz values converge toward *exterior* eigenvalues so that difficulties arise when *interior* eigenpairs (Ritz pairs) are sought. A wanted eigenvector may be very poorly represented in the Ritz vector corresponding to the Ritz value approximating a desired eigenvalue. In such circumstances this Ritz vector is not a good restarting choice when attempting to converge to this eigenvalue. In effect, interior Ritz pairs may be regarded as *bad* approximations to exterior Ritz pairs. One way to circumvent this problem is to compute the *harmonic* Ritz values rather than the Ritz values themselves and to use the *harmonic* Ritz vectors rather than Ritz vectors in restarting strategies.

Consider a set of j orthonomal vectors W_j generated by orthonormalizing the columns of AV_j where V_j is a Jacobi–Davidson orthonormal subspace. Then a vector $y \in V_j$ is a harmonic Ritz vector of A with corresponding harmonic Ritz value θ if (the Petrov–Galerkin condition) $w^T(A - \theta I)y = 0$ holds for all $w \in W_j$. Let $y = V_j s$, where $s \in \mathcal{R}^j$, then

$$W_j^T(A - \theta I)y = W_j^T A V_j s - \theta W_j^T V_j s = 0$$

or, since $V_j = A^{-1} W_j$ and $W_j^T W_j = I$, $s = \theta W_j^T A^{-1} W_j s$. Thus $W_j^T A^{-1} W_j s = 1/\theta s$, yielding an orthogonal projection process for computing eigenvalues of A. Note, however, that A need not be inverted explicitly since:

$$W_j^T A^{-1} W_j = V_j^T A V_j.$$

The harmonic Ritz values are reciprocals of the Ritz values of A^{-1} and converge, in general, to eigenvalues closest to zero — that is, to the interior eigenvalues of the eigenspectrum. To compute eigenvalues close to a specified value σ, the Jacobi–Davidson method may be applied to the shifted matrix $(A - \sigma I)$. In this adaptation of the method restarting can be carried out with vectors rich in the required eigenvectors. The need to maintain the orthonormality of the vectors of W_j incurs the usual computation penalty associated with the use of a Gram–Schmidt orthonormalization procedure. The techniques for computing harmonic Ritz pairs sketched here can also be incorporated in Lanczos methods.

Aspects of the parallel implementation of the Davidson and Jacobi–Davidson methods are discussed in Refs. [10, 51, 72].

5.9   THE UNSYMMETRIC EIGENPROBLEM

Iterative methods for the partial eigensolution of the unsymmetric eigenproblem have the same structure as those for the symmetric case discussed earlier. However, the symmetric problem is very well behaved — for example, all of the eigenvalues lie on the real line, coincident eigenvalues have distinct associated eigenvectors and the eigensystem of a slightly perturbed matrix is close to that of its parent. None of these properties is, in general, enjoyed by unsymmetric matrices.

The power method may be applied unaltered to real unsymmetric matrices provided that the eigenvalue of maximum modulus is real. Complex eigenvalues of such matrices occur as conjugate pairs and so have the same magnitude. Inverse iteration may be used as in the symmetric case to compute the eigenvector of an eigenpair with eigenvalue closest to a specified value, σ. The matrix $(A - \sigma I)$ will not, of course, be symmetric. This may affect the techniques used in solving the linear equations associated with the application of $(A - \sigma I)$. Subspace iteration with projection may also be applied almost without change [4,24,76].

The Lanczos method is a specialization for symmetric matrices of the Krylov space-based Arnoldi method [3,4,11,12,59,63,66], which is applicable to general matrices. In Arnoldi's method, for $A \in \mathcal{R}^{n \times n}$, j-step Arnoldi factorizations are generated according to:

$$A R_j = R_j H_j + r_j e_j^{\mathrm{T}}$$

where $R_j \in \mathcal{R}^{n \times j}$ (whose columns are referred to as Arnoldi vectors) is orthonormal, $R_j^T r_j = 0$, and $H_j \in \mathcal{R}^{j \times j}$ is a Hessenberg matrix with nonnegative subdiagonal elements. Applications of the algorithm proceeds as for the Lanczos method with the eigenvalues of the H_j tending toward Ritz values of A. Variations of the basic method incorporating explicit restarting and deflation are available to reduce storage requirements and to improve performance. Powerful implicitly restarted Arnoldi variants have also been developed and are incorporated in the widely used ARPACK (Arnoldi Package) software [40]. This software may be employed to solve both symmetric and unsymmetric problems. A parallel version of the package P_ARPACK (Parallel ARPACK) [44] is available for use on distributed memory machines. Message passing is realized by MPI [70] or BLACS [8].

The Jacobi–Davidson method may also be extended in a natural way for the unsymmetric eigenproblem [4,25,63,68]. Variants are available for computing the dominant eigenpair, a dominant set of eigenpairs and a set of interior eigenpairs. The structure of the algorithms mirrors that of their symmetric counterparts. The main difference in that M_j is unsymmetric. Rather than compute an eigensolution of M_j a Schur decomposition, $M_j = S_j T S_j^{\mathrm{T}}$, is computed where S_j is unitary and, in the general case where M_j may have complex eigenvalues, T is quasitriangular (perhaps having 2×2 blocks representing complex eigenvalues on its diagonal).

As for the symmetric case the crucial operations are vector dot products, matrix–vector products, Gram–Schmidt orthogonalization, eigensolution or Schur decomposition of relatively small matrices — as have compared with the order, n, of the problem solved — and solutions of sets of linear equations when shift and invert techniques are employed. Successful parallel implementation depends on the efficiency with which these operations can be realized on the given architecture. Often in subspace iteration methods, restarted Krylov space-based methods and Jacobi–Davidson methods eigensolutions of only very small matrices is required and these are computed using serial algorithms. In all methods orthonormalization of vectors should be kept to a minimum.

5.10   PARALLEL CONSIDERATIONS

It will be clear from the discussions of the iterative methods that all have in common a reliance on certain operations. These are:

5.10.1   MATRIX–VECTOR PRODUCTS

In each cycle of subspace iteration m products of this kind are required where m is the number of vectors in the subspace. In single vector Lanczos methods and in the Jacobi–Davidson method only one matrix–vector product is required in each iteration. For block Lanczos methods b products are needed, where b is the block size [30].

5.10.2   VECTOR ORTHONORMALIZATION

This operation is usually implemented using classical Gram–Schmidt method with orthogonalization refinement or using the MGS method with iterative refinement [4, 18, 32]. In subspace iteration m vectors need to be orthonormalized in each iterative cycle. In single vector Lanczos-based methods and in the Jacobi–Davidson method one newly generated vector is orthogonalized with respect to the members of an already orthonormalized set of vectors. Since this is an expensive operation several strategies have been proposed which relax the requirement for full orthonormalization in each iteration. For block methods in which two block Krylov spaces are iteratively extended the biorthogonality of two sets of block vectors must be maintained.

The MGS algorithm may be efficiently implemented on a wide range of parallel machines. In Tables 5.1 and 5.2 results are presented for the orthonormalization of a few long, randomly generated, vectors on a distributed memory machine — the Cray T3D [48]. It will be observed that, when using a small number of processors, the efficiency of the parallel implementation is impressive even for shorter vectors. The efficiency declines considerably for shorter vectors as the number of processors increases. This is because the ratio of computation time to communication time decreases. However, the implementation displays good scalability characteristics. For example, when executed on 16 processors with vectors of length 10^4 (see Table 5.1) the efficiency of the implementation is less than 50%; when the length of the vectors is increased to 3×10^5 (see Table 5.2), however, the efficiency rises to nearly 95%. A lot of computation is needed to offset the cost of interprocessor communication. Parallel implementation of dot products on distributed memory machines requires global communication. As a result, the scalability of software, which relies heavily on these operations, is limited.

Gram–Schmidt-based algorithms may be implemented with great efficiency on a wide range of parallel architectures [31, 52, 83, 87].

5.10.3   EIGENSOLUTION OF SMALL DENSE MATRICES

All of the methods require the computation of accurate eigensolutions of matrices whose orders are much smaller than those of the target matrices whose partial eigensolutions are sought. In subspace iteration and Jacobi–Davidson the matrices are symmetric and full. In Lanczos methods

TABLE 5.1

Time (in sec) on the Cray T3D for the Modified Gram–Schmidt Algorithm (10 Vectors of Length 10,000)

	Sequential	Parallel Implementation				
		1 Proc	2 Proc	4 Proc	8 Proc	16 Proc
Total time	0.4066	0.4068	0.2239	0.1240	0.0762	0.0539
Speedup		1.0	1.82	3.28	5.34	7.55
Efficiency (%)		100	91	82	67	47

TABLE 5.2
Time (in sec) on the Cray T3D for the Modified Gram–Schmidt
Algorithm (10 Vectors of Length 300,000)

			Parallel Implementation			
	Sequential	**1 Proc**	**2 Proc**	**4 Proc**	**8 Proc**	**16 Proc**
Total time	12.1799	12.1893	6.2189	3.1280	1.5851	0.8110
Speedup		1.0	1.96	3.89	7.68	15.02
Efficiency (%)		100	98	97	96	94

the matrices are symmetric and tridiagonal. In methods such as the classical Lanczos and Jacobi–Davidson methods in which fixed length runs and restarting are not employed, the order of the eigenproblems to be solved may become significantly large. In subspace iteration and fixed length restarting methods, the order of the eigenproblems are very small in relation to the order of the target eigenproblem.

If the small dense eigenproblems are sufficiently large then parallelism may be exploited. For tridiagonal systems this usually involves a multisection approach [42] (based on Sturm sequence properties) for the parallel computation of eigenvalues and the independent parallel computation (by inverse iteration) of batches of eigenvectors on the individual processors of the machine. If the matrices are full then an initial phase to reduce them to tridiagonal form is required. In this phase, which is based on Householder reflections, significant advantage may also be gained through parallel implementation. The parallel divide-and-conquer algorithms of Dongarra and Sorensen [21] and Tisseur and Dongarra [80] may also be considered for the solution of these eigenproblems.

In Tables 5.3 and 5.4 results are presented relating to the computation of complete eigensystems (of the size likely to arise in unconstrained Lanczos-like methods) on a distributed memory machine — the Cray T3D — using Householder reduction, Sturm sequence properties, and inverse interation [49]. The matrices used in the experiments are dense. In all three phases of the computation — reduction to tridiagonal form, isolation of eigenvalues, and computation of eigenvectors — parallel execution may be employed to advantage. However, as the number of processors increases the efficiency of the parallel implementation falls off. This is due to the increasing communication requirement and the declining amount of arithmetic computation

TABLE 5.3
Time (in sec) to Compute the Complete Eigensystem of a Random
Matrix of Order 1024 on the Cray T3D

	Number of Processors						
	1	**2**	**4**	**8**	**16**	**32**	**64**
Reduction	61.04	31.32	17.09	9.69	5.85	4.61	3.71
Eigenvalues	55.09	28.43	15.42	8.48	4.88	2.66	1.61
Eigenvectors	52.96	26.77	13.95	7.25	3.95	2.32	1.59
Total time	169.09	86.53	46.46	25.41	14.69	9.58	6.91
Speedup	1.0	1.95	3.64	6.65	11.51	17.65	24.47
Efficiency (%)	100	98	91	83	72	55	38

TABLE 5.4

Time (in sec) to Compute the Complete Eigensystem of a Random Matrix of Order 2048 on the Cray T3D

	Number of Processors						
	1	**2**	**4**	**8**	**16**	**32**	**64**
Reduction	224.97	114.99	62.07	34.38	22.77	16.05	15.42
Eigenvalues	111.13	58.63	33.36	19.80	11.92	7.35	4.31
Eigenvectors	227.30	104.45	54.46	28.22	15.345	8.95	5.93
Total time	563.399	278.07	149.90	82.40	50.039	32.35	25.65
Speedup	1.0	2.02	3.76	6.84	11.26	17.42	21.96
Efficiency(%)	100	101	94	86	70	54	34

performed on each processor. On 32 processors, for example, the efficiency falls to 54% for a matrix of order 1000. For larger matrices where the computational load on each processor is higher the efficiency rises. Thus, for a matrix of order 2000, the efficiency on 32 processors is 70%.

When embedded in a Lanczos process only p (the order of partial eigensolution sought) eigenpairs of these small matrices need be computed. In addition since, in Lanczos methods, the T_j are tridiagonal the reduction phase can be omitted. In subspace iteration methods and Jacobi–Davidson methods, however, the projected eigenproblem is dense. Efficient eigensolvers (and Schur decomposition algorithms) for (relatively) small dense matrices can also be constructed for other parallel architectures. In restarted methods the order of the projected eigenproblem is so small that its solution is computed on a single processor using a serial implementation.

5.10.4 SOLUTION OF LINEAR EQUATIONS

When sets of internal eigenpairs are to be computed there is a requirement to solve sets of linear equations of the form $(A - \sigma I)x = b$. If A is very sparse, then so also is $(A - \sigma I)$. The straightforward approach is to compute an LU factorization of $(A - \sigma I)$. However, in general this will yield factors L and U, which are not sparse, so that, when A is very large storage demands may be prohibitive. Methods are available which limit the degree of fill-in (the replacement of zero entries by non zero values) and recently, very efficient algorithms have been developed for sparse factorization [1, 41]. However, it is more usual to solve the equations using iterative methods [62] such as the CG algorithm. CG is itself a Krylov subspace orthogonal projection method so that its implementation shares the same operations as the iterative eigensolution methods being discussed. Thus heavy use is made of matrix–vector products, vector updates, and dot products.

To improve the convergence of these iterative methods they are usually applied in preconditioned form. Essentially, preconditioned methods require the construction of an approximate inverse, K, of the coefficient matrix $(A - \sigma I)$ and the solution of sets of linear equations involving K. The parallel generation of effective preconditioners is a difficult problem and is the subject of intense current research activity.

5.11 ITERATIVE METHODS AND PARALLEL MACHINES

The influence of the architecture of the machine on the performances of iterative algorithms of the sort discussed here is illustrated using experimental results obtained from executing several variants of the Lanczos algorithm on several parallel computers of different kinds. (All of the results presented stem from work carried out by the author and his collaborators.) Many of these machines are now obsolete but the results remain valuable in that they reflect the sorts of performances which might be expected of the algorithms on current machines having similar features. In addition, although results are presented for only Lanczos-based algorithms the reader is reminded that all of the other methods discussed share the same basic structure and employ the same operations.

5.11.1 CLASSICAL LANCZOS METHOD

The computational loads associated with the various basic operations in the classical Lanczos method and a slightly modified version (see below) are illustrated in Tables 12.5–12.8. These are: matrix–vector products, reorthogonalization, eigensystem computation, and convergence monitoring. The information in the tables relates to the computation of partial eigensolutions of [1,2,1] matrices of various sizes. A [1,2,1] matrix is tridiagonal with each of its diagonal elements being 2 and each of the elements on its main super- and sub-diagonal being 1. The target matrices in these experiments are already tridiagonal: however, the eigenvalues of each may be calculated using a trigonometric expression (so that they may readily be computed) and the purpose of the experiments is to estimate the division of work among the phases of the algorithms.

The results in Tables 5.5 and 5.7 detail the performance of the classical Lanczos algorithm; those in Tables 5.6 and 5.8 detail the performance of a variant in which parallelism is employed in monitoring convergence [77] without, however, computing accurately the eigensolutions of the T_j.

The results in Tables 5.5 and 5.7 were obtained using a Convex C3800 — series machine, a shared memory vector computer. The experiments were conducted on a machine with two C3840 processors. The results in Tables 5.6 and 5.8 were obtained using an Intel iPSC/860 distributed memory machine with 16 processors connected in a hypercube architecture. On this machine a significant amount of interprocessor communication is required in computing matrix–vector products and in vector orthonormalization. On the Convex an eigensolution of order $p = 3$ was sought; on the Intel an eigensolution of order $p = 2$ was sought. For the Convex, results corresponding to a series of [1,2,1] matrices are presented; for the Intel results for a [1,2,1] matrix of order 10,000 are given. To highlight the crucial importance of the communication load in distributed memory implementations the [1,2,1] matrix in the Intel experiments is regarded both as a dense matrix and a sparse (tridiagonal) matrix.

It will be noted that, in all cases, the combined arithmetic load associated with matrix–vector products and orthogonalization accounts for a very high percentage of time devoted to computation. The time devoted to full orthogonalization is, for the experiments carried out on the

TABLE 5.5
Convex C3840 (Classical Lanczos Method — Two Processors)

n	512	1,024	2,048	4,096	8,192	16,384	32,768
j	249	421	450	456	462	457	461
Matrix–Vector product(%)	23.80	19.26	17.58	15.91	15.28	15.63	15.76
cgce Monitor(%)	22.33	16.33	12.38	8.73	5.34	3.11	1.64
Orthogonalization (%)	52.37	63.20	67.701	73.69	77.42	79.06	80.33
Eigensystem Composition (%)	0.79	0.52	0.65	0.85	1.02	1.17	1.20
Total time (sec)	9.27	26.07	37.09	54.26	90.40	157.50	297.78
Speedup ratio	1.03	1.16	1.4	1.69	1.71	1.81	1.89

TABLE 5.6
Intel iPSC/860(Classical Lanczos Method — 16 Processors)

j	262 Dense Matrix		262 Tridiagonal Matrix		Speedup	
$n = 10,000$	Serial	Parallel	Serial	Parallel	Dense	Tridiagonal
Total Matrix–Vector product	85.80%	64.60%	0.68%	2.87%	15.79%	1.12
Matrix–Vector communication	—	0.84%	—	2.64%	—	—
cgce Monitor	0.12%	1.41%	0.83%	3.90%	1.0	1.0
Total orthogonalization	13.95%	32.35%	97.73%	88.91%	5.13	5.17
Orthogonalized communication	—	20.55%	—	55.92%	—	—
Total communication time (sec)	—	21.38%	—	58.56%	—	—
Total time (sec)	2304.00	193.80	329.04	69.967	11.89	4.70

TABLE 5.7
Convex C3840 (Classical Lanczos Method, Parallel cgce Monitoring — Two Processors)

n	512	1,024	2,048	4,096	8,192	16,384	32,768
j	249	421	450	456	462	457	461
Matrix–Vector product(%)	24.58	14.10	13.37	13.24	13.95	14.11	14.84
cgce Monitor (%)	0.82	0.29	0.20	0.13	0.07	0.04	0.02
Orthogonalization (%)	72.63	84.57	85.18	85.12	84.27	84.00	83.28
Eigensystem composition (%)	1.14	0.54	0.70	0.90	1.02	1.12	1.19
Total time (sec)	5.3	21.32	31.67	48.59	86.49	153.53	287.55
Speedup ratio	1.22	1.09	1.41	1.63	1.70	1.77	1.87

TABLE 5.8
Intel iPSC/860(Classical Lanczos Method with Parallel Monitor — 16 Processors)

j	234 Dense Matrix		234 Tridiagonal Matrix		Speedup	
$n = 10,000$	Serial	Parallel	Serial	Parallel	Dense	Tridiagonal
Total Matrix–Vector product (%)	87.00%	66.89%	0.77%	2.39%	15.70%	1.51%
Matrix–Vector communication	—	1.23%	—	2.12%	—	—
cgce Monitor	0.18%	1.24%	1.36%	3.60%	1.70	1.77
Total orthogonalization	12.81%	30.05%	97.23%	89.57%	5.15	5.12
Orthogonalized communication	—	17.34%	—	51.62%	—	—
Total communication time (sec)	—	19.52%	—	56.53%	—	—
Total time (sec)	2037.87	168.79	266.64	56.57	12.07	4.71

Convex, exceedingly high; this is also true in the Intel experiments when the sparsity of the target matrix is exploited thereby reducing very significantly the cost of a matrix–vector product. The cost of interprocessor communication in the Intel is dominated by the requirements for the orthonormalization process. The results show that it is imperative to orthogonalize as little as possible through the employment of selective orthogonalization and other techniques. It will be observed that, for the classical Lanczos method the length of the factorizations (denoted by j) is large enough to make the use of a parallel convergence monitoring routine worthwhile. For the shared memory machine the parallel efficiencies achieved on two processors exceeds 85%. For the distributed memory machine the parallel efficiencies are low when the sparsity of the target matrix is taken into account. This is due to the poor parallel performance of the matrix–vector products in which communication time swamps the very low cost of this operation when a tridiagonal matrix is involved. Note that, if the target matrix is taken to be dense, the parallel efficiency of this operation on the 16 processor machine is nearly optimal. On the Intel some slight advantage is also to be gained by using a parallel convergence monitoring procedure. On both types of machine the computational load associated with computing eigensolutions of the T_j is very small.

5.11.2 UNBOUNDED EXPLICITLY RESTARTED AND IMPLICITLY RESTARTED LANCZOS METHODS

In Tables 5.9 and 5.10 the performances of two unconstrained explicitly restarted Lanczos methods (Res1 and Res2) based on the approaches sketched above [78] are compared with that of an implicit, fixed length run restarted method taken from ARPACK (the routine SSAUPD) [38, 40]. In the implicit method a restart is made after min $(16, 2p)$ Lanczos steps where p is the order of the eigensolution sought. Two different architectures have been used in the experiments — a distributed memory SIMD machine (the Connection Machine CM-200 — Table 5.9) and a shared memory MIMD machine (the Cray J-90 — Table 5.10). The Connection Machine used provides a massively parallel computing environment having 8K processors. In the explicitly restarted methods all n-vectors are stored as distributed arrays to be processed in parallel. Fine grain parallelism is exploited in the massive n-vector operations (of type BLAS-1 and BLAS-2 [9]) associated with

TABLE 5.9

Time (in sec) on the CM-200 for Matrices BCSSTK17 and BCSSTK18 from the BCSSTRUC2 Collection (Figures in Brackets Show the Percentage of Execution Time Devoted to the Evaluation of Matrix–Vector Products Aq_j)

	BCSSTK17			BCSSTK18		
	$n = 10,974$ ($nz = 219,812, d = 20.3$) $\lambda_1 = 1.29606 \times 10^{10}$, $\lambda_{32} = 4.53540 \times 10^9$			$n = 11,948$ ($nz = 80,519, d = 6.74$) $\lambda_1 = 4.29520 \times 10^{10}$, $\lambda_{32} = 1.93316 \times 10^{10}$		
p	SSAUPD	Res1	Res2	SSAUPD	Res1	Res2
1	9.9 (42%)	4.4 (91%)	4.4 (91%)	8.3 (22%)	1.6 (79%)	1.6 (78%)
2	16.4 (48%)	6.2 (90%)	6.2 (90%)	8.7 (23%)	2.3 (80%)	2.3 (80%)
4	20.5 (42%)	14.6 (90%)	12.3 (91%)	21.3 (18%)	6.3 (78%)	6.1 (76%)
8	34.4 (29%)	31.3 (89%)	28.1 (90%)	36.9 (12%)	15.0 (75%)	13.5 (76%)
16	84.1 (15%)	68.4 (88%)	52.6 (83%)	245.5 (4%)	51.3 (67%)	44.8 (68%)
32	373.1 (6%)	172.8 (85%)	182.4 (85%)	737.6 (2%)	122.1 (64%)	108.4 (65%)

matrix–vector products, orthogonalization and the computation of Ritz vectors. Operations involving the small matrices T_j are carried out sequentially on the "front-end" machine. Thus, for example, appropriate routines from the EISPACK library [2] and BLAS library routines [9] may be employed in the implementation of these.

A consequence of the restarting mechanism used in these experiments is that the Lanczos factorizations are unlikely to become sufficiently large for the use of parallelism in the eigendecomposition of T_j to be profitable. All the methods have been implemented using a *reverse communication* strategy [40, 43], in which the implementer is responsible for providing an efficient realization of the massive matrix–vector product operation. This approach allows the employment of highly optimized machine specific sub operations in the implementation of this computationally expensive operation.

On the Cray J-90 the eigensolutions of the tridiagonal matrices T_j are computed (if they are sufficiently large) using parallel multisection [26] for the *isolation* of the eigenvalues followed by the independent extraction of roughly equal numbers of eigenvalues on the individual processors. The associated eigenvectors are then computed (again independently), by inverse iteration using the computed eigenvalues as shifts. This phase may be carried out using the appropriate EISPACK routines.

The test matrices have been selected from the Harwell–Boeing Sparse Matrix Collections [23]. The results show how the choice of which Lanczos-based method to use depends on the nature of the available parallel machine. Clearly in Table 5.9, on the massively parallel SIMD machine on which matrix–vector products may be computed cheaply, the explicitly restarted methods are superior to the implicit method. The figures in brackets indicate the number of matrix–vector products required. The explicit methods employ many more of these operations than does the implicit method. Nevertheless they significantly outperform the implicit method.

Conversely, on the shared memory MIMD machine (see Table 5.10) SSAUPD significantly outperforms the explicitly restarted methods. The reason is that on the Cray J-90 the cost of a matrix–vector product is much greater than on the CM-200 with the result that the execution time is dominated by the computational load associated with the large numbers of these operations required by the explicitly restarted methods.

TABLE 5.10

Time (sec) on the Cray J-90 with Eight Processors for Matrices BCSSTK17 and BCSSTK18 from the BCSSTRUC2 Collection (Figures in Brackets Show the Number of Matrix–Vector Products Aq_j Required)

	BCSSTK17			BCSSTK18		
	$n = 10,974$ ($nz = 219,812, d = 20.3$) $\lambda_1 = 1.29606 \times 10^{10}$, $\lambda_{32} = 4.53540 \times 10^9$			$n = 11,948$ ($nz = 80,519, d = 6.74$) $\lambda_1 = 4.29520 \times 10^{10}$, $\lambda_{32} = 1.93316 \times 10^{10}$		
p	SSAUPD	Res1	Res2	SSAUPD	Res1	Res2
1	1.30 (40)	1.37 (28)	1.37 (28)	0.92 (32)	1.10 (26)	1.14 (26)
2	1.58 (57)	1.67 (34)	1.67 (34)	1.54 (57)	1.44 (34)	1.43 (34)
4	2.33 (69)	3.83 (78)	2.82 (60)	2.47 (87)	4.6 (102)	3.56 (81)
8	2.83 (81)	8.01 (152)	5.11 (99)	2.76 (91)	10.33 (233)	7.5 (166)
16	3.53 (93)	19.46 (360)	13.09 (248)	7.76 (211)	32.17 (617)	28.15 (540)
32	3.21 (133)	46.13 (784)	39.04 (663)	7.57 (203)	74.29 (1380)	71.67 (1290)

In addition, it is not always possible to decide in advance which explicit restarting method will be most efficient. For example, for matrix BCSSTK17, Res2 performs at least as well on the SIMD machine as Res1 until a partial eigensolution of order 32 is sought. On the other hand on the MIMD machine Res2 always outperforms Res1. Note also that the numbers of matrix–vector products required varies across the machines, but not greatly.

5.11.3   FIXED LENGTH RUN EXPLICITLY RESTARTED AND IMPLICITLY RESTARTED LANCZOS METHODS

In Tables 5.11 and 5.12 the performances of two fixed length run, explicitly restarted, methods (viz., LRes1 and LRes2 — fixed length variants of Res1 and Res2, respectively) are compared with the same implicit, fixed length run, restarted method used in the experiments discussed earlier [79]. The computers used are (again) the CM-200 and a Cray T3D distributed memory machine with 32 processors.

Comparing the results in Tables 5.9 and 5.11 it is evident that, in general, the performances on the SIMD machine of LRes1 and LRes2 are less efficient than those of Res1 and Res2, respectively. This reflects the loss of information about the sought eigenspace arising from the curtailment of the Lanczos factorization and restarting with poorer information than is available in the unconstrained explicit restarting methods. Note that the number of matrix–vector products increases when using LRes1 and LRes2 and this can happen to such an extent that SSAUPD outperforms the explicit methods — see the results for BCSSTK17.

In Table 5.12 the performances on the CM-200 of LRes1 and LRes2 are compared with that of SSAUPD for two highly sparse matrices NOS7 and PLAT1919 from the LANPRO and the PLATZ collections. Their densities are 3.66 and 2.51, respectively. PLAT1919 is generated as the square of a skew-symmetric matrix. Thus its eigenvalues occur in pairs with the exception of a singleton zero eigenvalue. As the order of the eigensolution p increases the proportion of execution time devoted to matrix-vector products declines for all the methods. However, for the explicitly restarted methods the decrease is from a high threshold — for PLAT1919, 66% for LRes1 and 96% for LRes2 when a single eigenpair is sought — whereas, for SSAUPD, it is from a low threshold — only 18% for a single eigenpair. When $p = 64$ the number of such products is negligible for SSAUPD but

---

TABLE 5.11

Time (in sec) on the CM-200 for Matrices BCSSTK17 and BCSSTK18 from the BCSSTRUC2 Collection (Figures in Brackets Show the Number of Matrix–Vector Products Aq_j Required)

	BCSSTK17			BCSSTK18		
	$n = 10,974$ ($nz = 219,812, d = 20.3$) $\lambda_1 = 1.29606 \times 10^{10}$, $\lambda_{32} = 4.53540 \times 10^9$			$n = 11,948$ ($nz = 80,519, d = 6.74$) $\lambda_1 = 4.29520 \times 10^{10}$, $\lambda_{32} = 1.93316 \times 10^{10}$		
p	SSAUPD ($k = \max(16,2p)$)	LRes1 ($k = 32$)	LRes2 ($k = 32$)	SSAUPD ($k = \max(16,2p)$)	LRes1 ($k = 32$)	LRes2 ($k = 32$)
1	9.9 (24)	3.2 (21)	3.6 (25)	8.3 (24)	2.3 (32)	2.9 (47)
2	16.4 (43)	7.1 (47)	8.0 (54)	8.7 (30)	4.6 (64)	6.1 (97)
4	20.5 (48)	16.0 (103)	17.6 (119)	21.3 (55)	11.5 (160)	12.3 (197)
8	34.4 (57)	36.6 (237)	42.1 (282)	36.9 (63)	30.6 (416)	29.4 (431)
16	84.1 (71)	83.1 (529)	91.1 (615)	245.5 (154)	138.8 (1792)	116.5 (1783)
32	373.1 (114)	789.0 (4752)	511.0 (3316)	737.6 (175)	393.9 (4800)	374.3 (5517)

TABLE 5.12

Time (in sec) for Matrices NOS7 and PLAT1919 from the LANPRO and PLATZ Collections Using the Cray T3D — 32 Processors (Figures in Brackets Show the Percentage of Total Time for the Computation of Matrix-Vector Products Aq_j)

	NOS7			PLAT1919		
	$n = 729$ ($nz = 2673, d = 3.66$)			$n = 1919$ ($nz = 4831, d = 2.51$)		
	$\lambda_1 = 9.86403 \times 10^6$, $\lambda_{128} = 7.11115 \times 10^2$			$\lambda_1 = 2.92163$, $\lambda_{128} = 1.00529$		
p	SSAUPD ($k =$ max(16, 2p))	LRes1 ($k = 32$)	LRes2 ($k = 32$)	SSAUPD ($k =$ max(16, 2p))	LRes1 ($k = 32$)	LRes2 ($k = 32$)
1	0.7 (21%)	0.1 (62%)	0.1 (84%)	1.4 (18%)	0.3 (66%)	0.2 (96%)
2	1.4 (18%)	0.3 (50%)	0.2 (83%)	3.3 (22%)	0.9 (62%)	0.6 (90%)
4	4.2 (11%)	0.7 (47%)	0.5 (80%)	3.9 (17%)	2.0 (60%)	1.4 (89%)
8	5.1 (8%)	1.5 (44%)	1.2 (74%)	8.8 (10%)	4.9 (56%)	3.4 (86%)
16	8.5 (4%)	3.2 (39%)	3.0 (65%)	28.0 (4%)	13.0 (51%)	9.1 (80%)
32	18.9 (3%)	7.8 (31%)	8.9 (49%)	123.3 (1%)	38.3 (44%)	24.6 (69%)
64	192.8 (1%)	34.3 (19%)	27.1 (35%)	568.7 (0%)	125.2 (33%)	78.2 (55%)
128	1816.4 (0%)	302.75 (10%)	179.2 (21%)		438.4 (22%)	262.7 (40%)

remains significantly large for the explicitly restarted methods. In NOS7 SSAUPD failed to find λ_{128} whereas both LRes1 and LRes2 succeeded. This is one of the advantages of computing one eigenpair at a time.

The performances of the three methods on the SIMD machine depends, to some extent, on the density of the target matrix. In BCSSTK17, where the density, d, is highest, SSAUPD performs best, outstripping the other methods when p reaches 32.

In Table 5.13 the results of experiments on the CM-200 for the [1,2,1] matrix of order 10,000 are presented. For this matrix $d = 0.99$ and the same pattern as before is evident. The numbers of matrix–vector products for the explicit methods far exceeds that for the implicit method — when

TABLE 5.13

Time (in sec) for a [1,2,1] Matrix of Order 10,000 Using the Cray T3D — 32 Processors (Figures in Brackets Show the Number of Matrix-Vector Products Aq_j Required)

		$n = 10,000, nz = 9,999, d = 0.99$	
		$\lambda_1 = 3.9999959$, $\lambda_{16} = 3.9999380$	
p	SSAUPD ($k = 32$)	LRes1 ($k = 32$)	LRes2 ($k = 32$)
1	32.8 (64)	0.9 (66)	0.5 (70)
2	427.9 (1,310)	15.3 (1,021)	6.9 (903)
4	474.8 (1,242)	54.0 (3,453)	24.2 (3,081)
8	1114.0 (2,094)	150.4 (9,073)	68.7 (8,141)
16	2735.0 (2,763)	381.3 (21,129)	178.4 (18,840)

$p = 16$, by factors of more than 6. Nevertheless, the execution times for both LRes1 and LRes2 are much lower than that for SSAUPD — by a factor of more than 15 when $p = 16$ in the case of LRes2.

In Table 5.14 the performances of the three methods are compared on the distributed memory, Cray T3D computer. On this machine the explicitly restarted methods are competitive only when p is very small. The density of the target matrix does not appear to have a significant impact on the performances. The decline in the competitiveness of LRes1 and LRes2 is probably due to the increasing number of vector dot products required in the orthogonalization phase.

The results show that the architecture of the machine on which an iterative algorithm for computing a partial eigensolution may significantly affect its performance. The decision as to which algorithms to employ in a particular parallel environment depends critically on the efficiency with which matrix–vector products and vector dot products can be implemented in the environment and, to a lesser extent, on how efficiently eigensolutions of (perhaps, moderately large) dense matrices can be computed.

5.12 FURTHER READING

Solving sets of linear equations and computing eigensystems are essential requirements in methods for the numerical solution of many diverse problems in science and engineering. A wide range of literature is devoted to these topics — much of it given over to the exposition of efficient computational techniques.

Among the general texts which deal with algorithmic aspects of numerical linear algebra the following are recommended as background reading. *Numerical Linear Algebra* by Trefethen and Bau [81] is an elegantly written and gentle introduction to the theory on which the methods discussed here are based. *Matrix Computations* by Golub and van Loan [28] is an encyclopedic text offering deep insights into algorithmic techniques in numerical linear algebra and addressing issues of parallel implementation [73]. Stewart's classic *Introduction to Matrix Computations* offers a rigorous exposition of the foundations on which numerical methods are built. Watkins' [84] *Fundamentals of Matrix Computations* is also a fine introductory text. *Matrix Computations* by Jennings

TABLE 5.14

Time (in sec) for Matrices BCSSTK17 and BCSSTK18 from the BCSSTRUC2 Collection Using the Cray T3D — 32 Processors

	BCSSTK17			BCSSTK18		
	$n = 10,974$ ($nz = 219,812, d = 20.3$) $\lambda_1 = 1.29606 \times 10^{10}$ $\lambda_{32} = 4.53540 \times 10^9$			$n = 11,948$ ($nz = 80,519, d = 6.74$) $\lambda_1 = 4.29520 \times 10^{10}$ $\lambda_{32} = 1.93316 \times 10^{10}$		
p	SSAUPD ($k = \max(16, 2p)$)	LRes1 ($k = 32$)	LRes2 ($k = 32$)	SSAUPD ($k = \max(16, 2p)$)	LRes1 ($k = 32$)	LRes2 ($k = 32$)
1	53.06	1.68	1.13	18.54	1.24	0.55
2	50.77	3.63	2.42	18.30	3.31	1.32
4	55.87	8.27	5.47	19.51	8.40	3.98
8	53.04	21.49	13.86	19.00	25.00	12.93
16	49.17	54.51	34.58	26.62	217.57	113.29
32	50.69	1332.16	591.85	23.26	960.28	527.69

and McKeown [34] gives an easily assimilable treatment of many important methods written from a scientific practitioner's viewpoint.

Three important books dealing exclusively with the numerical solution of eigenproblems are Wilkinson's [85] monumental *The Algebraic Eigenvalue Problem*, Stewart's [74] recent *Matrix Algorithms, Vol. II: Eigensystems* (in which in-depth treatments of Lanczos, Arnoldi, and Jacobi–Davidson methods are presented) and Parlett's [57] *The Symmetric Eigenvalue Problem*, which is more restricted in scope but is an invaluable reference work for those interested in the special case of symmetric matrices.

For full and leisurely coverage of the use of iterative methods in all areas of numerical linear algebra Demmel's [20] *Applied Numerical Algebra* is to be recommended. Saad's [61] *Numerical Methods for Large Eigenvalue Problems* gives a well-motivated in-depth treatment of subspace iteration methods and Krylov space-based methods for both symmetric and nonsymmetric matrices. The same author's *Iterative Methods for Sparse Linear Systems* [62] provides excellent coverage of CG approaches to the iterative solution of sets of equations, which are integral components of many iterative eigensolvers: representation of sparse matrices and parallel implementation issues are also discussed. Cullum and Willoughby's [17] *Lanczos Algorithms for Large Symmetric Eigenvalue Computations* focuses specifically on Lanczos methods. H. van der Vorst [82] provides extensive coverage of methods for the eigensolution of large problems.

Templates for the Solution of Algebraic Eigenvalue Problems edited by Bai et al. [4] is a peerless resource for applications software developers who have a requirement to solve large eigenproblems. The book comprises contributions from leading scientists active at the forefront of research in numerical linear algebra. It is an indispensible reference work for those wishing to use the most efficient methods currently available for computing partial eigensolutions. It gives deep and extensive treatment of all of the methods discussed above and the book has an associated website giving details of how software for a wide range of methods may be accessed. *Numerical Linear Algebra for High-Performance Computers* by Dongarra et al. [22] includes a chapter on methods for the computation of eigensystems on parallel machines.

Among the most widely used software currently available for the partial solutions of eigenproblems is ARPACK [38,40]. This package supports implicitly restarted methods for both symmetric (Lanczos-based) and unsymmetric (Arnoldi-based) matrices. The routines in the package may be flexibly implemented on a variety of architectures using the technique of reverse communication which offers users the opportunity to implement key operations efficiently.

Implementations of iterative methods may be built from software components available from libraries such as ScaLAPACK [7]. ScaLAPACK is a development of the widely used library of numerical linear algebra routines, LAPACK [2], which supports parallel execution. These packages make heavy use of the BLAS, a three-level hierarchically structured collection of basic linear algebraic operations. At its lowest level (Level 1) the BLAS provides vector–vector operations such as dot products; at its highest level (Level 3) it provides matrix–matrix operations such as matrix products.

ScaLAPACK is designed for use in a distributed memory (message passing) computing environment which includes MIMD machines, networks of workstations and clusters of SMPs. It is primarily used wherever Level 3 BLAS operations are widely employed to ensure that the computation — communication ratio is high: this approach favors the use of block algorithms. Interprocessor communication is handled by BLACS [8], which rests on the MPI [70] and PVM [27] computational models. Building blocks for efficient implementations are also available from repositories such as the NAG Parallel Library [50] and the Harwell Subroutine Library [33].

Several subspace iteration packages for unsymmetric eigenproblems are available [39]. These may be used (with some degradation of performance) in the symmetric case but need to be adapted for parallel execution. They include Bai and Stewart's package SRRIT [5] and Stewart and Jennings' package LOPSI [75]. A routine (EA12) from the Harwell Subroutine Library by Duff and Scott [33] is also available and the influential recipe book *Handbook for Automatic Computation, Vol. II, Linear Algebra* [86] contains a wealth of material on which parallel implementations may be based.

REFERENCES

[1] P.R. Amestoy, I.S. Duff, J.Y. L'Excellent, J. Koster, and M. Tuma, Mumps 4.1.6 http://www.enseeiht.fr/lima/apo/MUMPS/, 2000.

[2] E. Anderson, Z. Bai, C. Bischof, J. Demmel, J. Dongarra, J. Croz, A. Greenbaum, S. Hammarling, A. McKenney, S. Ostrouchov, and D. Sorensen, *LAPACK User's Guide*, SIAM, Philadelphia, PA, 1992.

[3] W.E. Arnoldi, The principle of minimized iterations in the solution of the matrix eigenvalue problem, *Quart. Appl. Math.*, 9(1951) 17–29.

[4] Z. Bai, J. Demmel, J. Dongarra, A. Ruhe, and H. van der Vorst, *Templates for the Solutions of Algebraic Eigenvalue Problems – A Practical Guide*, SIAM, Philadelphia, PA, 2000.

[5] Z. Bai and G.W. Stewart, Algorithm 776: SRRIT – A FORTRAN subroutine to calculate the dominant invariant subspaces of a nonsymmetric matrix, *ACM Trans. Math. Software*, 23(1998) 494–513.

[6] M.W. Berry, D. Mezher, B. Philippe, and A. Sameh, Parallel Computation of the Singular Value Decomposition, Rapport de Recherche No. 4694, INRIA, 2003.

[7] L.S. Blackford, J. Choi, and A. Cleary, E.D' Azevedo, J. Demmel, I. Dhillon, J. Dongarra, S. Hammarling, G. Heary, A. Petitet, K. Stanley, D. Walker, and R.C. Whaley, *ScaLAPACK Users' Guide*, http://www.netlib.org/scalapack/slug/, 1997.

[8] BLACS, http://www.netlib.org/blacs/.

[9] BLAS, http://www.netlib.org/blas/.

[10] L. Borges and S. Oliviera, A parallel Davidson-type algorithm for several eigenvalues, *J. Comput. Phys.*, 144(2)(1998) 727–748.

[11] T. Braconnier, The Arnoldi–Tchebycheff Algorithm for Solving Large Nonsymmetric Eigenproblems, Technical Report TR/PA/94/08, CERFACS, Toulouse, France, 1994.

[12] F. Chatelin and D. Ho, Arnoldi–Tchebychev procedure for large scale nonsymmetric matrices, *Math. Model. Num. Anal.*, 24 (1990) 53–65.

[13] M. Clint and A. Jennings, The evaluation of eigenvalues and eigenvectors of real symmetric matrices by simultaneous iteration, *Comp. J.*, 13 (1970) 76–80.

[14] M. Crouzeix, B. Philippe, and M. Sadkane, The Davidson method, *SIAM J. Sci. Comput.*, 15 (1994) 62–76.

[15] J. Cullum and W.E. Donath, A block Lanczos algorithm for computing the q algebraically largest eigenvalues and a corresponding eigenspace for large, sparse symmetric matrices, in *Proc. 1974 IEEE Conf. Decision Control*, New York, 1974, pp. 505–509.

[16] J. Cullum, The simultaneous computation of a few of the algebraically largest and smallest eigenvalues and a corresponding eigenspace of a large, symmetric sparse matrix, *BIT*, 18 (1978) 265–275.

[17] J. Cullum and R.A. Willoughby, *Lanczos Algorithms for Large Symmetric Eigenvalue Computations, Theory*, Vol. 1, Birkhäuser, Boston, 1985.

[18] J. Daniel, W.B. Gragg, L. Kaufman and G.W. Stewart, Reorthogonalization and stable algorithms for updating the Gram–Schmidt QR factorization, *Math. Comput*, 30 (1976) 772–795.

[19] E.R. Davidson, The iterative calculation of a few of the lowest eigenvalues and corresponding eigenvectors of large real symmetric matrices, *J. Comput. Phys.*, 17 (1975) 87–94.

[20] J. Demmel, *Applied Numerical Linear Algebra*, SIAM, Philadelphia, PA, 1997.

[21] J. Dongarra and D. Sorensen, A fast algorithm for the symmetric eigenvalue problem, *SIAM J. Sci. Stat. Comput.*, 8(2)(1987) s139–s154.

[22] J.J. Dongarra, I.S. Duff, D.C. Sorensen, and H. van der Vorst, *Numerical Linear Algebra for High-Performance Computers*, SIAM, Philadelphia, PA, 1998.

[23] I.S. Duff, R.G. Grimes, and J.G. Lewis, *User's Guide for the Harwell–Boeing Sparse Matrix Collection* (release I), available online ftp orion.cerfacs.fr, 1992.

[24] I.S. Duff and J.A. Scott, Computing selected eigenvalues of large sparse unsymmetric matrices using subspace iteration, *ACM Trans. Math. Software*, 19(1993) 137–159.

[25] D.R. Fokkema, G.L.G. Sleijpen, and H.A. van der Vorst, Jacobi-Davidson style QR and QZ algorithms for the partial reduction of matrix pencils, *SIAM J. Sci. Comput.*, 20(1998) 94–125.

[26] T.L. Freeman and C. Phillips, *Parallel Numerical Algorithms*, Prentice Hall, New York, 1992.

[27] A. Geist, A. Beguelin, J. Dongarra, W. Jiang, R. Manchek, and V. Sunderam. *PVM 3 Users Guide and Reference Manual*. Oak Ridge National Laboratory, Oak Ridge, TN, 1994.

[28] G. Golub and C.F. van Loan, *Matrix Computations*, 3rd ed., Johns Hopkins University Press, Baltimore, 1996.

[29] G. Golub and R. Underwood, The block Lanczos method for computing eigenvalues, in *Mathematical Software III* J. Rice, Ed., Academic Press, New York, 1977, pp. 364–377.

[30] R.G. Grimes, J.G. Lewis, and H.D. Simon, A shifted block Lanczos algorithm for solving sparse symmetric generalized eigenproblems, *SIAM J. Matrix Anal. Appl.*, 15(1994) 228–272.

[31] B. Hendrickson, Parallel QR factorization using the torus–wrap mapping, *Parallel Comput*, 19(1993) 1259–1271.

[32] W. Hoffmann, Iterative algorithms for Gram–Schmidt orthogonalization, *Computing*, 41 (1989) 335–348.

[33] HSL (Harwell Subroutine Library), http://www.cse.clrc.ac.uk/nag/hsl/.

[34] A. Jennings and J. J. McKeown, *Matrix Computation*, 2nd ed., John Wiley & Sons, New York, 1992.

[35] W. Karush, An iterative method for finding characteristic vectors of a symmetric matrix, Pac. J. of Math. 1 (1951) 233–248.

[36] C. Lanczos, An iterative method for the solution of the eigenvalue problem of linear differential and integral operations, *J. Res. Nat. Bur. Stand.*, 45 (1950) 255–282.

[37] R.B. Lehoucq and D.C. Sorensen, Deflation techniques for an implicitly restarted Arnoldi iteration, *SIAM J. Matrix Anal. Appl.*, 17 (1996) 789–821.

[38] R.B. Lehoucq, D.C. Sorensen, and P.A. Vu, *SSAUPD: Fortran Subroutines for Solving Large Scale Eigenvalue Problems*, Release 2.1, available from netlib@ornl.gov in the scalapack directory, 1994.

[39] R.B. Lehoucq and J.A. Scott, An Evaluation of Software for Computing Eigenvalues of Sparse Nonsymmetric Matrices, Technical Report MCS - P 547-1195, Argonne National Laboratory, Argonne, IL, 1995.

[40] R.B. Lehoucq, D.C. Sorensen, and C. Young, *ARPACK Users' Guide: Solution of Large-scale Eigenvalue Problems with Implicitly Restarted Arnoldi Methods*, SIAM, Philadelphia, PA, 1998.

[41] X.S. Li, *Superlu version 2*, http://www.nersc.gov/xiaoye/SuperLU/, 1999.

[42] S. Lo, B. Philippe, and A.H. Sameh, A multiprocessor algorithm for the symmetric tridiagonal eigenvalue problem, *SIAM J. Sci. Stat. Comput.*, 8(2)(1987) s155–s165.

[43] O.A. Marques, *BLZPACK: Description and User's Guide*, CERFACS Report TR/PA/95/30, Toulouse, 1995.

[44] K.J. Maschhoff and D.C. Sorensen, P_ARPACK: an efficient portable large scale eigenvalue package for distributed parallel architectures, in *Applied Parallel Computing, Industrial Computation and Optimization*, J. Wasniewski, J. Dongarra, K. Madsen, and D. Olesen, Eds., LNCS 1184, Springer-Verlag, Heidelberg, 1996, pp. 651–660.

[45] R.B. Morgan, Generalizations of Davidson's method for computing eigenvalues of large nonsymmetric matrices, *J. Comput. Phys.*, 101 (1992) 287–291.

[46] R.B. Morgan and D.S. Scott, Generalizations of Davidson's method for computing eigenvalues of sparse symmetric matrices, *SIAM J. Sci. Stat. Comput.*, 7 (1986) 817–825.

[47] K. Murphy, M. Clint, M. Szularz, and J. Weston, The parallel computation of partial eigensolutions using a modified Lanczos method, *Parallel Algorithms Appl.*, 11 (1997) 299–323.

[48] K. Murphy, M. Clint, and R.H. Perrott, Re-engineering statistical software for efficient parallel execution. *Comput. Stat. Data Anal.*, 31(1999) 441–456.

[49] K. Murphy, M. Clint, and R.H. Perrott, Solving dense symmetric eigenproblems on the Cray T3D, in *High Performance Computing*, R.J. Allan, M.F. Guest, A.D. Simpson, D.S. Henty, and D.A. Nicole, Eds., Kluwer Academic Publishers, New York, 1999, pp. 69–78.

[50] NAG Parallel Library, http://www.nag.co.uk/numeric/fd/FDdescription.asp.

[51] M. Nool and A. van de Ploeg, A parallel Jacobi–Davidson-type method for solving large generalized eigenvalue problems in magnetohydrodynamics, *SIAM J. Sci. Stat. Comput.*, 22(1)(2000) 95–112.

[52] D. O'Leary and P. Whitman, Parallel QR factorization by householder and modified Gram–Schmidt algorithm, *Parallel Comput.*, 16(1990) 99–112.

[53] C.C. Paige, Error analysis of the Lanczos algorithm for tridiagonalizing a symmetric matrix, *J. Inst. Math. Appl.*, 18(1976) 341–349.

[54] C.C. Paige, Computational variants of the Lanczos method for the eigenproblem, *J. Inst. Math. Appl.*, 10 (1972) 373–381.

[55] C.C. Paige, Practical use of the symmetric Lanczos process with reorthogonalization methods, *BIT*, 10 (1970) 183–195.

[56] B.N. Parlett, H. Simon, and L.M. Stringer, On estimating the largest eigenvalue with the Lanczos algorithm, *Math. Comput.*, 38 (1982) 153–165.

[57] B.N. Parlett, *The Symmetric Eigenvalue Problem*, Prentice Hall, Englewood Cliffs, NJ, 1980. Reprinted as *Classics in Applied Mathematics*, Vol. 20, SIAM, Philadelphia, PA, 1997.

[58] B.N. Parlett and D.S. Scott, The Lanczos algorithm with selective orthogonalization, *Math. Comput.*, 33 (1979) 217–238.

[59] Y. Saad, Variations on Arnoldi's method for computing eigenelements of large unsymmetric matrices, *Linear Algebra Appl.*, 34 (1980) 269–295.

[60] Y. Saad, Chebyshev acceleration techniques for solving nonsymmetric eigenvalue problems, *Math. Comput.*, 42 (1984) 568–588.

[61] Y. Saad, *Numerical Methods for Large Eigenvalue Problems*, Wiley-Halsted Press, New York, 1992.

[62] Y. Saad, *Iterative Methods for Linear Systems*, PWS Publishing, Boston, 1996.

[63] M. Sadkane, Block-Arnoldi and Davidson methods for unsymmetric large eigenvalue problems, *Num. Math.*, 64 (1993) 195–211.

[64] A.H. Sameh and J.A. Wisniewski, A trace minimization algorithm for the generalized eigenvalue problem, *SIAM J. Num. Anal.*, 19 (1982) 1243–1259.

[65] A. Sameh and Z. Tong, The trace minimization method for the symmetric generalized eigenvalue problem, *J. Comput. Appl. Math.*, 123 (2000) 155–175.

[66] J.A. Scott, An Arnoldi code for computing selected eigenvalues of sparse real unsymmetric matrices, *ACM Trans. Math. Software*, 21 (1995) 432–475.

[67] H.D. Simon, Analysis of the symmetric Lanczos algorithm with reorthogonalization methods, *Linear Algebra Appl.*, 61 (1984) 101–131.

[68] G.L.G. Sleijpen and H.A. van der Vorst, Jacobi–Davidson iteration method for linear eigenvalue problems, *SIAM J. Matrix Anal. Appl.*, 17 (1996) 401–425.

[69] A. van der Sluis and H.A. van der Vorst, The convergence behaviour of Ritz values in the presence of close eigenvalues, *Linear Algebra Appl.*, 88/89 (1987) 651–694.

[70] M. Snir, S.W. Otto, S. Huss-Lederman, D.W. Walker, and J. Dongarra, *MPI: The Complete Reference*, The MIT Press, Cambridge, MA, 1996.

[71] D.C. Sorensen, Implicit application of polynomial filters in a K-step Arnoldi method, *SIAM J. Matrix Anal. Appl.*, 13 (1992) 357–385.

[72] A. Stathopoulos and C.F. Fischer, Reducing synchronization on the parallel Davidson method for the large, sparse eigenvalue problem, *Proceedings Supercomputing*, '93, ACM Press, Portland (1993) 172–180.

[73] G.W. Stewart, *Introduction to Matrix Computations*, Academic Press, New York, 1973.

[74] G.W. Stewart, *Matrix Algorithms, Vol. II: Eigensystems*, SIAM, Philadelphia, PA, 2001.

[75] W.J. Stewart and A. Jennings, Algorithm 570: LOPSI a simultaneous iteration method for real matrices, *ACM Trans. Math. Software*, 7 (1981) 230–232.

[76] W.J. Stewart and A. Jennings, A simultaneous iteration algorithm for real matrices, *ACM Trans. Math. Software*, 7 (1981) 184–198.

[77] M. Szularz, J. Weston, K. Murphy and M. Clint, Monitoring the convergence of the Lanczos algorithm in parallel computing environments, *Parallel Algorithms Appl.*, 6 (1995) 287–302.

[78] M. Szularz, J. Weston, and M. Clint, Restarting techniques for the Lanczos algorithm and their implementation in parallel computing environments: architectural influences, *Parallel Algorithms Appl.*, 14(1999) 57–77.

[79] M. Szularz, J. Weston, and M. Clint, Explicitly restarted Lanczos algorithms in an MPP environment, *Parallel Comput*, 25 (1999) 613–631.

[80] F. Tisseur, and J. Dongarra, A parallel divide-and-conquer algorithm for the symmetric eigenvalue problem on distributed memory architectures, *SIAM J. Sci. Stat. Comput.*, 20 (1999) 2223–2236.

[81] L.N. Trefethen and D. Bau, *Numerical Linear Algebra*, SIAM, Philadelphia, PA, 1997.

[82] H. van der Vorst, Computational methods for large eigenvalue problems, in *Handbook of Numerical Analysis*, P.G. Ciarlet and J.L. Lions, Ed., Vol. VIII, North-Holland, Amsterdam, 2002, pp. 3–179.

[83] L.C. Waring and M. Clint, Parallel Gram–Schmidt orthogonalization on a network of transputers, *Parallel Comput.*, 17 (1991) 1043–1050.

[84] D.S. Watkins, *Fundamentals of Matrix Computations*, Wiley, New York, 1991.

[85] J.H. Wilkinson, *The Algebraic Eigenvalue Problem*, Clarendon Press, Oxford, 1965.

[86] J.H. Wilkinson and C. Reinsch, *Handbook for Automatic Computation, Vol. II, Linear Algebra*, Springer-Verlag, Heidelberg, 1971.

[87] E. Zapata, J. Lamas, F. Rivera, and O. Plata, Modifed Gram–Schmidt QR factorization on hypercube SIMD computers, *J. Parallel Distrib. Comput.*, 12 (1991) 60–69.

Optimization

6 Parallel Optimization Methods*

*Yair Censor** and Stavros A. Zenios*

CONTENTS

---

* Portions of this chapter are reprinted from *Parallel Optimization: Theory, Algorithms, and Applications* by Yair Censor and Stavros A. Zenios, copyright 1997 by Oxford University Press. Used by permission of Oxford University Press.
** The work of Yair Censor was done at the Center for Computational Mathematics and Scientific Computation (CCMSC) at the University of Haifa and was supported by grant 522/04 of the Israel Science Foundation (ISF), and by grant No. HL 70472 by NIH and grant number 2003275 from the BSF.

ABSTRACT

Parallel computing technology is improving by a quantum leap the size and complexity of mathematical programming models that can be represented and solved on a computer. We discuss appropriate mathematical algorithms for structured optimization problems that utilize efficiently the parallel architectures. Parallel algorithms are presented that exploit parallelism either due to the mathematical structure of the algorithm's operations, or due to the structure of the optimization problem at hand.

6.1 INTRODUCTION

At the beginning of the 21st century we are witnessing an unprecedented growth in the power of computer systems. This growth has been brought about, to a large extent, by parallel computer architectures: multiple, semi-autonomous processors coordinating for the solution of a single problem. The late 1980s witnessed the advent of *massive parallelism* whereby the number of processors brought to bear on a single problem could be several thousands. The 1990s witnessed the proliferation of high-performance workstations, high-speed communications, and the rapid growth of the internet. Within this technological landscape the multicomputer — a computer system composed of heterogeneous, autonomous, processors, linked through a network and with an architecture that is mostly transparent to the user — gained widespread acceptance and this trend continues in the 2000s. Parallel computers are not found solely in research laboratories and academia but are emerging, in self-organizing forms through networking, in industry and commerce.

Parallelism is improving by quantum leaps the size and complexity of models that can be represented on and solved by a computer. But these improvements are not sustained by the architecture of the machines alone. Equally important is the development of appropriate mathematical algorithms and the proper decomposition of the problem at hand, to utilize effectively the parallel architectures. We introduce methods of parallel computing for large-scale constrained optimization and structured linear programs, drawing heavily from the recent book on this topic by Censor and Zenios [27], and using additional material that appeared since then. We take foremost an algorithmic approach, presenting specific implementable algorithms from three broad families of algorithms, and we discuss their suitability for parallel computations. The treatment is unavoidably not exhaustive of the vast landscape of parallel computing methods for optimization. Additional material can be found in Censor and Zenios [27], parallel methods for global optimization are presented in Pardalos [73], and numerical methods for parallel and distributed computing, including extensive coverage of optimization problems, are given in Bertsekas and Tsitsiklis [7]. Parallel methods for unconstrained optimization and for the solution of nonlinear equations are found in Schnabel [81]. A snapshot of current research in parallel optimization can be found in Butnariu, Censor, and Reich [15].

Several of the algorithms and their underlying theories have a long history that dates back to the 1920s. At that time the issue of computations — that is, implementation of the algorithm as a computer program — was irrelevant. (The calculations involved in an algorithm were in those days carried out by humans, or by humans operating mechanical calculators. The issues underlying the efficient execution of these calculations are different from the issues of computation, as we understand them today.) Here we present the algorithms not just with a view toward computations, but more importantly, toward parallel computations. It is when viewed within the context of parallel computing that these algorithms are understood to possess some important features. They are well-suited for implementation on a wide range of parallel architectures, with as few as two processors or as many as millions. It is precisely our aim to show that parallel optimization methods discussed here are implementable algorithms whose efficiency is independent of the specific parallel machine. The theory underlying parallel optimization algorithms and the algorithms themselves (and, of course, the applications) are independent of the rapidly changing landscape of parallel computer architectures.

6.2 CLASSIFICATION OF PARALLEL ALGORITHMS

A *parallel algorithm* is any algorithm that can be implemented on a parallel machine in a way that results in clock-time savings when compared to its implementation, for the solution of the same problem, on a serial computer. This property of an algorithm, i.e., of being parallel or not, stems, from either its mathematical structure, or from the features of the problem to which it is applied, or from both. Even an algorithm that is mathematically sequential can become, for certain decomposable problems, a parallel algorithm. We first look at several mathematical structures of algorithms and discuss their parallelism, and then turn to algorithms that become parallel due to problem structure.

We deal with algorithms drawn from three broad algorithmic classes: (i) *iterative projection algorithms*, (ii) *model decomposition algorithms*, and (iii) *interior point algorithms*. These classes of algorithms exploit parallelism in fundamentally different ways. Algorithms in the first class do so mostly due to their mathematical structure, whereas algorithms in the second and third classes do so for problems with special structure. Iterative projection methods are the most flexible in exploiting parallelism because they do so by suitable reorganization of their operations. The reorganization of operations can be accomplished in many different ways, and a classification of the alternative approaches is given. Model decomposition and interior point algorithms can exploit parallelism by taking advantage of the structure of the model of the problem that is being solved. A general preview of the potential parallelism of these two classes of algorithms is also given.

6.2.1 PARALLELISM DUE TO ALGORITHM STRUCTURE: ITERATIVE PROJECTION ALGORITHMS

We consider applications that can be modeled by a system of linear or nonlinear equations or inequalities, i.e., a system of the form

$$f_i(x) \star 0, \quad i = 1, 2, \ldots, m, \tag{6.1}$$

where $x \in \mathbb{R}^n$ ($\mathbb{R}^n$ is the n-dimensional Euclidean space), $f_i : \mathbb{R}^n \to \mathbb{R}$, for all i, are real-valued functions, and $\star$ stands for equality signs, inequality signs, or a mixture of such. This generic description allows us to classify a broad set of iterative procedures and discuss their suitability for parallel computations. The mathematical classification of algorithms we give next depends on how each algorithm makes use of *row-action* or *block-action* iterations. This classification is purely logical and independent of the architecture of the machine on which the algorithm will be implemented or of the structure of the problem to be solved. Solving a *feasibility problem* (6.1) may be attempted if the system is *feasible*, i.e., if the set $\{x \in \mathbb{R}^n \mid f_i(x) \star 0, \quad i = 1, 2, \ldots, m\}$ is not empty, or even if the system is infeasible. The mathematical model may also be set up as an *optimization problem* with an objective function $f_0 : \mathbb{R}^n \to \mathbb{R}$ imposed and (6.1) serving as constraints. The optimization problem can be written as:

$$\text{Optimize} \quad f_0(x) \tag{6.2}$$

$$\text{subject to} \quad f_i(x) \star 0, \quad i = 1, 2, \ldots, m. \tag{6.3}$$

Here *optimize* stands for either minimize or maximize. Additional constraints of the form $x \in Q$ can also be imposed, where $Q \subseteq \mathbb{R}^n$ is a given subset describing, for example, box constraints.

Special-purpose iterative algorithms designed to solve these problems may employ iterations that use in each iterative step a single row of the constraints system (6.1) or a group of rows. A *row-action iteration* has the functional form

$$x^{\nu+1} \doteq R_{i(\nu)}(x^\nu, f_{i(\nu)}), \tag{6.4}$$

where v is the iteration index, and $i(v)$ is the *control index*, $1 \leq i(v) \leq m$, specifying the row that is acted upon by the algorithmic operator $R_{i(v)}$. The algorithmic operator generates, in some specified manner, the new iterate x^{v+1} from the current iterate x^v and from the information contained in $f_{i(v)}$. $R_{i(v)}$ may depend on additional parameters that vary from iteration to iteration, such as relaxation parameters, weights, tolerances, etc. The system (6.1) may be decomposed into M groups of constraints (called *blocks*) by choosing integers $\{m_t\}_{t=0}^{M}$ such that

$$0 = m_0 < m_1 < \cdots < m_{M-1} < m_M = m, \tag{6.5}$$

and defining for each t, $1 \leq t \leq M$, the subset

$$I_t \doteq \{m_{t-1} + 1, \ m_{t-1} + 2, \ldots, m_t\}. \tag{6.6}$$

This yields a partition of the set:

$$I \doteq \{1, 2, \ldots, m\} = I_1 \cup I_2 \cup \cdots \cup I_M. \tag{6.7}$$

A *block-action iteration* has then the functional form

$$x^{v+1} \doteq B_{t(v)}(x^v, \{f_i\}_{i \in I_{t(v)}}), \tag{6.8}$$

where $t(v)$ is the control index, $1 \leq t(v) \leq M$, specifying the block which is used when the algorithmic operator $B_{t(v)}$ generates x^{v+1} from x^v and from the information contained in all rows of (6.1) whose indices belong to $I_{t(v)}$. Again, additional parameters may be included in each $B_{t(v)}$, such as those specified for the row-action operator $R_{i(v)}$. The special-purpose iterative algorithms that we consider in our classification scheme may address any problem of the form (6.1) or (6.2) and (6.3) and may be classified as having one of the following four basic structures.

Sequential Algorithms. For this class of algorithms we define a control sequence $\{i(v)\}_{v=0}^{\infty}$ and the algorithm performs, in a strictly sequential manner, row-action iterations according to (6.4), from an appropriate initial point until a stopping rule applies.

Simultaneous Algorithms. Algorithms in this class first execute simultaneously row-action iterations on all rows

$$x^{v+1,i} \doteq R_i(x^v, f_i), \quad i = 1, 2, \ldots, m, \tag{6.9}$$

using the same current iterate x^v. The next iterate x^{v+1} is then generated from all intermediate ones $x^{v+1,i}$ by an additional operation

$$x^{v+1} \doteq S(\{x^{v+1,i}\}_{i=1}^{m}). \tag{6.10}$$

Here S and R_i are algorithmic operators, the R_i's being all of the row-action type.

Sequential Block-Iterative Algorithms. Here the system (6.1) is decomposed into fixed blocks according to (6.7), and a control sequence $\{t(v)\}_{v=0}^{\infty}$ over the set $\{1, 2, \ldots, M\}$ is defined. The algorithm performs sequentially, according to the control sequence, block iterations of the form (6.8).

Simultaneous Block-Iterative Algorithms. In this class, block iterations are first performed using the same current iterate x^v, on all blocks simultaneously:

$$x^{v+1,t} \doteq B_t(x^v, \{f_i\}_{i \in I_t}), \quad t = 1, 2, \ldots, M. \tag{6.11}$$

The next iterate x^{v+1} is then generated from the intermediate ones $x^{v+1,t}$ by

$$x^{v+1} \doteq S\left(\{x^{v+1,t}\}_{t=1}^{M}\right). \tag{6.12}$$

Here S and B_t are algorithmic operators, the B_t's being block-iterative as in Equation (6.8).

An important generalization of block-iterative algorithms refers to *variable block-iterative* algorithms, wherein block iterations are performed according to formula (6.8). The blocks $I_{t(v)}$, however, are not determined *a priori* according to (6.7) and kept fixed, but the algorithm employs an infinite sequence $\{I_{t(v)}\}_{v=0}^{\infty}$ of nonempty blocks $I_{t(v)} \subseteq I$. This means that as the iterations proceed, both the sizes of the blocks and the assignment of constraints to blocks may change in a much more general manner than the scheme with fixed blocks. Of course, convergence theory of such an algorithm could still impose certain restrictions on how the sets $I_{t(v)}$ may be constructed.

6.2.1.1   Parallel Computing with Iterative Algorithms

We examine two approaches for introducing parallelism in both the *sequential* and the *simultaneous* algorithms described above. We address in particular the problem of task partitioning, that is, identifying operations of the algorithm that can be executed concurrently.

A sequential row-action algorithm uses information contained in a single row $f_{i(v)}(x) \star 0$ and the current iterate x^v to generate the next iterate x^{v+1}. The control sequence $\{i(v)\}_{v=0}^{\infty}$ specifies the order in which rows are selected. Is it possible to select rows in such a way that two (or more) rows can be operated upon simultaneously without altering the mathematical structure of the algorithm? This can be achieved if at every iteration only different components of the updated vector are changed. If the algorithmic operators $R_{i(v)}$ and $R_{i(v+1)}$, which are vector-to-vector mappings, use different components of x^v and x^{v+1} to operate upon and leave all other components unchanged, then the operations can proceed concurrently. Identifying row-indices $i(v)$ so that the operators can be applied concurrently on such rows depends on the structure of the family $\{f_i(x)\}_{i=1}^m$. Several important problems have a structure that allows us to identify such row-indices. The parallel algorithm is, with this approach, mathematically identical to the serial algorithm.

The parallelism in the simultaneous algorithms is obvious. The iterations (6.9) can be executed concurrently using up to m parallel processors. The step (6.10) is a synchronization step where the processors must cooperate, and exchange information contained in the m vectors $\{x^{v+1,i}\}_{i=1}^m$ to compute the next iterate x^{v+1}.

The block-iterative algorithms — sequential or simultaneous — lead to parallel computing identical to the two ways outlined above. However, the introduction of blocks permits more flexibility in handling the task scheduling problem. The blocks can be chosen so as to ensure that all processors receive tasks of the same computational difficulty. Hence, all processors complete their work in roughly the same amount of time. Delays while processors wait for each other are minimized. The specification of block-sizes will typically depend on the computer architecture. For fine-grain massively parallel systems it is preferable to create a very large number of small blocks. Fewer, but larger, blocks are needed for coarse-grain parallel computers.

6.2.2   PARALLELISM DUE TO PROBLEM STRUCTURE: MODEL DECOMPOSITION AND INTERIOR POINT ALGORITHMS

We consider now the solution of *structured* optimization problems. We introduce first the following definition.

Definition 6.1 (Block Separability) *A function F is block-separable if it can be written as the sum of functions of subsets of its variables. Each subset is called a* block.

Consider, for example, the vector $x \in \mathbb{R}^{nK}$, which is the concatenation of K subvectors, each of dimension n,

$$x = \left((x^1)^\mathrm{T}, (x^2)^\mathrm{T}, \ldots, (x^K)^\mathrm{T} \right)^\mathrm{T},$$

where $x^k \in \mathbb{R}^n$, for all $k = 1, 2, \ldots, K$. (boldface letters denote vectors in the product space $\mathbb{R}^{nK}$). Then $F : \mathbb{R}^{nK} \to \mathbb{R}$ is block-separable if it can be written as $F(x) \doteq \sum_{k=1}^{K} f_k(x^k)$, where $f_k : \mathbb{R}^n \to \mathbb{R}$. We consider now the following constrained optimization problem: **Problem** $[\mathcal{P}]$

$$\text{Optimize} \quad F(x) \doteq \sum_{k=1}^{K} f_k(x^k) \tag{6.13}$$

$$\text{subject to} \quad x^k \in X_k \subseteq \mathbb{R}^n, \qquad \text{for all } k = 1, 2, \ldots, K, \tag{6.14}$$

$$x \in \Omega \subseteq \mathbb{R}^{nK}. \tag{6.15}$$

The block-separability of the objective function and the structure of the constraint sets of this problem can be exploited for parallel computing. Two general algorithms that exploit this structure are described below.

6.2.2.1 Model Decomposition Algorithms

If the product set $X_1 \times X_2 \times \cdots \times X_K$ of problem $[\mathcal{P}]$ is a subset of Ω, then the problem can be solved by simply ignoring the constraints $x \in \Omega$. A solution can be obtained by solving K independent subproblems in each of the x^k vector variables. These subproblems can be solved in parallel.

In general, however, the *complicating constraints* $x \in \Omega$ cannot be ignored. A model decomposition applies a *modifier* to problem $[\mathcal{P}]$ to obtain a problem $[\mathcal{P}']$ in which the complicating constraints are not explicitly present. It then employs a suitable algorithm to solve $[\mathcal{P}']$. Since the modified problem does not have complicating constraints it can be decomposed into K problems, each one involving only the x^k vector and the constraint $x^k \in X_k$, for $k = 1, 2, \ldots, K$. These problems are not only smaller and simpler than the original problem $[\mathcal{P}]$, but they are also independent of each other, and can be solved in parallel. If the solution to the modified problem $[\mathcal{P}']$ is "sufficiently close" to a solution of the original problem $[\mathcal{P}]$ then the process terminates. Otherwise the current solution is used to construct a new, modified problem, and the process repeats until some stopping criterion is satisfied (see Figure 6.1). A wide variety of algorithms fall under this broad framework, and specific instances are discussed in Section 6.4.

Of particular interest is the application of model decomposition algorithms to problems with linear constraints, with a constraint matrix that is block-angular or dual block-angular. For example, the constraint matrix of the classic *multicommodity network flow* problem found in Rockafellar [78] and Censor and Zenios [27] is of the following, block-angular form

$$A = \begin{pmatrix} A_1 & & & \\ & A_2 & & \\ \vdots & & \ddots & \\ & & & A_k \\ I & I & \ldots & I \end{pmatrix}. \tag{6.16}$$

Here $\{A_k\}$, $k = 1, 2, \ldots, K$, are $m \times n$ network flow constraints matrices, and I denotes the $n \times n$ identity matrix. The constraint set of this problem can be put in the form of problem $[\mathcal{P}]$ by defining X as the product set $X \doteq \prod_{k=1}^{K} X_k$, where $X_k = \{x^k \mid A_k x^k = b^k, \ 0 \le x^k \le u^k\}$,

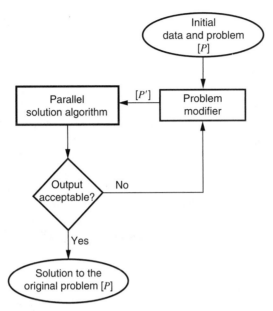

FIGURE 6.1 Structure of model decomposition algorithms: Modifications $[\mathcal{P}']$ of the original problem $[\mathcal{P}]$ are solved repeatedly using a parallel solution algorithm. The solution of each modified problem is used to construct the next modification until an acceptable solution to the original problem is reached.

where b^k denotes the right-hand side of the equality constraints and u^k are vectors of upper bounds on the variables, and by defining $\Omega \doteq \{x \mid x \in X, \; Ex \leq U\}$, where $E = [I \mid I \mid \ldots \mid I]$. A model decomposition algorithm, found in Censor and Zenios [27], can now be applied to solve the multicommodity flow problem. The constraint matrix of linear programming formulations of two-stage stochastic programming problems, Birge and Louveaux [9], Censor and Zenios [27], and Kall and Wallace [55] is of the dual block-angular form:

$$
A = \begin{pmatrix}
A_0 & & & \\
T_1 & W_1 & & \\
\vdots & & \ddots & \\
T_N & & & W_N
\end{pmatrix}.
\tag{6.17}
$$

The matrix A_0 is of dimension $m_0 \times n_0$. The matrices W_l are of dimensions $m_l \times n_l$, for $l = 1, 2, \ldots, N$. The matrices T_l are of dimensions $m_l \times n_0$. It is assumed that A_0 and W_l have full row rank, with $m_l \leq n_l$ for all $l = 1, 2, \ldots, N$, and that $n_0 \leq \sum_{l=1}^{N} n_l$.

The dual formulation of this problem, which is formulated using the transpose A^T of the constraint matrix, can be put in the form of $[\mathcal{P}]$ using definitions for X and Ω similar to those introduced above for the multicommodity flow problem.

6.2.2.2 Interior Point Algorithms

Consider now special cases of problem $[\mathcal{P}]$, when the objective function is linear or separable quadratic, and the constraints sets X and Ω are defined by linear equalities and inequalities. Then we can solve the problem directly using a general-purpose interior point algorithm for linear or quadratic programs (Kojima et al. [59], Lustig et al. [64], and Censor and Zenios [27]).

The major computational effort of these algorithms appears in the factorization and solution of a system of equations, for the unknown vector Δy, of the form

$$(A\Theta A^{\mathrm{T}})\Delta y = \psi, \tag{6.18}$$

where A is the constraint matrix of the problem, ψ is a given vector, Θ is a given diagonal matrix, and Δy denotes a dual step direction of the interior point algorithm.

When the constraint matrix A is of the structure shown in (6.16) it is possible to design special-purpose procedures that solve the system of equations (6.18) by solving subsystems involving the smaller matrices $A_1, \ldots, A_K$, or $A_0, W_1, \ldots, W_N, T_1, \ldots, T_N$ (see Section 6.5). These subsystems are independent of each other and can be solved in parallel. Hence, while interior point algorithms are not parallel algorithms *per se*, they can exploit parallel architectures when executing the numerical linear algebra calculations for problems with special structures.

6.3 GENERAL FRAMEWORK OF PROJECTION ALGORITHMS

6.3.1 ORTHOGONAL PROJECTIONS

Projection algorithms employ projections onto convex sets (POCS) in various ways. They may use different kinds of projections and, sometimes, even use different projections within the same algorithm. They serve to solve a variety of problems which are either of the feasibility or the optimization types. They have different algorithmic structures, of which some are particularly suitable for parallel computing, and they demonstrate nice convergence properties and/or good initial behavior patterns. This class of algorithms has witnessed great progress in recent years and its member algorithms have been applied with success to fully discretized models of problems in image reconstruction and image processing.

The *convex feasibility problem* is to find a point (any point) in the nonempty intersection $C := \bigcap_{i=1}^{m} C_i \neq \emptyset$ of a family of closed convex subsets $C_i \subseteq \mathbb{R}^n$, $1 \leq i \leq m$, of the n-dimensional Euclidean space. It is a fundamental problem in many areas of mathematics and the physical sciences and has been used to model significant real-world problems in image reconstruction from projections, in radiation therapy treatment planning, and in crystallography, to name but a few, and has been used under additional names such as *set theoretic estimation* or the *feasible set approach*. A common approach to such problems is to use projection algorithms which employ *orthogonal projections* (i.e., nearest point mappings) onto the individual sets C_i. The orthogonal projection $P_\Omega(z)$ of a point $z \in \mathbb{R}^n$ onto a closed convex set $\Omega \subseteq \mathbb{R}^n$ is defined by

$$P_\Omega(z) := \operatorname{argmin}\{\|z - x\|_2 \mid x \in \Omega\}, \tag{6.19}$$

where $\| \cdot \|_2$ is the Euclidean norm in $\mathbb{R}^n$. Frequently a *relaxation parameter* is introduced so that

$$P_{\Omega,\lambda}(z) := (1 - \lambda)z + \lambda P_\Omega(z) \tag{6.20}$$

is the *relaxed projection* of z onto Ω with relaxation λ.

Another problem that is related to the convex feasibility problem is the *best approximation problem* of finding the projection of a given point $y \in \mathbb{R}^n$ onto the nonempty intersection $C := \bigcap_{i=1}^{m} C_i \neq \emptyset$ of a family of closed convex subsets $C_i \subseteq \mathbb{R}^n$, $1 \leq i \leq m$. In both problems the convex sets $\{C_i\}_{i=1}^{m}$ represent mathematical constraints obtained from the modeling of the real-world problem. In the convex feasibility approach any point in the intersection is an acceptable solution to the real-world problem whereas the best approximation formulation is usually appropriate if some point $y \in \mathbb{R}^n$ has been obtained from modeling and computational efforts which initially did not take into account the constraints represented by the sets $\{C_i\}_{i=1}^{m}$ and now one wishes to incorporate them by seeking a point in the intersection of the convex sets which is closest to the point y.

Iterative projection algorithms for finding a projection of a point onto the intersection of sets are more complicated then algorithms for finding just any feasible point in the intersection. This is so because they must have, in their iterative steps, some built-in "memory" mechanism to remember the original point whose projection is sought after. We do not deal with these algorithms here although many of them share the same algorithmic structural features described below.

6.3.2 BREGMAN PROJECTIONS

Bregman projections onto closed convex sets were introduced by Censor and Lent [25], based on Bregman's seminal work [11] and were subsequently used in a plethora of research works as a tool for building sequential and parallel feasibility and optimization algorithms.

A *Bregman projection* of a point $z \in \mathbb{R}^n$ onto a closed convex set $\Omega \subseteq \mathbb{R}^n$ with respect to a, suitably defined, *Bregman function f* is denoted by $P_\Omega^f(z)$. It is formally defined as

$$P_\Omega^f(z) := \text{argmin}\{D_f(x, z) | x \in \Omega \cap clS\} \tag{6.21}$$

where clS is the closure of the open convex set S, which is the *zone* of f, and $D_f(x, z)$ is the so-called *Bregman distance*, defined by

$$D_f(x, z) := f(x) - f(z) - \langle \nabla f(z), x - z \rangle, \tag{6.22}$$

for all pairs $(x, z) \in clS \times S$, where $\langle \cdot, \cdot \rangle$ is the standard inner product in $\mathbb{R}^n$. If $\Omega \cap clS \neq \emptyset$ then (6.21) defines a unique $P_\Omega^f(z) \in clS$, for every $z \in S$. If, in addition, $P_\Omega^f(z) \in S$ for every $z \in S$, then f is called *zone consistent* with respect to Ω.

Orthogonal projections are a special case of Bregman projections, obtained from (6.21) by choosing $f(x) = (1/2) \parallel x \parallel^2$ and $S = \mathbb{R}^n$. Bregman generalized distances and generalized projections are instrumental in several areas of mathematical optimization theory. They were used, among others, in special-purpose minimization methods, in the proximal point minimization method, and for stochastic feasibility problems. These generalized distances and projections were also defined in non-Hilbertian Banach spaces, where, in the absence of orthogonal projections, they can lead to simpler formulas for projections.

Bregman's method for minimizing a convex function (with certain properties) subject to linear inequality constraints employs Bregman projections onto the half-spaces represented by the constraints. Recently [12] the extension of this minimization method to nonlinear convex constraints has been identified with the Han–Dykstra projection algorithm for finding the projection of a point onto an intersection of closed convex sets. It looks as if there might be no point in using nonorthogonal projections for solving the convex feasibility problem in $\mathbb{R}^n$ since they are generally not easier to compute. But this is not always the case. Shamir and coworkers [60] have used a multiprojection method of Censor and Elfving [21] to solve filter design problems in image restoration and image recovery posed as convex feasibility problems. They took advantage of that algorithm's flexibility to employ Bregman projections with respect to *different* Bregman functions within the same algorithmic run. Another example is the seminal paper by Csiszár and Tusnády [33], where the central procedure uses alternating entropy projections onto convex sets. In their "alternating minimization procedure," they alternate between minimizing over the first and second arguments of the Kullback–Leibler divergence. This divergence is nothing but the generalized Bregman distance obtained by using the negative of Shannon's entropy as the underlying Bregman function. Recent studies about Bregman projections, Bregman/Legendre projections and averaged entropic projections — and their uses for convex feasibility problems in $\mathbb{R}^n$ — attest to the continued theoretical and practical interest in employing Bregman projections in projection methods for convex feasibility problems.

6.3.3   PROJECTION ALGORITHMIC SCHEMES

Projection algorithmic schemes for the convex feasibility problem and for the best approximation problem are, in general, either *sequential* or *simultaneous* or *block-iterative*. In the following, we explain and demonstrate these structures along with the recently proposed *string-averaging* structure. The philosophy behind these algorithms is that it is easier to calculate projections onto the individual sets C_i then onto the whole intersection of sets. Thus, these algorithms call for projections onto individual sets as they proceed sequentially, simultaneously or in the block-iterative or the string-averaging algorithmic models.

6.3.3.1   Sequential Projections

The well-known POCS algorithm [82] for the convex feasibility problem is a *sequential* projection algorithm. Starting from an arbitrary initial point $x^0 \in \mathbb{R}^n$, the POCS algorithm's iterative step is

$$x^{k+1} = x^k + \lambda_k (P_{C_{i(k)}}(x^k) - x^k), \tag{6.23}$$

where $\{\lambda_k\}_{k \geq 0}$ are relaxation parameters and $\{i(k)\}_{k \geq 0}$ is a *control sequence*, $1 \leq i(k) \leq m$, for all $k \geq 0$, which determines the individual set $C_{i(k)}$ onto which the current iterate x^k is projected. A commonly used control is the *cyclic control* in which $i(k) = k \mod m + 1$, but other controls are also available. Bregman's projection algorithm allowed originally only unrelaxed projections, i.e., its iterative step is of the form

$$x^{k+1} = P_{C_{i(k)}}^f(x^k), \quad \text{for all } k \geq 0. \tag{6.24}$$

For the Bregman function $f(x) = (1/2) \parallel x \parallel^2$ with zone $S = \mathbb{R}^n$ and for unity relaxation ($\lambda_k = 1$, for all $k \geq 0$), (6.24) coincides with (6.23).

6.3.3.2   The String Averaging Algorithmic Structure

The *string-averaging* algorithmic scheme works as follows. Let the *string* σ_t, for $t = 1, 2, \ldots, M$, be an ordered subset of $\{1, 2, \ldots, m\}$ of the form

$$\sigma_t = (i_1^t, i_2^t, \ldots, i_{\mu(t)}^t), \tag{6.25}$$

with $\mu(t)$ denoting the number of elements in σ_t. Suppose that there is a set $S \subseteq \mathbb{R}^n$ such that there are operators $R_1, R_2, \ldots, R_m$ mapping S into S and an operator R which maps $S^M = S \times S \times \cdots \times S$ (M times) into S. Initializing the algorithm at an arbitrary $x^0 \in S$, the iterative step of the string-averaging algorithmic scheme is as follows. Given the current iterate x^k, calculate, for all $t = 1, 2, \ldots, M$,

$$T_t x^k = R_{i_{\mu(t)}^t} \ldots R_{i_2^t} R_{i_1^t} x^k, \tag{6.26}$$

and then calculate

$$x^{k+1} = R(T_1 x^k, T_2 x^k, \ldots, T_M x^k). \tag{6.27}$$

For every $t = 1, 2, \ldots, M$ this algorithmic scheme applies to x^k successively the operators whose indices belong to the tth string. This can be done in parallel for all strings and then the operator R maps all end points onto the next iterate x^{k+1}. This is indeed an algorithm provided that the operators $\{R_i\}_{i=1}^m$ and R all have algorithmic implementations. In this framework we get a *sequential* algorithm by the choice $M = 1$ and $\sigma_1 = (1, 2, \ldots, m)$ and a *simultaneous* algorithm by the choice $M = m$ and $\sigma_t = (t), t = 1, 2, \ldots, M$.

  We demonstrate the underlying idea of the string-averaging algorithmic scheme with the aid of Figure 6.2. For simplicity, we take the convex sets to be hyperplanes, denoted by H_1, H_2, H_3, H_4, H_5,

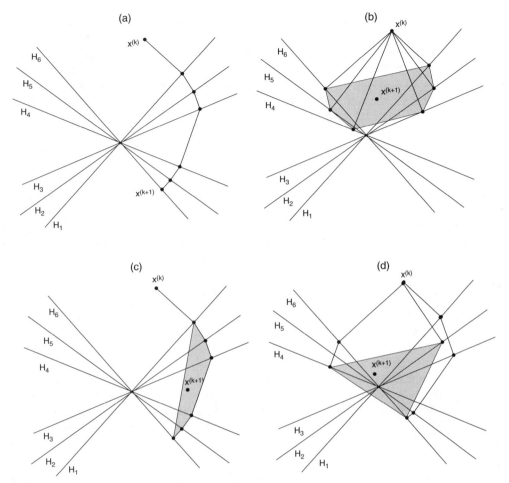

FIGURE 6.2 (a) Sequential projections. (b) Fully simultaneous projections. (c) Averaging of sequential projections. (d) String averaging. (Reproduced from Censor et al. [22].)

and H_6, and assume all operators R_i to be orthogonal projections onto the hyperplanes. The operator R is taken as a convex combination

$$R(x^1, x^2, \ldots, x^M) = \sum_{t=1}^{M} \omega_t x^t, \qquad (6.28)$$

with $\omega_t > 0$, for all $t = 1, 2, \ldots, M$, and $\sum_{t=1}^{M} \omega_t = 1$.

Figure 6.2(a) depicts the purely sequential algorithm. This is the so-called POCS algorithm which coincides, for the case of hyperplanes, with the Kaczmarz algorithm, see, e.g., Algorithms 5.2.1 and 5.4.3, respectively, in Censor and Zenios [27]. The fully simultaneous algorithm appears in Figure 6.2(b). With orthogonal reflections instead of orthogonal projections it was first proposed by Cimmino [30] for solving linear equations. Here the current iterate x^k is projected on all sets simultaneously and the next iterate x^{k+1} is a convex combination of the projected points. In Figure 6.2(c) we show how a simple averaging of *successive* projections (as opposed to averaging of parallel projections in Figure 6.2(b)) works. In this case $M = m$ and $\sigma_t = (1, 2, \ldots, t)$ for $t = 1, 2, \ldots, M$. This scheme, inspired the formulation of the general string-averaging algorithmic scheme whose action is demonstrated in Figure 6.2(d). It averages, via convex combinations, the end points obtained from strings of sequential projections and in this figure the strings are $\sigma_1 = (1, 3, 5, 6)$, $\sigma_2 = (2)$, and $\sigma_3 = (6, 4)$. Such schemes offer a variety of options for steering the iterates toward a solution of the convex feasibility problem. It is an *inherently parallel* scheme in that its mathematical formulation is parallel (like the fully simultaneous method mentioned above). We use this term to contrast such algorithms with others which are sequential in their mathematical formulation but can, sometimes, be implemented in a parallel fashion based on appropriate model decomposition (i.e., depending on the structure of the underlying problem). Being inherently parallel, this algorithmic scheme enables flexibility in the actual manner of implementation on a parallel machine.

At the extremes of the "spectrum" of possible specific algorithms, derivable form the string-averaging algorithmic scheme, are the generically sequential method, which uses one set at a time, and the fully simultaneous algorithm, which employs all sets at each iteration. The "block-iterative projections" (BIP) scheme also has the sequential and the fully simultaneous methods as its extremes in terms of block structures. Nevertheless, the string-averaging algorithmic structure gives users a tool to design many new inherently parallel computational schemes.

6.3.4 BLOCK-ITERATIVE COMPONENT AVERAGING

A recent member of the powerful family of block-iterative projection algorithms is the (BICAV) *block-iterative component averaging* algorithm of Censor et al. [24], which was applied to a problem of image reconstruction from projections. The BICAV algorithm is a block-iterative companion to the *component averaging* (CAV) method for solving systems of linear equations. In these methods the sparsity of the matrix is explicitly used when constructing the iteration formula.

In Cimmino's simultaneous projections method, with relaxation parameters and with equal weights $w_i = 1/m$, the next iterate x^{k+1} is the average of the orthogonal projections of x^k onto the hyperplanes H_i defined by the ith row of the linear system $Ax = b$ and has, for every component $j = 1, 2, \ldots, n$, the form

$$x_j^{k+1} = x_j^k + \frac{\lambda_k}{m} \sum_{i=1}^{m} \frac{b_i - \langle a^i, x^k \rangle}{\| a^i \|_2^2} a_j^i, \tag{6.29}$$

where a^i is the ith column of the transpose A^T of A and b_i is the ith component of the vector b and λ_k are relaxation parameters. When the $m \times n$ system matrix $A = (a_j^i)$ is sparse, only a relatively small number of the elements $\{a_j^1, a_j^2, \ldots, a_j^m\}$ of the jth column of A are nonzero, but in (6.29) the sum of their contributions is divided by the relatively large m. This observation led to the replacement of the factor $1/m$ in (6.29) by a factor that depends only on the *nonzero* elements in the set $\{a_j^1, a_j^2, \ldots, a_j^m\}$. For each $j = 1, 2, \ldots, n$, denote by s_j the number of nonzero elements of column j of the matrix A, and replace (6.29) by

$$x_j^{k+1} = x_j^k + \frac{\lambda_k}{s_j} \sum_{i=1}^{m} \frac{b_i - \langle a^i, x^k \rangle}{\| a^i \|_2^2} a_j^i. \tag{6.30}$$

Certainly, if A is sparse then the s_j values will be much smaller than m. The iterative step (6.29) is a special case of

$$x^{k+1} = x^k + \lambda_k \sum_{i=1}^{m} \omega_i \frac{b_i - \langle a^i, x^k \rangle}{\| a^i \|_2^2} a^i, \tag{6.31}$$

where the fixed weights $\{w_i\}_{i=1}^{m}$ must be positive for all i and $\sum_{i=1}^{m} \omega_i = 1$. The attempt to use $1/s_j$ as weights in (6.30) does not fit into the scheme (6.31), unless one can prove convergence of the iterates of a fully simultaneous iterative scheme with component-dependent (i.e., j-dependent) weights of the form

$$x_j^{k+1} = x_j^k + \lambda_k \sum_{i=1}^{m} \omega_{ij} \frac{b_i - \langle a^i, x^k \rangle}{\| a^i \|_2^2} a_j^i, \tag{6.32}$$

for all $j = 1, 2, \ldots, n$. To formalize this consider a set $\{G_i\}_{i=1}^m$ of real diagonal $n \times n$ matrices $G_i = \mathrm{diag}(g_{i1}, g_{i2}, \ldots, g_{in})$ with $g_{ij} \geq 0$, for all $i = 1, 2, \ldots, m$ and $j = 1, 2, \ldots, n$, such that $\sum_{i=1}^m G_i = I$, where I is the unit matrix. Referring to the sparsity pattern of A one needs the following definition.

Definition 6.2 (sparsity pattern oriented) *A family $\{G_i\}_{i=1}^m$ of real diagonal $n \times n$ matrices with all diagonal elements $g_{ij} \geq 0$ and such that $\sum_{i=1}^m G_i = I$ is called sparsity pattern oriented (SPO, for short) with respect to an $m \times n$ matrix A if, for every $i = 1, 2, \ldots, m$, $g_{ij} = 0$ if and only if $a_j^i = 0$.*

The CAV algorithm combines three features: (i) each orthogonal projection onto H_i is replaced by a *generalized oblique projection with respect to G_i*, denoted below by $P_{H_i}^{G_i}$. (ii) The scalar weights $\{\omega_i\}$ in (6.31) are replaced by the diagonal weighting matrices G_i. (iii) The actual weights are set to be inversely proportional to the number of nonzero elements in each column, as motivated by the discussion preceding Equation (6.30). The iterative step resulting from the first two features has the form:

$$x^{k+1} = x^k + \lambda_k \sum_{i=1}^{m} G_i (P_{H_i}^{G_i}(x^k) - x^k). \tag{6.33}$$

The basic idea of the BICAV algorithm is to break up the system $Ax = b$ into "blocks" of equations and treat each block according to the CAV methodology, passing cyclically over all the blocks. Throughout the following, T will be the number of blocks and, for $t = 1, 2, \ldots, T$, let the block indices $B_t \subseteq \{1, 2, \ldots, m\}$, be an ordered subset of the form $B_t = \{i_1^t, i_2^t, \ldots, i_{\mu(t)}^t\}$ where $\mu(t)$ is the number of elements in B_t, such that every element of $\{1, 2, \ldots, m\}$ appears in at least one of the sets B_t. For $t = 1, 2, \ldots, T$, let A_t denote the matrix formed by taking all the rows of A whose indices belong to the block of indices B_t, i.e.,

$$A_t := \begin{bmatrix} a^{i_1^t} \\ a^{i_2^t} \\ \vdots \\ a^{i_{\mu(t)}^t} \end{bmatrix}, \quad t = 1, 2, \ldots, T. \tag{6.34}$$

The iterative step of the BICAV algorithm uses, for every block index $t = 1, 2, \ldots, T$, generalized oblique projections with respect to a family $\{G_i^t\}_{i=1}^m$ of diagonal matrices which are SPO with respect to A_t. The same family is also used to perform the diagonal weighting. The resulting iterative step has the form:

$$x^{k+1} = x^k + \lambda_k \sum_{i \in B_{t(k)}} G_i^{t(k)} (P_{H_i}^{G_i^{t(k)}}(x^k) - x^k), \tag{6.35}$$

where $\{t(k)\}_{k \geq 0}$ is a *control sequence* according to which the $t(k)$-th block is chosen by the algorithm to be acted upon at the kth iteration, thus, $1 \leq t(k) \leq T$, for all $k \geq 0$. The real numbers $\{\lambda_k\}_{k \geq 0}$ are user-chosen *relaxation parameters*. Finally, in order to achieve the acceleration, the diagonal matrices $\{G_i^t\}_{i=1}^m$ are constructed with respect to each A_t. Let s_j^t be the number of nonzero elements $a_j^i \neq 0$ in the jth column of A_t and define

$$g_{ij}^t := \begin{cases} 1/s_j^t, & \text{if } a_j^i \neq 0, \\ 0, & \text{if } a_j^i = 0. \end{cases} \tag{6.36}$$

It is easy to verify that, for each $t = 1, 2, \ldots, T$, $\sum_{i=1}^m G_i^t = I$ holds for these matrices. With these particular SPO families of G_i^t's one obtains the block-iterative algorithm:

Algorithm 6.1 BICAV

Step 0: (Initialization.) $x^0 \in \mathbb{R}^n$ is arbitrary.

Step 1: (Iterative Step.) Given x^k, compute x^{k+1} by using, for $j = 1, 2, \ldots, n$, the formula:

$$x_j^{k+1} = x_j^k + \lambda_k \sum_{i \in B_{t(k)}} \frac{b_i - \langle a^i, x^k \rangle}{\sum_{l=1}^n s_l^{t(k)} (a_i^i)^2} a_j^i, \tag{6.37}$$

where λ_k are relaxation parameters, $\{s_l^t\}_{l=1}^n$ are as defined above, and the control sequence is cyclic, i.e., $t(k) = k \bmod T + 1$, for all $k \geq 0$.

6.4 GENERAL FRAMEWORK OF MODEL DECOMPOSITIONS

Coming back to the problem $[\mathcal{P}]$, defined by (6.13)–(6.15), we assume that the function F is convex and continuously differentiable. The sets $X_k, k = 1, 2, \ldots, K$, and Ω are assumed to be closed and convex. Figure 6.3 illustrates the structure of this problem in two dimensions.

If the product set $X_1 \times X_2 \times \cdots \times X_K \subseteq \Omega$, then problem $[\mathcal{P}]$ can be solved by simply ignoring the constraints $x \in \Omega$ and solving K independent subproblems in each of the x^k vector variables. In this respect, the constraints $x \in \Omega$ are *complicating* (or *coupling*) constraints. When the complicating constraints cannot be ignored, a model decomposition applies a *modifier* to problem $[\mathcal{P}]$

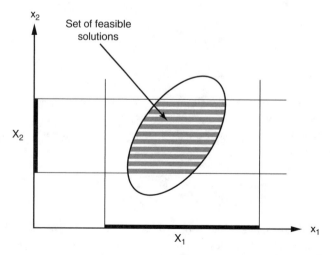

FIGURE 6.3 Constraint sets and set of feasible solutions of problem $[\mathcal{P}]$ in $\mathbb{R}^2$ with $k = 2$.

to obtain a problem [$\mathcal{P}'$] in which the complicating constraints are not explicitly present. It then employs a suitable algorithm to solve [$\mathcal{P}'$]. If the solution to [$\mathcal{P}'$] is *sufficiently close* to a solution of the original problem [$\mathcal{P}$] then the process terminates. Otherwise, the current solution is used to construct a new modified problem and the process repeats (see Figure 6.1).

It is the judicious combination of a modifier and a suitable algorithm for the solution of the modified problem [$\mathcal{P}'$] that leads to a decomposition of the original problem [$\mathcal{P}$] suitable for parallel computations. We discuss in this section modifiers and algorithms suitable for solving the modified problems.

6.4.1 PROBLEM MODIFIERS

We present now two modifiers for problem [$\mathcal{P}$], drawing on the theory of penalty and barrier methods, and augmented Lagrangians (Fiacco and McCormick [41], Bertsekas [6], and Censor and Zenios [27]).

Definition 6.3 (Penalty Functions) *Given a nonempty set $\Omega \subseteq \mathbb{R}^n$, a function $p \colon \mathbb{R}^n \to \mathbb{R}$ is called an (exterior) penalty function with respect to Ω if the following hold: (i) $p(x)$ is continuous, (ii) $p(x) \geq 0$, for all $x \in \mathbb{R}^n$, and (iii) $p(x) = 0$ if and only if $x \in \Omega$.*

Definition 6.4 (Barrier Functions) *Given a robust set $\Omega \subseteq \mathbb{R}^n$, a function $q \colon \mathbb{R}^n \to \mathbb{R}$ is called a barrier function with respect to Ω (also called an interior penalty function) if the following hold: (i) $q(x)$ is continuous on int Ω, (ii) $q(x) \geq 0$, for all $x \in$ int Ω, and (iii) $\lim q(x) = +\infty$, as x tends to any boundary point of Ω.*

6.4.1.1 Modifier I: Penalty or Barrier Functions

The first modifier eliminates the complicating constraints $x \in \Omega$ by using a *penalty* or a *barrier* function. Using a penalty function $p \colon \mathbb{R}^{nK} \to \mathbb{R}$ with respect to the set Ω, defining the complicating contraints, the modified problem can be written as: **Problem [$\mathcal{P}'$]:**

$$\text{Minimize} \quad F(x) + cp(x) \tag{6.38}$$

$$\text{subject to} \quad x^k \in X_k, \quad \text{for all } k = 1, 2, \ldots, K. \tag{6.39}$$

It is possible to construct penalty functions that are *exact*, i.e., there exists some constant $\bar{c} > 0$, such that for $c > \bar{c}$ any solution of [$\mathcal{P}'$] is also a solution to [$\mathcal{P}$]. Hence, a solution to [$\mathcal{P}$] can be obtained by solving [$\mathcal{P}'$]. Note that [$\mathcal{P}'$] has a simpler constraint set than [$\mathcal{P}$] because the complicating constraints $x \in \Omega$ have been removed. However, problem [$\mathcal{P}'$] still cannot be solved by solving K independent subproblems, since the function p is not necessarily block-separable; (see Definition 6.1). The next section explores algorithms that induce separability of this function.

Consider now situations when the set Ω has a nonempty interior. Such sets arise when inequality constraints are used to define them, e.g.,

$$\Omega = \{x \mid g_l(x) \leq 0, \quad \text{for all } l = 1, 2, \ldots, L\}, \tag{6.40}$$

where $g_l \colon \mathbb{R}^{nK} \to \mathbb{R}$. In this case we can use a *barrier function* (see Definition 6.4) to establish a barrier on the boundary of Ω so that the iterates of an algorithm that starts with an interior point remain in the interior of the set, therefore, satisfying the constraints $x \in \Omega$. For example, a barrier function for the set Ω defined by (6.40) can be constructed with the aid of Burg's entropy as:

$$q(x) = -\sum_{l=1}^{L} \log g_l(x),$$

where g_l for, $l = 1, 2, \ldots, L$, are the functions used in the definition of Ω. With the use of such a barrier function the modified problem is written as: **Problem [$\mathcal{P}'$]:**

$$\text{Minimize} \quad F(x) + cq(x) \tag{6.41}$$

$$\text{subject to} \quad x^k \in X_k, \quad \text{for all } k = 1, 2, \ldots, K. \tag{6.42}$$

A solution to the problem [$\mathcal{P}$] can be approximated by solving the modified problem with the barrier function for a sufficiently small value of the parameter c. It is also possible to solve the barrier-modified problem repeatedly for a sequence of barrier parameters $\{c_\nu\}$, such that $c_\nu > c_{\nu+1} > 0$. If $\{x^\nu\}$ denotes the sequence of solutions of these barrier problems, then it is known that $\{x^\nu\}$ converges to a solution of [$\mathcal{P}$] as $c_\nu \to 0$.

Like the penalty-modified problem [$\mathcal{P}'$], the barrier-modified problem has a simpler constraint structure than problem [$\mathcal{P}$]. However, it still cannot be decomposed into independent components since the barrier function is not necessarily block-separable. The algorithms of the next section can be used to induce separability of penalty and barrier functions.

6.4.1.2 Modifier II: Variable Splitting and Augmented Lagrangian

The second modifier first *replicates* (or *splits*) the components x^k of the vector x into two copies, one of which is constrained to belong to the set X_k and the other constrained to satisfy the complicating constraints. Let $z \in \mathbb{R}^{nK}$ denote the replication of x, where $z = \left((z^1)^\mathsf{T}, (z^2)^\mathsf{T}, \ldots, (z^K)^\mathsf{T}\right)^\mathsf{T}$ and the vector $z^k \in \mathbb{R}^n$, for all $k = 1, 2, \ldots, K$. Consider now the equivalent split-variable formulation of [$\mathcal{P}$]: **Problem [Split-$\mathcal{P}$]:**

$$\text{Minimize} \quad \sum_{k=1}^{K} f_k(x^k) \tag{6.43}$$

$$\text{subject to} \quad x^k \in X_k, \quad \text{for all } k = 1, 2, \ldots, K, \tag{6.44}$$

$$z \in \Omega, \tag{6.45}$$

$$z^k = x^k, \quad \text{for all } k = 1, 2, \ldots, K. \tag{6.46}$$

The constraints $z^k = x^k$ link the variables that appear in the constraint sets X_k with the variables that appear in the set Ω. An augmented Lagrangian formulation is now used to eliminate these complicating constraints. We let $\pi \doteq \left((\pi^1)^\mathsf{T}, (\pi^2)^\mathsf{T}, \ldots, (\pi^K)^\mathsf{T}\right)^\mathsf{T}$ where $\pi^k \in \mathbb{R}^n$ denotes the Lagrange multiplier vector for the complicating constraints $z^k = x^k$, and let $c > 0$ be a constant. Then a partial augmented Lagrangian for (6.43)–(6.46) can be written as:

$$\mathcal{L}_c(x, z, \pi) = \sum_{k=1}^{K} f_k(x^k) + \sum_{k=1}^{K} \langle \pi^k, z^k - x^k \rangle + \frac{c}{2} \sum_{k=1}^{K} \| z^k - x^k \|^2. \tag{6.47}$$

A solution to problem [Split-$\mathcal{P}$] can be obtained by solving the dual problem: **Problem [$\mathcal{P}''$]:**

$$\underset{\pi \in \mathbb{R}^{nK}}{\text{Maximize}} \quad \varphi_c(\pi), \tag{6.48}$$

where $\varphi_c(\pi) \doteq \min_{x^k \in X_k, \, z \in \Omega} \mathcal{L}_c(x, z, \pi)$. This is the modified problem whose solution yields a solution of [$\mathcal{P}$].

An algorithm for solving convex optimization problems using augmented Lagrangians is the *method of multipliers*, which is an instance of the augmented Lagrangian algorithmic scheme found in Bertsekas [6] and Censor and Zenios [27]. It proceeds by minimizing the augmented Lagrangian for a fixed value of the Lagrange multiplier vector, followed by a simple update of this vector. Using the method of multipliers to solve the dual problem $[\mathcal{P}'']$ we obtain the following algorithm:

Algorithm 6.2: Method of Multipliers for Solving the Modified Problem $[\mathcal{P}'']$

Step 0: (Initialization.) Set $\nu = 0$. Let $\boldsymbol{\pi}^0$ be an arbitrary Lagrange multiplier vector.

Step 1: (Minimizing the augmented Lagrangian.)

$$\left(x^{\nu+1}, z^{\nu+1}\right) = \operatorname*{argmin}_{x^k \in X_k,\ z \in \Omega} \mathcal{L}_c\left(\boldsymbol{x}, z, \boldsymbol{\pi}^\nu\right) \tag{6.49}$$

Step 2: (Updating the Lagrange multiplier vector.) For $k = 1, 2, \ldots, K$, update:

$$(\pi^k)^{\nu+1} = (\pi^k)^\nu + c((z^k)^{\nu+1} - (x^k)^{\nu+1}). \tag{6.50}$$

Step 3: Replace $\nu \leftarrow \nu + 1$ and return to Step 1.

The minimization problem in Step 1 has a block-decomposable constraint set. The problem, however, still cannot be decomposed into K independent subproblems since the augmented Lagrangian is not block-separable due to the cross-products $\langle z^k, x^k \rangle$ in the quadratic term of (6.47). The next section explores algorithms that induce separability of this term. Step 2 consists of simple vector operations that can be executed very efficiently on parallel architectures.

6.4.2 SOLUTION ALGORITHMS

Both modified problems $[\mathcal{P}']$ and $[\mathcal{P}'']$ have a block-decomposable constraint set, but the objective function is not block-separable. These problems can be written in the general form:

$$\text{Minimize} \quad \Phi(\boldsymbol{x}) \doteq \Phi(x^1, x^2, \ldots, x^K) \tag{6.51}$$

$$\text{subject to} \quad \boldsymbol{x} \in X \doteq X_1 \times X_2 \times \cdots \times X_K. \tag{6.52}$$

We consider in this section two solution algorithms that — when applied to problems of this form — give rise to block-separable functions and thus decompose the problem into K subproblems, which can then be solved in parallel. The first algorithm uses linear approximations to the nonlinear function Φ. These linear approximations are block-separable. The second algorithm uses a diagonal approximation for cases where Φ is a nonseparable quadratic function, such as the last summand appearing in the augmented Lagrangian (6.47).

6.4.2.1 Solution Algorithms Based on Linearization

One of the earliest algorithms suggested for the solution of nonlinear optimization problems using linear approximations is the Frank–Wolfe algorithm [44]. It uses Taylor's expansion formula to obtain a first-order approximation $\hat{\Phi}$ to Φ around the current iterate $\boldsymbol{x}^\nu$, i.e.,

$$\hat{\Phi}(\boldsymbol{x}) = \Phi(\boldsymbol{x}^\nu) + \langle \nabla\Phi(\boldsymbol{x}^\nu), \boldsymbol{x} - \boldsymbol{x}^\nu \rangle,$$

ignoring second- and higher-order terms. The Frank–Wolfe algorithm now minimizes this linear function, subject to the original constraints. The solution of this linear program, $\overline{\boldsymbol{y}}$, is a vertex of the constraint set, which determines a direction of descent for the original nonlinear function, given by $\boldsymbol{p} = \overline{\boldsymbol{y}} - \boldsymbol{x}^\nu$. The algorithm then performs a one-dimensional search along this direction to determine the step length where the nonlinear function attains its minimum. Figure 6.4 illustrates the algorithm.

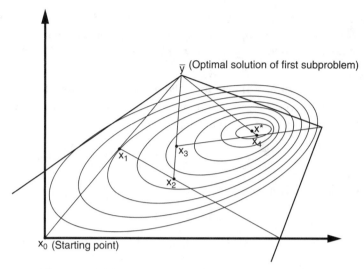

FIGURE 6.4 Illustration of the Frank–Wolfe linearization algorithm for a function $\Phi(x_1, x_2)$. (The curves denote level sets.)

6.4.2.2 Solution Algorithms Based on Diagonalization

Consider now a special structure of the objective function (6.51) arising from the modified problem $[\mathcal{P}'']$. In particular, we assume that $\Phi(\cdot)$ can be written as

$$\Phi(\mathbf{x}, \mathbf{z}) \doteq \sum_{k=1}^{K} \| x^k - z^k \|^2 . \tag{6.53}$$

This is the structure of the quadratic terms of the augmented Lagrangian (6.47). (We consider only the quadratic terms of the augmented Lagrangian, since these are the nonseparable terms that prevent us from decomposing the minimization of Step 1 of Algorithm 6.2 into K independent subproblems.) We will approximate this nonseparable function using a separable quadratic function. The term *diagonal quadratic approximation* is also used, which indicates that the Hessian matrix of $\Phi(\mathbf{x}, \mathbf{z})$ is approximated by a diagonal matrix.

The terms in (6.53) can be expanded as

$$\| x^k - z^k \|^2 = \| x^k \|^2 + \| z^k \|^2 - 2\langle x^k, z^k \rangle,$$

and we discuss only the cross-product terms $\langle x^k, z^k \rangle$ for $k = 1, 2, \ldots, K$. Using Taylor's expansion formula we obtain a first-order approximation of the cross-product terms around the current iterate $(\mathbf{x}^\nu, \mathbf{z}^\nu)$ as:

$$\langle x^k, z^k \rangle \approx \langle x^k, (z^k)^\nu \rangle - \langle (x^k)^\nu, (z^k)^\nu \rangle + \langle (x^k)^\nu, z^k \rangle,$$

for $k = 1, 2, \ldots, K$. With this approximation of its cross-product terms, the function (6.53) is approximated by the expression:

$$\sum_{k=1}^{K} \| x^k - (z^k)^\nu \|^2 - \sum_{k=1}^{K} \| (x^k)^\nu - (z^k)^\nu \|^2 + \sum_{k=1}^{K} \| (x^k)^\nu - z^k \|^2 .$$

A solution algorithm, based on diagonalization solves problem $[\mathcal{P}'']$ using the method of multipliers (Algorithm 6.2) but instead of minimizing the augmented Lagrangian in Step 1 it minimizes the diagonal quadratic approximation. This approximation is block-separable into the variable blocks $x^k, k = 1, 2, \ldots, K$, and the minimization is decomposed into K independent problems. These problems can be solved in parallel.

6.5 PARALLEL MATRIX FACTORIZATION PROCEDURES FOR THE INTERIOR POINT ALGORITHM

We turn now to the parallelization of interior point algorithms. In particular, we consider the solution of the linear systems (6.18) for structured problems. We assume that the constraint matrix A has the block-angular structure (6.17), which arises in the deterministic equivalent formulation of two-stage stochastic programs with recourse and in the dual formulation of multicommodity network flow problems. The stochastic programming problems and multicommodity network flow problems encountered in practical applications are of extremely large size, and matrices of this form occasionally have millions of columns and hundreds of thousands of rows. Furthermore, while these matrices are sparse the product matrix $A\Theta A^{\mathrm{T}}$ could be completely dense due to the presence of the matrices T_l, $l = 1, 2, \ldots, N$.

This section develops a matrix factorization procedure for solving the linear systems of equations for the calculation of Δy, that exploits the special structure of the constraint matrix A. The vectors Δy and ψ in Equation (6.18) are written as concatenations of subvectors, i.e., as $((\Delta y^0)^{\mathrm{T}}, (\Delta y^1)^{\mathrm{T}}, \ldots, (\Delta y^N)^{\mathrm{T}})^{\mathrm{T}}$, and $((\psi^0)^{\mathrm{T}}, (\psi^1)^{\mathrm{T}}, \ldots, (\psi^N)^{\mathrm{T}})^{\mathrm{T}}$, respectively, with Δy^l, $\psi^l \in \mathbb{R}^{m_l}$ for $l = 0, 1, \ldots, N$. Θ is assumed to be diagonal, as is the case for linear programs and for separable quadratic programs.

6.5.1 THE MATRIX FACTORIZATION PROCEDURE FOR THE DUAL STEP DIRECTION CALCULATION

The procedure for solving (6.18) is based on the use of a generalized version of the Sherman–Morrison–Woodbury formula. This formula is stated in the following lemma, where I denotes the identity matrix.

Lemma 6.1 *For matrices $A \in \mathbb{R}^{n \times n}$, $U \in \mathbb{R}^{n \times K}$, and $V \in \mathbb{R}^{n \times K}$ such that both A and $(I + V^{\mathrm{T}}A^{-1}U)$ are invertible,*

$$(A + UV^{\mathrm{T}})^{-1} = A^{-1} - A^{-1}U(I + V^{\mathrm{T}}A^{-1}U)^{-1}V^{\mathrm{T}}A^{-1}.$$

Proof. See, for example, Householder [50] or Censor and Zenios [27] (p. 228). □

The following theorem provides the foundation for the matrix factorization routine that solves the system (6.18) for Δy.

Theorem 6.1 *Let $M \doteq A\Theta A^{\mathrm{T}}$, where Θ is a diagonal matrix, and let $S \doteq \mathrm{diag}(S_0, S_1, S_2, \ldots, S_N)$, $S_l \doteq W_l \Theta_l W_l^{\mathrm{T}} \in \mathbb{R}^{m_l \times m_l}, l = 1, 2, \ldots, N$, where the matrix $S_0 \doteq I$ is an $m_0 \times m_0$ identity matrix, and for $l = 0, 1, 2, \ldots, N$, $\Theta_l \in \mathbb{R}^{n_l \times n_l}$ is the diagonal submatrix of Θ corresponding to the lth block. Also, denoting $(\Theta_0^{-1})^2$ by Θ_0^{-2}, let*

$$G_1 \doteq \Theta_0^{-2} + A_0^{\mathrm{T}}A_0 + \sum_{l=1}^{N} T_l^{\mathrm{T}} S_l^{-1} T_l, \tag{6.54}$$

$$G \doteq \begin{pmatrix} G_1 & A_0^{\mathrm{T}} \\ -A_0 & 0 \end{pmatrix}, \quad U \doteq \begin{pmatrix} A_0 & I \\ T_1 & 0 \\ \vdots & \vdots \\ T_N & 0 \end{pmatrix}, \quad V \doteq \begin{pmatrix} A_0 & -I \\ T_1 & 0 \\ \vdots & \vdots \\ T_N & 0 \end{pmatrix}.$$

If A_0 and W_l, $l = 1, 2, \ldots, N$, have full row rank then the matrices M and $G_2 \doteq -A_0 G_1^{-1} A_0^{\mathrm{T}}$ are invertible, and

$$M^{-1} = S^{-1} - S^{-1} U G^{-1} V^{\mathrm{T}} S^{-1}. \tag{6.55}$$

Proof. See Birge and Qi [10] or Censor and Zenios [27] (p. 229). $\qquad\qquad\square$

It is easy to verify, using equation (6.55), that the solution of the linear system $(A \Theta A^{\mathrm{T}}) \Delta y = \psi$ is given by $\Delta y = p - r$ where p solves the system $Sp = \psi$, and r is obtained from the system

$$Gq = V^{\mathrm{T}} p \quad \text{and} \quad Sr = Uq. \tag{6.56}$$

The vector p can be computed componentwise by solving $S_l p^l = \psi^l$, for $l = 1, 2, \ldots, N$. To solve for q we exploit the block structure of G and write:

$$Gq = \begin{pmatrix} G_1 & A_0^{\mathrm{T}} \\ -A_0 & 0 \end{pmatrix} \begin{pmatrix} q^1 \\ q^2 \end{pmatrix} = \begin{pmatrix} \hat{p}^1 \\ \hat{p}^2 \end{pmatrix} \quad \text{where} \quad \begin{pmatrix} \hat{p}^1 \\ \hat{p}^2 \end{pmatrix} \doteq V^{\mathrm{T}} p. \tag{6.57}$$

Hence, we get

$$q^2 = -G_2^{-1}(\hat{p}^2 + A_0 G_1^{-1} \hat{p}^1), \tag{6.58}$$

$$q^1 = G_1^{-1}(\hat{p}^1 - A_0^{\mathrm{T}} q^2). \tag{6.59}$$

Once q is known, r can be computed componentwise from (6.56). The procedure for calculating Δy can be summarized as follows. (We adopt here the notation A_i for the ith row of A, and A_j for the jth column of A.)

Procedure 6.1 Matrix factorization for dual step direction calculation

Step 1: *Solve $Sp = \psi$.*
Step 2: *(Solve $Gq = V^{\mathrm{T}} p$.)*

    a. *For all $l = 1, 2, \ldots, N$, solve $S_l(u^l)^i = (T_l)._i$, for $(u^l)^i$, $i = 1, 2, \ldots, n_0$ thus computing the columns of the matrix $S_l^{-1} T_l$.*
    b. *For all $l = 1, 2, \ldots, N$ multiply $T_l^{\mathrm{T}}(u^l)^i$, for $i = 1, 2, \ldots, n_0$ to form $T_l^{\mathrm{T}} S_l^{-1} T_l$. Form G_1 (Equation (6.54)). Compute $\hat{p}^1$, $\hat{p}^2$ (Equation (6.57)).*
    c. *Solve $G_1 u = \hat{p}^1$ for u and set $v = \hat{p}^2 + A_0 u$ (Equation (6.58)).*
    d. *Form G_2 by solving $G_1 w^i = (A_0^{\mathrm{T}})_i$ for w^i, for $i = 1, 2, \ldots, m_0$, and setting $G_2 = -A_0[w^1 \, w^2 \cdots w^{m_0}]$.*
    e. *Solve $G_2 q^2 = -v$ for q^2, and solve $G_1 q^1 = \hat{p}^1 - A_0^{\mathrm{T}} q^2$ for q^1, (Equation (6.59)).*

Step 3: *(Solve $Sr = Uq$.) Set $r^0 = A_0 q^1 + q^2$ and for all indices $l = 1, 2, \ldots, N$ solve $S_l r^l = T_l q^1$ for r^l.*
Step 4: *(Form Δy.) For $l = 1, 2, \ldots, N$, set $\Delta y^l = p^l - r^l$.*

6.5.1.1 Decompositions for Parallel Computing

Procedure 6.1 is well-suited for parallel implementation because it relies largely on matrix (sub)block computations that can be performed independent of one another. The computation begins with the submatrices T_l, W_l, and Θ_l and the vector segment ψ^l located at the lth processor. Processor l can compute S_l and proceed independently with all computations involving only local data.

Interprocessor data communication is necessary at only three instances in the parallel algorithm. After forming the terms $T_l^{\mathrm{T}} S_l^{-1} T_l$ in Steps 2a–2b of Procedure 6.1 in parallel, the processors communicate to form the matrix G_1 and the vectors $\hat{p}^1$ and $\hat{p}^2$ in Step 2b. The results can be accumulated at a single processor, designated as the *master*. The computations involving the dense matrix G_1 can be done serially on the master processor. Steps 2d–2e, which involve the dense matrix G_2, are also done serially on the master processor.

The master processor must then broadcast the computed vector q to all other nodes. The remaining Steps 3 and 4 require only the distributed data S_l, T_l, and p^l on the lth processor and may be carried out with full parallelism. A final communication step accumulates the partial vectors Δy^l at the master processor. This vector can then be made available to all processors for use in the calculation of the directions Δx and Δz of an interior point algorithm.

An alternative parallel implementation of Procedure 6.1 would distribute the matrices G_1 and G_2 to all processors and let the processors proceed locally (and redundantly) with all calculations involving these matrices. The approach based on a master processor described above uses an *all-to-one* communication step to accumulate the dense matrices at the master, followed by a *one-to-all* communication step to distribute the results to the processors. The alternative approach suggested here combines these two communication steps into a single *all-to-all* communication step that distributes the dense matrices to all processors. Both of these alternatives are very efficient on present day distributed memory machines with high-bandwidth communication networks.

Yet another alternative approach is to distribute the dense matrices across processors and use parallel dense linear algebra techniques for all calculations that involve these matrices. This approach is more suitable to shared-memory, tightly coupled multiprocessors, or when the dense matrices G_1 and G_2 are large.

6.6   NOTES AND REFERENCES

The development of programs for parallel machines is discussed by Carriero and Gelernter [18]. For alternative viewpoints on parallel computing in general, see Buzbee and Sharp [17], Deng et al. [35], and the special issue of the journal *Daedalus* [54]. A collection of references on parallel computing has been compiled by the Association for Computing Machinery [43]. Parallel computer architectures are treated by Hwang and Briggs [52] and Hwang [51]. Kumar et al. [61] also contains material on hardware aspects. A historical note on the development of parallel architectures is given by Desrochers [36]. A taxonomy of parallel architectures based on the interaction of instructions and data streams is due to Flynn [42].

For perspectives on parallel computing for optimization problems, see Kindervater et al. [56] and Zenios [86]. Advanced treatment of numerical methods for parallel and distributed computing, including extensive coverage of optimization problems, is given in Bertsekas and Tsitsiklis [7]. Special issues of journals focusing on parallel optimization were edited by Meyer and Zenios [71], Mangasarian and Meyer [66–68], and Rosen [79]. Pardalos et al. [74] treat some aspects of mathematical programming with respect to their potential for parallel implementations. Extensive lists of references on parallel computing for optimization are collected in Schnabel [80] and Zenios [85].

Methods of parallel computing for global optimization problems are collected in the volume by Pardalos [73] (see also Pardalos [72]). A discussion of parallel methods for unconstrained optimization and for the solution of nonlinear equations is found in Schnabel [81].

The use of array processors for linear programming was first discussed by Pfefferkorn and Tomlin [75], and the use of vector computing for the solution of large-scale nonlinear optimization problems by Zenios and Mulvey [87, 88].

Extensive discussion on the use of block-separability for the solution of large-scale optimization problems can be found in Griewank and Toint [46]. Block-separability was exploited, in the context of parallel computing, by Lescrenier [62], Lescrenier and Toint [63], and Zenios and Pinar [89]. Problems with block- or dual block-angular structure appear in many application domains.

For a general introduction to interior point algorithms, see Goldfarb and Todd [45], Kojima et al. [59], and Marsten et al. [70].

For a general discussion of direct matrix inversion methods see Householder [50]. Parallel implementation of interior point algorithms for dense problems is discussed in Qi and Zenios [77] and Eckstein et al. [38]. The exploitation of matrix structure in computing the dual step direction for interior point algorithms has been addressed in Choi and Goldfarb [29] and Czyzyk et al. [34]. The use of the Sherman–Morrison–Woodbury updating formula for computing dual steps for stochastic programs was suggested by Birge and Qi [10], and was further extended by Birge and Holmes [8] who also performed extensive numerical experiments with this method. The extension to multistage stochastic programming problems is discussed in Holmes [49]. Its use for solving multicommodity network flow problems has been suggested by Choi and Goldfarb [28, 29]. The parallel matrix factorization procedure and its implementation on hypercubes and other parallel machines, was developed by Jessup et al. [53]. Yang and Zenios [83] discuss the application of the parallel matrix factorization procedure within an interior point algorithm for the solution of large-scale stochastic programs.

For references on the convex feasibility problem see, for example, Stark and Yang [82], Censor and Zenios [27] and Combettes [31, 32] and references therein. See Herman [48] for applications in image reconstruction from projection, Censor et al. [20] and Censor [19] for applications in radiation therapy treatment planning, and Marks et al. [69] for an application in crystallography. Bauschke and Borwein [3] published a review paper on projection algorithms. For the best approximation problems, consult Deutsch [37] and for projection algorithms for its solution see, for example, Bregman et al. [12] and Bauschke [2].

Censor and Lent [25] introduced Bregman projections based on Bregman [11]. See, for example, Censor and Elfving [21], Censor and Reich [26], Censor and Zenios [27], De Pierro and Iusem [76], Kiwiel [57, 58] and Bauschke and Borwein [4], and references therein for further developments. For a treatments in non-Hilbertian Banach spaces see Butnariu and Iusem [16]. The multiprojection method, which appears in Censor and Elfving [21], was used by Shamir and coworkers [60,65]. The method of alternating Bregman projections appears in Csiszár and Tusnády [33]. Further studies on Bregman projections appear in Kiwiel [58], Bauschke and Borwein [4], and Butnariu et al. [14].

For sequential projection methods, see Bregman [11], Gubin et al. [47], and Youla [84]. The string averaging algorithmic scheme was proposed by Censor et al. [22]. For Cimmino's original algorithm see Ref. [30]. The BIP method was proposed by Aharoni and Censor [1], see also Butnariu and Censor [13], Bauschke and Borwein [3], Bauschke et al. [5], Elfving [40], and Eggermont, Herman and Lent [39]. The CAV and BICAV algorithms appear in Censor et al. [23, 24].

REFERENCES

[1] R. Aharoni and Y. Censor. Block-iterative projection methods for parallel computation of solutions to convex feasibility problems. *Linear Algebra and Its Applications*, 120:165–175, 1989.

[2] H.H. Bauschke. The approximation of fixed points of compositions of nonexpansive mappings in Hilbert space. *Journal of Mathematical Analysis and Applications*, 202:150–159, 1996.

[3] H.H. Bauschke and J.M. Borwein. On projection algorithms for solving convex feasibility problems. *SIAM Review*, 38:367–426, 1996.

[4] H.H. Bauschke and J.M. Borwein. Legendre functions and the method of random Bregman projections. *Journal of Convex Analysis*, 4:27–67, 1997.

[5] H.H. Bauschke, J.M. Borwein, and A.S. Lewis. The method of cyclic projections for closed convex sets in Hilbert space. *Contemporary Mathematics*, 204:1–38, 1997.

[6] D.P. Bertsekas. *Constrained Optimization and Lagrange Multipliers Methods*. Academic Press, New York, NY, USA, 1982.

[7] D.P. Bertsekas and J.N. Tsitsiklis. *Parallel and Distributed Computation: Numerical Methods*. Prentice Hall, Englewood Cliffs, NJ, USA, 1989.

[8] J.R. Birge and D.F. Holmes. Efficient solution of two-stage stochastic linear programs using interior point methods. *Computational Optimization and Applications*, 1:245–276, 1992.

[9] J.R. Birge and F. Louveaux. *Introduction to Stochastic Programming*. Springer-Verlag, Heidelberg, Germany, 1997.

[10] J.R. Birge and L. Qi. Computing block-angular Karmarkar projections with applications to stochastic programming. *Management Science*, 34:1472–1479, 1988.

[11] L.M. Bregman. The relaxation method of finding the common point of convex sets and its application to the solution of problems in convex programming. *USSR Computational Mathematics and Mathematical Physics*, 7:200–217, 1967.

[12] L.M. Bregman, Y. Censor, and S. Reich. Dykstra's algorithm as the nonlinear extension of Bregman's optimization method. *Journal of Convex Analysis*, 6:319–333, 1999.

[13] D. Butnariu and Y. Censor. Strong convergence of almost simultaneous block-iterative projection methods in Hilbert spaces. *Journal of Computational and Applied Mathematics*, 53:33–42, 1994.

[14] D. Butnariu, Y. Censor, and S. Reich. Iterative averaging of entropic projections for solving stochastic convex feasibility problems. *Computational Optimization and Applications*, 8:21–39, 1997.

[15] D. Butnariu, Y. Censor, and S. Reich, Eds. *Inherently Parallel Algorithms in Feasibility and Optimization and Their Applications*. Elsevier Science Publisher, Amsterdam, The Netherlands, 2001.

[16] D. Butnariu and A.N. Iusem. *Totally Convex Functions for Fixed Points Computation and Infinite Dimensional Optimization*. Kluwer Academic Publishers, Dordrecht, The Netherlands, 2000.

[17] B.L. Buzbee and D.H. Sharp. Perspectives on supercomputing. *Science*, 227:591–597, 1985.

[18] N. Carriero and D. Gelernter. *How to Write Parallel Programs: A First Course*. The MIT Press, Cambridge, MA, USA, 1992.

[19] Y. Censor. Mathematical optimization for the inverse problem of intensity modulated radiation therapy. In C.R. Palta and T.R. Mackie, Eds. *Intensity-Modulated Radiation Therapy: The State of the Art*, American Association of Physicists in Medicine, Medical Physics Monograph No. 29. Medical Physics Publishing, Madison, WI, USA, 2003, pp. 25–49.

[20] Y. Censor, M.D. Altschuler, and W.D. Powlis. On the use of Cimmino's simultaneous projections method for computing a solution of the inverse problem in radiation therapy treatment planning. *Inverse Problems*, 4:607–623, 1988.

[21] Y. Censor and T. Elfving. A multiprojections algorithm using Bregman projections in a product space. *Numerical Algorithms*, 8:221–239, 1994.

[22] Y. Censor, T. Elfving, and G.T. Herman. Averaging strings of sequential iterations for convex feasibility problems. In Y. Censor, D. Butnariu, and S. Reich, Eds. *Inherently Parallel Algorithms in Feasibility and Optimization and Their Applications*. Elsevier Science Publishers, Amsterdam, The Netherlands, 2001, pp. 101–114.

[23] Y. Censor, D. Gordon, and R. Gordon. Component averaging: An efficient iterative parallel algorithm for large and sparse unstructured problems. *Parallel Computing*, 27:777–808, 2001.

[24] Y. Censor, D. Gordon, and R. Gordon. BICAV: A block-iterative, parallel algorithm for sparse systems with pixel-related weigting. *IEEE Transactions on Medical Imaging*, 20:1050–1060, 2001.

[25] Y. Censor and A. Lent. An iterative row-action method for interval convex programming. *Journal of Optimization Theory and Applications*, 34:321–353, 1981.

[26] Y. Censor and S. Reich. Iterations of paracontractions and firmly nonexpansive operators with applications to feasibility and optimization. *Optimization*, 37:323–339, 1996.

[27] Y. Censor and S.A. Zenios. *Parallel Optimization: Theory, Algorithms, and Applications*. Oxford University Press, New York, NY, USA, 1997.

[28] I. Choi and D. Goldfarb. Solving multicommodity network flow problems by an interior point method. In T. Coleman and Y. Li, Eds. *Large Scale Numerical Optimization*. Society for Industrial and Applied Mathematics (SIAM), Philadelphia, PA, USA, 1990, pp. 58–69.

[29] I.C. Choi and D. Goldfarb. Exploiting special structure in a primal-dual path-following algorithm. *Mathematical Programming*, 58:33–52, 1993.

[30] G. Cimmino. Calcolo approssimato per le soluzioni dei sistemi di equazioni lineari. *La Ricerca Scientifica*, XVI, Series II, Anno IX, 1:326–333, 1938.

[31] P.L. Combettes. The foundations of set-theoretic estimation. *Proceedings of the IEEE*, 81:182–208, 1993.

[32] P.L. Combettes. The convex feasibility problem in image recovery. *Advances in Imaging and Electron Physics*, 95:155–270, 1996.

[33] I. Csiszár and G. Tusnády. Information geometry and alternating minimization procedures. *Statistics and Decisions*, (Suppl. 1):205–237, 1984.

[34] J. Czyzyk, R. Fourer, and S. Mehrotra. A study of the augmented system and the column-splitting approaches for solving two-stage stochastic linear programs by interior point methods. *ORSA Journal on Computing*, 7:474–490, 1995.

[35] Y. Deng, J. Glimm, and D.H. Sharp. Perspectives on parallel computing. *Daedalus*, 121:31–52, 1992.

[36] G. R. Desrochers. *Principles of Parallel and Multi-Processing*. McGraw-Hill, New York, NY, USA 1987.

[37] F. Deutsch. *Best Approximation in Inner Product Spaces*. Springer-Verlag, New York, NY, USA, 2001.

[38] J. Eckstein, R. Qi, V.I. Ragulin, and S.A. Zenios. Data-parallel implementations of dense linear programming algorithms. Report 92-05-06, Decision Sciences Department, The Wharton School, University of Pennsylvania, Philadelphia, PA, USA, 1992.

[39] P.P.B. Eggermont, G.T. Herman, and A. Lent. Iterative algorithms for large partitioned linear systems, with applications to image reconstruction. *Linear Algebra and its Applications*, 40:37–67, 1981.

[40] T. Elfving. Block-iterative methods for consistent and inconsistent linear equations. *Numerische Mathematik*, 35:1–12, 1980.

[41] A.V. Fiacco and G.P. McCormick. *Nonlinear Programming*: *Sequential Unconstrained Minimization Techniques*. In *Classics in Applied Mathematics*, Vol. 4. Society for Industrial and Applied Mathematics (SIAM), Philadelphia, PA, USA, 1990 (Republication of the work first published by Wiley, NY, USA, 1968).

[42] M.J. Flynn. Some computer organizations and their effectiveness. *IEEE Transactions on Computers*, C-21:948–960, 1972.

[43] Association for Computing Machinery. *Resources in Parallel and Concurrent Systems*. ACM Press, New York, NY, USA, 1991.

[44] M. Frank and P. Wolf. An algorithm for quadratic programming. *Naval Research Logistics Quarterly*, 3:95–110, 1956.

[45] D. Goldfarb and M.J. Todd. Linear programming. In G.L. Nemhauser, A.H.G. Rinnooy Kan, and M.J. Todd, Eds. *Handbooks in Operations Research and Management Science*, Vol. 1. North-Holland, Amsterdam, The Netherlands, 1989, pp. 73–170.

[46] A. Griewank and Ph.L. Toint. On the unconstrained optimization of partially separable functions. In M.J.D. Powell, Ed., *Nonlinear Optimization 1981*. Academic Press, New York, NY, USA, 1982, pp. 301–312.

[47] L. Gubin, B. Polyak, and E. Raik. The method of projections for finding the common point of convex sets. *USSR Computational Mathematics and Mathematical Physics*, 7:1–24, 1967.

[48] G.T. Herman. *Image Reconstruction from Projections: The Fundamentals of Computerized Tomography*. Academic Press, New York, NY, USA, 1980.

[49] D. Holmes. An explicit factorization for solving multistage stochastic linear programs using interior point methods. Report 93-18, Department of Industrial and Operations Engineering, The University of Michigan, Ann Arbor, MI, USA, 1993.

[50] A.S. Householder. *The Theory of Matrices in Numerical Analysis*. Dover Publications, New York, NY, USA, 1975.

[51] K. Hwang. *Advanced Computer Architecture: Parallelism, Scalability, Programmability.* McGraw-Hill, New York, NY, USA, 1987.

[52] K. Hwang and F.A. Briggs. *Computer Architecture and Parallel Processing.* McGraw-Hill, New York, NY, USA, 1984.

[53] E.R. Jessup, D. Yang, and S.A. Zenios. Parallel factorization of structured matrices arising in stochastic programming. *SIAM Journal on Optimization*, 4:833–846, 1994.

[54] Journal of the American Academy of Arts and Sciences. *Daedalus*, Winter 1992.

[55] P. Kall and S.W. Wallace. *Stochastic Programming.* John Wiley & Sons, New York, NY, USA, 1994.

[56] G.A.P. Kindervater, J.K. Lenstra, and A.H.G. Rinnooy Kan. Perspectives on parallel computing. *Operations Research*, 37:984–990, 1989.

[57] K. Kiwiel. Free-steering relaxation methods for problems with strictly convex costs and linear constraints. *Mathematics of Operations Research*, 22:326–349, 1997.

[58] K. Kiwiel. Generalized Bregman projections in convex feasibility problems. *Journal of Optimization Theory and Applications*, 96:139–157, 1998.

[59] M. Kojima, N. Megiddo, T. Noma, and A. Yoshise. A unified approach to interior point algorithms for linear complementarity problems. *Lecture Notes in Computer Science*, Springer-Verlag, New York, NY, USA, 1991.

[60] T. Kotzer, N. Cohen, and J. Shamir. A projection-based algorithm for consistent and incosistent constraints. *SIAM Journal on Optimization*, 7:527–546, 1997.

[61] V. Kumar, A. Grama, A. Gupta, and G. Karypis. *Introduction to Parallel Computing.* The Benjamin/Cummins Publishing Company, Redwood City, CA, USA, 1994.

[62] M. Lescrenier. Partially separable optimization and parallel computing. *Annals of Operations Research*, 14:213–224, 1988.

[63] M. Lescrenier and Ph. L. Toint. Large scale nonlinear optimization on the FPS164 and CRAY X-MP vector processors. *International Journal of Supercomputer Applications*, 2:66–81, 1988.

[64] I.J. Lustig, R.E. Marsten, and D.F. Shannon. Computational experience with a primal–dual interior point method for linear programming. *Linear Algebra and Its Applications*, 152:191–222, 1991.

[65] D. Lyszyk and J. Shamir. Signal processing under uncertain conditions by parallel projections onto fuzzy sets. *Journal of the Optical Society of America*, A 16:1602–1611, 1999.

[66] O.L. Mangasarian and R.R. Meyer, Eds. *Parallel Optimization*, Vol. 42, No. 2 of *Series* B, 1988. Special Issue of *Mathematical Programming*.

[67] O.L. Mangasarian and R.R. Meyer, Eds. *Parallel Optimization II*, Vol. 1, No. 4, 1991. Special Issue of *SIAM Journal on Optimization*.

[68] O.L. Mangasarian and R.R. Meyer, Eds. *Parallel Optimization III*, Vol. 4, No. 4, 1994. Special Issue of *SIAM Journal on Optimization*.

[69] L.D. Marks, W. Sinkler, and E. Landree. A feasible set approach to the crystallographic phase problem. *Acta Crystallographica*, A 55:601–612, 1999.

[70] R. Marsten, R. Subramanian, M. Saltzman, and D. Shanno. Interior point methods for linear programming: Just call Newton, Lagrange, Fiacco and McCormick! *Interfaces*, 20(4):105–116, 1990.

[71] R.R. Meyer and S.A. Zenios, Eds. *Parallel Optimization on Novel Computer Architectures*, Vol. 14 of *Annals of Operations Research*, 1988.

[72] P.M. Pardalos. *Proceedings of the 24th Annual Allerton Conference on Communication, Control and Computing.* University of Illinois, Urbana, IL, USA, 1986, pp. 812–821.

[73] P.M. Pardalos, Ed., *Advances in Optimization and Parallel Computing.* North-Holland, Amsterdam, The Netherlands, 1992.

[74] P.M. Pardalos, A.T. Phillips, and J.B. Rosen. *Topics in Parallel Computing in Mathematical Programming.* Science Press, New York, NY, USA, 1992.

[75] C.E. Pfefferkorn and J.A. Tomlin. Design of a linear programming system for the Illiac IV. Technical report, Department of Operations Research, Stanford University, Stanford, CA, USA, April 1976.

[76] A.R. De Pierro and A.N. Iusem. A relaxed version of Bregman's method for convex programming. *Journal of Optimization Theory and Applications*, 51:421–440, 1986.

[77] R.-J. Qi and S.A. Zenios. On the scalability of data-parallel decomposition algorithms for stochastic programs. *Journal of Parallel and Distributed Computing*, 22:565–570, 1994.

[78] R.T. Rockafellar. *Network Flows and Monotropic Programming.* John Wiley & Sons, New York, NY, USA, 1984.

[79] J.B. Rosen, Ed., *Supercomputers and Large-Scale Optimization: Algorithms, Software and Applications,* Vol. 22 of *Annals of Operations Research,* 1990.

[80] R.B. Schnabel. Parallel computing in optimization. In K. Schittkowski, Ed., *Computational Mathematical Programming.* Springer-Verlag, Berlin, Germany, 1985, pp. 357–382.

[81] R.B. Schnabel. A view of the limitations, opportunities and challenges in parallel nonlinear optimization. *Parallel Computing,* 21:875–905, 1995.

[82] H. Stark and Y. Yang. *Vector Space Projections: A Numerical Approach to Signal and Image Processing, Neural Nets and Optics.* John Wiley, New York, NY, USA, 1998.

[83] D. Yang and S.A. Zenios. A scalable parallel interior point algorithm for stochastic linear programming and robust optimization. *Computational Optimization and Applications,* 7:143–158, 1997.

[84] D.C. Youla. Mathematical theory of image restoration by the method of convex projections. In H. Stark, Ed., *Image Recovery: Theory and Applications,* Academic Press, Orlando, FL, USA, 1987, pp. 29–77.

[85] S.A. Zenios. Parallel numerical optimization: Current status and an annotated bibliography. *ORSA Journal on Computing,* 1:20–43, 1989.

[86] S.A. Zenios. Parallel and super-computing in the practice of management science. *Interfaces,* 24(5):122–140, 1994.

[87] S.A. Zenios and J.M. Mulvey. Nonlinear network programming on vector supercomputers: A study on the CRAY X-MP. *Operations Research,* 34:667–682, 1986.

[88] S.A. Zenios and J.M. Mulvey. Vectorization and multitasking of nonlinear network programming algorithms. *Mathematical Programming,* 42:449–470, 1988.

[89] S.A. Zenios and M.C. Pinar. Parallel block-partitioning of truncated newton for nonlinear network optimization. *SIAM Journal on Scientific and Statistical Computing,* 13:1173–1193, 1992.

7 Parallel Computing in Global Optimization

Marco D' Apuzzo, Marina Marino, Athanasios Migdalas, Panos M. Pardalos, and Gerardo Toraldo

CONTENTS

ABSTRACT

Global optimization problems (GOPs) arise in a wide range of real-world problems. They include applications in operations research, engineering, biological sciences, and computer science. The effective use of parallel machines in global optimization is a very promising area of research since, due to inherent difficulty of problems it studies, only instances of limited dimension can be solved in reasonable computer time on conventional machines. However, the use of parallel and distributed processing can substantially increase the possibilities for the success of the global optimization approach in practice. A survey on parallel algorithms proposed for the solution of GOPs is given. A majority of these algorithms belongs to the class of heuristic techniques. This is because such methods easily parallelize according to general principles. Even deterministic optimizing methods, such as branch-and-bound (BB) and interval methods, tend to be excellent candidates for parallel computing, although the effort required to achieve both efficient parallelization and to preserve their convergence properties is certainly greater than in the case of heuristic methods. That is why, until very recently, virtually no work (or a very little) had be done on parallel deterministic methods, although the effective parallelization of the search process appear the only way to make solution of large-scale problems practical.

7.1 INTRODUCTION

Global optimization problems (GOPs) are widespread in the mathematical modeling of real-world systems for a very broad range of applications. Such applications include engineering design and control, structural optimization, VLSI chip design and database problems, parameter estimation, molecular conformation problems, allocation and location problems, economies of scale, problems on graphs, and other combinatorial problems, etc. (see Refs. [1–10]).

The general GOPs to be considered is to find a function value $f^\star$ and an associated feasible point $x^\star$ such that

$$\text{(GOP)} \quad f^\star = f(x^\star) = \text{(global) } \min_{x \in \mathcal{S}} f(x)$$

where $\mathcal{S} \subset \mathbb{R}^n$ is some not necessarily convex compact set.

This is in general a very difficult problem since it may have many local minima and other stationary points. Although standard nonlinear programming (NLP) techniques (see Ref. [1, 11]) may obtain a local optimum or a stationary point to a global optimization problem, such a point will only be global when certain conditions are satisfied (such as quasiconvexity of the objective function in case of minimization). The problem of designing algorithms to compute global solutions is very difficult, since in general there are no local criteria in deciding whether a local solution is global. In fact, the task of determining the global (or even a local) minimum is $\mathcal{NP}$-hard, even for special cases of GOP [12, 13].

To find the global minima of large-scale problems in engineering and business applications, it is necessary to design algorithms which take advantage of the special structures usually present in such applications. Often this involves functions of a particular type, e.g., separable, bilinear, biconvex, Lipschitzian, monotonic, differential convex (d.c.), fractional, economy-of-scale (concave minimization), minimax, multiplicative, etc. Also the constraints may be simple, sparse, equilibration of network flows, reverse convex, linear, or even other optimization or complementarity problems as in hierarchical and equilibrium problems. For detailed description of most of these problem types, as well as their interrelations (see Refs. [14–19]) the textbooks and monographs on the subject [7, 8, 20–23], the area handbook [6], and the references therein.

Global optimization algorithms (both parallel and serial) can be categorized into two very broad groups, *stochastic* and *deterministic*, with respect to their implementation characteristics. Moreover, the algorithms can be classified as *optimizing* or *(meta-)heuristic* depending on whether there are or not *a priori* guarantees on the global (ϵ-) optimality of the furnished final solution. Optimizing deterministic algorithms are typically based either on successive approximation techniques [6, 8] that utilize the concepts of cuts, minorants and decomposition, or "enumerative" approaches that adaptively perform partition, search and bounding as in *branch-and-bound methods* (BB) [6–8, 21, 24–28] or in *interval methods* [29–42]. Heuristic algorithms, on the other hand, are typically based on stochastic search and randomization, as well as on grid search. Solution techniques such as random, grid, and multistart local search belong to this category [1, 2, 6, 9, 11, 21, 45, 46]. Adaptive stochastic search algorithms, such as clustering methods and Bayesian heuristics [1, 11, 21, 43, 47, 59–61], simulated annealing (SA) [62–67], evolutionary and genetic algorithms (GAs) [68–72], the greedy randomized adaptive search (GRASP) [9, 46, 73], as well as deterministic adaptive search techniques, such as the tabu search (TS) algorithm [67, 74, 79], are all of heuristic nature in practice, despite the theoretical results that hold for some of them. The optimizing deterministic methods pose prerequisites on the analytic properties of the functions involved as well as on the size and the general structure of the problems to which they can be applied. On the other hand, the heuristic approaches are less demanding of such properties. Thus, while optimizing methods are applicable to specific problem classes, heuristics can be used in a more general framework. Both are computing intensive; the latter methods more flexible in this respect.

A large number of serial deterministic optimizing methods for global optimization have been developed, but until very recently, virtually no work had been done on *parallel* deterministic methods for global optimization. On the other hand, most heuristic methods are very similar in their basic structure, and easily parallelize according to general principles. The deterministic optimizing algorithms of practical importance are usually BB approaches or interval methods and hence tend to be excellent candidates for parallel computing.

7.2 BRANCH-AND-BOUND ALGORITHMS AND INTERVAL ANALYSIS METHODS

BB is a deterministic method aimed at obtaining optimal solutions with prescribed accuracy. It was originally proposed for integer and combinatorial optimization problems, however, it can be applied to a wide class of problems, including GOPs with convex or nonconvex domains, with convex or nonconvex objective functions for which no particular continuity information is required. The basic idea is that the solution space is subdivided recursively into even smaller parts, a process called *branching* and represented by a *branching tree*. Large numbers of these smaller sets can be discarded from further exploration by utilizing lower *bounds* on the objective function. Hence, hopefully only a small part of the branching tree will be generated and explored. Thus, BB is classified as an *implicit tree search*, or *implicit enumeration* method. The lower bounds can be obtained using linear or convex nonlinear underestimation (convex envelope), Lagrangian duality, piecewise linear approximation, majorization and minorization methods based on the Lipschitzian properties of f, d.c. methods, interval analysis, and in particular the interval Newton method, etc. The partitioning can be rectangular, simplicial, and conical. Generally, rectangular partitions are used with convex envelopes and interval analysis methods, whereas simplicial and conical partitions are used to solve concave minimization problems (see Refs. [8, 21, 29, 32]). Of special interest here are techniques derived from interval analysis.

In recent years the use of techniques derived from interval analysis in GOP has become very popular, motivated both from computational and theoretical reasons, because they lead to very effective algorithms, for many of which the convergence to the global solution can be shown. Interval arithmetic operates on interval values rather than points and it allows to examine large areas of a space. The idea that is common to all interval-BB algorithms is to partition the set S into a collection of boxes for which the lower bound on the objective function, higher derivatives, Lipschitz constants, can be calculated by using interval techniques. A prototype pseudocode for the BB method, particularly suitable for a naive description of an interval analysis BB, is given below:

algorithm `Interval-BB`

Put S into an ordered list $\mathcal{L}$

$i \longleftarrow 1$

$f_{\text{best}} \longleftarrow +\infty$! Initial upper bound

$\mathcal{F} \longleftarrow \emptyset$! Fully explored subdomains

<u>do while</u> $\mathcal{L} \neq \emptyset$

    Pull a subdomain S^i from $\mathcal{L}$

    <u>if</u> `size`$(S^i) \leq \epsilon$ <u>then</u>

        $\mathcal{F} \longleftarrow \mathcal{F} \cup S^i$

        $i \longleftarrow i + 1$

        <u>cycle</u>

    <u>endif</u>

Compute lower and upper bounds: $f_{lbd} \leq f(x)|_{x \in S^i} \leq f_{ubd}$
if $f_{lbd} > f_{best}$ then
    Discard S^i from further exploration
    $i \longleftarrow i + 1$
    cycle
endif
if $f_{ubd} \leq f_{best}$ then
    $f_{best} \longleftarrow f_{ubd}$
endif
Branch S^i into Q subdomains $[S^i_k]^Q_{k=1}$
$\mathcal{L} \longleftarrow \mathcal{L} \cup [S^i_k]^Q_{k=1}$
$i \longleftarrow i + 1$
enddo
return $f_{best}, \mathcal{F}$

If S is a (box-shaped) region and the objective function is formulated by an analytic expression, then the subdomains S^i are subboxes in which bounds are calculated through interval analysis methods such as interval-Newton/generalized-bisection procedures. Detailed exposition of the interval analysis BB as well as of the necessary interval analysis is given in Refs. [29, 32, 39, 40]. Publicly available implementations of interval analysis and/or BBs based on it are discussed in Refs. [38, 42, 33–39]. Of these the former are implemented in C++ and the latter in Fortran 77 and 90, respectively. Additional packages are available from Ref. [80].

Among the most widely used BB algorithms in global optimization are the codes of αBB by Floudas and coworkers [24, 25] and BARON by Sahinidis and associates [26, 28]. The first is based on d.c. transformations. It identifies the structures of nonconvexities and uses customized lower bounding functions for special structures, for example bilinearities, whereas for general nonconvexities it uses the parameter α to derive lower bounding functions. It can be used on both constrained and unconstrained GOPs, and requires MINOS and/or NPSOL [81] to solve the resulting linear or convex optimization subproblems. It is callable both from C and Fortran. On the other hand, BARON combines interval analysis and duality with other enhanced concepts. It is particularly well-designed for structured nonconvex problems, able to solve large-scale instances. It can also handle less structured nonconvexities as far as the user supplies routines for underestimation. It can handle integer variables, it can be called from Fortran or GAMS, and requires MINOS or the OSL library for its specialized modules. For additional BB codes, consult the survey of Pintér [82] and the websites [80, 81].

There are several possible sources of parallelism in a BB algorithm including parallel expansion of the branching tree, expansion of a single node in parallel, and parallel evaluation of subproblems. In general, a BB can be parallelized in one of two ways; by using either low- or high-level parallelism. In *low-level parallelism* only parts of the sequential algorithm are parallelized and in such a manner that the sequential characteristics of the algorithm are not altered, for instance, the parallel algorithm will branch from the same subproblems in the same order, that is, the size of the branching tree is not affected. Examples of sources for low-level parallelism are the computation of the bounds, the selection of the subproblem to branch from, or the application of the subproblem elimination rule. In *high-level parallelism*, a coarse-grained paradigm is followed, for instance, several iterations of the main algorithm loop could be run in parallel. The resulting algorithm will be essentially different from its sequential counterpart; the order in which the work is performed by the parallel algorithm does not necessarily equal to the work done by the sequential one, and actually the branching trees may be totally different. Thus, unless caution is paid, the parallel BB may actually expand a much larger tree and consequently it may end up taking more time than its sequential counterpart. Such anomalies are well-known and well-discussed by several researchers [83, 88]; the *detrimental anomaly* refers to the case where the parallel BB is slower than the sequential counterpart, the *speedup anomaly* refers to the case where the parallel algorithm has superlinear speedup over the sequential one, the *efficiency anomaly* refers to the fact that the larger amount of memory in a parallel system allows often the processors to reduce the amount of work necessary to perform expansions. The work distribution and the load balancing adopted in an implementation of a parallel BB is important. For instance, parallel evaluation of subproblems could be done in one of several ways; one is to split the search tree statically between available processors to keep communication costs low, another is a dynamic distribution with farming of available work to distribute the work more evenly between the available processors, or a dynamic distribution with farming of large tasks, etc. It is important that the ratio of communication to calculation is kept as low as possible and that the algorithm is scalable, i.e., that it maintains constant performance as problem size and hardware size increase. In Refs. [88, 89], a portable parallel BB library is presented that will easily parallelize sequential BB algorithms for several architectures. The library overtakes the management of the subproblems and provides load balancers for balanced distribution of the subproblems. It can be used either for automatic parallelization of a sequential BB, but also to implement and test new load balancing strategies. An extensive survey of parallel BB algorithms for combinatorial optimization problems is given in Ref. [90].

The general idea in the parallelization of interval analysis BB is that each processor applies the algorithm to a box or hyperrectangle (subdomain), independently of all others. The upper bound found by any one of the processors is communicated immediately to others so that unnecessary box subdivisions are avoided. To keep all the processors busy all the time, *dynamic load balancing* is needed. The processors should always be assigned the "best" boxes to avoid work on boxes that the serial version would not consider. Three parallelization models have been used in the past: a *master–slave paradigm*, an *asynchronous multiple pool paradigm*, and a *combined paradigm*. In the master–slave paradigm, a master processor keeps a global list and a global upper bound. It sends boxes to slaves and receives their results. To reduce the communication, the BB tree is treated in a *depth-first order* [91]. There are a few important drawbacks associated with this approach: (i) the master becomes a bottleneck as the number of processors increases, and (ii) the length of the list kept is limited by the memory of the master. In the asynchronous multiple pool paradigm, Eriksson and coworkers [31, 37] uses a *processor farm* linked in a ring on an iPSC/2 hypercube. Each processor has its own list and the *best-first strategy* is used. New upper bounds are broadcasted asynchronously. Slightly super linear speedups are reported with this approach. Moore et al.[92] use a similar approach. However, with two substantial differences: (i) the *oldest-first strategy* is used and (ii) while Eriksson transmits requests through the ring sending the answers directly to the requesting processor, Leclerc et al. send requests to a randomly chosen processor. Later Leclerc [93] obtains a faster version by utilizing the *best-first strategy*. Berner [94] combines the master–slave and processor farm paradigms into a new scheme, where each processor keeps its own list and works essentially independent of the others. Upper bounds are distributed in asynchronous broadcasts and the best-first strategy is used. The master processor is used for dynamic load balancing and does not work on boxes. Only a few speedups were obtained on a CM5. Parallel BB algorithms based on interval analysis and dynamic load balancing with applications to GOPs in chemical engineering are the subject of Gau and Stadtherr [95–97].

Publicly available parallel code for interval analysis BB is found in Ref. [98]. It solves GOPs using generalized bisection and interval-Newton methods. It is highly scalable, coarse-grained code based on the single program multiple data (SPMD) paradigm (see Section 7.3) with balanced workload.

One of the most extensively tested parallel deterministic approaches is concerned with the global minimization of *partially separable functions* [2, 99–101]. The general problem considered is to find

$$(\text{SP}) \; f^\star = (\text{global}) \min_{(x,y,z) \in \, \mathcal{S}} \phi(x) + d^\mathrm{T} y + s(z)$$

where $\mathcal{S}$ is a nonempty bounded polyhedral set in $\mathbb{R}^{n+k+p}$. The nonlinear term $\phi(x)$ is assumed to be separable and concave whereas the nonlinear term $s(z)$ is assumed only to be convex (but not necessarily separable). A large number of functions can be expressed in this form including the very important classes of quadratic and synomial functions. The general parallel algorithm for solving the partially separable global optimization problem is a BB based on hyperrectangle branching and heuristic computations of upper and lower bounds for $f^\star$ based on convex problems [101].

A parallel BB for a nonconvex network design problem based on lower bounds obtained from the convex envelope of the objective function is proposed in Ref. [102]. Although the lower bounds are quite weak, the master–slave implementation of the algorithm on a 10-processor sequent balance proved to result in good speedups and good efficiencies.

Finally, the implementation of BB algorithms on computational grids for the solution of large challenging problems is considered. A computational grid (or metacomputer) is a collection of loosely coupled, geographically distributed, and heterogeneous computers. A computational grid enable the sharing, selection, and aggregation of computational sources by presenting them as a single resource for solving large-scale and data-intensive computing applications. The choice of the parallel strategy is strongly influenced by the computational grid characteristics, mainly:

- Heterogeneity of the computational resources available
- Low (and often uneven) bandwidth
- Dynamical availability of the computational resources

These points make a computational grid very different from a parallel machines so that very specific issues must be addressed to achieve high efficiency:

- A load balance that takes into account of not having well-balanced and time-constant computing resources available
- Availability of fault-tolerant algorithms
- Minimization of work at master level and maximizing the usage of worker machines

All these issues have been successfully addressed some recent works [103, 104] dealing with the parallel implementation of BB techniques on metacomputers, mainly for nonlinear problems. As an example, Goux and Leyffer [103] consider the solution of large mixed integer nonlinear programming (MINLP) problems

$$(\text{MINLP}) \; f^\star = \min f(x, y)$$
$$\text{subject to} \;\; g(x, y) \leq 0$$
$$x \in X, \; y \in Y \;\; \text{integer}$$

Note that the objective function is assumed to be convex (otherwise, just a local minimum will be found). The BB technique is a quite classical one for such kind of problems. Initially integer restrictions are ignored, the resulting nonlinear NLP is solved (in Ref. [103] by using a sequential quadratic programming algorithm), and let $(\bar{x}, \bar{y})$ be the solution; then if some of the components of $\bar{y}$ take an integer value at the solution, for one of such variables, say y_i, the algorithm branches two subproblems, respectively, with $y_i \leq \lfloor \bar{y}_i \rfloor$ and $y_i \geq \lfloor \bar{y}_i \rfloor + 1$ where $\lfloor \alpha \rfloor$ is the largest integer not greater than α. In a recursive way, therefore, a tree is generated in which each node corresponds to a NLP problem, as an edge indicates the addition of a branching bound. A node is fully explored when some fathoming rule is satisfied. What is specifically tailored for the computational environment, however, is the parallelization strategy. A *master–worker* strategy is adopted (as suggested in Ref. [105]); in order for the master to avoid too much communication, instead of receiving a single node, each processor handles a subtree (nested tree search). Then, no further communication between master and worker is needed, until the worker gets the partial result of its tree search. Great care must be taken in the implementation in order for the algorithm not to become very unbalanced. More details about the implementation based on Condor [106] as resource manager and MW [107] to build the parallel master–worker application is described in Ref. [105].

7.3   HEURISTICS AND METAHEURISTICS

Compared to the optimizing methods of the previous section, the methods examined here do not attempt to explore systematically the entire solution space of GOP. Instead, they can be considered as iterative methods that move from a given iterate to another within a neighborhood of the former, i.e., within a region specified by some distance function. The distances between such points can be measured either in combinatorial terms or in terms of certain norms, e.g., the Euclidean norm, depending on the nature of the problem under consideration and the representation or encoding of points. In this setting, these methods attempt to explore only promising parts of the solution space and the effort is placed in the selection of such exploration areas, in avoiding reexploration of the same areas and in avoiding getting stuck in nonpromising local optima.

Heuristics and metaheuristics constitute an increasingly essential component of solution approaches intended to tackle difficult problems, in general, and global and combinatorial problems in particular. Parallel versions of such methods are proposed with increasing frequency mainly for difficult combinatorial problems but also for GOPs to certain extent. Similar to BB methods, heuristic and metaheuristic techniques for combinatorial and GOPs share a few common basic parallelization ideas. Thus, in describing these ideas we examine a few parallel implementations across both fields.

Parallel heuristics and metaheuristics are designed along three parallelization strategies, independently of their inherent characteristics:

Fine-grained parallelization attempts to speedup the computations of the algorithm by identifying operations that can be carried in parallel, without altering the sequential character of the method, that is, it will produce the same solution, albeit faster. No attempt is made to diversify the exploration of the solution space. This is also known as *low-level parallelization*.

Single program multiple data (SPMD) parallelization attempts to increase the exploration of the solution space by initiating multiple searches that simultaneously, but not necessarily synchronously, move through the solution space. In multistart local search (see Section 7.3.1), as well as of other methods based on this concept, the SPMD parallelization is translated into parallelization of the outer loop. In this sense, these methods conform to *embarrassing parallelism*. However, various degrees of synchronization and cooperation between the search threads can be devised.

Domain decomposition attempts to solve in parallel a set of smaller problems and to compose an improved overall solution out of their solutions. The resulting algorithm is different from its sequential counterpart. SPMD and domain decomposition are both approaches of *coarse-grained parallelization*.

7.3.1 LOCAL SEARCH AND MULTISTART LOCAL SEARCH TECHNIQUES

Heuristics based on different variations of local search have been successfully applied to a variety of difficult combinatorial and global optimization optimization problems (see Refs. [9, 45, 108]). The idea behind these heuristics are quite simple; starting with a tentative solution point $x^0 \in S$, a neighborhood $\mathcal{N}(x^0)$ of x^0 is searched for a new point x^1 that is *better* than x^0 with respect to some predefined measure, e.g., the objective function f. A pseudocode for such a scheme is given below:

<u>method</u> LOCAL_SEARCH
<u>input</u> $x \in S$
$k \longleftarrow 0$
<u>do</u> ! for ever
        Search $\mathcal{N}(x^k)$ for a point x^{k+1} with $f(x^{k+1}) < f(x^k)$
        <u>if</u> no such point exists <u>then</u>
            <u>exit</u> ! point x^k cannot be improved
        <u>else</u>
            $x^k \longleftarrow x^{k+1}$
            $k \longleftarrow k + 1$
        <u>endif</u>
<u>enddo</u>

For combinatorial problems and certain global optimization problems, discrete or combinatorial neighborhoods are utilized [9, 45]. These are defined in terms of distance functions that measure distance in terms of elementary transformations or operations needed to obtain a point in S from another point in S. Otherwise, in GOPs, neighborhoods based on the Euclidean norms are in use. In Ref. [9], a classification of local search implementations is proposed based on the nature of the neighborhood, on the type of neighborhood search, and other factors. The choice of neighborhood and of the starting point is crucial for the performance of the algorithm. Typically, the algorithm is run from several different starting points, the final result of each run is compared to the others and the best point is selected as an "answer" to problem. This approach constitutes the *multistart local search* method and is the basis upon which *coarse-grained parallelization* of local search is implemented. The multistart local search is simple and easy to implement either sequentially or in parallel, and usually combines random sampling with a local minimization algorithm. Each of the randomly sampled points constitute a starting point from which a local minimizer is sought via a local minimization algorithm (Section 7.3.1.3).

The multistart *selective search method* (MSSS) is introduced in Ref. [109] and applied to *nonlinear least-square problems* (NLSP) in Ref. [48]. It is observed there that, under certain conditions, the methods for NLSP, such as Gauss–Newton and the Levenberg–Marquardt methods, possess a so-called selective minimization property. That is, for zero or very small residual least-square problems, iterates generated by these methods are locally attracted to local minima with sufficiently low function values and repelled from local minima with high function values. This property leads to the idea of modifying the multistart local search in a manner that instead of searching for a local minimizer from every starting point, a selective minimization algorithm is employed to try to find a minimizer that is better than the closest local minimizer. The problem considered by Velázquez et al. [48] is a GOP with

$$f(x) = \frac{1}{2} R(x)^\mathrm{T} R(x),$$
$$S \equiv \mathbb{R}^n,$$

where $R(x) = [r_1(x), \ldots, r_m(x)]^\mathrm{T} \in \mathbb{R}^m$ for some $m > n$. The gradient of $f(x)$ is $\nabla f(x) = J(x)^\mathrm{T} R(x)$, where $J(x) \in \mathbb{R}^{m \times n}$ is the Jacobian. If $x^\star$, is the global minimum, the problem for which $R(x^\star) = 0$, and hence $f(x^\star) = 0$, is called a zero-residual problem, whereas if these quantities are small but not zero, the problem is called a small-residual problem.

The *Levenberg–Marquardt method* is a classic technique for searching local minima to the NLSP. In this, given a current iterate x^k, the next iterate is calculated as

$$x^{k+1} = x^k - \alpha (J(x^k)^\mathrm{T} J(x^k + \mu_k I))^{-1} J(x^k)^\mathrm{T} R(x^k),$$

for some $\mu_k > 0$ and some step length $\alpha \in (0, 1]$, where I is the unit matrix. For $\mu_k = 0$, the method reduces to that of Gauss–Newton, which is numerically less stable than the former. In Ref. [48], the step length is fixed to 1, regardless of whether the function values decrease or not. This is considered as a key future in their implementation of the MSSS method, as it prevents the algorithm to be trapped easily at local optima. Hence, the algorithm will more likely locate a global minimum. The computational results presented are for test problems from Floudas and Pardalos [14], and the efficiency of the method is compared to the SA Fortran code `Simann` [110], to the tunneling method Fortran code `Tunnel` [111], and to a classical multistart local search Matlab code developed by Velázquez et al. [48]. They conclude that the numerical results indicate that the MSSS method is a viable and promising approach to solve the NLSP. Although they do not present results for a parallel implementation, their method easily parallelizes, as can be seen below, in the coarse-grain manner of the classical multistart local search method.

algorithm `multi-start_Levenberg-Marquardt`
input $\{y^i\}_{i=1}^m \in S, \tau, \epsilon, k_{\max}$
parallel do i=1,m
$\quad\quad x^0 \longleftarrow y^i; k \longleftarrow 0$
$\quad\quad$do
$\quad\quad\quad\quad$if $||J(x^k)^\mathrm{T} R(x^k)|| \le \epsilon$ or $k > k_{\max}$ then
$\quad\quad\quad\quad\quad\quad f^i \longleftarrow f(x^k)$
$\quad\quad\quad\quad\quad\quad x^i \longleftarrow x^k$
$\quad\quad\quad\quad\quad\quad$exit
$\quad\quad\quad\quad$endif
$\quad\quad\quad\quad \mu_k \longleftarrow \tau ||R(x^k)||$
$\quad\quad\quad\quad s^k \longleftarrow -(J(x^k)^\mathrm{T} J(x^k) + \mu_k I)^{-1} J(x^k)^\mathrm{T} R(x^k)$
$\quad\quad\quad\quad x^{k+1} \longleftarrow x^k + s^k$
$\quad\quad\quad\quad x^k \longleftarrow x^{k+1}; k \longleftarrow k+1$
$\quad\quad$enddo
parallel enddo
$f^\star \longleftarrow f(x^\star) = \min_{\{i=1,m\}}\{f^i\}$
return $f^\star, x^\star$

7.3.1.1   Tunneling Methods for Bounded Constrained Problems

There are other methods that can be considered of hybrid nature, as although the main structure of the method is deterministic in the sense that they approach the global solution in a sequential fashion, but they have also a stochastic element to explore the space. In this category we consider the tunneling methods (TM) [112, 113].

The TM are of hybrid nature, as their general design is deterministic in the sense that they find a sequence of local minima with monotonically decreasing objective function, by a two-phase process of finding a local minimum and then tunneling to find a point in another valley, that serve as the initial point to find the next local minimum. It ignores all the minima above the best already found; they can find minima at the same level and never find again the same minimum. They use a local method for both minimization and tunneling phases. But they also have a stochastic element, as the search for points in other valley is started in random directions. It is this stochastic element that can be exploited to perform a smart exploration of the space in parallel processors, to improve the performance and the speed of the method. A detailed description of the sequential methods can be found in Refs. [112, 113].

A description of the parallel method follows [114]. Once a local minimum has been obtained using any local method, to be able to tunnel from one valley to another using gradient-type methods, it is necessary to *destroy* the minimum, placing a pole at the minimum point x^* and generating directions that would move the iterates away from it. To find a point x^{tun} in another valley with less or equal value than $f(x^*) = f^*$, one has to solve the inequality

$$T(x) = f(x) - f(x^*) \leq 0 \tag{7.1}$$

and to place a pole at x^* any of the *tunneling functions*

$$T_e(x) = \left(f(x) - f^* \right) \exp\left(\frac{\lambda^*}{\| x - x^* \|} \right), \qquad T_c(x) = \frac{f(x) - f^*}{\| x - x^* \|^{\lambda^*}}$$

can be used.

Solving problem (7.1) is then equivalent to finding a x^{tun} such that

$$T_e(x^{\text{tun}}) \leq 0 \qquad \text{or} \qquad T_c(x^{\text{tun}}) \leq 0. \tag{7.2}$$

To solve this inequality problem, the descent directions algorithm adopted to find the local minimum can be used, as long as appropriate stopping conditions checking convergence for problem (7.2) are employed. Because, in many applications, there may exist multiple local minima at the same level, equality would then arise in (7.2). Thus, to avoid finding repeatedly minima at the same level, the corresponding destruction poles have to be active, implying the use of the following tunneling function:

$$T_e(x) = \left(f(x) - f^* \right) \prod_{i=1}^{t} \exp \frac{\lambda_i^*}{\| x - x_i^* \|}$$

where $t = 1$ is set as soon as an x_{t+1}^* is found with value strictly smaller than $f(x^*)$.

To achieve the parallelization of the method, the choice of initial points is exploited. The maximum number of initial points allowed for the tunneling phase, serves also to control the total amount of computing effort. The idea is the efficient exploration of the solution space in parallel. In Ref. [114], the parallelization of the method is accomplished along the following lines:

- There is a central processor P_0 that controls the process. It finds the first local minima and broadcasts the initial data and the first minimum to all processors P_i. When it receives a new minimum found by any processor, checks if the new minimum is the best found so far and proceeds to send this information to all processors. It keeps the information of all different local minima found. When the global solution is found, or the general stopping conditions have been satisfied, it stops the whole process.

- Every processor will perform the tunneling phase from the last (and best) local minimum it has obtained, in different random directions, and proceeds to find a local minimum when it finds a point in another valley. When a local minimum has been found, this information is sent to the central processor.

- Each processor checks if there is a message from the central processor only at the tunneling phase, as it is not wise to interrupt the minimization phase once a new valley has been found.

Because all the processors are searching for points in different valleys during the tunneling phase, starting from different initial points (first located in a neighborhood of the last local minimum and subsequently in the entire feasible region), they explore efficiently several regions of the feasible space simultaneously. Sequential implementations of the tunneling method have been used successfully to solve several academic and industrial problems, see for instance Refs. [115–118]. Description of sequential and parallel codes based on two different local methods are found in Refs. [119, 120].

7.3.1.2 Multistart Local Search for Nonconvex Minimization over Polytopes

The motivation behind such methods is the fact that a concave function, when minimized over a set of linear constraints, achieves its optimal solution at an extreme point of the feasible region. Therefore, there exists a linear cost function that yields the same optimal solution as the concave function. The concept is based on Euclidean distance neighborhoods.

Local search based on such noncombinatorial neighborhoods for concave network flow problems (NCFP) are explored by Minoux [121], Yaged [122], and Pardalos and coworkers [2, 124]. Assume we have a set $\{y^i\}_{i=1}^m$ of starting points in the feasible region $\mathcal{S}$. A multistart local search for NCFP can be based on the well-known *Frank–Wolfe method*.

algorithm `multi-start_Frank-Wolfe`
input $\{y^i\}_{i=1}^m \in \mathcal{S}$
parallel do i=1,m
  $\quad x^0 \longleftarrow y^i; k \longleftarrow 0$
  do
    $\quad\quad x^{k+1} \in \arg\min_{x \in \mathcal{S}} \nabla f(x^k)^\mathrm{T} x$
    $\quad\quad$ if $x^{k+1} = x^k$ then
      $\quad\quad\quad f^i \longleftarrow f(x^k)$
      $\quad\quad\quad x^i \longleftarrow x^k$
      $\quad\quad\quad$ exit
    $\quad\quad$ endif
    $\quad\quad x^k \longleftarrow x^{k+1}; k \longleftarrow k+1$
  enddo
parallel enddo
$f^\star \longleftarrow f(x^\star) = \min_{\{i=1,m\}}\{f^i\}$
return $f^\star, x^\star$

A major advantage of the algorithm is that it parallelizes easily with respect to the outer loop. In particular no communication is required between the different processes so generated. It is important to notice that such a parallelization is intended mainly to increase the probability of locating a good suboptimal solution by running from many starting points rather than to achieve speedup according to its classical definition. However, good efficiencies can be obtained and linear speedups have been observed [44, 124]. The choice of the set of starting points is crucial in this respect, because in applying the multistart local search we generate a set of local optima $\{x^i\}_{i=1}^m$, only n of which are distinct in general ($1 \leq n \leq m$). An important question that arises here is to find conditions that guarantee that $x^\star$ is a good approximate solution, or even a global optimum. Pardalos and coworkers [2, 123] derive theoretical conditions that guarantee the ϵ-optimality for Lipschitzian functions and relate the approach above to the space covering techniques [1, 11, 125].

Linearization of the original problem can be achieved not only by replacing the original objective function by its tangential approximation as in the Frank–Wolfe method above; Kim and Pardalos [126], for instance, devise a *dynamic slope scaling procedure* (DSSP) for the NCFP, where again a series of linear network flow problems are solved. A major advantage of this approach is that it can be applied to nondifferentiable objective functions, for instance, to the case of fixed-costs.

A version of multistart local search is applied to the *linear complementarity problem* (LCP) by Pardalos et al. [2, 20]. The LCP is concerned with finding an $x \geq 0$ such that

$$Mx + q \geq 0$$
$$x^{\mathrm{T}}(Mx + q) = 0,$$

where $M \sim n \times n$ and $q \in \mathbb{R}^n$, or proving that such an x does not exist. The problem is restated as a GOP with

$$f(x) = \frac{1}{2}x^{\mathrm{T}}(M + M^{\mathrm{T}})x + q^{\mathrm{T}}x,$$
$$\mathcal{S} = \{x : Mx + q \geq 0, x \geq 0\}.$$

It can be shown that if LCP has a solution, then it has a solution that occurs at a vertex of $\mathcal{S}$. However, LCP may lack solution even if the feasible region is nonempty. Assuming that $\mathcal{S}$ is bounded, the following two-phase parallel multistart local search algorithm based on linearizations is proposed for the LCP:

algorithm `multi-start_LCP`
input $\{u^i\}_{i=1}^m$ orthogonal directions, for instance the eigenvectors of $M + M^{\mathrm{T}}$
parallel do i=1,m
    Solve the multiple row linear program
      $v^i_{\pm} = \arg\min_{x \in \mathcal{S}} \pm u^{i^{\mathrm{T}}}x$
    to generate a set $\mathcal{V}$ of $k < 2n$ vertices v^i of $\mathcal{S}$.
    if $f(v^i) = 0$ for some vertex v^i then
      $f^\star \longleftarrow f(v^i)$
      $x^\star \longleftarrow v^i$
      return $f^\star, x^\star$
      terminate
    endif
  parallel enddo
Select a subset $\mathcal{W} \subseteq \mathcal{V}$ of vertices with the smallest objective function values.

forall $v^i \in \mathcal{W}$ parallel do

    $x^i \longleftarrow$ LOCAL_SEARCH (v^i)

    if $f(x^i) = 0$ then

        $f^\star \longleftarrow f(x^i)$

        $x^\star \longleftarrow x^i$

        return $f^\star, x^\star$

        exit

    endif

parallel enddo

Computational results on Cray 1S and X-MP/48 with this algorithm are reported in Refs. [20, 127], where the MINOS code is used in place of the LOCAL_SEARCH procedure, and where most of the test problems are solved within an average time of $O(n^4)$. Whenever the multi-start-LCP fails to solve a problem, LCP is transformed into a mixed binary linear program and solved as such with increasing computational effort.

7.3.1.3 Grids and Clusters

The main difficulty in obtaining with high probability a global optimum using the multistart local search is that the number of starting points chosen should be much larger than the usually unknown number of local minimizers relative to the neighborhood chosen. If one can assume that the local minimizers are "far" from each other, then the basic version of the multistart algorithm is often modified in one of two ways [11]:

The first is to assign each evaluated local optimum a neighborhood and whenever a search point enters this neighborhood it is moved to the local optimum. The crucial point is that the chosen neighborhood should be subset of the attraction neighborhood of the local optimum. Grid and covering methods are based on this.

The second variation consists in simultaneous hill climbs from many starting points and joining neighborhood points. Joining here means that all such points are replaced by one of them, the one with the lowest function value. Cluster analysis is used for this purpose leading to clustering methods.

Grid algorithms, also known as passive covering methods, are based on deterministic construction of a *grid* of points $\mathcal{G} = \{x^1, \ldots, x^n\}$ independently of the function values $f(x^i)$, $i = 1, \ldots, n$. In its most simple form, it will return the pair $(x^\star, f(x^\star))$, where $x^\star = \arg\min_{x^i \in \mathcal{G} \cap S}\{f(x^i)\}$, as the "answer" to the problem. In multistart local search the points in $\mathcal{G}$ are used as starting points for hill climbs to local optima. The Frank–Wolfe based algorithm of Section 7.3.1.2 belongs to the latter.

Active covering methods on the other hand use the information obtained to exclude noninteresting regions of S from farther search, whereas intensifying the search in other regions. Such methods are surveyed in Refs. [11, 125] and their suitability for parallel implementations is also discussed there.

Pure random search is the probabilistic analogy of the simple grid algorithm. The set $\mathcal{G}$ is generated by choosing randomly from a uniform distribution over S. The smallest value of f calculated is the candidate for global minimum value, and the point at which it is calculated is the candidate for the global optimal point.

In practice one would usually perform a multistart local search from several points with the lowest function values obtained. If additional information, which could enable biases to certain subsets of S, is available, then nonuniform sampling based on some probability distribution on S can be used.

Clustering methods are stochastic methods based on refinement of the multistart local search and random search. They are all very similar to each other in their basic structure: at each iteration, sample points are randomly generated over the feasible region S. A subset of those points are then selected as starting points for local searches.

A sample point is selected as a starting point if and only if it has the lowest function value of all points within some "critical distance" of it, for example, the usual Euclidean distance and certain threshold values can be used to determine if a point is close enough to be included into a particular cluster. The local minimization is finally begun from each start point and eventually terminates providing a local minimizer.

A probabilistic stopping rule is then applied to determine if another iteration is required. If so, the process is repeated by adding new sample points to the already existing sample (without the previously selected starting points). If not, then the local minimizer achieving the least function value is reported as the global minimizer.

A *Bayesian stopping rule* is usually used to terminate the procedure. In particular, let q denote the number of distinct local minimizers found after k iterations, and let $\sigma = \gamma \cdot k \cdot m$, where $\gamma \in (0, 1]$ is a fixed constant and m is the number of starting points in the sample. It has been shown that a Bayesian estimate of the total number of local minimizers is given by $(q(\sigma - 1))/(\sigma - q - 2)$ and that a Bayesian estimate of the portion of S covered by the attraction neighborhoods of the local minimizers found so far is given by $((\sigma - q - 1)(\sigma + q))/(\sigma(\sigma - 1))$. One such stopping rule used is that the first estimate must exceed q by not more than 0.5 and the second estimate must exceed it by 0.995.

Thorough treatment of stochastic methods and statistical inference methods for global optimization can be found in the books of Refs. [1, 11, 43, 47] and the surveys of [51, 54, 55, 128]. For additional information see Refs. [59, 60, 129].

A parallel algorithm based on these ideas is easily obtained:

- The feasible region S is first partitioned into p subregions, where p is the number of processors.
- Each processor generates sample points in its own subregion, and tentatively selects the candidate starting points. If any two neighboring regions contain candidate start points which are within the critical distance of each other, then the one with the higher function value is discarded.
- All starting points are then collected centrally and distributed to the processors to begin the local searches. If there are more starting points than processors, then each processor will begin another local minimization from a new starting point as soon as it has obtained a local minimizer for the previous point.

Phillips et al. [130] present a parallel implementation for NCFP on a Cray X-MP using multistart local search based on the Frank–Wolfe method (see Section 7.3.1.2), random initialization and the Bayesian stopping rule. Dixon and Jha [44] implemented in parallel on a transputer system tree four methods; the pure random search, the multistart local search initialized with points randomly generated from a uniform distribution, a clustering method, and the interval method (see Section 7.2), which is a deterministic exact method. The master–slave paradigm was used for the multistart and the clustering methods. They found out that the three variations of the local search theme were much faster than the exact approach, although the latter was shown very robust. The cluster method showed high overheads as the number of starting points increased, however, these overheads decrease superlinearly with the number of processors. The multistart method outperforms pure random search, whose speedup, however, appears to be perfect as the number of processors increases. Byrd and coworkers [53, 57, 58] developed several parallel implementations of clustering methods. The parts of the algorithm that execute in parallel between synchronization or communication points are based on local search and are therefore very computing intensive. Thus, possibly some processors remain idle for significant time between synchronization points due to the time irregularity of the local search, and also because the number of local searches is not necessarily a multiple of the number of processors. It is clear that this effect would be more pronounced whenever the number of processors is large. On relatively small problems, on the other hand, such a method performs quite an effective use of processors, as long as their number is small.

Also the algorithms by Byrd et al. [131–134] and by Smith and Schnabel [56, 135] can be classified as clustering techniques which attempt to overcome the limitations of the straightforward parallelization adopted in the standard approach above. In addition, the nonadaptive partition in subdomains causes the standard algorithm to put equal sampling effort in each subdomain, regardless of the importance of the subdomain with respect to global optimum. Byrd et al. propose asynchronous, adaptive, two-phase methods, the purpose of which are to concentrate firstly on the sampling and the minimization effort in the most promising subdomains, and secondly, to improve the load balancing to distribute more evenly the work among the processors and reduce substantially or eliminate completely processor idle time. During the first phase, points are randomly sampled in each subdomain, and starting points for local search are selected among them. Moreover, adaptive decisions are made that concern whether the subdomain should be splitted into smaller subdomains, what the new density of sample points for the subdomains should be, and what the relative priority of continuing to process the assigned subdomain should be. In the second phase, as processors become available, local search is initiated from each starting point selected in the first phase. Different scheduling strategies for such entirely asynchronous, dynamic parallel algorithms are possible. Smith and Schnabel [56, 135] have investigated such strategies thoroughly and found that *fully centralized* and *fully distributed* scheduling strategies are not comparable, with respect to scalability and ease of implementation, to the *centralized mediator* strategy that they devised. Computational results in Refs. [57, 136] obtained on network of workstations show that the adaptivity introduced in the first phase considerably improve the efficiency of the algorithm compared to its nonadaptive counterpart, particularly for problems with unevenly distributed local optima.

In Refs. [58, 137, 138] the previous algorithm is further improved to handle large-scale problems of molecular configuration and clustering. The first phase of this approach is essentially running an algorithm of the type discussed earlier. That is, sampling of points, selection of starting points, and local search from everyone of them is performed on each processor during the first phase. This phase is concerned with small subproblems, i.e., sampling and local search is performed only for a small subset of the decision variables, whereas the remaining variables are kept fixed (c.f., domain decomposition in Section 7.3). The second phase is performed on group rather than individual processors, and is concerned with the improvement of a subset of the best local minimizers found in the first phase; parallelism within each processor group is utilized, and fully dimensional local searches are launched from each such point to obtain new, improved local minima. Clearly, the second phase accounts for the overwhelm of the computational effort. In Ref. [134], this coarse-grained parallelization approach is described in detail. The approach uses *multiple levels of parallelism* and the task scheduling is performed in asynchronous fashion during some portions of the algorithm. The implementation on an Intel iPSC/860 and an Intel Delta demonstrated capabilities for effective use of massively parallel distributed memory architectures. The first phase is easily parallelized according to the lines discussed above; the initial sampling of points are distributed among the available processors, and then each local search is performed by a processor. If the number of the selected starting points is greater than the number of processors, local searches are assigned to processors as they become available. In the second phase, several levels of parallelism are exploited, and asynchronous scheduling techniques are used. In particular, they have found that for the molecular configuration problem it was beneficial to divide the processors in groups, each working asynchronously from the other groups and solving, using a parallel algorithm, the GOP assigned. The number of processors in each group is chosen to be the same as the number of second-phase fully dimensional local minimizations done per iteration for efficiency purposes, as this step is computationally the most demanding step of the second phase. The computational tests show that the asynchronous version of the implementation is more efficient in terms of processor utilization and also in terms of elapsed time.

7.3.1.4   Greedy Randomized Adaptive Search Procedure

The GRASP is a two-phase multistart local search method that has gained considerable popularity in combinatorial optimization (see Refs. [139–141]). In the first phase, a *randomized greedy technique* provides feasible solutions incorporating both greedy and random characteristics. In the second phase local searches are initialized from these points, and the final result is simply the best solution found over all searches (multistart local search). Typically, the sequential approach is thought of as an iterative process. That is, in each iteration, phase one is used to generate a starting point, then the local search of phase two is applied, before proceeding to the next iteration. The GRASP algorithm is described by the pseudocode below:

> algorithm GRASP
> do while stopping criteria not satisfied
>   call GREEDY_RANDOM_SOLUTION(Solution)
>   call LOCAL_SEARCH(Solution)
>   if Solution is better than Best_Solution_Found then
>   Best_Solution_Found ⟵ Solution
>   endif
> enddo
> return Best_Solution_Found

When implementing a GRASP for a particular problem the procedure for constructing the initial (feasible) solution must be supplied. Briefly, the construction phase can be described as stepwise adding one element at a time to the partial (incomplete) solution. The strategy for choosing the next element is based on randomly choosing the element from a list which is built up with regard to a greedy function. The heuristic is adaptive in the sense that the effect of already chosen elements are considered in this strategy.

The neighborhood mapping used in the local search phase of a GRASP must also be defined. Of course, different problems require different construction and local search strategies. The advantage of GRASP compared to other heuristics is that there are only two parameters to tune (the size of the candidate list and the number of GRASP iterations). Compared to TS, SA, and GA GRASP appears to be competitive with respect to the quality of the produced solutions and the efficiency and is easier to implement and tune. GRASP has been used extensively to solve difficult combinatorial optimization problems [46, 71, 139] but also GOP with combinatorial neighborhoods, such as the NCFP [9, 142] and the quadratic assignment problem (QAP) [72].

An obvious (and trivial) way of parallelizing GRASP is by its outer *do while* loop according to the SPMD strategy used in the multistart local search (see Section 7.3.1.2). In Ref. [73] the method is implemented in this fashion on a Kendal Square KRS-1 and excellent speedup is reported. A similar approach is adopted by Aiex et al. [143] to solve the three index assignment problem. In Ref. [144], a domain decomposition approach is adopted to parallelize GRASP; the original problem solved is decomposed into many smaller problems, and each processor is given a set of these to be solved with GRASP. The authors were able to solve large instances of the maximum independent set problem obtaining almost linear speedup on an eight processor Alliant FX/80 MIMD computer. Close to linear speedup is also obtained in Ref. [145] following a similar approach.

7.3.2 Simulated Annealing

Since first proposed several years ago, SA has attracted considerable attention as a solution technique for combinatorial and GOP (see Refs. [4, 11, 55, 62–66, 110, 146, 147]). It is a Monte Carlo method in the sense that it utilizes random number generation in some aspect of its search for a global optimum. It is based on an analogy between statistical mechanics and optimization. The term annealing refers to the process of a thermal system by first melting at high temperatures and then cooling based on an annealing schedule, until the vicinity of the solidification temperature is reached, where the system is allowed to reach its lowest energy state, the ground state. A general description of the method is given by the pseudocode below:

algorithm SA
$\overline{\text{input } x^0 \in \mathcal{S}}$
   $t > 0$! called the initial temperature
   $\alpha \in (0, 1)$! temperature reduction parameter
   $k_{\max}$! maximum number of internal iterations
$f_{\text{best}} \longleftarrow f^0 \longleftarrow f(x^0)$
$x_{\text{best}} \longleftarrow x^0$
$\underline{\text{do}}$
   $k \longleftarrow 0$
   $\underline{\text{do}}$
     Select $x^{k+1} \in \mathcal{N}(x^k)$
     Select randomly $r \in (0, 1)$
     $\underline{\text{if}}\ f^k \geq f^{k+1} \longleftarrow f(x^{k+1})\ \underline{\text{or}}\ r \leq \exp(\frac{f^k - f^{k+1}}{t})\ \underline{\text{then}}$
       $f^k \longleftarrow f^{k+1}$
       $x^k \longleftarrow x^{k+1}$
       $\underline{\text{if}}\ f_{k+1} < f_{\text{best}}\ \underline{\text{then}}$
         $f_{\text{best}} \longleftarrow f^{k+1}$
         $x_{\text{best}} \longleftarrow x^{k+1}$
       $\underline{\text{endif}}$
     $\underline{\text{endif}}$
     $k \longleftarrow k + 1$
   $\underline{\text{until}}\ k > k_{\max}$
   $t \longleftarrow \alpha t$! temperature reduction
$\underline{\text{until}}$ some stopping criterion is met
$\underline{\text{return}}\ f_{\text{best}}, x_{\text{best}}$

There are quite a few crucial decisions to be made to tune the algorithm to a successful implementation. The initial temperature must be chosen carefully, as a too high value slows down the algorithm, whereas a value too close to 0 will exclude the uphill moves, reducing thus the approach to a pure local search (a possible choice for initial temperature calculation is suggested in Ref. [64]). For the same reasons, the reduction of the temperature should not be either too fast or too slow. The choice of $k_{\max}$ as well as the stopping criterion have great influence on the behavior of the algorithm.

The crucial step in any SA algorithm is the decision whether or not to accept the "trial" point x^{k+1} in place of the current iterate x^k. The acceptance is made with probability $\min\{1, \exp((f^k - f^{k+1})/t)\}$, which implies that every downhill movement is accepted, but it is also possible to accept uphill moves, albeit to a limited extent.

The general concepts of parallel SA techniques are discussed in Refs. [63, 65]. A major issue in these discussions is whether the adopted parallelization strategy affects the statistical convergence properties of SA. Several parallel SA algorithms have been implemented for combinatorial [46, 67, 71] and global optimization problems [62, 66, 110, 147]. The parallel implementations follow the principles outlined in Section 7.3.

Fine-grained parallelization of SA can be realized mainly in two ways: *single-trial* parallelism, where only one move is examined at a time, and *multiple-trial* parallelism, where several moves are evaluated simultaneously. In the former case, *functional parallelism* is exploited, that is, the sequential character of the method is preserved but intensive tasks are split into subtasks and assigned to other processes. Although the approach preserves the original convergence properties of the sequential SA, it is strongly application dependent and it also leads to moderate speedups. The multiple-trial approach on the other hand executes one main iteration (outer **do** loop above) on separate processors. However, this raises a considerable problem; caution, possibly by synchronization or error control, must be taken when f_{best} and x_{best} are replaced. Synchronization effectively imposes the sequential character of the original method and therefore preserves the convergence properties of SA. If multiple-trial parallelization is combined with data partition, parallelization according to the principle of domain decomposition is obtained. Preserving the convergence properties of the sequential SA can be done by strict partition of the moves and restoration of the global state, i.e., evaluation of f, following each synchronization. Unless the underlying hardware is a shared memory system, this may cause considerable communication degrading thus the performance of the algorithm. A second approach would then be to accept computation errors and place instead the effort on controlling the amount of such errors. Finally, parallelization according to the SPMD principle can be done in the fashion of the multistart local search and the GRASP algorithms of the previous sections. Attempts have been made to reduce the number of iterations in each search by introducing synchronous cooperation with exchange of information at the end of each iteration. However, unless this is done very carefully, it may significantly damage the convergence properties of the resulting parallel SA. For a detailed survey on this subject see Crainic and Toulouse [67]. In Ref. [62] a parallelization of SA that consists in mapping the algorithm onto a dynamically structured tree of processors is proposed.

A hybrid algorithm, called the *two-level SA* and based on the combination of the algorithm above with pure local search, has shown much more efficient in practice [4, 148]. For any given feasible point $x \in \mathcal{S}$, let $\underline{x}$ be the local optimum with respect to neighborhood mapping $\mathcal{N}(\cdot)$ computed using local search with starting point x. Suppose further that $y \in \mathcal{N}(x)$. In the algorithm above we determine whether to accept or reject y by comparing $f(y)$ to $f(x)$. However, since we are mainly interested in local minimizers, it seems more appropriate to perform the test for acceptance or rejection of y by comparing the values $f(\underline{y})$ to $f(\underline{x})$ instead. This is the main idea of the two-level SA which is stated below:

algorithm two-level_SA
input $x^0 \in \mathcal{S}$
       $t > 0$! called the initial temperature

$\alpha \in (0, 1)$! temperature reduction parameter
$k_{\max}$! maximum number of internal iterations
$\underline{x}^0 \longleftarrow$ LOCAL_SEARCH(x^0)
$f_{\text{best}} \longleftarrow f^0 \longleftarrow f(\underline{x}^0)$
$x_{\text{best}} \longleftarrow \underline{x}^0$
<u>do</u>
    $k \longleftarrow 0$
    <u>do</u>
        Select $x^{k+1} \in \mathcal{N}(x^k)$
        $\underline{x}^{k+1} \longleftarrow$ LOCAL_SEARCH(x^{k+1})
        Select randomly $r \in (0, 1)$
        <u>if</u> $f^k \geq f^{k+1} \longleftarrow f(\underline{x}^{k+1})$ <u>or</u> $r \leq \exp(\frac{f_k - f_{k+1}}{t})$ <u>then</u>
            $f^k \longleftarrow f^{k+1}$
            $x^k \longleftarrow x^{k+1}$
            $\underline{x}^k \longleftarrow \underline{x}^{k+1}$
            <u>if</u> $f^{k+1} < f_{\text{best}}$ <u>then</u>
                $f_{\text{best}} \longleftarrow f^{k+1}$
                $x_{\text{best}} \longleftarrow \underline{x}^{k+1}$
            <u>endif</u>
        <u>endif</u>
        $k \longleftarrow k + 1$
    <u>until</u> $k > k_{\max}$
    $t \longleftarrow \alpha t$! temperature reduction
<u>until</u> some stopping criterion is met
<u>return</u> $f_{\text{best}}, x_{\text{best}}$

As can be seen from this pseudocode, the two-level simulated annealing operates on two sequences of points, i.e., $\{x^k\}$ and $\{\underline{x}^k\}$. The first sequence constitutes the upper level whereas the second is designated as the lower level for the simple reason that the objective function of $\underline{x}^k$ is lower than or equal to that of x^k. The selection of neighboring points is made on the upper level whereas the decision of accepting or rejecting a move is based on comparison of the objective function values on the lower level. This reflects also the naming of the algorithm. A synchronized master–slave implementation of the two-level variant is given by Xue [4] for the molecular conformation problem.

7.3.3 TABU SEARCH

TS was introduced for $\mathcal{NP}$-hard combinatorial problems by Glover [74, 75] who considered it as a *meta-heuristic*, that is, a guiding technique over imposed on a local search. In this sense, TS provides a stepwise search where each step is proceeded according to a local search method. Because the local search gets trapped at local optima, the purpose of the TS is to provide means to avoid getting stuck at such points and instead be able to both intensify the search around a local optimum and diversify the search to new regions of the feasible solution space. To reach these goals, TS uses flexible memory—as opposed to memoryless local search—based on local search history, tabu restrictions, aspiration criteria, intensification criteria, and diversification criteria. *Flexible memory* permits evaluation criteria and historical search information to be exploited more thoroughly. *Tabu restrictions* and *aspiration criteria* implement conditions that strategically constrain the search process, whereas *intensification* and *diversification criteria* are used in reinforcing the attributes historically found good and driving the search into new regions.

Conceptually, TS is an iterative procedure that transforms one point in the solution space to the next one which is obtained from the previous by performing a set of moves. In order for the search to move out from the attraction neighborhood of a local optimum, it occasionally accepts a move even if such a move fails to be of descent nature. To avoid cycling, i.e., regeneration of moves, a phenomenon associated with the occasional nondescent moves taken, flexible memory is utilized. Typically, this is implemented as a list, called the *tabu list*, of forbidden moves. Moves get the tabu status whenever they satisfy a set of tabu restrictions. Usually the tabu list is restricted to the last few points (specified number) visited by the search. The length of the tabu list is a factor of great importance for the performance of the algorithm and is usually decided through experimentation. Note that the use of tabu lists restricts but does not prevent the occurrence of cycles of larger length. Therefore, the procedure is equipped with a termination criterion for such a case. A prototype for the method is given in the code below, where instead of moves, points enter the tabu list. It is an idealized presentation of the method, where tabu lists of fixed size are assumed.

$$
\begin{aligned}
&\underline{\text{method}}\ \texttt{TS} \\
&\underline{\text{input}}\ x^0 \in \mathcal{S} \\
&\qquad k_{\max}\ !\ \text{max number of nonproductive moves} \\
&\qquad q_{\max}\ !\ \text{max number of tabu elements} \\
&f_{\text{best}} \longleftarrow f^0 \longleftarrow f(x^0) \\
&x_{\text{best}} \longleftarrow x^0 \\
&k_{\text{iter}} \longleftarrow 0\ !\ \text{overall iteration counter} \\
&\ell_{\text{iter}} \longleftarrow 0\ !\ \text{record of last improvement move} \\
&\mathcal{T} \longleftarrow \emptyset\ !\ \text{tabu list} \\
&\underline{\text{do}} \\
&\qquad \underline{\text{if}}\ k_{\text{iter}} - \ell_{\text{iter}} \geq k_{\max}\ \underline{\text{then}} \\
&\qquad\qquad \underline{\text{exit}}\ !\ \text{No progress} \\
&\qquad \underline{\text{else}} \\
&\qquad\qquad \text{Select a neighborhood}\ \mathcal{N}(x^k) \\
&\qquad\qquad x^{k+1} \longleftarrow \arg\min_{y \in \mathcal{N}(x^k)} f(y) \\
&\qquad\qquad f^{k+1} \longleftarrow f(x^{k+1}) \\
&\qquad\qquad \underline{\text{if}}\ x^{k+1} \notin \mathcal{T}\ \underline{\text{or}}\ \text{aspiration criteria satisfied}\ \underline{\text{then}} \\
&\qquad\qquad\qquad \ell_{\text{iter}} \longleftarrow k_{\text{iter}}\ !\ \text{Accept the move} \\
&\qquad\qquad\qquad \underline{\text{if}}\ f^{k+1} < f_{\text{best}}\ \underline{\text{then}} \\
&\qquad\qquad\qquad\qquad f_{\text{best}} \longleftarrow f^{k+1} \\
&\qquad\qquad\qquad\qquad x_{\text{best}} \longleftarrow x^{k+1} \\
&\qquad\qquad\qquad \underline{\text{endif}} \\
&\qquad\qquad\qquad \underline{\text{if}}\ |\mathcal{T}| = q_{\max}\ \underline{\text{then}} \\
&\qquad\qquad\qquad\qquad \mathcal{T} \longleftarrow (\mathcal{T} \setminus \{\texttt{oldest}(y \in \mathcal{T})\}) \bigcup \{x^{k+1}\} \\
&\qquad\qquad\qquad \underline{\text{else}} \\
&\qquad\qquad\qquad\qquad \mathcal{T} \longleftarrow \mathcal{T} \bigcup \{x^{k+1}\} \\
&\qquad\qquad\qquad \underline{\text{endif}} \\
&\qquad\qquad\qquad f^k \longleftarrow f^{k+1} \\
&\qquad\qquad\qquad x^k \longleftarrow x^{k+1} \\
&\qquad\qquad \underline{\text{endif}} \\
&\qquad\qquad k_{\text{iter}} \longleftarrow k_{\text{iter}} + 1 \\
&\qquad \underline{\text{endif}}
\end{aligned}
$$

<u>enddo</u>
<u>return</u> f_{best}, x_{best}

The tabu status of moves are overridden whenever at least one of a set of aspiration criteria are satisfied. Intensification of the search in a promising region is usually enforced through penalization. Restarting the procedure from different initial points is typical in an effort not to neglect entirely regions of the solution space (multistart search). A more detailed description of the main features, aspects, applications and extensions of TS can be found in Ref. [76].

TS has been found to compare advantageously with popular Monte Carlo methods, such as the SA method, and has therefore attracted considerable applications in combinatorial optimization running both sequentially and in parallel (see Refs. [67, 71, 79, 149]). For continuous optimization problems, TS based on continuous neighborhoods has been proposed only recently by Cvijović and Klinowski [78]. They introduce the notion of *conditional neighborhood*, a discretized neighborhood, for global optimization problems with simple constraints, i.e., S is a hyperrectangle in $\mathbb{R}^n$. The solution space is partitioned into disjunct cells by division of the coordinate intervals along the $x_1, \ldots, x_n$ axes into $p_1, \ldots, p_n$ parts. The partition parameters p_i are problem-specific and empirical. They induce a unique partition of S into cells with unique "addresses." At each iteration, n_s sample points are drawn from a uniform distribution over n_c randomly chosen cells. These points are the neighbors of the current iterate x^k. The size of this neighborhood is $n_s \times n_c$ and remains constant whereas the content of the neighborhood changes at each iteration. At each iteration, the TS accepts a nontabu move with the lowest value, not necessarily to a cell that is a neighbor of the current solution. Limited computational comparisons with SA, pure random search, and a clustering method on a small set of standard test functions gave the indication that the continuous TS is reliable and efficient.

Battiti and Tecchiolli [77] introduce the notion of *reactive* TS for combinatorial optimization, where they explicitly check for the repetition of point configurations and compute the appropriate size of the tabu list automatically by reacting to occurrence of cycles. In addition, if the search appears to be repeating an excessive number of points excessively often, then the search is diversified by making many random moves proportional to a moving average of the cycle length. Moreover, they extend the approach to GOPs with simple constraints by combining the combinatorial optimization algorithm with a stochastic local minimizer in a hybrid scheme. The initial hyperrectangle is partitioned in boxes and a tree of boxes is formed. The tree has 2^n leaves of equal size which are obtained by dividing in half the initial range of each variable. Each box is then subdivided into 2^n equal-sized children, as soon as two different local optima are found in it. The reactive TS locates the most promising boxes where starting points for the local minimizer are generated. The algorithm is compared on a set of standard benchmarks with some existing algorithms, and in particular with the optimizing method of Strongin and coworkers [150, 151], which was proposed for Lipschitz functions.

For structured GOPs on the other hand, combinatorial neighborhoods enter in a natural way. For the production-transportation problem (PTP), for instance, Ghannadan et al. [152, 153] derived a TS based on a combinatorial neighborhood. With this approach they were able to solve instances of the problem for which an exact approach would require more than 7 days in just 40 min.

Parallelization of TS follows the general strategies of Section 7.3. Fine-grained parallelism executes the TS in serial manner but evaluates the possible moves in the neighborhood of the current iterate in parallel. The approach can utilize slave processors to achieve this. An implementation that does not require a specific master processor is also possible. In this case, each processor broadcasts its best move to all other processors, which then perform all the tasks of a master. In Ref. [154], a third approach, called *probing* is introduced. In this, the slaves also perform a few local search iterations. Parallelization based on domain decomposition can be based on partitioning the vector x of decision variables and perform a TS on each subset. This is the approach taken in Ref. [154]. SPMD parallelization of TS in the sense of multistart search is the most straightforward approach. The technique of Battiti and Tecchiolli [155] is along these lines. Cooperation between the parallel processes is also a possibility. Computational results with different parallel implementation of TS for the QAP and for combinatorial optimization problems are surveyed in Refs. [67, 79], where also extensive and systematic taxonomies of the parallel TS algorithms are presented.

7.3.4 CONTROLLED RANDOM SEARCH METHODS

The controlled random search algorithm (CRSA) is a stochastic direct search global optimization algorithm proposed by Price [156], which is applicable both to constrained and unconstrained problems. Chronologically, it is among the first methods to employ population of candidate points. It is based on clustering techniques and is considered robust and very reliable for problems with noise in the object function (noise tolerant). Although the algorithm has been proven as such in practice, it lacks theoretical results that support its behavior. Direct search algorithms tend to require many more function evaluations compared to gradient-based algorithms to converge to the minimum. The latter, however, cannot be used in situations where the analytical derivatives are unavailable and where numerical derivatives cannot be employed due to the noises in the objective function, as for instance, whenever the objective function is available in the form of a "black-box" code.

algorithm Controlled Random Search **CRSA**
input $\{x^i\}_{i=1}^m \in \mathcal{S}$ (random sample points in $\mathcal{S}$)
repeat
       STEP 1 Determine the points x^B and x^W in $\mathcal{S}$ with
                 the lowest and the highest $f(x)$ value.
       STEP 2 Generate a new point x^N in $\mathcal{S}$
       STEP 3 If $x^N \in \mathcal{S}$ and $f(x^N) < f(x^W)$ then
$\mathcal{S} = \mathcal{S} \cup \{x^N\} - \{x^W\}$
            go to STEP 2
until termination conditions are not satisfied
return $f(x^B), x^B$

The core operation in CRSA is obviously the way in which STEP 2 is performed. In the original formulation of Price,

$$x^N = \frac{1}{m-1} \sum_{i=1}^{m-1} x^{P_i} + x^{P_m},$$

where $\{x^{P_i}, x^{P_i}, \ldots, x^{P_m}\}$ is a set of points randomly chosen in STEP 2. Different modifications, mainly different crossover strategies, have been proposed by Price [157, 158], Ali and coworkers [159, 160] and Mohan and Shanker [161, 162], which improved the effectiveness of the original algorithm.

In their original formulations the CRSAs do not show the same parallelism of GAs (see Section 7.3.5). Nevertheless, interesting and quite effective parallelizations of the CRSA, have been proposed by Price and coworkers [158, 163] for CAD acceleration, by Garcia et al. [164–166] in GOPs with computationally expensive objective function in image processing, and by McKeqwon [167]. Parallelism can be basically introduced in generating the initial set $\mathcal{S}$ in STEP 1 and in generating several offspring in STEP 2 performing crossover operations on each processor.

7.3.5 EVOLUTIONARY AND GENETIC ALGORITHMS

GAs are based on an imitation of the biological process in which new and better populations among different species are developed during evolution. Thus, unlike most standard heuristics, GAs use information of a *population* of solutions, called *individuals*, when they search for better solutions. A GA aims at computing suboptimal solutions by letting a randomly generated population undergo a series of unary and binary transformations governed by a selection scheme biased towards high-quality solutions. A GA is a stochastic iterative procedure that maintains the population size constant in each iteration, called a *generation*. To form a new population, a binary operator called *crossover*, and a unary operator, called *mutation*, are applied. Crossover takes two individuals, called parents, and produces two new individuals, called *offsprings*, by swapping parts of the parents. The mutation operator is usually thought of as introduction of noise to prevent premature convergence to local optima. The individuals involved in crossover and mutation are selected differently. For crossover, individuals are selected with a probability proportional to their relative fitness to ensure that the expected number of times an individual is chosen is approximately proportional to its relative performance in the population, i.e., so that good individuals have more chances of mating. Each individual represents a feasible solution in some problem space through a suitable encoding (mapping), often *strings* of finite length. This encoding is done through a structure called *chromosome*, where each chromosome is made up of units named *genes*. Typically, the values of each gene are binary, and are often called *alleles*. The mutation operator is applied by flipping such bits at random in a string with a certain probability called *mutation rate*. There are no convergence guarantees in GAs. Termination is based either on the maximum number of generations or by qualifying the acceptance of the best solution found. The following pseudocode summarizes a standard GA:

algorithm GA
Produce an initial population $\mathcal{P}^0$ of individuals
$k \longleftarrow 0$! Generation counter
do while termination conditions are not satisfied
      Evaluate the fitness of all individuals in $\mathcal{P}^k$
      Select the fittest individuals in $\mathcal{P}^k$ for reproduction
      Produce new individuals $\overline{\mathcal{P}}^k$
      Generate a new population $\underline{\mathcal{P}}^{k+1}$ by discarding
      some bad old individuals from $\mathcal{P}^k$ and by inserting
      a corresponding number of good individuals from $\overline{\mathcal{P}}^k$
      Mutate some individuals in $\underline{\mathcal{P}}^{k+1}$, and let $\mathcal{P}^{k+1}$
      be the resulting population
      $k \longleftarrow k + 1$
enddo
return Best Solution Found

Related to GAs are the evolutionary algorithms EAs which are based on evolution strategies. The main difference is with respect to encoding. EAs in their original form [168] work with continuous parameters represented as floating point numbers, rely on mutation as the only genetic operator, and use a population of only two members, the parent and the offspring. Subsequent development lead essentially towards a merge of the two fields; because as multimembered populations were introduced in the EAs, the latter could be thought of as GAs using different encoding concept and different genetic operators [69, 169, 170].

Evolutionary and genetic algorithms have attracted immense attention and the number of publications is overwhelming. For their applications in combinatorial and global optimization consult the books by Refs. [69, 72, 168], the proceedings by Refs. [70, 171], and the surveys by Refs. [46, 68, 71, 155, 169].

Low-level parallelism can be used to obtain performance improvements without altering the sequential character of the algorithm. For instance, if the calculation of each individual fitness is computational heavy, then low-level parallelism will divide the individuals among the available processors to perform the fitness computations in parallel, whereas the other parts of the algorithm remain sequential. Such an implementation is achieved following the synchronous master–slave paradigm, where the master processor performs the sequential part of the algorithm and is responsible to allocate individuals to slave processors to be evaluated. The communication is only between the master and the slaves at the beginning and the end of each computation. Implementations of parallel GAs of this type are given in Refs. [172–176].

algorithm <u>Simple_parallel_GA</u>
<u>Produce</u> an initial population $\mathcal{P}^0$ of individuals
$k \longleftarrow 0$! Generation counter
<u>do while</u> termination conditions are not satisfied
      <u>parallel do</u>
           Evaluate the fitness of all individuals in $\mathcal{P}^k$
      <u>parallel enddo</u>
      Select the fittest individuals in $\mathcal{P}^k$ for reproduction
      Produce new individuals $\overline{\mathcal{P}}^k$
      Generate a new population $\underline{\mathcal{P}}^{k+1}$ by discarding
      some bad old individuals from $\mathcal{P}^k$ and by inserting
      a corresponding number of good individuals from $\overline{\mathcal{P}}^k$
      Mutate some individuals in $\underline{\mathcal{P}}^{k+1}$, and let $\mathcal{P}^{k+1}$
      be the resulting population
      $k \longleftarrow k + 1$
<u>enddo</u>
<u>return</u> Best Solution Found

The simplest way of achieving parallelism of the SPMD paradigm would be to run several copies of the algorithm on the available processors simultaneously and independently, and for different initial populations. As in the multistart local search, the best individual from all independent runs is selected as the required result. Due to the stochastic character of the GAs, several runs are required even in the sequential case to draw statistically significant conclusions. Thus, these runs can be done in parallel. Alternatives to this embarrassing parallel GA are based on the subdivision of the population in *subpopulations* allocated across the available processors. Such algorithms are based on the observation that several subpopulations evolve in parallel tending to possess a spatial structure. As a result, *demes* make their appearance, i.e., semi-independent subpopulations that are only loosely coupled to neighboring demes. This coupling may take the form of migration, crossover or diffusion of some individuals from one deme to another.

The so-called *parallel island* GA model [177–179] is based on these concepts. *Migration* is used to revive diversity into an otherwise converging subpopulation. Migration takes place according to a specified frequency and various patterns. Within each subpopulation, a sequential GA is run. A possible description of such a generic scheme is given below:

algorithm <u>Parallel_island_GA</u>
Produce K initial subpopulations $[\mathcal{P}^0]_{i=1}^K$ of N individuals each
$k \longleftarrow 0$! Generation counter
<u>do while</u> termination conditions are not satisfied
      <u>parallel do</u> i=1,K
           Evaluate the fitness of all individuals in $\mathcal{P}_i^k$
           Select the fittest individuals in $\mathcal{P}_i^k$ for reproduction and migration
           <u>if</u> frequency condition is met <u>then</u>
                Let a number ($q < N$) of best individuals to migrate to a
                neighboring subpopulation
                Receive q immigrants from a neighboring subpopulation and
                let them replace the q worst individuals in the subpopulation
           <u>endif</u>
           Produce new individuals $\overline{\mathcal{P}}_i^k$
           Generate a new population $\underline{\mathcal{P}}_i^{k+1}$ by discarding some bad old
           individuals from $\mathcal{P}_i^k$ and by inserting a corresponding
           number of good individuals from $\overline{\mathcal{P}}_i^k$
           Mutate some individuals in $\underline{\mathcal{P}}_i^{k+1}$,
           and let $\mathcal{P}_i^{k+1}$ be the resulting subpopulation
      <u>parallel enddo</u>
      $k \longleftarrow k + 1$
<u>enddo</u>
<u>return</u> Best Solution Found

Substantial experimentations have been done with nonconvex test problems to study how migration parameters and subpopulation sizes affect the performance of the algorithm [178, 180, 181]. When the migration takes place between nearest neighbor subpopulations, the algorithm is also known as *stepping stone*. Computational results for this approach on optimization of difficult functions are reported in Refs. [182, 183]. It has been found that parallel GAs, apart from being significantly faster, help in preventing premature convergence and that, when equipped with local search capabilities, they are effective for multimodal function optimization.

7.4 CONCLUSION

Many parallel algorithms for the solution of global optimization problems have been proposed in recent years. A majority of these algorithms belongs to the class of heuristic methods. Such methods seem to fit naturally in the third option of Schnabel's taxonomy [184] who identified three possible level for introducing parallelism into optimization:

1. Parallelize the computation of objective function, gradient, constraints
2. Parallelize the linear algebra
3. Parallelize the algorithm at high level

This does not necessarily mean that the other two options should be discarded. However, high-level parallel restatement of the methods seems to be definitely the most natural way to proceed. The computational experience reported for the parallel versions of algorithms such as GA or GRASP show very high speedups for a large range of different optimization problems.

On the other hand, the most recent experience shows that exact optimization deterministic algorithms, such as BB and interval methods, tend to be excellent candidates for parallel computing, although the effort required to achieve both efficient parallelization and to preserve their convergence properties is certainly greater than for heuristic methods.

Finally, the potential utilization of computational grids is probably of the highest priority in the future development of global optimization algorithms.

ACKNOWLEDGMENT

M. D'Apuzzo, M. Marino, and G. Toraldo were supported by the MIUR FIRB project "Large Scale Nonlinear Optimization," No. RBNE01WBBB, ICAR-CNR, branch of Naples.

REFERENCES

[1] Törn, A. and Zilinskas, A. (1989). *Global Optimization*, Lecture Notes in Computer Science 350. Berlin: Springer-Verlag.

[2] Pardalos, P.M., Phillips, A.T., and Rosen, J.B. (1992). *Topics in Parallel Computing in Mathematical Programming*. New York: Science Press.

[3] Pardalos, P.M., Xue, G., and Shalloway, D. (1994). Optimization methods for computing global minima of nonconvex potential energy functions. *Journal of Global Optimization* 4: 117–133.

[4] Xue, G. (1994). Molecular conformation on the CM-5 by parallel two-level simulated annealing. *Journal of Global Optimization* 4: 187–208.

[5] Gu, J. (1995). Parallel algorithms for satisfiability (SAT) problem. In Pardalos, P.M., Resende, M.G.C., and Ramakrishnan, K.G., Eds., *Parallel Processing of Discrete Optimization Problems*. DIMACS Series in Discrete Mathematics and Theoretical Computer Sciences 22, New York: American Mathematical Society.

[6] Horst, R. and Pardalos, P.M., Eds. (1995). *Handbook of Global Optimization*. Dordrecht: Kluwer Academic Publishers.

[7] Horst, R., Pardalos, P.M., and Thoai, N.V. (1995). *Introduction to Global Optimization*. Dordrecht: Kluwer Academic Publishers.

[8] Horst, R. and Tuy, H. (1996). *Global Optimization—Deterministic Approaches*, 3rd ed. Berlin: Springer-Verlag.

[9] Holmqvist, K., Migdalas, A., and Pardalos, P.M. (1997). Parallel continuous non-convex optimization. In Migdalas, A., Pardalos, P.M., and Starøy, S., Eds., *Parallel Computing in Optimization*. Dordrecht: Kluwer Academic Publishers.

[10] Abello, J., Butenko, S., Pardalos, P.M., and Resende, M.G.C. (2001). Finding independent sets in a graph using continuous multivariable polynomial formulations. *Journal of Global Optimization* 21: 111–137.

[11] Zhigljavsky, A.A. (1991). *Theory of Global Random Search*. Dordrecht: Kluwer Academic Publishers.

[12] Pardalos, P.M. and Schnitger, G. (1998). Checking local optimality in constrained quadratic programming is NP-hard. *Operations Research Letters* 7: 33–35.

[13] Vavasis, S.A. (1995). Complexity issues in global optimization: a survey. In Horst, R., and Pardalos, P.M., Eds., *Handbook of Global Optimization*. Dordrecht: Kluwer Academic Publishers, pp. 27–41.

[14] Floudas, C.A. and Pardalos, P.M., Eds. (1992). *Recent Advances in Global Optimization*. Princeton, NJ: Princeton University Press.

[15] Floudas, C.A. and Pardalos, P.M., Eds. (1996). *State of the Art in Global Optimization*. Dordrecht: Kluwer Academic Publishers.

[16] Grossmann, I.E., Ed. (1996). *Global Optimization in Engineering Design*. Dordrecht: Kluwer Academic Publishers.

[17] Migdalas, A., Pardalos, P.M., and Storøy, S., Eds. (1997). *Parallel Computing in Optimization*. Dordrecht: Kluwer Academic Publishers.

[18] Migdalas, A., Pardalos, P.M., and Värbrand, P., Eds. (2001). *From Local to Global Optimization*. Dordrecht: Kluwer Academic Publishers.

[19] Pardalos, P.M., Migdalas, A., and Burkard, R.E. (2002). *Discrete and Global Optimization*. Singapore: World Scientific.

[20] Pardalos, P.M. and Rosen, J.B. (1987). *Global Optimization: Algorithms and Applications*, Lecture Notes in Computer Science 268. Berlin: Springer-Verlag.

[21] Pintér, J.D. (1996). *Global Optimization in Action*. Dordrecht: Kluwer Academic Publishers.

[22] Konno, H., Thach, P.T., and Tuy, H. (1997). *Optimization on Low Rank Nonconvex Structures*. Dordrecht: Kluwer Academic Publishers.

[23] Tuy, H. (1998). *Convex Analysis and Global Optimization*. Dordrecht: Kluwer Academic Publishers.

[24] Androulakis, I.P., Maranas, C.D., and Floudas, C.A. (1995). αBB: a global optimization method for general constrained nonconvex problems. *Journal of Global Optimization* 7: 337–363.

[25] Adjiman, C.S., Androulakis, I.P., and Floudas, C.A. (1996). A global optimization method, αBB, for process design. *Computers and chemical Engineering Suppl.* 20: S419–S424.

[26] Tawarmalini, M. and Sahinidis, N.V. (1999). Global Optimization of Mixed Integer Nonlinear Programs: A Theoretical and Computational Study. Technical report, Department of Chemical Engineering, University of Illinois, Urbana-Champaign, IL.

[27] Bomze, I.M. (2002). Branch-and-bound approaches to standard quadratic optimization problems. *Journal of Global Optimization* 22: 17–37.

[28] http://archimedes.scs.uiuc.edu/baron/baron.html

[29] Ratschek, H. and Rokne, J.G. (1988). *New Computer Methods for Global Optimization*. Chichester: Ellis Horwood.

[30] Neumaier, A. (1990). *Interval Methods for Systems of Equations*. Cambridge, MA: Cambridge University Press.

[31] Eriksson, J. (1991). Parallel Global Optimization Using Interval Analysis. Research report UMINF-91.17, Department of Computing Science, Institute of Information Processing, University of Umeå, Ume.

[32] Hansen, E.R. (1992). *Global Optimization Using Interval Analysis*. New York: Marcel Dekker.

[33] Knüppel, O. (1993). BIAS — Basic Interval Arithmetic Subroutines. Berichte des Forschungsschwerpunktes Informations- und Kommunikationstechnik, Bericht 93.3, Technische Universität, Hamburg, Germany.

[34] Knüppel, O. (1993). PROFIL — Programmer's Runtime Optimized Fast Interval Library. Berichte des Forschungsschwerpunktes Informations- und Kommunikationstechnik, Bericht 93.4, Technische Universität, Hamburg, Germany.

[35] Knüppel, O. (1993). PROFIL/BIAS Extensions. Berichte des Forschungsschwerpunktes Informations- und Kommunikationstechnik, Bericht 93.5, Technische Universität, Hamburg, Germany.

[36] Knüppel, O. (1993). A Multiple Precision Arithmetic for PROFIL. Berichte des Forschungsschwerpunktes Informations- und Kommunikationstechnik, Bericht 93.6, Technische Universität, Hamburg, Germany.

[37] Eriksson, J. and Lindström, P. (1995). A parallel interval method implementation for global optimization using dynamic load balancing. *Reliable Computing* 1: 77–91.

[38] Holmqvist, K. and Migdalas, A. (1996). A C++ class library for interval arithmetic in global optimization. In Floudas, C.A. and Pardalos, P.M., Eds., *State of the Art in Global Optimization*. Dordrecht: Kluwer Academic Publishers.

[39] Kearfott, R.B. (1996). *Rigorous Global Search: Continuous Problems*. Dordrecht: Kluwer Academic Publishers.

[40] Kearfott, R.B. and Kreinovich, V., Eds., *Applications of Interval Computations*. Dordrecht: Kluwer Academic Publishers.

[41] Neumaier, A. (2002). Grand Challenges and Scientific Standards in Interval Analysis. Technical note, Institut für Mathematik, Universität Wien, Austria.

[42] Leclerc, A.P. (1992). Efficient and Reliable Global Optimization. Ph.D. thesis, Graduate School, Ohio State University.

[43] Mockus, J. (1989). *Bayesian Approach to Global Optimization*. Dordrecht: Kluwer Academic Publishers.

[44] Dixon, L.C.W. and Jha, M. (1993). Parallel algorithms for global optimization. *Journal of Optimization Theory and Application* 79: 385–395.

[45] He, L. and Polak, E. (1993). Multistart method with estimation scheme for global satisfycing problems. *Journal of Global Optimization* 3: 139–156.

[46] Holmqvist, K., Migdalas, A., and Pardalos, P.M. (1997). Parallelized heuristics for combinatorial search. In Migdalas, A., Pardalos, P.M., Størøy, S., Eds., *Parallel Computing in Optimization*. Dordrecht: Kluwer Academic Publishers.

[47] Mockus, J., Eddy, W., Mockus, A., Mockus, L., and Reklaitis, G. (1997). *Bayesian Heuristic Approach to Discrete and Global Optimization*. Dordrecht: Kluwer Academic Publishers.

[48] Velázquez, L., Phillips, G.N., Jr., Tapia, R.A., and Zhang, Y. (1999). Selective Search for Global Optimization of Zero or Small Residual Least-Squares Problems: A Numerical Study. Technical report, September 1999, Revised May 2000, Francis Bitter Magnet Laboratory, Massachusetts Institute of Technology, Cambridge, MA.

[49] de Angelis, P.L., Bomze, I.M., and Toraldo, G. (2004). Ellipsoidal approach to box-constrained quadratic problems. *Journal of Global Optimization* 28(1): 1–15

[50] Boender, C.G., Rinnoy Kan, A.H.G., Stougie, L., and Timmer, G.T. (1982). A stochastic method for global optimization. *Mathematical Programming* 22: 125–140.

[51] Rinnooy Kan, A.H.G., and Timmer, G.T. (1987). Stochastic global optimization methods. Part I: Clustering methods; Part II: Multi level methods. *Mathematical Programming* 39: 27–78.

[52] Bertocchi, M. (1990). A parallel algorithm for global optimization. *Optimization* 21: 379–386.

[53] Byrd, R.H., Dert, C.L., Rinnooy Kan, A.H.G., and Schnabel, R.B. (1990). Concurrent stochastic methods for global optimization. *Mathematical Programming* 46: 1–29.

[54] Betró, B. (1991). Bayesian methods in global optimization. *Journal of Global Optimization* 1: 1–14.

[55] Schoen, F. (1991). Stochastic techniques for global optimization: a survey of recent advances. *Journal of Global Optimization* 1: 207–228.

[56] Smith, S.L. and Schnabel, R.B. (1992). Dynamic Scheduling Strategies for an Adaptive, Asynchronous Parallel Global Optimization Algorithms. Research report CU-CS-652-93, Department of Computer Science, University of Colorado at Boulder, Campus Box 430, Boulder, Colorado.

[57] Byrd, R.H., Eskow, E., Schnabel, R.B., and Smith, S.L. (1993). Parallel global optimization: numerical methods, dynamic scheduling methods, and application to molecular configuration. In Ford, B. and Fincham, A., Eds., *Parallel Computation*. Oxford: Oxford University Press, pp. 187–207.

[58] Byrd, R.H., Derby, T., Eskow, E., Oldenkamp, K.P.B., and Schnabel, R.B. (1994). A new stochastic/perturbation method for large-scale global optimization and its application to water cluster problems. In Hager, W., Hearn, D., and Pardalos, P.M., Eds., *Large-Scale Optimization: State of the Art*. Dordrecht: Kluwer Academic Publishers, pp. 71–84.

[59] Ritter, K. and Schäffler, S. (1994). A stochastic method for constraint global optimization. *SIAM Journal on Optimization* 4: 894–904.

[60] Schoen, F. (1994). On an new stochastic global optimization algorithm based on censored observations. *Journal of Global Optimization* 4: 17–35.

[61] Boender, C.G.E. and Romeijn, H.E. (1995). Stochastic methods. In Horst, R. and Pardalos, P.M., Eds., *Handbook of Global Optimization*. Dordrecht: Kluwer Academic Publishers, pp. 829–869.

[62] Chamberlain, R.D., Edelman, M.N., Franklin, M.A., and Witte, E.E. (1988). Simulated annealing on a multiprocessor. In *Proceedings of 1988 IEEE International Conference on Computer Design*, pp. 540–544.

[63] Aarts, E. and Korst, J. (1990). *Simulated Annealing and Boltzmann Machines*. Chichester: John Wiley & Sons.

[64] Dekker, A. and Aarts, E. (1991). Global optimization and simulated annealing. *Mathematical Programming* 50: 367–393.

[65] Azencott, R. (1992). *Simulated Annealing — Parallelization Techniques*. New York: John Wiley & Sons.

[66] Desai, R. and Patil, R. (1996). SALO: combining simulated annealing and local optimization for efficient global optimization. In *Proceedings of the Ninth Florida AI Research Symposium (FLAIRS-96)*. Key West, FL, pp. 233–237.

[67] Crainic, T.G. and Toulouse, M. (1997). Parallel Metaheuristics. Technical report, Centre de recherche sur les transports, Université de Montréal, Montréal, Québec, Canada.

[68] Tomassini, M. (1993). The parallel genetic cellular automata: application to global function optimization. In *Proceedings of the International Conference on Artificial Neural Networks and Genetic Algorithms*. Wien: Springer-Verlag.

[69] Michalewicz, Z. (1994). *Genetic Algorithms + Data Structures = Evolution Programs*, Second, Extended Edition. Berlin: Springer-Verlag.

[70] Alander, J.T. (1995). *Proceedings of the First Nordic Workshop on Genetic Algorithms and their Applications*. Vaasa: Vaasan yliopisto.

[71] Pardalos, P.M., Pitsoulis, L., Mavridou, T., and Resende, M.G.C. (1995). Parallel search for combinatorial optimization: genetic algorithms, simulated annealing, tabu search and GRASP. In Ferreira, A. and Rolim, J., Eds., *Proceedings of Workshop on Parallel Algorithms for Irregularly Structured Problems*, Lecture notes in Computer Science 980. Berlin: Springer-Verlag, pp. 317–331.

[72] Osyczka, A. (2002). *Evolutionary Algorithms for Single and Multicriteria Design Optimization*. Heidelberg: Physica-Verlag.

[73] Pardalos, P.M., Pitsoulis, L. and Resende, M.G.C. (1995). A parallel GRASP implementation for the quadratic assignment problem. In Ferreira, A. and Rolim, J., Eds., *Solving Irregular Problems in Parallel — State of the Art*. Dordrecht: Kluwer Academic Publishers.

[74] Glover, F. (1989). Tabu search — Part I. *ORSA Journal on Computing* 1: 190–206.

[75] Glover, F. (1990). Tabu search — Part II. *ORSA Journal on Computing* 2: 4–31.

[76] Glover, F., Taillard, E., and De Werra, D. (1993). A user's guide to tabu search. In Glover, F., Laguna, M., Taillard, E., and De Werra, D., Eds., *Tabu Search*. Annals of Operations Research 41. J.C. Baltzer AG, Basel, Switzerland, pp. 2–28.

[77] Battiti, R. and Tecchiolli, G. (1994). The Continuous Reactive Tabu Search: Blending Combinatorial Optimization and Stochastic Search for Global Optimization. Preprint UTM 432, Dipartimento Di Matematica Universitá di Trento, Trento, Italy.

[78] Cvijović, D., and Klinowski, J. (1995). Tabu search: an approach to the multiple minima problem. *Science* 267: 664–666.

[79] Crainic, T.G., Toulouse, M., and Gendreau, M. (1997). Towards a taxonomy of parallel tabu search algorithms. *INFORMS Journal on Computing* 9: 61–72.

[80] http://plato.la.asu.edu/gom.html

[81] http://www. mat.univie.ac.at/~neum/glopt/software_g.html

[82] Pintér, J.D. (1996). Continuous global optimization software: a brief review. *Optima* 52: 1–8.

[83] Trienekens, H.W.J.M. (1990). Parallel Branch and Bound Algorithms. Ph.D. thesis, Department of Computer Science, Erasmus University, Rotterdam, The Netherlands.

[84] Kronsjö, L., and Shumsheruddin, D., Eds., *Advances in Parallel Algorithms*. London: Blackwell Scientific Publications.

[85] de Bruin, A., Kindervater, G.A.P., and Trienekens, H.W.J.M. (1995). Asynchronous Parallel Branch and Bound and Anomalies. Lyon, France, September 4–6.

[86] Clausen, J. (1997). Parallel branch and bound — principles and personal experiences. In Migdalas, A., Pardalos, P.M., and Värbrand, P., Eds., *From Local to Global Optimization*. Dordrecht: Kluwer Academic Publishers, pp. 239–267.

[87] http://liinwww.ira.uka.de/bibliography/Parallel/par.branch.and.bound.html

[88] Tschöke and S., and Polzer, T. (1999). Portable Parallel Branch-and-Bound Library: PPBB-Lib, User Manual, Library Version 2.0. Department of Computer Science, University of Paderborn, D-33095 Paderborn, Germany.

[89] http://www.uni-paderborn.de/~ppbb-lib

[90] Gendron, B. and Crainic, T.G. (1994). Parallel branch-and-bound algorithms: survey and synthesis. *Operations Research* 42: 1042–1066.

[91] Henriksen, T. and Madsen, T. (1992). Use of a Depth-First Strategy in Parallel Global Optimization. Research report 92-10, Technical University of Denmark, Lyngby, Denmark.

[92] Moore, R.E., Hansen, E., and Leclerc, A. (1992). Rigorous methods for global optimization. In Floudas, C.A. and Pardalos, P.M., Eds., *Recent Advances in Global Optimization*. Princeton, JN: Princeton University Press.

[93] Leclerc, A.P. (1993). Parallel interval global optimization and its implementation in C++. *Interval Computations* 3: 148–163.

[94] Berner, S. (1996). Parallel methods for verified global optimization — practice and theory. *Journal of Global Optimization* 9: 1–22.

[95] Gau, C.Y. and Stadtherr, M.A. (1999). A systematic analysis of dynamic load balancing strategies for parallel interval analysis. In *AIChE Annual Meeting*. Dallas, TX, October 31–November 5.

[96] Gau, C.Y. and Stadtherr, M.A. (2000). Trends in parallel computing for process engineering. In *AspenWorld 2000*. Orlando, FL, February 6–11.

[97] Gau, C.Y. and Stadtherr, M.A. (2000). Parallel branch-and-bound for chemical engineering applications: load balancing and scheduling issues. In *VECPAR 2000*. Porto, Portugal, June 21–23.

[98] http://happy.dt.uh.edu/~ParaGlobSol.html

[99] Phillips, A.T. and Rosen, J.B. (1989). Guaranteed ϵ-approximate solution for indefinite quadratic global minimization. *Naval Research Logistics Quarterly* 37: 499–514.

[100] Pardalos, P.M. and Rosen, J.B., Eds. (1990). *Computational Methods in Global Optimization*. Annals of Operations Research 25.

[101] Phillips, A.T. and Rosen, J.B. (1990). A parallel algorithm for partially separable non-convex global minimization. In Pardalos, P.M., and Rosen, J.B., Eds., *Computational Methods in Global Optimization* 25. J.C. Baltzer AG, Basel, Switzerland.

[102] Boffey, T.B. and Saeidi, P. (1996). A parallel branch-and-bound method for a network design problem. *Belgian Journal of Operations Research* 32: 69–83.

[103] Goux, J.P. and Leyffer, S. (2001). Solving large MINLPs on computational grids. University of Dundee Report NA200.

[104] Anstreicher, K., Brixius, N., Linderoth, J., and Goux, J.P. (2002). Solving large quadratic assignment problems on computational grids. *Mathematical Programming Series B* 91: 563–588.

[105] Goux, J.P., Linderoth, J., and Yoder, M. (2002). Metacomputing and the master/worker paradigm. Preprint MCS/ANL P792/0200, Computer Science Division, Argonne National Laboratory, USA.

[106] Livny, M., Basney, J., Raman, R., and Tannebaum, T. (1997). *Mechanisms for High-throughput Computing*. Speedup 11.

[107] Goux, J.P., Kulkarni, L., Linderoth, J., and Yoder, M. (2000). An enabling framework for master–worker applications on computational grid. *Cluster Computing* 43–50.

[108] Pardalos, P.M., Resende, M.G.C., and Ramakrishnan, K.G., Eds. (1995). *Parallel Processing of Discrete Optimization Problems*. DIMACS Series in Discrete Mathematics and Theoretical Computer Science: 22. New York American Mathematical Society.

[109] Zhang, Y., Tapia, R.A., and Velázquez, L. (1999). On Convergence of Minimization Methods: Attraction, Repulsion and Selection. Technical report TR 99-12, Department of Computational and Applied Mathematics, Rice University, Housto, TX.

[110] Goffe, B. (1994). Global optimization of statistical functions with simulated annealing. *Journal of Econometrics* 30: 65–100.

[111] Castellanos, L. and Gómez, S. (1998). Tunnel: A Fortran Subroutine for Global Optimization with Bounds on the Variables using the Tunneling Method. Technical report, IIMAS-UNAM.

[112] Levy, A.V. and Gomez, S. (1985). The tunneling method applied to global optimization. In Boggs, P.T., Byrd, R.H., and Schnabel, R.B., Eds., *Numerical Optimization*. Philadelphia: SIAM, pp. 213–244.

[113] Levy, A.V. and Montalvo, A. (1985). The tunneling algorithm for the global minimisation of functions. *SIAM Journal of Scientific and Statistical Computing* 6(1):15–29.

[114] Gomez, S., Del Castillo, N., Castellanos, L., and Solano, J. (2003). The parallel tunneling method. *Journal of Parallel Computing* 9: 523–533.

[115] Gomez, S. and Romero, D. (1994). Two global methods for molecular geometry optimization. *Progress in Mathematics*. Vol. 121. Boston: Birkhauser, pp. 503–509.

[116] Gomez, S., Gosselin, O., and Barker, J. (2001). Gradient-based history-matching with a global optimization method. *Society of Petroleum Engineering Journal* 200–208.

[117] Nichita, D.V., Gomez, S., and Luna, E. (2002). Multiphase equilibria calculation by direct minimization of Gibbs free energy with a global optimization method. In *AICHE Spring Meeting Proceedings*. Computers and Chemical Engineering.

[118] Nichita, D.V., Gomez, S., and Luna, E. (2002). Phase stability analysis with cubic equations of state using a global optimization method. *Fluid Phase Equilibria* 194–197: 411-437.

[119] Gomez, S., Castellanos, L., and Del Castillo, N. (2003). *Users Manuals for the Sequential Tunneling Methods, TunTGNS and TunLBFS*. Serie Manuales, IIMAS-UNAM, No. 12.

[120] Gomez, S., Castellanos, L., and Del Castillo, N. (2003). *Users Manuals for the Parallel Tunneling Methods, TunTGNP and TunLBFP*. Serie Manuales, IIMAS-UNAM, No. 13.

[121] Minoux, M. (1976). Multiflots de Coût Minimal Avec Fonctions de Coût Concaves. *Annals of Telecommunication* 31(3–4): 77–92.

[122] Yaged, B. Jr. (1971). Minimum cost routing for static network models. *Networks* 1: 139–172.

[123] Li, Z., Pardalos, P.M., and Levine, S.H. (1992). Space-covering approach and modified Frank–Wolfe algorithm for optimal nuclear reactor reload design. In Floudas, C.A., and Pardalos, P.M., Eds., *Recent Advances in Global Optimization*. Princeton, NJ: Princeton University Press.

[124] Ten Eikelder, H.M.M., Verhoeven, M.G.A., Vossen, T.W.M., and Aarts, E.H.L. (1996). A probabilistic analysis of local search. In Osman, I.H., and Kelly, J.P., Eds., *Meta-Heuristics: Theory and Applications*. Boston: Kluwer Academic Publishers, pp. 605–618.

[125] Evtushenko, Y.G. (1985). Numerical Optimization Techniques. New York. *Optimization Software*.

[126] Kim, D. and Pardalos, P.M. (2000). Dynamic slope scaling and trust interval techniques for solving concave piecewise linear network flow problems. *Networks* 35: 216–222.

[127] Phillips, A.T. and Rosen, J.B. (1988). A parallel algorithm for the linear complementarity problem. *Annals of Operations Research* 14: 77–104.

[128] Mockus, J. (1994). Application of Bayesian approach to numerical methods of global and stochastic optimization. *Journal of Global Optimization* 4: 347–365.

[129] Mockus, J. (2002). Bayesian heuristic approach to global optimization and examples. *Journal of Global Optimization* 22: 191–203.

[130] Phillips, A.T., Rosen, J.B., and van Vliet, M. (1992). A parallel stochastic method for solving linearly constrained concave global minimization problems. *Journal of Global Optimization* 2: 243–258.

[131] Byrd, R.H., Derby, T., Eskow, E., Oldenkamp, K.P.B., and Schnabel, R.B. (1994). A new stochastic/perturbation method for large-scale global optimization and its application to water cluster problems. In Hager, W., Hearn, D., and Pardalos, P.M., Eds., *Large-scale Optimization: State of the Art*. Dordrecht: Kluwer Academic Publishers, pp. 71–84.

[132] Byrd, R.H., Eskow, E., van der Hoek, A., Schnabel, R.B., and Oldenkamp, K.P.B. (1995). A parallel global optimization method for solving molecular cluster and polymer conformation problems. In Bailey, D.H. et al., Eds., *Proceedings of the Seventh SIAM Conference on Parallel Processing for Scientific Computing*, Philadelphia: SIAM, pp. 72–77.

[133] Byrd, R.H., Eskow, E., van der Hoek, A., Schnabel, R.B., Shao, C.S., and Zou, Z. (1995). Global optimization methods for protein folding problems. In Pardalos, P.M., Shalloway, D., and Xue, G., Eds., *Proceedings of the DIMACS Workshop on Global Minimization of Nonconvex Energy Functions: Molecular Conformation and Protein Folding*. New York: American Mathematical Society.

[134] Byrd, R.H., Eskow, E., and Schnabel, R.B. (1999). A Large-Scale Stochastic-Perturbation Global Optimization Method for Molecular Cluster Problems. Research report, Department of Computer Science, University of Colorado Boulder, CO.

[135] Smith, S.L. and Schnabel, R.B. (1992). Centralized and distributed dynamic scheduling for adaptive parallel algorithms. In Mehrotra, P., Saltz, J., and Voigt, R., Eds., *Unstructured Scientific Computation on Scalable Multiprocessors*. Cambridge, MA: MIT Press, pp. 301–321.

[136] Smith, S.L., Eskow, E., and Schnabel, R.B. (1990). Adaptive, asynchronous stochastic global optimization for sequential and parallel computation. In Coleman, T. and Li, Y., Eds., *Large Scale Numerical Optimization*. Philadelphia: SIAM, pp. 207–227.

[137] Byrd, R.H., Eskow, E., and Schnabel, R.B. (1992). A New Large-Scale Global Optimization Method and its Application to Lennard–Jones Problems. Technical report CU-CS-630-92, Department of Computer Science, University of Colorado, Boulder, CO.

[138] Byrd, R.H., Derby, T., Eskow, E., Oldenkamp, K., Schnabel, R.B., and Triantafillou, C. (1993). Parallel global optimization methods for molecular configuration problems. In *Proceedings of the Sixth SIAM Conference of Parallel Processing for Scientific Computation*. Philadelphia: SIAM, pp. 165–169.

[139] Feo, T.A. and Resende, M.G.C. (1995). Greedy randomized adaptive search procedures. *Journal of Global Optimization* 6: 109–133.

[140] Festa, P. and Resende, M.G.C. (2001). GRASP: an annoted bibliography. In Hansen, P. and Ribeiro, C.C., Eds., *Essays and Surveys on Metaheuristics*. Dordrecht: Kluwer Academic Publishers.

[141] Resende, M.C.C., Pardalos, P.M., and Ekşioğlu, S.D. (2001). Parallel metaheuristics for combinatorial optimization. In Correa, R. et al., Eds., *Advanced Algorithmic Techniques for Parallel Computation with Applications*. Dordrecht: Kluwer Academic Publishers.

[142] Holmqvist, K., Migdalas, A., and Pardalos, P.M. (1998). A GRASP algorithm for the single source uncapacitated minimum concave-cost network flow problem. *DIMACS Series in Discrete Mathematics and Theoretical Computer Science* 40: 131–142.

[143] Aiex, R., Pardalos, P.M., Resende, M., and Toraldo, G. (2005). GRASP with path relinking for the three-index assignment problem. *INFORMS Journal on Computing*, 17(2): 224–247

[144] Feo, T.A., Resende, M.G.C., and Smith, S.H. (1994). A greedy randomized adaptive search procedure for maximum independent set. *Operations Research* 42: 860–878.

[145] Murphey, R.A., Pardalos, P.M., and Pitsoulis, L. (1998). A parallel GRASP for the data association multidimensional assignment problem. In *Parallel Processing of Discrete Problems*. IMA Volumes in Mathematics and its Applications 106. Heidelberg: Springer-Verlag, pp. 159–180.

[146] Ingber, L. (1993). Simulated annealing: practice versus theory. *Journal of Mathematical and Computer Modelling* 18: 29–57.

[147] Romeijn, H.E. and Smith, R.L. (1994). Simulated annealing for constrained global optimization. *Journal of Global Optimization* 5: 101–126.

[148] Pardalos, P.M., Xue, G., and Panagiotopoulos, P.D. (1995). Parallel algorithms for global optimization. In Ferreira, A. and Pardalos, P.M., Eds., *Solving Irregular Problems in Parallel: State of the Art*. Berlin: Springer-Verlag.

[149] Glover, F., Laguna, M., Taillard, E., De Werra, and D., Eds., (1993). *Tabu Search*. Annals of Operations Research 41.

[150] Gergel, V.P., Sergeyev, Ya.D., and Strongin, R.G. (1992). A parallel global optimization method and its implementation on a transputer system. *Optimization* 26: 261–275.

[151] Strongin, R.G. and Sergeyev, Y.D. (1992). Global multidimensional optimization on parallel computer. *Parallel Computing* 18(11): 1259–1273.

[152] Ghannadan, S., Migdalas, A., Tuy, H., and Värbrand, P. (1994). Heuristics based on tabu search and lagrangean relaxation for the concave production–transportation problem. *Studies in Regional and Urban Planning* 3: 127-140.

[153] Ghannadan, S., Migdalas, A., Tuy, H., and Värbrand, P. (1996). Tabu meta-heuristic based on local search for the concave production–transportation Problem. *Studies in Location Analysis*. Special Issue Edited by C. R. Reeves. 8: 33–47.

[154] Crainic, T.G., Toulouse, M., and Gendreau, M. (1995). Synchronous tabu search parallelization strategies for multicommodity location–allocation with balancing requirements. *OR Spectrum* 17: 113–123.

[155] Battiti, R. and Tecchiolli, G. (1992). Parallel biased search for combinatorial optimization: genetic algorithms and tabu. *Microprocessors and Microsystems* 16: 351–367.

[156] Price, W.L. (1978). A controlled random search procedure for global optimization. In Dixon, L.C.W. and Szegö,G.P., Eds., *Towards Global Optimization 2*. Amsterdam: North-Holland, pp. 71–84.

[157] Price, W.L. (1979). A controlled random search procedure for global optimization. *The Computer Journal* 20: 367–370.

[158] Price, W.L. (1983). Global optimization algorithms for a CAD workstation. *Journal of Optimization Theory and Applications* 55: 133–146.

[159] Ali, M.M. and Storey, C. (1994). Modified controlled random search algorithms. *International Journal of Computational Mathematics* 53: 229–235.

[160] Ali, M.M., Törn, A., and Viitanen, S. (1997). A numerical comparison of some modified controlled random search algorithms. *Journal of Global Optimization* 11: 377–385.

[161] Mohan, C. and Shanker, K. (1988). A numerical study of some modified versions of controlled random search method for global optimization. *International Journal of Computer Mathematics* 23: 325–341.

[162] Mohan, C. and Shanker, K. (1994). A controlled random search technique for global optimization using quadratic approximation. *Asia-Pacific Journal of Operational Research* 11: 93–101.

[163] Woodhams, F.W.D. and Price, W.L. (1988). Optimizing accelerator for CAD workstation. *IEEE Proceedings, Part E* 135: 214–221.

[164] Garcia, I., Ortigosa, P.M., Casado, L.G., Herman, G.T., Matej, S. (1995). A parallel implementation of the controlled random search algorithm to optimize an algorithm reconstruction from projections. In *Third Workshop on Global Optimization*. Szeged, Hungary, December, pp. 28–32.

[165] Garcia, I. and Herman, G.T. (1996). Global optimization by parallel constrained biased random search. In Floudas, C.A. and Pardalos, P.M., Eds., *State of the Art in Global Optimization: Computational Methods and Applications*. Dordrecht: Kluwer Academic Publishers, pp. 433–455.

[166] Garcia, I., Ortigosa, P.M., Casado, L.G., Herman, G.T., and Matej, S. (1997). Multi-dimensional optimization in image reconstruction from projections. In Bomze, L.M., Csendes, T., Horst, R., and Pardalos, P.M., Eds., *Developments in Global Optimization*. Dordrecht: Kluwer Academic Publishers, pp. 289–300.

[167] McKeown, J.J. (1980). Aspects of parallel computations in numerical optimization. In Arcetti, F. and Cugiarni, M., Eds., *Numerical Techniques for Stochastic Systems*, pp. 297–327.

[168] Schwefel, H.P. (1981). *Numerical Optimization for Computer Models*. Chichester: John Wiley & Sons.

[169] Bäck, T., Hoffmeister, F., and Schwefel, H.P. (1991). A survey of evolution strategies. In *Proceedings of the Fourth International Conference on Genetic Algorithms*. Los Altos, CA: Morgan Kaufmann.

[170] Schwefel, H.P. and Männer, R., Eds. (1991). *Parallel Problem Solving from Nature*. Lecture notes in Computer Science 496. Berlin: Springer-Verlag.

[171] Schaffer, J., Ed. (1989). *Proceedings of the Third International Conference on Genetic Algorithms*. Los Altos, CA: Morgan Kaufmann Publishers.

[172] Fogerty, T.C. and Haung, R. (1991). Implementing the genetic algorithm on transputer based parallel systems. In Schwefel, H.P. and Männer, R., Eds., *Parallel Problem Solving from Nature*. Lecture notes in Computer Science 496. Berlin: Springer-Verlag, pp. 145–149.

[173] Abramson, D. and Abela, J. (1992). A parallel genetic algorithm for solving the school timetabling problem. In Gupta, G. and Keen, C. Eds., *15th Australian Computer Science Conference*. Department of Computer Science, University of Tasmania, Australia, pp. 1–11.

[174] Cui, J. and Fogarty, T.C. (1992). Optimization by using a parallel genetic algorithm on a transputer computing surface. In Valero, M. et al., Eds., *Parallel Computing and Transputer Applications*. Amsterdam: IOS Press, pp. 246–254.

[175] Abramson, D., Millis, G., and Perkins, S. (1993). Parallelization of a genetic algorithm for the computation of efficient train schedules. In Arnold, D. et al., Eds., *Proceedings of the 1993 Parallel Computing and Transputers Conference*. IOS Press, pp. 139–149.

[176] Hauser, R. and Männer, R. (1994). Implementation of standard genetic algorithms on MIMD machines. In *Parallel Problem Solving from Nature III*. Lecture notes in Computer Science 866. Berlin: Springer-Verlag, pp. 504–514.

[177] Cohoon, J.P., Hedge, S.U., Martin, W.N., and Richards, D. (1987). Punctuated equilibria: a parallel genetic algorithm. In Grefenstette, J.J., Ed., *Proceedings of the Second International Conference on Genetic Algorithms*. Hillsdale, NJ: Lawrence Erlbaum Associates, pp. 148–154.

[178] Tanese, R. (1987). Parallel genetic algorithm for a hypercube. In Grefensette, J.J., Ed., *Proceedings of the Second International Conference on Genetic Algorithms*. Hillsdale, NJ: Lawrence Erlbaum Associates, pp. 177–183.

[179] Whitley, D. (1989). GENITOR II: A distributed genetic algorithm. *Journal of Experimental and Theoretical Artificial Intelligence* 2: 189–214.

[180] Starkweather, T., Whitney, D. and Mathias, K. (1991). Optimization using distributed genetic algorithms. In Schwefel, H.P. and Männer, R., Eds., *Parallel Problem Solving from Nature*. Lecture notes in Computer Science 496. Berlin: Springer-Verlag, pp. 176–185.

[181] Belding, T. (1995). The distributed genetic algorithms revised. In Eshelman, L.J., Eds., *Proceedings of the Sixth International Conference on Genetic Algorithms*. San Mateo, CA: Morgan Kaufmann, pp. 114–121.

[182] Mühlenbein, H. (1989). Parallel genetic algorithms, population genetics and combinatorial optimization. In Schaffer, J., Ed., *Proceedings of the Third International Conference on Genetic Algorithms*. Los Altos, CA: Morgan Kaufmann Publishers, pp. 416–421.

[183] Mühlenbein, H., Schomisch, M., and Born, J. (1991). The parallel genetic algorithm as function optimizer. *Parallel Computing* 7: 619–632.

[184] Schnabel, R.B. (1995). A view of the limitations, opportunities, and challenges in parallel nonlinear optimization. *Parallel Computing* 21(6): 875–905.

8 Nonlinear Optimization: A Parallel Linear Algebra Standpoint

Marco D'Apuzzo, Marina Marino, Athanasios Migdalas, Panos M. Pardalos, and Gerardo Toraldo

CONTENTS

ABSTRACT

Parallel computing research in the area of nonlinear optimization has been extremely intense during the last decade. One of the perspectives for this research is to parallelize the linear algebra operations that arise in numerical optimization, and thus to consider the impact that the most recent advances in parallel linear algebra can have in building efficient high-performance software for nonlinear optimization. Bringing the high-quality parallel linear algebra algorithms and software, produced in the last decade, into the field of nonlinear optimization is really a crucial issue to deal with. This is because the linear algebra needs in optimization strongly depend on the kind of problem and/or method one considers, and therefore, for each algorithm, it is necessary to thoroughly analyze the parallel numerical linear algebra strategy to be adopted. By considering a few general classes of methods, well representative for the different aspects of the overall problem, some of the crucial linear algebra kernels, some related existing parallel software and/or algorithms, as well as some specific linear algebra parallelization issues, unresolved or little considered, will be pointed out. Finally, some issues related to parallel preconditioning will also be addressed.

8.1   INTRODUCTION

Due to the increasing availability of low-cost high-performance computers (see e.g. http://www.top500.org) and the ever-increasing need to solve large-scale problems drawn from different application fields [1,2], parallel computing research in the areas of the linear and nonlinear optimization has been intense during the last decade. This effort has followed different directions [3]. Thus, parallel optimization algorithms have been *ad hoc* devised or tuned for specific applications [4–6] that require huge computational work. Completely new algorithms, based on parallel strategies easily scalable and with very low interprocessor interactions, have also been proposed for many problem classes [7–9] to take full advantage of new computer architectures. Multidirectional line search strategies, which belong to the latter case, despite of their very slow convergence properties, represent, at this stage, one of the most innovative trend in designing efficient parallel optimization algorithms suitable for very large number of processors. New algorithmic strategies are probably the only way to fully exploit the computational power of massively parallel MIMD computers. A different general approach consists in parallelizing existing sequential algorithms. The advantage of such an approach is that it can make use of a very large number of highly sophisticated algorithms developed in the last 20 years, for which reliable and well-established software has already been produced and theory supporting their computational efficiency has been developed. The goal is to generate general purpose parallel software, with high level of robustness, portability, and reliability with the highest possible degree of scalability, at least on moderately large parallel MIMD machines. However, this generality may end up less suitable for specific large-scale problems that arise in certain applications, and may fail to take advantage of massively parallel computational environments. Indeed, on massively parallel machines the use of new parallel algorithms is likely to be more efficient than any parallelization strategy of existing methods. However, parallelized version of sequential algorithms, especially of efficient and robust sequential codes, can be seen as part of a general framework in which parallelization is achieved at a higher level. Multidimensional search strategies are good examples of this.

Schnabel [10] gives an excellent survey of perspectives and challenges for parallel computing in nonlinear optimization; he identifies three possible levels of parallelization:

1. Parallelization of the function and/or the derivative evaluations in the algorithm
2. Parallelization of the linear algebra operations
3. Parallelization through modifications of the basic algorithm by performing, for instance, multiple function and/or derivative evaluations to increase the degree of intrinsic parallelism

The third approach is very close to the design of new algorithms, and is the one that mainly involves researchers from the optimization community; it has been especially pursued in global optimization [11] where much of the most recent progress has been driven by the availability of new parallel machines (see Refs. [12–14] and Chapter 7). On the other hand, the first and the second approaches require the interaction with researchers either from the field of numerical linear algebra or the parallel computing communities. The first option, in particular, is the only one available for problems usually derived from modeling and simulation of physical processes, in which the number of variables is small but the objective function evaluation is very expensive because its computation requires some subproblem to be solved. Thus, parallelizing the linear algebra kernels would not result in any meaningful benefit in this case. Naturally, the three different approaches are not mutually exclusive, and actually they are strongly complementary. Merging them together is a very challenging task that can lead to efficient parallel software. A very interesting example of such a merging is given by Hough and Meza [15], who incorporate a multilinear parallel direct search method within a trust-region Newton framework, so that the inherent parallelism of the former can be exploited jointly with the robustness of the latter. The combination of the two methods gives rise to a robust and accurate solution procedure, where the use of the parallel direct search in solving the standard trust-region subproblem takes advantage of parallel architectures.

This chapter will focus on the second of Schnabel's parallelization strategies. That is, it is focused on the impact that the most recent advances in parallel linear algebra can have in building efficient parallel software for nonlinear optimization. The target machines are mainly distributed memory MIMD computers, ranging from multiprocessors to clusters of workstations and/or PCs, because such architectures represent today the majority of the computational power available.

Several software packages for solving numerical linear algebra problems in high-performance computer environments either by direct or iterative methods are now freely available on the Web for both dense and sparse systems.[1]

For the dense case, the ScaLAPACK project [16] is undoubtedly the most elaborated and permanent effort, involving several institutions and people. Its purpose is to provide the scientific community with robust, efficient, and scalable routines to tackle standard medium-size linear algebra problems and to obtain good performances on a variety of parallel machines.

In recent years, some research efforts have been devoted to develop the so-called "infrastructures" for coding and tuning linear algebra algorithms at a high level of abstraction [17, 18]. For this kind of approach, performance is sacrificed, but by coding at a higher level of abstraction, more sophisticated algorithms can be implemented. Moreover, an abstract interface for linear algebra should allow existing software, for instance, an optimization code, to utilize many different representations of the linear algebra objects, from simple serial vectors to distributed objects. The use of such a linear algebra abstract interface should also allow linear algebra systems to perform operations on remote and geographically distributed machines. One of the most exciting projects in this direction is NetSolve [19], which aims at bringing together disparate computational resources connected by computer networks, allowing one to remotely access both hardware and software components. One of the first packages integrated in this project is ScaLAPACK.

For the sparse case, the use of suitable graph coloring and ordering strategies, as well as multifrontal schemes, seems to represent a promising and consistent way in developing efficient parallel distributed direct solvers able to solve large-scale problems in many application fields [20–22].

For the iterative solution of linear systems of equations several packages are currently freely available.[2] The most important features of such packages concern the iterative methods and preconditioners that they supply. Moreover, the interface they present to the user's data structures is also an important defining characteristic. In the context of parallel processing, the most implemented iterative methods, based on the Krylov subspaces, allow for a natural and easy parallelization of their basic computational kernel, i.e., the matrix–vector products. This drastically changes if a suitable preconditioner, necessary for the convergence of the iterative methods for difficult (ill-conditioned) problems, is used. Good preconditioners are hard to design, and this is especially true in parallel environments. Furthermore, the choice of a preconditioner, as well as the strategy for its implementation, have to take into account the structure and special properties of the system to be solved. Incomplete factorizations, developed for symmetric sparse problems, are among the most successful preconditioners. Unfortunately, they are intrinsically sequential and cannot be immediately used on parallel machines. For this reason, few parallel implementations of incomplete factorizations, especially for systems derived from finite element methods, are actually available, and most packages focus on block Jacobi and/or Additive Schwarz preconditioners, which are of very parallel nature, but generally provide only a limited improvement in the number of iterations. On the other hand, the use of domain decomposition and multicoloring strategies should allow to perform global parallel incomplete factorizations with good performance on systems with special structure [23–30].

Bringing the high-quality parallel linear algebra algorithms and software, produced in the last decade, into the field of nonlinear optimization is really a crucial issue to deal with. This represents a research challenge, to be pursued in the future by the people in the numerical linear algebra

---

[1] See for instance: http://www.netlib.org/utk/people/JackDongarra/la-sw.html, http://www.netlib.org/utk/papers/iterative-survey/, for a quite exhaustive list.

[2] See for instance: http://www.netlib.org/utk/papers/iterative-survey/, and references therein.

community, aiming at the fulfillment of the needs in nonlinear optimization. The linear algebra requirements in optimization are quite varying, depending on the kind of problem and/or algorithm one considers, and therefore, for each algorithm, it is necessary to thoroughly analyze the parallel numerical linear algebra strategy to be adopted. Nevertheless, a quite clear picture can be drawn by referring to a few general classes of methods, well representative for the different aspects of the overall problem. In Sections 8.2 and 8.3 *Newton* and *active set* and *interior point* (IP) methods will be considered, and for each of them a very general algorithmic framework will be supplied, pointing out the crucial linear algebra kernels, some related existing parallel software and/or algorithms, as well as some specific linear algebra parallelization issues, unresolved or little considered in the case of optimization methods. Finally, some issues related to parallel preconditioning are also addressed.

8.2 NEWTON METHODS

The most well-established methods for local minimization, both constrained and unconstrained, belong to the very large family of methods derived from the classical Newton method for the minimization of a twice continuously differentiable function $f: R^n \to R$. Given a current iterate x_k, the following steps must be performed to compute the next iterate x_{k+1}:

1. Compute the search direction p_k by solving the system

$$H_k p_k = -\nabla f(x_k), \tag{8.1}$$

2. Make a line search to compute a stepsize α_k such that

$$f(x_k + \alpha_k p_k) < f(x_k),$$

3. $x_{k+1} \leftarrow x_k + \alpha_k p_k$; calculate $\nabla f(x_{k+1})$ and decide whether to stop.
4. Compute H_{k+1}.

The methods of the Newton family are basically differentiated by the way they compute the stepsize and the matrix H. From the point of view of numerical linear algebra, the computation of the matrix H, together with the solution of the linear system (8.1), constitute the computational kernel.

In the original method, the matrix H is the Hessian matrix of the function at the current iterate and assumptions about convexity are needed for convergence. The trust-region strategies [31], however, enable the indefinite case to be elegantly handled so as to ensure global convergence. The computation of the n gradient components and of the n^2 second derivatives offers the opportunity for a quite easy and very efficient parallelization, because all derivatives can be independently computed if either a finite difference scheme or forward automatic differentiation [32] is used. The numerical linear algebra involved in the algorithm concerns the solution of the system (8.1). This can be done in parallel by using well-established parallel computational tools (see Chapter 4). A quite common claim is that Newton's method, although very powerful and robust, because of the need for computing first- and second-order derivatives is suitable only for problems "not too large." However, an answer to the question about which are the problems having this characteristic is not straightforward. Indeed, because of the nonlinearity, the concept of "large problems" may be misleading. Computationally, for a problem with relatively few variables, say 100 or less, but with very expensive objective function evaluations, the use of the Newton method could be a reasonable choice. For this kind of problems, a parallelization strategy is proposed for the trust-region algorithm by Hough and Meza [15]. In contrast, for problems with a large number of variables and sparsity, the use of Newton method can be prohibitive, and quasi-Newton techniques appear much more convenient.

In quasi-Newton methods, step (4) of the algorithm uses an approximation H_{k+1} of the Hessian matrix. This is computed, using an updating process, from H_k, as for example, in the Broyden, Fletcher, Goldfarb, and Shanno direct BFGS method [31]:

$$H_{k+1} = H_k - \frac{H_k s_k s_k^T H_k}{s_k H_k s_k} + \rho_k y_k y_k^T, \tag{8.2}$$

where $\nabla f(x)$ is the gradient of $f(x)$, $s_k = x_{k+1} - x_k$, $y_k = \nabla f(x_{k+1}) - \nabla f(x_k)$, and $\rho_k = 1/y_k^T s_k$, i.e., H_{k+1} is obtained from H_k through a rank-two modification process. An alternative, the inverse BFGS, because a low-rank modification on a matrix implies a low-rank modification on its inverse, would be to store and maintain the inverse $\overline{H}_{k+1}$ of H_{k+1}. In this case, step (1) is substituted by

$$p_k = -\overline{H}_k \nabla f(x_k), \tag{8.3}$$

where the following updating formula is used:

$$\overline{H}_{k+1} = V_k^T \overline{H}_k V_k + \rho_k s_k s_k^T, \tag{8.4}$$

with $V_k = I - \rho_k y_k s_k^T$.

Using (8.2) requires the solution of a linear system at each step, whereas the use of (8.3) only requires a matrix–vector product. On sequential machines, the seeming computational cost disadvantage of using (8.2) can be eliminated by keeping and updating the Cholesky factor of H_k; the Cholesky factor L_{k+1} of H_{k+1}, can be computed by making a rank-two update on the Cholesky factor L_k of H_k [33] through a sequence of Given's rotations. This makes the computational cost comparable to the one computing $\overline{H}_{k+1}$. Because the numerical behavior of this approach is considered superior to the one computing the approximation of the inverse, which might lose positive definiteness because of finite precision, (8.2) is preferred in practical implementations [34, 35]. Moreover, very sophisticated sparse techniques have also been developed for the modification of sparse Cholesky factorizations [33, 36–38].

Unfortunately, the updating process, based on vector–vector operations performs very badly on parallel machines [39], also for dense problems. Indeed, the LINPACK (40) package, in contrast to its evolutions LAPACK [41], and ScaLAPACK (16), includes multiple rank modifications, based only on level 1 BLAS [42] of dense Cholesky factorizations. The efficient implementation of such either dense or sparse linear algebra operations is a well-understood and quite stable process. Nonetheless, the availability of a general purpose, high-performance parallel either dense or sparse implementation is severely hampered by the intrinsic sequential character of such operation. Recently, some parallel strategies have been presented [43], but a parallel direct BFGS still looks like a very difficult task to be achieved.

In contrast, the inverse BFGS algorithm is, from the parallel point of view, a much more appealing strategy. Such considerations led Byrd et al. [44] to make numerical performance comparisons between the direct and the inverse BFGS methods through extensive computational tests, which show that the features of the two approaches are basically the same. Motivating their parallel quasi-Newton method they write, "This is an interesting example where the consideration on parallelism has led to the reexamination of a part of a basic optimization algorithm, with interesting and surprising conclusions" [10].

The main drawback in (8.2) and (8.4) is that H_{k+1} and $\overline{H}_{k+1}$ do not keep the sparsity pattern of $\nabla^2 f$. In the last two decades, several limited BFGS methods (L-BFGS) have been proposed [45], in which the approximate inverse Hessian is not formed explicitly, and only a certain number of pairs (y_k, s_k) is stored; the idea is to keep the most recent information about the curvature of the function, taking into account the storage available. In other words, while $\overline{H}_{k+1}$ is computed in the BFGS method as

$$\overline{H}_{k+1} = F(\overline{H}_0, y_0, s_0, y_1, s_1, \ldots, y_{k-1}, s_{k-1}, y_k, s_k),$$

it is computed in the L-BFGS method as

$$\overline{H}_{k+1} = F(\overline{H}_0, y_0, s_0, y_{k-m}, s_{k-m}, \ldots, y_{k-1}, s_{k-1}, y_k, s_k). \tag{8.5}$$

This approach does not take directly into consideration the sparsity pattern of the Hessian matrix. However, compared to BFGS, it allows dealing with much larger problems. Byrd et al. [46] use the compact form described in Ref. [47] for bound-constrained nonlinear problems, using a projected gradient strategy similar to the one described in Ref. [48] to deal with the constraints. Extensive numerical tests show (see Ref. [49] and references therein) that the algorithm works well in practice for $m < 30$ and that it compares favorably to the Newton method in terms of computing time. Due to the poor quality of the Hessian approximation, the number of iterations is greater than that of the Newton method, but the overall computational costs are comparable. On the NEOS Server (see http://www-neos.mcs.anl.gov/), the code L-BFGS-B [50], enhanced by the ADIFOR [51] and ADOLC [52] automatic differentiation tools, is available for solving large problems. Further details on limited memory quasi-Newton strategies are found in Ref. [53]. Sequential and parallel implementations have found several important applications, for example in weather forecast problems (see Refs. [54, 55] and references therein). Efficient parallelizations can be also obtained by computing concurrently the function value and the derivatives, and by parallelizing the matrix–vector product that arises in (8.5).

An efficient alternative for large-scale optimization problems is given by the truncated Newton methods in which the Newton equation is approximately solved by an iterative method (inner algorithm), whereas a trust-region strategy usually ensures global convergence. The first convergence results for this approach are given in the pioneering work of Dembo et al. [56], whereas a recent, extensive survey is given in Ref. [57]. Usually, the inner iterations utilize some variant of the *conjugate gradient* (CG) method. Scalar and vector operations do not allow a very efficient parallelization, and therefore a parallelization strategy should concentrate on the function and gradient evaluations as well as on block linear algebra. In the parallel BTN package by Nash and Sofer [58, 59], a block CG method is implemented in the inner iteration; moreover, parallel computation can be performed in the gradient evaluation too.

In general, the parallelization of the Newton family methods seems to be achievable according to a two-level strategy, namely high-level parallelization of the algorithm and low-level parallelization, with respect to the linear algebra operations, of the Newton equation in step (1). Example of the first approach is the parallel variable metric algorithm in Ref. [60], in which, at each step, parallel search directions are computed, whose number depends on the number of available processors, and then a parallel line search is performed. A natural form of parallelism arises in the partitioned quasi-Newton methods for large-scale optimization. This is based on the observation that every function with sparse Hessian is partially separable [61], i.e., the objective function can be written (separated) as sum of several internal functions, and therefore the computation of the gradient and the quasi-Newton updating can be split into element-by-element operations. A sophisticated strategy is used in the LANCELOT package [62], which is able to fully exploit the partial separability of the problem.

The common feature of almost all Newton methods is the solution of one or more linear system (more or less exactly) at each iteration; it is worth noting that, in contrast to what happens in interior point strategies (see Section 8.3), very often, at least in the early stages of the algorithms, little precision in the solution is required. The CG algorithm is a quite popular choice in this respect. In several algorithms, the CG method is used not only to approximate a system solution, but also to access the information supplied by the intermediate iterates and residuals. This is the case in the iterated subspace minimization method [63], in which, at each iterate x_k, $\nabla^2 f(x_k)$-conjugate directions are used to define a manifold in which a multidimensional search is performed to determine the next iterate. Similarly, in the quasi-Newton preconditioner by Morales and Nocedal [64], and particularly in their enriched method [65], L-BFGS and Hessian Free Newton (HFN) methods are interlaced, so that the best features of both methods are combined by using the quasi-Newton preconditioner computed during a L-BFGS step in the next HFN iterate.

Hence, CG methods for solving sparse systems of linear equations are playing an ever important role in numerical methods for nonlinear optimization as well as in solving discretized partial differential equations. Actually, it is in this field that the most recent results on efficient parallelization and preconditioning techniques of the CG method have appeared [66].[1]

8.3 INTERIOR POINT VERSUS ACTIVE SET

In this section, the *box-constrained quadratic programming* problem is considered to compare two major classes of constrained optimization methods, namely the interior point and active set methods, from the parallel linear algebra point of view. The focus is on the parallelization of the linear algebra kernels in each iteration. The box-constrained quadratic programming problem is of very general interest because it arises naturally in a number of different applications in engineering, economics, physics and computer science and it is a basic kernel for many nonlinear programming algorithms. It can be stated as follows:

$$
\begin{aligned}
\text{minimize} \quad & f(x) = \tfrac{1}{2}x^{\mathrm{T}}Ax - b^{\mathrm{T}}x \\
\text{subject to} \quad & x \in \Omega = \{x : l \leq x \leq u\}.
\end{aligned}
\tag{8.6}
$$

Here A is a $n \times n$ symmetric, positive definite matrix, b, l, and u are known n-vectors. Moreover, all inequalities involving vectors are interpreted componentwise, and $\nabla f(x) = Ax - b$. For the sake of simplicity we assume that the bounds, l and u, are finite.

Solution algorithms for (8.6) are essential subroutines in many nonlinear optimization codes. *Sequential quadratic programming* (SQP) algorithms have been among the most successful methods for solving general nonlinear programming problems. The basic idea is to linearize the constraints and to consider, in a neighborhood of the current iterate, a quadratic approximation of the Lagrangian function. Therefore, a new iterate is computed by solving a quadratic programming problem, actually more general than (8.6), possibly with some restriction on the step length but for which most of the computational analysis in this section, can easily be extended. The description of the computational kernels, as well as the results of extensive numerical tests for SQP methods, for sparse or dense problems can be found, for example in Refs. [67–70]. Moreover, some of the state-of-the-art codes, such as SNOPT [71] or NLPQL(P) [70, 72] for general nonlinear optimization implement SQP methods. Unfortunately the iterative process of an SQP algorithm is highly sequential. Schittkowski [70] suggests to bring parallelism in the NLPQL code through the computation of first and second derivatives by using finite differences; then (at least) n-function evaluations are needed, which can be performed in parallel. Of course, this approach is reasonable when the derivatives cannot be computed analytically. A more general parallelization strategy would be to use an efficient parallel quadratic programming solver at each iteration.

---

[1] See for example the PARFEM project at http://www.fz-juelich.de/zam/RD/coop/parfem.html

Until the advent of IP algorithms in the mid-1980s [73], the computational scene was dominated by the active set algorithms, implemented in many mathematical software libraries such as the NAG [34] and the IMSL [35] libraries; they have been deeply analyzed and still represent a very popular and reliable approach for constrained optimization [74]. In contrast to active set algorithms which generate a sequence of extreme points, a generic IP method generates a sequence of points in the interior of Ω. Several IP algorithms have been proposed with very attractive computational efficiency and excellent theoretical convergence properties as they have polynomial complexity. In particular, the *potential reduction* (PR) algorithm, proposed by Han et al. in [75], extends the primal-dual PR algorithm for the convex linear complementarity problem, developed by Todd and Ye [76].

Problem (8.6) can be transformed to the standard form

$$
\begin{aligned}
&\text{minimize} \quad f(x) = \tfrac{1}{2}x^{\mathrm{T}}Ax - b^{\mathrm{T}}x \\
&\text{subject to} \quad x + z = e \text{ and } x, z \geq 0,
\end{aligned}
\tag{8.7}
$$

where $z \in R^n$ is the slack variable, e is the vector of ones. The dual problem of (8.7) is stated as

$$
\begin{aligned}
&\text{maximize} \quad e^{\mathrm{T}}y - \tfrac{1}{2}x^{\mathrm{T}}Ax \\
&\text{subject to} \quad s_x = Ax - b - y \geq 0 \text{ and } s_z = -y \geq 0,
\end{aligned}
\tag{8.8}
$$

where $y \in R^n$ is the dual vector. The duality gap of (8.7) and (8.8) is given by

$$
\Delta = \frac{1}{2}x^{\mathrm{T}}Ax - b^{\mathrm{T}}x - \left(e^{\mathrm{T}}y - \frac{1}{2}x^{\mathrm{T}}Ax\right) = x^{\mathrm{T}}s_x + z^{\mathrm{T}}s_z \geq 0.
$$

The interior points $x, z, s_x,$ and s_z are characterized by using the primal–dual potential function of Todd and Ye [76]

$$
\phi(x, z, s_x, s_z) = \rho \ln(x^{\mathrm{T}}s_x + z^{\mathrm{T}}s_z) - \sum_{i=1}^{n} \ln(x_i(s_x)_i) - \sum_{i=1}^{n} \ln(z_i(s_z)_i),
$$

where $\rho \geq 2n + \sqrt{2n}$. The relationship between the duality gap and the PR function was shown by Ye [77], who proved that

$$
\text{if } \phi(x, z, s_x, s_z) \leq -O((\rho - 2n)L) \text{ then } \Delta < 2^{-O(L)},
$$

where L is the size of input data. Therefore, the PR method is based on the minimization of the potential function, to minimize the duality gap. Kojima et al. [78] showed that the Newton direction of the nonlinear system

$$
\begin{cases}
XS_xe = \dfrac{\Delta}{\rho}e \\[2mm]
ZS_ze = \dfrac{\Delta}{\rho}e,
\end{cases}
\tag{8.9}
$$

with fixed Δ guarantees a constant reduction in the potential function. In (8.9), the uppercase letters (X, Z, S_x, S_z) designate the diagonal matrices of the vectors (x, z, s_x, s_z), respectively, and the slack vectors are given by

$$
z = e - x, \quad s_x = Ax - b - y, \quad s_z = -y.
\tag{8.10}
$$

Following the Newton method, given the current iterates $\bar{x}$ and $\bar{y}$, which are interior feasible points for the primal (8.7) and the dual (8.8), respectively, it can be verified that the search direction $\{\delta x, \delta z, \delta y\}$ can be computed by solving the following linear system:

$$
\begin{bmatrix} B & E \\ -E & F \end{bmatrix} \begin{bmatrix} \delta x \\ -E^{-1}\delta y \end{bmatrix} = \begin{bmatrix} c \\ d \end{bmatrix},
\tag{8.11}
$$

where $B = A + \bar{X}^{-1}S_x$, $E = S_z$, $F = \bar{Z}E$, c and d are right-hand side vectors, and $\delta z = -\delta x$, $\delta s_x = A\delta x - \delta y$, $\delta s_z = -\delta y$. One observes that as the current iterate approaches the boundary, the diagonal elements of B can become very large, producing an increasing ill-conditioning. To tackle this numerical difficulty, a suitable diagonal scaling can be used.

Define the diagonal matrix D with

$$
D_{i,i} = \sqrt{\bar{x}_i \bar{z}_i}, \quad \text{for } i = 1, ..., n.
$$

Then, the matrix D can be used to transform the system (8.11) to

$$
\begin{bmatrix} DBD & DED \\ -DED & DFD \end{bmatrix} \begin{bmatrix} D^{-1}\delta x \\ -D^{-1}E^{-1}\delta y \end{bmatrix} = \begin{bmatrix} Dc \\ Dd \end{bmatrix}.
\tag{8.12}
$$

Once the direction is obtained by solving (8.12), the next iterate is generated as

$$
\bar{x} \leftarrow \bar{x} + \bar{\theta}\delta x \quad \text{and} \quad \bar{y} \leftarrow \bar{y} + \bar{\theta}\delta y,
$$

where $\bar{\theta}$ is the step length chosen to minimize the potential function, and the slack vectors are updated according to (8.10). Han et al. [75] give an easy computational procedure for $\bar{\theta}$, and they observe that, for ρ large enough, $\bar{x}$ and $\bar{y}$ guarantee a constant reduction in the potential function.

The dominant computational task in each step of the PR algorithm is the solution of the linear system (8.12), which can be further decomposed as

$$
(DBD + G)(D^{-1}\delta x) = Dc - Qd
$$
$$
-D\delta y = G(D^{-1}\delta x) + Qd,
\tag{8.13}
$$

where $G = E\bar{X}$ and $Q = \bar{Z}^{-1}D$.

An active set strategy for problem (8.6) is implemented by the *projected gradient* algorithm BCQP [79]. BCQP uses the standard active set strategy [38] to explore the face of the feasible region defined by the current iterate, and the projected gradient method to move to a different face. The face of the feasible set which contains the current iterate can be defined in terms of the active set. The set of the *active constraints* at x is defined by

$$
\text{Act}(x) = \left\{ i \colon x^i \in \left\{ l^i, u^i \right\} \right\},
$$

whereas the set of the *binding constraints* at x is defined by

$$
\text{Bind}(x) = \left\{ i \colon \left(x^i = l^i \text{ and } [\nabla f(x)]^i \geq 0 \right) \text{ or } \left(x^i = u^i \text{ and } [\nabla f(x)]^i \leq 0 \right) \right\}.
$$

Note that x_* is a stationary point if and only if $\text{Act}(x_*) = \text{Bind}(x_*)$ and

$$
[\nabla f(x)]^i = 0 \quad \text{if } i \notin \text{Act}(x_*).
$$

Given a current iterate $\overline{x}$, a Newton step of the BCQP algorithm computes

$$\overline{y} = \operatorname{argmin} f(\overline{x} + y), \ y_i = 0, \ \forall i \in \operatorname{Act}(\overline{x}), \qquad (8.14)$$

and then

$$\overline{x} \leftarrow \overline{x} + \alpha \overline{y},$$

where the scalar α is defined by:

$$\alpha = \max\{\alpha \in [0, 1] : l \leq \overline{x} + \alpha \overline{y} \leq u\}.$$

If $\alpha < 1$, the active set is updated by adding at least one of the constraints that becomes active at the new iterate $\overline{x}$; in this case, a new Newton step can be performed. If $\alpha = 1$, then $\overline{x}$ is a global minimizer of subproblem (8.14). If $\operatorname{Act}(\overline{x}) = \operatorname{Bind}(\overline{x})$, then the solution of (8.6) has been found and the algorithm stops, otherwise a sequence of projected gradient steps is performed:

$$\overline{x} \leftarrow P_\Omega (\overline{x} - \alpha \nabla f(\overline{x})),$$

where $P_\Omega(x)$ indicates the orthogonal projection of x onto the set Ω, and the positive stepsize α is chosen so that some sufficient decrease condition [79] is met. The sequence of projected gradient steps stops when either the active set settles down (in two consecutive projected gradient steps) or more than *maxitp*-projected gradient steps are performed, for some prefixed value of *maxitp*. The first of these two conditions is guaranteed to be satisfied in a finite number of steps only for nondegenerate problems; however, the number of steps may be very large if the current iterate is far from the solution. Therefore, the condition on the maximum number of consecutive projected gradient steps is justified both computationally and theoretically. In case of nondegeneracy, the algorithm stops in a finite number of iterations. The solution of (8.14) is the computational kernel of the algorithm and can be easily shown that it requires the solution of the linear system

$$A_{\mathrm{W}} y_{\mathrm{W}} = -g_{\mathrm{W}}, \qquad (8.15)$$

where g_{W} and A_{W} are respectively the gradient in x and the Hessian reduced to the variables which are not active (working set). An active set strategy basically performs a sequence of unconstrained minimizations of the objective function in the affine subspace defined by the current active set (more generally, a subset of it). It iterates as long as the current active set does not match the active set of the solution of (8.6) (identification of the face in which the solution lies). Note that, at each projected gradient step, the problem is reduced to the nonbinding variables, whereas in the Newton step the problem is reduced to the inactive variables.

A comparison between the two algorithms with respect to the main computational issues involved must be done on the basis of the systems (8.13) and (8.15). First of all, it has been observed that in practice the number of iterations in interior point algorithms is almost independent of the number of variables; in contrast, the number of iterations in an active set strategy strongly depends on the problem size (see for example the discussion and the computational results in Ref. [79]). The systems (8.15) to be solved in an active set strategy are strongly related to each other: in two consecutive systems to be solved, the matrix of the second system is obtained from the matrix of the first one through a rank-one update; then, if, for instance the Cholesky factorization is used, its efficient implementation strongly relies on efficient factorization updating strategies [33], so that the cost per iteration can be reduced from $O(m^3)$, when the Cholesky factorization is recomputed, to $O(m^2)$, where m is the number of free variables. For a more detailed discussion about the updating process see Ref. [79]. It is worth to note that the above complexity estimates represent upper bounds to the effective measuring of the computational effort on today's computers. Indeed, floating point operations counting is just a crude accounting approach, which ignores several other issues related to the practical algorithm and software efficiency. Among them, one mentions the problem's sparsity, the block algorithms, the vectorization and the memory hierarchy. By suitably taking into account such aspects, it is possible to have implementations that reduce the "observed" complexity. Nevertheless, in the sequel the simple operations count is used to compare the considered algorithms.

In the PR algorithm, the systems (8.13) are positive definite, unrelated to each other (the diagonal entries change at each iteration) but all with the sparsity pattern of the matrix A (if a direct sparse solver is used, one can take advantage of this). Unfortunately, these systems are very ill-conditioned, at least at the end-stages of the process, when the iterates approach the boundary. Because of this, great care must be taken in the numerical solution of such systems.

Roughly speaking, the PR method requires the solution of few quite ill-conditioned systems of size n with fixed sparsity pattern but unrelated, whereas the BCQP method requires the solution of many more stable systems of varying sizes, whose matrices do not need to be computed from scratch, and whose conditioning is related to that of the Hessian of the objective function. If the matrix A is dense and a direct solver is used in the inner iterations, then each iteration of the PR algorithm has $O(n^3)$ computational complexity, and therefore much higher than in an active set strategy, where, as already noted, the computational cost per iteration varies between $O(m^3)$ and $O(m^2)$, with m equal to the number of free variables. A comparison between the two algorithms shows that BCQP algorithm is the most reliable for ill-conditioned problems and a careful implementation of updating strategies allows to make the algorithm very competitive also in terms of computational cost. Because of this, for small- to medium-size problems, an active set strategy may be superior to PR algorithm.

Because the computational kernels of the two described approaches are quite different, so are their parallel features. The parallel implementation of the BCQP method has been recently discussed by D'Apuzzo et al. [80], who pointed out that the main drawbacks in implementing an active set strategy are the linear algebra operations that must be performed, in each iteration, on matrices and vectors reduced with respect to the current set of the free variables. In theory, working only on the free variables presents an advantage, because if few variables are free at the optimal solution, eventually very small problems must be solved. Unfortunately, neither the standard linear algebra package LAPACK [41] nor BLAS [42] can be used on submatrices and on subvectors. In addition, the very poor parallel performance, due to interprocessor communication, of the factorization updating process, joined with the possibly unbalanced data distribution, due to the modification of the active set in each iteration, represents a serious obstacle for an efficient parallel implementation. Linear algebra operations on matrices and vectors whose size changes dynamically in each iteration and drastically reduces in the last iterations are not suitable for parallel implementation. This is also confirmed from the fact that these nonstandard operations are not implemented in the most popular high-performance mathematical libraries. Because of this, the parallel performance of the BCQP method in Ref. [80] appears to be unsatisfactory.

The key issues arising in the development of a parallel version of the active set algorithm are common to any nonlinear optimization algorithms based on active set strategies: matrix updating processes, linesearches, operations with subvectors, etc. In fact, the general quadratic programming problem

$$\text{minimize} \quad f(x) = \tfrac{1}{2}x^T A x - b^T x$$

$$\text{subject to} \quad Bx \leq u,$$

(8.16)

with $B \in R^{m \times n}$, requires the solution of a sequence of problems of the form

$$\underset{y \in R^n}{\text{minimize}} \quad g(y) = \tfrac{1}{2} (\overline{x} + y)^T A (\overline{x} + y) - b^T (\overline{x} + y)$$

$$\text{subject to} \quad \overline{B} y = 0$$

or, equivalently of the form

$$\underset{y \in R^n}{\text{minimize}} \quad \tfrac{1}{2} y^T A y - \nabla f(\overline{x})^T y$$

$$\text{subject to} \quad \overline{B} y = 0, \tag{8.17}$$

where $\overline{x}$ is the current iterate, $\overline{B}$ is the submatrix of B whose rows are indexed by the working set $\overline{W}$, some subset of the set of indices corresponding to the active constraints in (8.16). The linear constraint in (8.17) guarantees feasibility for $\overline{x} + \alpha y$ for any small enough stepsize $\alpha > 0$. The first-order critical conditions of (8.17) require the existence of Lagrange multipliers λ such that

$$\begin{pmatrix} A & \overline{B}^T \\ \overline{B} & 0 \end{pmatrix} \begin{pmatrix} y \\ \lambda \end{pmatrix} = \begin{pmatrix} \nabla f(\overline{x}) \\ 0 \end{pmatrix}. \tag{8.18}$$

If $\overline{N}$ is a basis for the null-space of $\overline{B}$, i.e., a basis for the tangent subspace defined by the active constraints, (8.17) can be formulated as the following reduced unconstrained problem

$$\underset{y \in R^n}{\text{minimize}} \quad \tfrac{1}{2} y^T \left(\overline{N}^T A \overline{N} \right) y - \nabla f(\overline{x})^T \overline{N} y. \tag{8.19}$$

In principle, the problem (8.17) can be attacked by solving (8.18) or (8.19), and therefore (8.16) requires the solution of a sequence of closely related problems (major iterations) that basically depend on the changes on the matrix of the system (8.18). Hence as for the case of simple constraints in a general active set algorithm the data structure changes dynamically. On parallel computing environment the problem of the data redistribution among processors is added to the intrinsic linear algebra complications that naturally arise in sequential operations such as matrix projections and matrix factorization updates.

In contrast, the usually few but computationally expensive iterations of the interior point algorithms turn out to be an advantage, because they can be efficiently parallelized by using standard parallel linear algebra software tools. For the dense case, a parallel version of the potential reduction algorithm, based on ScaLAPACK [16], for distributed memory MIMD machines is described in Ref. [81], where it is assumed that the computational model consists of a two-dimensional grid of processes, where each process stores and operates on blocks of block-partitioned matrices and vectors. The blocks are distributed in a cyclic block fashion over the processes, and their size represents both the distribution block size and the computational blocking factor used by the processes to perform most of the computations. In particular, the ScaLAPACK routines are based on block-partitioned algorithms to minimize the frequency of data movement between different levels of the memory hierarchy. In Ref. [81] parallel implementation details are presented, showing how a very efficient parallel PR code can be obtained in a very straightforward way. Therefore, for the dense case the existing parallel linear algebra software tools appear quite satisfactory.

Moving to large and sparse problems, things can drastically change. Moré and Toraldo [48] present the algorithm GPCG, i.e., a version of the BCQP algorithm in which the system (8.15) is approximately solved by using the CG method. Encouraging computational results are shown on some classical, structured test problems arising from the discretization of partial differential equations (PDE) problems. The same test problems are used in Ref. [75] for the PR algorithm, in which the system (8.13) is solved by using a CG method without any preconditioning techniques.

In Ref. [82], the parallel performance of GPCG, within the Toolkit for Advanced Optimization [83] environment, is carefully analyzed. Although iterative methods appear more suitable than direct methods for dealing with systems of the form (8.15), some difficulties still arise when working with reduced Hessian and gradient. These difficulties are accentuated in parallel environments since keeping the data evenly distributed among processors is quite a hard task; moreover, the scalability of the algorithm is strongly affected by the varying size of the matrix A_W, and this is a critical issue in any implementation of active set strategies. From the numerical point of view, the choice of an efficient preconditioner is crucial to reduce the overall computing time and to deal with ill-conditioning. The implementation in Ref. [82] is based on object-oriented techniques and leverages the parallel computing and linear algebra environment of PETSc [84].

The computational results in Ref. [82] surprisingly show that an efficient implementation can keep the percentage of time needed to extract and distribute the reduced Hessian among the processors very low (around 2% of the total computing time). This is a very encouraging result, since it has been obtained using standard and widely available software toolkits such as PETSc. These results show that active set strategies using iterative solvers in each step can be a reasonable choice to build reliable and efficient parallel software for large and sparse quadratic problems. Regarding this aspect, note that if $\overline{N}$ is an orthogonal matrix, it can be shown [85] that the projected Hessian $\overline{N}^T A \overline{N}$ is not more ill-conditioned than A. Unfortunately, this nice feature of the orthogonal choice can oppose the goal of preserving the sparsity in $\overline{N}^T A \overline{N}$. Some suitable choices for $\overline{N}$ based on variable reduction strategies can be found in Ref. [85]. In large sparse problems, however, a reasonable strategy should not explicitly form $\overline{N}^T A \overline{N}$. Several *ad hoc* preconditioning techniques have been proposed (see Ref. [85]) for linear systems of the form

$$\left(\overline{N}^T A \overline{N} \right) y = g,$$

which is an obvious generalization of (8.15). In Ref. [74], a preconditioned CG algorithm is used to solve the major iteration, and the Schur complement method is used to update the factorization matrix. Nevertheless, the linear algebra framework for general quadratic programming is harder to parallelize than for the simple bounds. A parallel iterative approach to the linear systems arising in general quadratic programming problems is investigated in Ref. [86], where the authors split the matrix of the linear complementarity subproblems into p blocks, where p is the number of processors used. A parallel iterative solver based on the SOR method is applied to solve the linear subproblems by exploiting the splitting of the matrix. The main issue related to this approach is that the number of inner iterations increases with p, leading to increasing communication overhead. Nevertheless, the authors present promising results, in terms of parallel performance on a distributed memory MIMD machine. Much work is still needed to obtain highly efficient software for quadratic programming, and eventually to use it in a more general framework based on SQP [31] strategies.

The sequential and parallel implementation of IP algorithms is still a very open research field which deserves more attention. In Ref. [87] a sequential implementation of PR algorithm for general sparse problems is presented, in which a *preconditioned conjugate gradient* (PCG) solver is used to solve the system (8.13). As preconditioner, the incomplete Cholesky factorization package IFCS by Lin and Moré [88] is used. The same method has been successfully applied by the same authors [89] in a Newton method for nonlinear box-constrained problems. As its main feature, IFCS allows to specify the amount of additional memory available, without the need to use a drop tolerance.

In each step, the $n_k + p$ largest elements are retained, where p is the fill-in parameter, i.e., the additional memory of the Cholesky factor, and n_k is the number of nonzero elements in the kth column of the original matrix. Furthermore, in Ref. [87], the solution of (8.13) is computed with an accuracy that grows higher as the iterates approach the boundary. The basic idea is to relate the accuracy requirement of the PCG iterations to the value of the duality gap at the current iterate to use an adaptive termination rule for the PCG method that enables computing timesavings in the early steps of the outer scheme. More specifically, the PCG iterations are stopped when the following condition is satisfied:

$$\frac{\|r^l\|}{\|r^0\|} \leq \text{tol}_{\text{cg}},$$

where

$$\text{tol}_{\text{cg}} = \begin{cases} 10^{-5} & \text{if } 10^{-5}\Delta^k > 10^{-5} \\ 10^{-9} & \text{if } 10^{-5}\Delta^k < 10^{-9} \\ 10^{-5}\Delta^k & \text{otherwise,} \end{cases}$$

with r^l being the residual at the lth PCG iteration and Δ^k the current duality gap. Furthermore, the fill-in parameter p is incremented by 1 every 5 outer iterations, starting from $p = 1$. Such a strategy allows reduction in the overall number of CG iterations without increasing the number of outer steps, leading to significant timesavings. Computational results show that the described approach works quite well, and that it compares favorably with well-established software as LOQO [90] and MOSEK [91]. A drawback of such an approach is that the preconditioning techniques based on incomplete factorizations are hard to parallelize. In most cases, as for instance in the solution of the linear systems in the PR algorithm, the use of good preconditioners is crucial for the convergence of the iterative method used. Efficient preconditioners for general linear systems are hard to design in parallel environments. In Ref. [92] a parallel version of the PR method is presented, in which a block diagonal preconditioner is considered, and in each block the incomplete Cholesky factorization is applied. Such a strategy allows to combine the inherent parallelism of diagonal preconditioners with the effectiveness of incomplete factorization techniques. The computational results in Ref. [92] show that, for general sparse problems, the described approach allows a constant reduction of the execution time when the number of processors used increases. For structured optimization problems, the choice of a preconditioner, as well as its parallel implementation strategies, are generally based on the special structure of the system being solved.

The discussion can be generalized to any IP method for linear and nonlinear optimization. In fact, most of the IP methods for linear, quadratic, and nonlinear programming require in each iteration the solution of a linear system of the form

$$\begin{bmatrix} -(H + D) & A^{\text{T}} \\ A & E \end{bmatrix} \begin{bmatrix} \Delta x \\ \Delta y \end{bmatrix} = \begin{bmatrix} c \\ d \end{bmatrix}, \tag{8.20}$$

where H is the Hessian matrix, and D and E are some diagonal matrices. Computational issues related to the solution of (8.20) have been thoroughly analyzed in Refs. [90, 91, 93–96]. A possible approach is to find Δx from the first equation of (8.20), and then eliminate it from the second equation to compute Δy. Similarly, one can proceed in the opposite way by first computing Δy, and then Δx. Focusing for instance on linear programming problems, A is the $m \times n$ constraint matrix, where m is the number of linear constraints, H is zero, as is the matrix E. By deriving Δx from the first block equation of (8.20) and substituting it, one has the symmetric, positive definite system

$$(AD^{-1}A^{T})\Delta y = AD^{-1}c + d. \tag{8.21}$$

Direct solution methods applied to either (8.20) or (8.21) cannot in general take advantage of the special structure of the coefficient matrix and can be numerically and computationally unsuitable because of the fill-in that they introduce. In LOQO, a modified Cholesky factorization for semidefinite systems, using a heuristic strategy to minimize the fill-in, is utilized to solve the large systems. In CPLEX,[1] the smaller systems are similarly solved.

For the system (8.21), the use of an implicit inverse representation of the normal equation matrix $AD^{-1}A^{T}$ can offer some advantages over the explicit Cholesky factorization. Among them, it is worth to mention a significant reduction of the memory used, a better exploitation of the sparsity of the matrix, and considerable timesavings. Furthermore, the implicit representation offers the opportunity to exploit block matrix operations and leads to natural and efficient parallel implementations. As noted before, because many real-life linear and nonlinear optimization problems display some particular block structures either in the objective function or in the constraint matrix, several implementations of IP methods exploit such structures, mainly in the linear algebra context, and are therefore suitable for parallel computational environments.

Another recently proposed approach [97] for developing parallel IP solvers for structured linear programs is based on an object-oriented implementation of a set of routines for matrix operations, that can support any structure. This approach is based on the definition of an abstract matrix class containing a set of virtual functions that provide all the linear operations required by an IP method. From the defined basic matrix class, other classes can be derived, as for instance classes for block-structured matrices. For such matrices, the structure and the hierarchy of blocks can be represented as a tree, whose root is the whole matrix. Thus, the tree provides an exhaustive description of how the matrix is block-partitioned, and almost all linear algebra operations involving the matrix can be recursively performed by operations on the blocks of the tree. Moreover, block-matrix operations are natural candidates for a coarse-grained parallelization. In Ref. [97], a parallel implementation of an objected-oriented strategy for solving large-scale linear programs with primal and dual block-angular structures is presented.

As largely documented, IP methods are the state-of-the-art in linear programming, and they are going to have a great impact on nonlinear programming too [14, 90, 91, 94, 98, 99]. Extensive computational comparisons between state-of-the-art software [93, 100] show that IP-based codes are very reliable, and that they outperform active set algorithms such as SNOPT on many problems. Active set methods are very competitive for problems with few degrees of freedom, i.e., for problems with most of the constraints active at the solution, where IP strategies are unable to exploit such problem features. Because of several new results [100–102] that allow one to overcome some of the structural weakness of the IP approach, it seems reasonable to expect in the near future the development of far more reliable general purpose and robust software for nonlinear programming problems. It is a well-established claim [91] that "the practical implementation of any IP method depends on the efficiency and the reliability of the linear algebra kernel in it."

Actually, it is a common belief that, due to ill-conditioning of the matrices in (8.9), direct methods should be used [90, 91, 103] for the solution of the Newton systems which arise in IP methods. However, a very enlightening recent paper by Wright [101] gives an elegant and sharp error analysis, which shows why the effect of ill-conditioning in IP has much less dramatic effect

---

[1] See http://www.ilog.com/products/cplex/barrier.cfm

on iterative approaches than one would expect. Wang and O'Leary [96] also propose the use of iterative techniques in IP methods for linear programming. KNITRO [98], which is an interesting and innovative software, implements a primal–dual IP method with trust region and uses an iterative solver, namely a projected CG, in its computational kernel.

Parallel implementations based on direct methods are very desirable, because of their robustness and generality. Therefore, the few available[1] scalable parallel sparse direct methods constitute a promising starting point for further research. With regard to this, it is worth mentioning MUMPS [20–22], a recently developed library for solving large sparse systems of equations on distributed memory machines. MUMPS uses a multifrontal approach and achieves high performance by exploiting parallelism based on both problem sparsity and matrix density. The implementation is based on the decomposition of large computational tasks into subtasks, and on a distributed task scheduling technique to dynamically distribute the workload between processors.

The use of iterative approaches appears to be an important alternative because of their inherent parallelism. The critical computational issues concern mainly the development of efficient and reliable parallel preconditioners. A great deal of work has been produced in this direction [23, 26, 28, 30, 66, 104]. In Section 8.4, a quick look at the current trends in the development and use of parallel preconditioners is given.

8.4　PARALLEL PRECONDITIONING ISSUES

In the solution process of many large-scale optimization problems, time and memory constraints do not permit a direct factorization of the linear systems arising in each step of the solution algorithm, and therefore an iterative solver must be used instead. As already mentioned, such systems may be highly ill-conditioned, particularly when they arise in the context of IP algorithms, and therefore iterative methods may fail to converge, or they may converge very slowly. Preconditioning techniques are consequently highly required.

Given a system $Ax = b$, the basic concept behind all the preconditioning techniques is to find a nonsingular matrix K, called the preconditioner, such that the matrix $K^{-1}A$ has "nicer" properties than A, in the sense that a specified iterative method converges faster for the preconditioned system $K^{-1}Ax = K^{-1}b$ than for the original one. Generally, K is chosen so that it is a good approximation of A in certain sense, and the computational cost for constructing it is limited. It is important to notice that the use of preconditioners adds to the computational complexity of the iterative method, and consequently their use is motivated only if it leads to a sufficient reduction in the number of iterations. By focusing on the currently most popular iterative methods, i.e., the Krylov subspace methods [105], one observes that the system matrix A is only needed in matrix–vector products. Thus, forming $K^{-1}A$ explicitly can be avoided. However, a system of the form $Ky = c$ has still to be solved in each step of the iterative method. Consequently, the solution of such a system must be less demanding than the solution of the original one.

As observed in Ref. [88], the use of preconditioning techniques, and mainly those based on incomplete Cholesky factorizations, in the solution of large-scale optimization problems continues to remain quite unexplored. This is due to the fact that the linear systems arising in optimization problems are not guaranteed to be positive definite. Furthermore, from an optimization point of view it is desirable that a preconditioner shows good performance, and that it has predictable and limited memory requirements. The preconditioners currently used in iterative optimization algorithms are in many cases "problem dependent," that is, they try to fully exploit the special structures present in a problem class or the special features present in a particular algorithm. Although problem-

---

[1]　See for instance http://www-users.cs.umn.edu/ mjoshi/pspases/index.html

dependent preconditioners have demonstrated their effectiveness in several cases, there still is a practical need for efficient preconditioners for many problem classes.

The current trends in the development of parallel iterative solvers are discussed in detail by Duff and van der Vorst [106], where they distinguish between iterative methods and preconditioning techniques. The parallelization strategies for the former rely mainly on the parallelization of the computationally dominant kernels, namely sparse inner and matrix–vector products. Such computations require communication between processors in distributed memory MIMD machines, and the amount of such an overhead depends strictly on the matrix distribution. For this reason, research has been focused on the use of adequate schemes for distributing the nonzero elements of the matrix, and on the use of suitable blocking decomposition strategies, to both minimize the interprocessor communication and to reduce memory references in processors with hierarchical local memory. The iterative algorithms developed for currently available parallel machines are based on coarse-grained parallelism with very low interprocessor communication. These algorithms use large matrix–vector operations as building blocks to fully exploit memory hierarchy. However, the current approaches for developing efficient parallel iterative methods can only lead to modest improvements, whereas the use of good preconditioners may drastically reduce computing time. This observation tends to shift the current research focus from iterative methods to preconditioners of high quality.

It has already been pointed out that the construction of an efficient preconditioner for certain problem classes is a hard problem, and even more so if it has to be a parallelizable design. One of the most simple preconditioners is $K = D$, where D is some diagonal matrix. If for instance $D = \text{diag}(A)$, the Jacobi preconditioning results, which is easily parallelizable but leads to poor improvements in terms of convergence speed. Block diagonal preconditioning strategies, in which the blocks are distributed among the available processors and a direct, possibly sparse, approach is locally used by each processor, should lead to a more significant reduction in the number of iterations. However, such an approach really works only on block-structured matrices.

The most general purpose and efficient preconditioners are based on incomplete factorizations of the coefficient matrix [105], and good results have been obtained by using them for a wide range of symmetric problems. Because the main goal of preconditioning is to obtain a matrix $K^{-1}A$ which is close to the identity matrix, it is quite natural to look at a factorization of A as $A = LU$, with L lower-triangular and U upper-triangular matrices. Because for sparse matrices the number of nonzero entries in these factors is generally greater than that in the original matrix, one resorts to incomplete factorization strategies to preserve the factors artificially sparse by retaining, during the factorization process, only some nonzero entries on the basis of a prefixed dropping rule. Let $\bar{L}$ and $\bar{U}$ denote the incomplete factors, then the result of the preconditioner application, that is $y = K^{-1}c$, can be computed by solving the system $\bar{L}\bar{U}y = c$, which in turn involves solving $\bar{L}z = c$ by forward substitutions and computing subsequently y from $\bar{U}y = z$ by back substitutions. Such processes are inherently sequential, because they are based on recursions. Therefore, it is a very difficult task to parallelize incomplete factorizations. To make such preconditioners more suitable for parallel architectures, reformulations of incomplete factorizations with reordering, and domain decomposition techniques have been proposed [23–30, 107]. Such attempts have demonstrated their parallel efficiency mainly on structured linear systems arising from finite difference or finite element discretization of PDE in two or three dimensions. Due to their special structure, valuable savings in the communication overhead can be obtained.

Standard reordering techniques consist in grouping the unknowns together according to their relationships. Each group is labeled by a color name, and the unknowns having the same color have almost no direct connections with the unknowns of all other colors. Using few colors, such as in the well-known red-black ordering, mainly applied to elliptic PDE, results in a very high degree of parallelism but also in increased number of iterations. With a larger number of colors, the global dependence between unknowns increases, and so does the convergence speed. However, the communication among processors also increases, leading thus to degrading of the parallel performance. A suitable balance between increasing convergence speed and increasing interprocessor communication is therefore a critical issue in the design of parallel algorithms.

The basic idea of domain decomposition methods is to split the given domain into subdomains, and to compute an approximate solution in each one of them. Parallel implementations can be obtained by distributing the subdomains among the available processors. To solve each subproblem, a direct method is commonly used for medium-size subdomains, whereas an incomplete version of it is used otherwise. The interaction between the subproblems represents the communication step and it is handled by an iterative scheme. As observed in Ref. [108], preconditioners based on domain decomposition can lead to improved convergence rates when the number of subdomains is not too large. This represents a serious obstacle for the scalability of parallel techniques. The number of neglected fill-in entries increases with the subdomains, degrading the convergence. Accepting more fill-in entries will avoid the deterioration of the preconditioner, but will increase instead the communication between processors. Recent results have shown that an increase in the number of subdomains does not affect the effectiveness of incomplete factorization preconditioners, at least for discretized elliptic PDE equations [26], as long as information from neighboring subdomains is utilized in their derivation.

Attractive alternatives to incomplete factorization have appeared recently in attempts for parallel implementations of approximate inverse preconditioners [109, 110]. Such techniques are based on the computation and the use of a sparse approximation of the inverse of the coefficient matrix, which is obtained by solving

$$\min_{K} \| I - AK \|,$$

where the Frobenius norm is usually considered. This approach results in a set of independent least-squares problems, each involving few variables, which can be solved in parallel. The approximate inverse approach may represent a valid alternative for incomplete LU [111]. However, it generally requires more time and storage than incomplete LU, even if it offers more room for parallelism.

A clear current trend is focusing on methods based on coarse-grained parallelism. Techniques like domain decomposition and sparse approximate inverses give the opportunity to split the overall computation into large tasks. Distribution of data among the processors still requires proper load balancing and reduction in communication costs for sparse matrix computations. Although such techniques have proven to be successful, mainly for PDE, more work is still needed to construct robust and efficient parallel software, based on preconditioned iterative methods, which can be used on both general and special-structured systems.

8.5 CONCLUSIONS

High-quality algorithms and software blend several ingredients, which are becoming more complex with the rapid progress in processor technology and the evolution of computer architectures. At the same time, the design of such algorithms and software poses a challenging task for researchers of different fields. The historically well-established symbiotic relationship between linear algebra and numerical optimization [103] is playing a fundamental role in bringing together the most efficient computational methodologies in mathematical programming with the new computer technologies. New algorithmic developments in *Optimization* are going not only to embody existing parallel linear algebra software in their codes, but also to stimulate new research in *Numerical Linear Algebra*.

Eventually, such an interaction between the two fields will lead to the production of high-quality parallel software.

ACKNOWLEDGMENTS

M. D'Apuzzo, M. Marino, and G. Toraldo were supported by the M.I.U.R. FIRB project "Large Scale Nonlinear Optimization," No. RBNE01WBBB, ICAR-CNR, branch of Naples.

REFERENCES

[1] de Leone, R., Murli, A., Pardalos, P.M., and Toraldo, G. (1998). *High Performance Algorithms and Software in Nonlinear Optimization*. Boston: Kluwer Academic Publishers.

[2] Pardalos, P.M. and Resende, M.G.C. (2002). *Handbook of Applied Optimization*. Oxford: Oxford University Press.

[3] Migdalas, A., Toraldo, G., and Kumar, V. (2003). Nonlinear optimization and parallel computing. *Parallel Comput.* 29(4) Special Issue: *Parallel Computing in Numerical Optimization*:375–391.

[4] Jooyounga, L., Pillardy, J., Czaplewski, C., Arnautova, Y., Ripoll, D., Liwo, A., Gibson, K., Wawak, R., and Scheraga, H. (2000). Efficient parallel algorithms in global optimization of potential energy functions for peptides, proteins, and crystals. *Comput. Phys. Commun.*. 128(1–2):399–411.

[5] Resende, M.G.C. and Veiga, G. (1993). In Johnson, D.S. and McGeoch, C.C., Eds., *Network Flows and Matching: First DIMACS Implementation Challenge. DIMACS Series on Discrete Mathematics and Theoretical Computer Science,* 12: 299–348.

[6] Wang, D.Z., Droegemeier, K.K., and White, L. (1998). The adjoint Newton algorithm for large-scale unconstrained optimization in meteorology applications. *Comput. Optim. Appl.* 10:281–318.

[7] Lewis, R.M. and Torczon, V.J. (1999). Pattern search algorithms for bound constrained minimization. *SIAM J. Optim.* 9(4):1082–1099.

[8] Lewis, R.M. and Torczon, V.J. (2000). Pattern search methods for linearly constrained minimization. *SIAM J. Optim.* 10:917–941.

[9] Pardalos, P.M., Pitsoulis, L., Mavridou, T., and Resende, M.G.C. (1995). Parallel search for combinatorial optimization: genetic algorithms, simulated annealing, tabu search, and GRASP. In Ferreira, A. and Rolim, J., Eds., *Lecture Notes in Computer Science — Proceedings of the Second International Workshop — Irregular'95*, Vol. 980. Berlin: Springer-Verlag, pp. 317–331.

[10] Schnabel, R.B. (1995). A view of the limitations, opportunities, and challenges in parallel nonlinear optimization. *Parallel Comput.* 21(6):875–905.

[11] Migdalas, A., Pardalos, P.M., and Värbrand, P., Eds. (2001). *From Local to Global Optimization*. Dordrecht: Kluwer Academic Publishers.

[12] Migdalas, A., Pardalos, P.M., and Storøy, S., Eds. (1997). *Parallel Computing in Optimization*. Dordrecht: Kluwer Academic Publishers.

[13] Holmqvist, K., Migdalas, A., and Pardalos, P.M. (1997). Parallel Continuous Non-Convex Optimization. In: Migdalas, A., Pardalos, P.M., and Storoy, S., Eds., *Parallel Computing in Optimization*. Dordrecht: Kluwer Academic Publishers.

[14] Pardalos, P.M. and Resende, M.G.C. (1996). Interior point methods for global optimization. In Terlaky, T., Ed., *Interior Point Methods in Mathematical Programming*. Boston: Kluwer Academic Publishers, pp. 467–500.

[15] Hough, P.D. and Meza, J.C. (1999). A Class of Trust-Region Methods for Parallel Optimization. Technical report SAND99-8245, Sandia National Laboratories, Livermore, CA.

[16] Blackford, L.S., Choi, J., Cleary, A., D'Azevedo, E., Demmel, J., Dhillon, I., Dongarra, J., Hammarling, S., Henry, G., Petitet, A., Stanley, K., Walker, D., and Whaley, R.C. (1997). *ScaLAPACK Users' Guide*. Philadelphia: SIAM.

[17] van de Geijn, R.A. (1997). *Using PLAPACK: Parallel Linear Algebra Package*. Cambridge, MA: MIT Press.

[18] Whaley, R.C., Petitet, A.A., and Dongarra, J.J. (2001). Automated empirical optimizations of software and the ATLAS project. *Parallel Comput.* 27(1–2):3–35.

[19] Agrawal, S., Arnold, D., Blackford, S., Dongarra, J., Miller, M., Sagi, K., Shi, Z., Seymour, K., and Vahdiyar, S. (2002). User's Guide to NetSolve V1.4.1. ICL Technical Report, ICL-UT-02-05, Tennessee, University of Tennessee, Knoxville.

[20] Amestoy, P.R., Duff, I.S., and L'Excellent, J.-Y. (2000). Multi-frontal parallel distributed symmetric and asymmetric solvers. *Comput. Meth. Appl. Mech. Eng.* 184:501–520.

[21] Amestoy, P.R., Duff, I.S., L'Excellent, J.-Y., and Koster J. (2001). A fully asynchronous multifrontal solver using distributed dynamic scheduling. *SIAM J. Matrix Anal. Appl.* 23(1):15–41.

[22] Amestoy, P.R., Duff, I.S., L'Excellent, J.-Y., and Li, X.S. (2001). Analysis, tuning and comparison of two general sparse solvers for distributed memory computers. *ACM Trans. Math. Softwares.* 27(4): 388–421.

[23] Corral, C., Giménez, I., Marín, J., and Mas, J. (1999). Parallel m-step preconditioners for the conjugate gradient method. *Parallel Comput.* 25(3):265–281.

[24] Doi, S. (1991). On parallelism and convergence of incomplete LU factorizations. *Appl. Numer. Math.* 7:417–436.

[25] Doi, S. and Hoshi, A. (1992). Large numbered multicolor MILU preconditioning on SX-3/14. *Int. J. Comput. Math.* 44:143–152.

[26] Haase, G. (1998). Parallel incomplete Cholesky preconditioners based on the non-overlapping data distribution. *Parallel Comput.* 24(11):1685–1703.

[27] Hysom, D. and Pothen, A. (2001) A scalable parallel algorithm for incomplete factor preconditioning. *SISC* 22(6):2194–2215.

[28] Magolu Monga Made, M. and van der Vorst, H. (2001). A generalized domain decomposition paradigm for parallel incomplete LU factorization preconditionings. *Future Generation Comput. Syst.* 17(8): 925–932.

[29] Magolu Monga Made, M.and van der Vorst, H. (2001). Parallel incomplete factorizations with pseudo-overlapped subdomains. *Parallel Comput.* 27(8):989–1008.

[30] Pakzad, M., Lloyd, J.L., and Phillips, C. (1997). Independent columns: a new parallel ILU preconditioner for the PCG method. *Parallel Comput.* 23(6):637–647.

[31] Dennis, J.E. and Schnabel, R.B. (1996). *Numerical Methods for Unconstrained Optimization and Nonlinear Equations.* Philadelphia: SIAM.

[32] Griewank, A. (2000). *Evaluating Derivatives: Principles and Techniques of Algorithmic Differentiation.* Philadelphia: SIAM.

[33] Gill, P.E., Murray, W., Saunders, M.A., and Wright, M.H. (1984). Procedure for optimization problems with a mixture of bounds and general linear constraints. *ACM Trans. Math. Software* 10(3):282–298.

[34] The Numerical Algorithms Group Ltd (2000). *e04 — Minimizing or Maximizing a Function.* Oxford, UK.

[35] *QPROG/DQPROG,* IMSL Math/Library Online User's Guide. Vol. 1–2.

[36] Davis, T.A. and Hager W.W. (2001). Multiple rank modifications of a sparse Cholesky factorization. *SIAM J. Matrix Anal. Appl.* 22(4):997–1013.

[37] Davis, T.A. and Hager W.W. (1999). Modifying a sparse Cholesky factorization. *SIAM J. Matrix Anal. Appl.* 20:606–627.

[38] Gill, P.E., Murray, W., and Wright, M.H. (1981). *Practical Optimization.* London: Academic Press.

[39] D'Apuzzo, M., De Simone, V., Marino, M., and Toraldo, G. (2000). Parallel computing in bound constrained quadratic programming. *Ann. Univ. Ferrara - Sez. VII - Sc. Mat..* XLV:479–491.

[40] Dongarra, J.J., Bunch, J.R., Moler, C.B., and Stewart, G.W. (1978). *LINPACK Users Guide.* Philadelphia: SIAM.

[41] Anderson, E., Bai, Z., Bischof, C., Demmel, J., Dongarra, J., Du Croz, J., Greenbaum, A., Hammarling, S., McKenney, A., Ostrouchov, S., and Sorensen, D. (1995). *LAPACK Users' Guide,* 2nd ed. Philadelphia: SIAM.

[42] Dongarra, J.J., Du Croz, J., Hammarling, S., and Duff, I. (1990). Algorithm 679: level 3 BLAS. *ACM TOMS* 16:18–28.

[43] Kontoghiorhes, E.J. (2000). Parallel strategies for rank-k updating of the QR decomposition. *SIAM J. Matrix Anal. Appl.* 22(3):714–725.

[44] Byrd, R.H., Schnabel, R.B., and Schultz, G.A. (1988). Parallel quasi-Newton methods for unconstrained optimization. *Math. Program.* 42:273–306.

[45] Liu, D.C. and Nocedal, J.(1989). On the limited memory BFGS method for large scale optimization. *Math. Program.* 45:503–528.

[46] Byrd, R.H., Lu, P., and Nocedal, J. (1995). A limited memory algorithm for bound constrained optimization. *SIAM J. Sci. Stat. Comput.* 16(5):1190–1208.

[47] Byrd, R.H., Nocedal, J., and Schnabel, R.B. (1994). Representation of quasi–Newton matrices and their use in limited memory methods. *Math. Program.* 63(4):129-156.

[48] Moré, J.J. and Toraldo, G. (1991). On the solution of quadratic programming problems with bound constraints. *SIAM J. Optim.* 1:93–113.

[49] Nocedal, J. (1997). Large scale unconstrained optimization. In Watson, A. and Duff, I., Eds., *The State of the Art in Numerical Analysis*. Oxford: Oxford University Press, pp. 311–338.

[50] Zhu, C., Byrd, R., Nocedal, J., and Lu, P. (1997). Algorithm 778: L-BFGS-B, FORTRAN routines for large scale bound constrained optimization. *ACM Trans. Math. Software* 23(4):550–560.

[51] Bischof, C., Carle, A., Khademi, P., Mauer, A., and Hovland, P. (1998). ADIFOR 2.0 User's Guide (Revision D). Technical report ANL/MCS-TM-192, Argonne National Laboratory, Argonne, IL.

[52] Griewank, A., Juedes, D., Mitev, H., Utke, J., Vogel, O., and Walther, A. (1996). Algorithm 755: ADOL-C. A package for the automatic differentiation of algorithms written in C/C++. *ACM TOMS* 22(2):131–167.

[53] Kolda, T., O'Leary, D.P., and Nazareth, L. (1998). BFGS with update skipping and varying memory. *SIAM J. Optim.* 8(4):1060–1083.

[54] Kalnay, E., Park, S.K., Pui, Z.X., and Gao, J. (2000). Application of the quasi-inverse method to data assimilation. *Monthly Weather Rev.* 128:864–875.

[55] Vogel, C.R. (2000). A limited memory BFGS method for an inverse problem in atmospheric imaging. In Hansen, P.C., Jacobsen, B.H., and Mosegaard, K., Eds., *Methods and Applications of Inversion*, Lecture notes in Earth Sciences 92. Heidelberg: Springer-Verlag, pp. 292–304.

[56] Dembo, R., Eisenstat, S., and Steihaug, T. (1982). Inexact Newton methods. *SIAM J. Numer. Anal.* 9:400–408.

[57] Nash, S.G. (2000). A survey of truncated-Newton methods. *J. Comp. Appl. Math.* 124:45–59.

[58] Nash, S.G., and Sofer, A. (1991). A general-purpose parallel algorithm for unconstrained optimization. *SIAM J. Optim.* 1:530–547.

[59] Nash, S.G., and Sofer, A. (1992). BTN: software for parallel unconstrained optimization. *ACM Trans. Math. Software* 18:414–448.

[60] Phua, P.K.-H., Fan Weiguoa, W., and Zeng, Y. (1998). Parallel algorithms for large-scale nonlinear optimization. *Int. Trans. in Oper. Res.* 5(1):67–77.

[61] Griewank, A. and Toint, Ph.L. (1982). Partitioned variable metric updates for large structured optimization problems. *Numer. Math.* 39:119–137.

[62] Conn, A.R., Gould, N.J., and Toint, Ph.L. (1992). *LANCELOT: A Fortran Package for Large-Scale Nonlinear Optimization (Release A)*. Berlin: Springer-Verlag.

[63] Conn, A.R., Gould, N.J., Sartenaer, A., and Toint, Ph.L. (1996). On the iterated-subspace minimization methods for nonlinear optimization with a combination of general equality and linear constraints. In *Proceedings on Linear and Nonlinear Conjugate Gradient-Related Methods*, Seattle, USA., July.

[64] Morales, J.L. and Nocedal, J. (2001). Algorithm PREQN: FORTRAN subroutines for preconditioning the conjugate gradient method. *ACM Trans. Math. Software* 27(1):83–91.

[65] Morales, J.L. and Nocedal, J. (2002). Enriched methods for large-scale unconstrained optimization. *Comput. Optim. Appl (COAP)* 21(2):143–154.

[66] Basermann, A., Reichel, B., and Schelthoff, C. (1997). Preconditioned CG methods for sparse matrices on massively parallel machines. *Parallel Comput.* 23(3):381–398.

[67] Boggs, P.T. and Tolle, J.W. (2000). Sequential quadratic programming for large-scale nonlinear optimization. *J. Comput. Appl. Math.* 124:123–137.

[68] Benson, H.Y., Shanno, D.F., and Vanderbei, R.J. (2003). A comparative study of large-scale nonlinear optimization algorithms. In Di Pillo, G. and Murli, A., Eds., *High Performance Algorithms and Software for Nonlinear Optimization*. Boston: Kluwer Academic Publishers.

[69] Chauvier, L., Fuduli, A., and Gilbert, J.C. (2003). A truncated SQP algorithm for solving nonconvex equality constrained optimization problems. In Di Pillo, G. and Murli, A., Eds., *High Performance Algorithms and Software for Nonlinear Optimization*. Boston: Kluwer Academic Publishers, pp. 149–176.

[70] Schittkowski, K. (2002). Implementation of a Sequential Quadratic Programming Algorithm for Parallel Computing. Report, Department of Mathematics, University of Bayreuth, Germany.

[71] Gill, P.E., Murray, W., and Saunders, M.A. (2002) SNOPT: an SQP algorithm for large-scale constrained optimization. *SIAM J. Optim.* 12:979–1006.

[72] Schittkowski, K. (1986). A FORTRAN subroutine for solving constrained nonlinear programming problems. *Ann. Oper. Res* 5:485–500.

[73] Karmarkar, N. (1984). A new polynomial time algorithm for linear programming. *Combinatorica* 4:373–395.

[74] Gould, N.I.M., and Toint, Ph.L. (2002). Numerical method for large-scale nonconvex quadratic programming. In Siddiqi, A.H. and Kocvara M., Eds., *Trends in Industrial and Applied Mathematics*. Boston: Kluwer Academic Publishers, pp. 149–179.

[75] Han, C.G., Pardalos, P.M., and Ye, Y. (1990). Computational aspects of an interior point algorithm for quadratic problems with box constraints. In Coleman, T. and Li, Y., Eds., *Lage-Scale Numerical Optimization*. Philadelphia: SIAM, pp. 92–112.

[76] Todd, M.J. and Ye, Y. (1990). A centered projective algorithm for linear programming. *Math. Oper. Res.* 15:508–529.

[77] Ye, Y. (1991). An $O(n^3L)$ potential reduction algorithm for linear programming. *Math. Program.* 50:239–258.

[78] Kojima, M., Megiddo, N., and Yoshise, A. (1991). An O($\sqrt{n}L$) iteration potential reduction algorithm for linear complementarity problems. *Math. Program.* 50:331–342.

[79] Moré, J.J. and Toraldo, G. (1989). Algorithms for bound constrained quadratic programming problems. *Numer. Mathe.* 55:377–400.

[80] D'Apuzzo, M., De Simone, V., Marino M., and Toraldo, G. (1998). A parallel algorithm for box-constrained convex quadratic programming. *AIRO* 1:57–80.

[81] D'Apuzzo, M., Marino M., Pardalos, P.M., and Toraldo, G. (2000). A parallel implementation of a potential reduction algorithm for box-constrained quadratic programming. In Bode, A., Ludwig, T., Karl, W., and Wismller, R., Eds., *Lecture Notes in Computer Science — Euro-Par 2000 Parallel Processing*. Berlin: Springer-Verlag, pp. 839–848.

[82] Benson, S.J., McInnes, L.C., and Moré, J.J. (2001). A case study of the performance and scalability of optimization algorithms. *ACM Trans. Math. Software* 27(3):361–376.

[83] Benson, S.J., McInnes, L.C., Moré, J.J., and Sarich, J. (2003). TAO Users Manual. Technical report ANL/MCS-TM-242-Revision 1.5. Argonne National Laboratory, Argonne, IL.

[84] Balay, S., Buschelman, K., Gropp, W.D., Kaushik, D., Knepley, M., McInnes, L.C., Smith, B.F., and Zhang, H. (2003). PETSc 2.0 Users manual. Technical report ANL-95/11 Revision 2.1.6, Argonne National Laboratory, Argonne, IL.

[85] Nash, S.G. and Sofer, A. (1996). Preconditioning reduced matrices. *SIAM J. Matrix Anal. Appl.* 17(1):47–68.

[86] Galligani, E., Ruggiero, V., and Zanni, L. (1998) Parallel solution of large scale quadratic programs. In De Leone, R., Murli, A., Pardalos, P.M., and Toraldo, G., Eds., *High Performance Algorithms and Software in Nonlinear Optimization*. Boston: Kluwer Academic Publishers, pp. 189–205.

[87] D'Apuzzo, M., De Cesare, M.L., Marino, M., and Toraldo, G. (2001). IP Software for BCQP Problems. Technical report TR-13-01, CPS, Naples, Italy.

[88] Lin, C.-J. and Moré, J.J. (1999). Incomplete Cholesky factorizations with limited memory. *SIAM J. Sci. Comput.* 21:24–45.

[89] Lin, C.-J. and Moré, J.J. (1999). Newton's method for large bound-constrained optimization problems. *SIAM J. Optim.* 9(4):1100–1127.

[90] Vanderbei, R.J. and Shanno, D.F. (1999). An interior point algorithm for nonconvex nonlinear programming. *Comput. Optim. Appl.* 13:231–252.

[91] Andersen, E.D., Gondzio, J., Mészáros, Cs., and Xu, X. (1996). Implementation of interior point methods for large scale linear programming. In Terlaky, T., Ed., *Interior-Point Methods of Mathematical Programming*. Dordrecht: Kluwer Academic Publishers, pp. 189–252.

[92] D'Apuzzo, M. and Marino, M. (2003). Parallel computational issues of an interior point method for solving large bound constrained quadratic programming problems. *Parallel Comput.* 29(4):467–483.

[93] Benson, H.Y., Shanno, D.F., and Vanderbei, R.J. (2004). Interior-point methods for nonconvex nonlinear programming: jamming and numerical testing. *Math. Program.* 99(1):35–48.

[94] Shanno, D.F. and Vanderbei, R.J. (2000). Interior point methods for nonconvex nonlinear programming: orderings and higher-order methods. *Math. Program.* 87(2):303–316.

[95] Shi, Y. (1995). Solving linear systems involved in constrained optimization. *Linear Algebra Appl.* 229:175–189.

[96] Wang, W. and O'Leary, D.P. (2000). Adaptive use of iterative methods in predictor–corrector interior point methods for linear programming. *Numer. Algorithms* 25(1–4):387–406.

[97] Gondzio, J. and Sarkissian, R. (2003). Parallel interior point solver for structured linear programs. *Math. Program.* 96(3):561–584.

[98] Byrd, R.H., Hribar, M.E., and Nocedal, J. (1999). An interior point algorithm for large-scale nonlinear programming. *SIAM J. Optim.* 9(4):877–900.

[99] Byrd, R.H., Gilbert, J.C., and Nocedal J. (2000). A trust region method based on interior point techniques for nonlinear programming. *Math. Program. A* 89:149–185.

[100] Morales, J.L. and Nocedal, J. (2003). Assessing the potential of interior methods for nonlinear optimization. In Biegler, L.T. et al., Eds., *Large-Scale PDE-Constrained Optimization*. Berlin: Springer-Verlag.

[101] Wright, M.H. (1998). Ill-conditioning and computational error in interior methods for nonlinear programming. *SIAM J. Optim.* 9(1):84–111.

[102] Wright, S.J. (2001). Effects of finite precision arithmetic on interior-point methods for nonlinear programming. *SIAM J. Optim.* 12(1):36–78.

[103] O'Leary, D.P. (2000). Symbiosis between linear algebra and optimization. *J. Comput. Appl. Math.* 123:447–465.

[104] Durazzi, C., Ruggiero, V., and Zanghirati, G. (2001). Parallel interior point method for linear and quadratic programs with special structure. *J. Optim. Theory Appl.* 110(2):289–313.

[105] Saad, Y. (1996). *Iterative Methods for Sparse Linear Systems*. Boston: PWS Publishing Company.

[106] Duff, I.S. and van der Vorst, H.A. (1999). Developments and trends in the parallel solution of linear systems. *Parallel Comput..* 25(13–14):1931–1970.

[107] Hu, F., Maguire, K.C.M., and Blake, R.J. (1999). Ordering unsymmetric matrices into bordered block diagonal form for parallel processing. In Amestoy, P., Berger, P., Daydé, M., Duff, I., Frayssé, V., Giraud, L., and Ruiz, D., Eds., *Lecture Notes in Computer Science, 1685*. Berlin: Springer-Verlag, pp. 295–302.

[108] Chan, T.F. and Goovaerts, D. (1990). A note on the efficiency of domain decomposed incomplete factorizations. *SIAM J. Sci. Stat. Comput.* 11:794–803.

[109] Barnard, S.T. and R.L. Clay, R.L. (1997). A portable MPI implementation of the SPAI preconditioner in ISIS++. In *Proceedings of the Eighth SIAM Conference on Parallel Processing for Scientific Computing. Minneapolis*, MN, March 14–17.

[110] Benzi M., Marin, J., and Tuma, M. (1999). A two-level parallel preconditioner based on sparse approximate inverse. In Kincaid, D.R. and Elster, A.C., Eds., *Iterative Methods in Scientific Computing IV*. IMACS Series in Computational and Applied Mathematics, Vol. 5. New Brunswick, NJ IMACS, pp. 167–178.

[111] Gould, N.I.M. and Scott, J.A. (1998). Sparse approximate inverse preconditioners using norm-minimization techniques. *SIAM J. Sci. Comput.* 19(2):605–625.

Statistical Applications

9 On Some Statistical Methods for Parallel Computation

Edward J. Wegman

CONTENTS

9.1 INTRODUCTION

The history of computing has been marked by the inevitable march toward higher performance computing devices. As the physical limits of electronic semiconductors reach submicron sizes and current movement is limited to a few hundreds of electrons, the search for increased speeds and performance in computing machines has shifted focus from high-performance specialized processors once found in past generations of supercomputers to so-called massively parallel computers built with commodity CPU chips. These commodity chips have evolved much more rapidly than specialized processors because the large scale of the installed base allows economies of scale in their development. Parallel computation is an efficient form of computing, which emphasizes the exploitation of concurrent events. Even the simplest microcomputers to some extent have elements of concurrency such as simultaneous memory fetches, arithmetic processing and input/output. However, generally the term, parallel computation, is reserved for machines or clusters of machines, which have multiple arithmetic/logical processing units and which can carry on simultaneous arithmetic/logical operations. The architectures that, when implemented, lead to parallel machines can be thought of in three generic forms: pipeline processors, array processors, and concurrent multiprocessors.

Conventional supercomputers were an evolution of mainframe computers that focused on the development of very powerful custom arithmetic processors. The CRAY Y-MP supercomputer, for example, at most had 16 processors whereas its successors and similar competitors had as many as 100 processors. The individual arithmetic processors were so physically small and operated at sufficiently high current levels that heat dissipation was a major problem and the systems required freon-type or other exotic coolants to allow them to function. In a general sense, while they were, for their era, extremely powerful number crunchers, the number of megaflops per byte of input data had to be comparatively large to keep their arithmetic processors operating at full potential. Calculations involving the solutions to partial differential equations (PDE) typically have such a character. Thus applications involving aerodynamic and fluid dynamics, high-energy physics calculations, computational chemistry, cryptographic analysis and seismic exploration were the primary beneficiaries of supercomputing. Most statistical calculations involve only a relatively modest number of megaflops per byte of data and so do not demand traditional supercomputing capability. Ironically, the current emphasis on data mining only became practical with improvements in data storage and parallel computation. Indeed, data mining applications such as clustering with large data sets demand the highest capabilities of modern computing. See Wegman [1] for a discussion of computational feasibility.

9.2   THE FORMS OF PARALLEL COMPUTERS

The three generic forms of parallel computing devices are pipeline processors, array processors, and concurrent multicomputers (either as a single machine with many processors or as a cluster of linked independent machines).

9.2.1   PIPELINE PROCESSORS

To understand a pipeline processor, it is important to understand that the process of executing an instruction consists of four major steps: the instruction fetch, the instruction decoding, the operand fetch, and the execution. In a nonpipeline computer these steps must be completed before the next instruction can be issued. In a pipeline machine, successive instructions are executed in an overlapped fashion. An instruction cycle is the amount of time it takes to execute a single instruction. The instruction cycle can be thought of as being made up of multiple pipeline cycles, a pipeline cycle being equal to the time required to complete the (four) stages of instruction execution. In the nonpipeline computer, a single instruction will require four pipeline cycles, whereas, in a pipeline computer, once the pipeline is filled up, an instruction is executed at each pipeline cycle. Therefore, for this example, the pipeline machine is four times as quick as a nonpipeline machine.

Actually a machine need not be limited to a four-stage pipeline, but could be extended to a k-stage pipeline depending on the machine designer's desire. Theoretically, a k-stage linear pipeline processor could be at most k times as fast. However, because of memory bank conflicts, conflicts due to simultaneous memory access, data dependency, branching and interrupts, this theoretical speedup will never in practice be achieved. Nonetheless, clearly considerable improvement in efficiency can be achieved if the pipeline is tuned to performing a specific operation repeatedly. That is to say, whenever there is a change in operation, for example, from addition to multiplication, the arithmetic pipeline must be drained and reconfigured. Therefore, most pipelines are tuned to specific operations, usually addition and multiplication, and used in situations where these operations are performed repeatedly. The prime example of these are vector processors used in traditional supercomputers in which the same operations are repeatedly applied to components of a vector.

9.2.2 ARRAY PROCESSORS

An array processor in contrast is a synchronous parallel computer with multiple arithmetic/logic units (processing elements) that operate simultaneously in a lockstep fashion. By replicating the processing elements in an appropriately configured and connected geometric array, it is possible to achieve a spatial distribution of processing elements, a spatial parallelism. An appropriate data routing mechanism must be constructed for the processing elements. The processing elements may be either special purpose processing elements or general purpose. Machines in this category often contain very large numbers of processors. In the case of special purpose machines such as the systolic array for multiplying $n \times n$ matrices, approximately n^2 processing elements are required. The general purpose case was illustrated by the massively parallel bit-slice processors; a commercial example was the Connection Machine. The Connection Machine could have been purchased with between 16,000 and 64,000 processing elements. Machines with a large number of simple processors were generally known as fine-grain machines. Pipeline processors have become the standard for commodity chips such as those manufactured by Intel, AMD, and Motorola and are no longer the province of supercomputers only. General and special purpose fine-grain machines have fallen out of favor and are no longer commercially produced.

9.2.3 CONCURRENT MULTICOMPUTERS

In contrast with the fine-grain machines are coarse-grain machines, which, as one would guess, have a more limited number of high-performance processors together with appropriate memory. Memory may either be shared or local to the processor. In any case the processors are configured with some connection topology that in simplest form may be a ring network, a lattice, a star network, a tree-structured graph, and, in most complicated form, a complete graph. A machine with a ring structure economizes on the total number of interconnects required, but exchanges this for distance between nodes. By distance between nodes we mean the number of intervening nodes through which a message must pass to go from source node to destination node. If communication between nodes is important for a given application, typically a ring-connected architecture will have bad performance because distance between nodes is a surrogate for communication time. At the other extreme is a machine with a complete graph. Here every node is connected to every other node, leading to fast internode communications but also cumbersome interconnect wiring and clumsy control mechanisms. Some intermediate configurations include the butterfly, the binary tree, and the hypercube configuration. In the mid-1980s to early 1990s, many commercial parallel machines were sold. Intel introduced a hypercube-based machine in 1985 known as the iPSC (Intel Personal Super Computer) based on the 80286 chip with ethernet connections between nodes. This was later replaced with the iPSC/2 using 80386 chips and the iPSC/860 and Paragon using 80860 risc chips. The later machines had special purpose network interface chips (NICs) that provided direct routing and improved communication overhead dramatically. More recently, networking individual PCs using ethernet connections (Beowulf clusters) have become popular.

In a multiprocessor environment one would like to achieve linear speedup, i.e., for k-processors, one would like a k-fold speedup. However, because of communication overhead and inherent nonparallelizability, this is never achieved. More processors make the communication overhead more intense so that there is a practical reason of diminishing returns for limiting total number of processors.

There are several taxonomies of parallel machines emphasizing different features. One classification involves instruction and data streams. An instruction stream is a sequence of instructions
as executed by a machine. A data stream is a sequence of data including input, intermediate or
partial results called for by the instruction scheme. A machine may involve a single or multiple
instruction stream and a single or multiple data stream. The single instruction single data (SISD) is
the mode of most commonly available serial machines. The machine may be pipelined, however,
so that there can be an element of parallelism even in a SISD machine. In the single instruction
multiple data (SIMD) architecture all processing elements receive the same instruction broadcast
from the control unit, but have their own unique data sets flowing from different data streams. Typically an array processor is an SIMD. The multiple instruction single data (MISD) involves multiple
processors receiving a distinct instruction set but operating on a common data set. No practical
examples of this style of computing architecture exists. The multiple instruction multiple data
(MIMD) uses multiple instruction streams and multiple data streams. Inherent in this organization
is the implication that the multiple processors have substantial interactions and usually share intermediate results. Most parallel computers have this architecture although they are often used in a
SIMD mode.

Another classification that is usually applied to multiprocessor machines is the shared memory
versus local memory dichotomy. In a shared memory machine, each of the processors accesses the
same memory. This may be done through a time-shared common bus, through a crossbar switch
network or by a multiport memory. The shared memory is the vehicle not only for storage of intermediate results, but also a vehicle for internode communications. In a local memory machine, each
processor owns and accesses only its own memory. Hence, internode communication must take
place along the communication linkages established between nodes. Such machines are frequently
referred to as message-passing machines.

9.2.4 SOME CONTEMPORARY ARCHITECTURES

Probably the most important high-performance computing development for statisticians in the
past ten years is the development of the message-passing Beowulf cluster concept (see http://
www.beowulf.org/). The idea began to take shape when Donald Becker and Thomas Sterling of
the Center of Excellence in Space Data and Information Sciences (CESDIS) began outlining a
commodity-based cluster system in late 1993 as an alternative to traditional supercomputers. By
this time many of the vendors who had developed and marketed commercial supercomputer systems
had ceased business due to the increasing power of the commodity CPUs. CESDIS was a division
of the nonprofit University Space Research Association and was located at Goddard Space Flight
Center in Greenbelt, Maryland.

The success of the Beowulf concept relies on three more or less concurrent development.
First, the cost of commodity chips has been driven down and their capabilities have been markedly
improved in the competitive environment of personal computers. Moreover, the commercial off-
the-shelf (COTS) industry provides highly reliable, fully assembled subsystems such as motherboards, graphics subsystems, disks subsystems, and network interface cards. Thus the hardware is
in place. Second, the rise of publicly available software, in particular, the Linux operating system
(see http://www.linux.org/), the GNU compilers and programming tools (see http://gcc.gnu.org/),
and the MPI (see http://www.mpi-forum.org/), and PVM (see http://www.csm.ornl.gov/pvm/
pvm_home.html) message-passing libraries, created a suite of hardware independent software,
which could be used as companions to the COTS hardware. Finally, of equal importance to the
considerable improvement in microprocessor speed and reliability, is the concomitant improvement
in network technology. In earlier commercial and academic systems, there was a dependence on
special purpose interconnect chips and interconnect schemes. The emergence of fast ethernet and

other commodity-type interconnect technology opened the door for the use of scalable and reusable message-passing software.

A Beowulf cluster is distinguished from a network of workstations (NOW) by several subtle, but real differences. All the workstations in a Beowulf cluster are dedicated to the cluster, which mitigates the effect of load imbalance due to external factors. The interconnection network is isolated from the external network so that the network load is determined only by the application run on the cluster. This minimizes problems associated with unpredictable latency. Finally all nodes in a Beowulf cluster are under the jurisdiction of a single administrator and the individual workstations are not visible to the outside world. This reduces the need for authentication between processors to only that needed for system integrity.

Of course, not all commercial vendors have exited the supercomputer market. IBM, for example, has adopted the cluster concept in its Cluster 1600 and its RS/6000 SP products. NEC has relatively recently concluded an arrangement with Cray Research to sell its SX-6 and other supercomputers in the US. The SX-6 is based on a proprietary single-chip vector processor and has up to eight or more of these processors. Compaq also markets a multiprocessor supercomputer based on the alpha chip.

Because of the popularity of linking independent PCs in Beowulf-type clusters, message passing has become the dominant style of parallel computing. Gropp and Lusk [2], Gropp et al. [3], and Snit et al. [4] are excellent references for parallel computation using message passing. Dongarra et al. [5] and Foster [6] are useful references for general parallel programming. Much of the recent work on parallel computing has focused on linear algebra. See for example Gatu and Kontoghiorghes [7] and Kontoghiorghes [8].

9.3 STOCHASTIC DOMAIN DECOMPOSITION

Contemporary statistical computations often focus the analysis of massive data sets with complex algorithms. Consequently, efforts to speedup the calculations are extremely important even with the impressive computational power available today. Parallel computation techniques are an important technology that is used to achieve speedup. Historically in the 1960s and early 1970s, parallelism was also used in the design of both hardware and software to enhance the reliability of systems through redundancy. In such a design, components (either hardware or software) run in parallel performing the same task. Each of the parallel processors process the same data, with a voting procedure used to determine the reported outcome of the computation. That is, the outcome with the most votes is the one thought to be correct. The object of the redundancy in this case is fault tolerance. Of course, this type of parallelism leads to no inherent speedup in the computations.

One may use parallelism in achieving speedup by sending different data to different processors. This can result in substantial speedup, depending on communication overhead and the details of the implementation of parallelism. However, in this mode of operation there is no mechanism for achieving fault detection. For example, the decomposition of an integral and assignment of portions of that integral to processors in a numerical quadrature algorithm is an illustration of this sort of parallelism. It would usually be impossible to know whether one processor returned an incorrect value for its portion of the integral. An interesting question then is whether or not there are situations where I can use parallelism for speedup and still maintain some of the properties of redundancy for my reliability checks. In fact, this is possible in some situations, as will be described below.

An important style of parallel computation is to exploit a cluster of machines in a SIMD mode. The same program is sent to each processor, but different elements of the data are sent to each processor. The art of dividing up the data in such a way that the results of the computation may be reassembled into a useful answer is known as domain decomposition. Xu et al. [9] suggested a form of stochastic domain decomposition. The general idea of domain decomposition is to parse the data so that each processor has as nearly as possible data sets of the same size and which will take as near as possible the same amount of time to compute (load balancing). In the numerical quadrature example mentioned above, if there are k processors and say, for example, it is intended to approximate the integral by $k \times 1000$ rectangular strips, then the domain decomposition strategy might be to assign the computations for the first strip, the $(k + 1)$st strip, the $(2k + 1)$st strip, ... to the first processor, the second, the $(k + 2)$nd, the $(2k + 2)$nd, ... to the second processor and so on. Thus every processor would have the area of 1000 strips to compute, which then could be summed to approximate the integral.

In a setting in which data may be assumed to be generated from some probabilistically homogeneous structure, the use of statistical hypothesis tests in place of voting procedures to compare results from different nodes is suggested. Data are assigned to nodes by parsing in an appropriate manner. As a first step, I assume that the data may be parsed into random samples. Because I began with stochastically homogeneous data and parsed it into random samples, the only variation in the output that I should expect to see from the different nodes is stochastic variation. Hence, I can use statistical tests to check the results for deviations from homogeneity. These tests yield a stochastic measure of redundancy for my parallel implementation. I can, thus, use the tests for fault detection in either of the node hardware or software. This form of domain decomposition is stochastic domain decomposition. I will describe several examples of using this methodology.

9.3.1 GENERAL PURPOSE STOCHASTIC DOMAIN DECOMPOSITION

The general purpose stochastic domain decomposition relies on the Central Limit Theorem and so is appropriate for relatively massive data sets. See, for example, Wegman [1]. Consider a random sample $X_1, \ldots, X_n$ where n is a large number, for example 10^6 to 10^{12}. Suppose also that I have k nodes. Typically k might be somewhere about 64 to say 1000, but very much smaller than n. The basic idea is to operate the machine in a SIMD fashion so that each node is running the same algorithm, but applying it to stochastically equivalent sets of data. The idea is to spawn m processes on each node where m is a relatively large number. Then $m \cdot k$ is the total number of processes and $n/(m \cdot k)$ is the number of observations per node. For sake of simplicity, let us temporarily assume that $n/(m \cdot k)$ is an integer. If not $[(n/(m \cdot k)) + 1]$ will be an integer, where $[\cdot]$ is the greatest integer function, and the algorithm will require a slight modification.

If the computation is a linear functional, e.g., mean, sum of observations, sum of squared observations (hence variance), kernel density estimator, or similar linear computation, then the final computation can be made by summing the components from each process on each node. Hence there is a possibility of substantial speedup based on this stochastic domain decomposition. Further, because each process on each node is operating on stochastically equivalent independent data with the same algorithm, I essentially have a sample of size m from each of the k nodes, so that I could form the mean value $\widehat{\mu}_i, i = 1, \ldots, k$ for each of the k nodes. By the CLT, these $\widehat{\mu}_i$ will be approximately normally distributed and so I can apply a standard ANOVA to test $H_0 : \mu_1 = \cdots = \mu_k$ against $H_1 : \mu_i \neq \mu_j$ for some i and j. By judiciously selecting the size of the test, I can adjust the sensitivity of the test. Clearly if I fail to reject the null, I would believe that all nodes are operating properly. If, however, I reject the null, I would believe that one or more of the nodes is delivering faulty results. Thus I can achieve speedup as well as achieving an indication of faulty nodes.

In this approach, I would want the number of processes, m, on each node to be sufficiently large that CLT approximations would be fairly accurate justifying the use of ANOVA calculations. Thus the sample size also needs to be fairly large. Obviously, this ties in nicely with the need for parallel processing because if the sample size is not large, there would be no need to parallelize. Clearly the same approach could be used for testing equivalence of medians or testing equivalence of order statistics including maximum and minimum values.

9.3.2 STOCHASTIC DOMAIN DECOMPOSITION FOR QUADRATURE

Let me consider the quadrature (numerical integration) of a function, say f. As I described before, I could divide the domain up into strips and, as suggested above, I could systematically assign the strips to each processor. However, if I randomly assign the strips to each processor, then I could reasonably expect the sum of rectangles, $\sum_i (f(X))$ for each $i = 1, \ldots, k$ to be approximately the same, where $\sum_i$ is the sum of all strips for the ith node. Again I could use an ANOVA test to examine whether the results for the nodes were statistically equivalent or not.

9.3.3 STOCHASTIC DOMAIN DECOMPOSITION FOR MULTIPLE LINEAR REGRESSION

The multiple linear regression application is one of particular interest, which I will examine in more detail. I have selected multiple regression because the procedure is well understood, the computations are straightforward, and the statistical tests of homogeneity are easily developed. The implementation of the parallelization of multiple linear regression was done using a hypercube configured with 16 nodes, a direct routing module for communication via message passing, and an additional vector pipeline coprocessor. In addition to the 16 nodes, there was also a host node, which "directs" the activity of the other nodes. The hypercube system had a distributed, message-passing architecture. Data passes through the host to the nodes and the results are gathered from the nodes back to the host. Given the message-passing nature of the architecture, communication overhead typically plays a significant part in the overall effectiveness of any algorithm.

In general, computational problems which require comparatively little internode communication are the most effective ones on the message-passing architectures. Bootstrapping and kernel smoothing operations are examples of computationally intensive tasks that fall into this category. Whereas multiple linear regression is comparatively communications intensive, it does admit a very effective parallel implementation and, moreover, elegantly illustrates our point so that I felt it was quite worth developing. Also, it allowed me to investigate the effect of varying the communication packet size on the potential speedup.

I often have the need to study a system in which the changes in several variables may affect the dependent variable. I may know or be willing to assume that the model is expressed as a linear model or I may use a linear model as an approximation to some unknown, more complex model. In either case, least squares estimation yields a computational technique generally known as regression. I distribute the computations necessary for multiple linear regression over several node processors. I then use statistical tests for homogeneity as a redundancy check for hardware and software faults. The tests used in this discussion depend on the assumption of normally distributed residuals for their complete validity although, of course, their nonparametric analogs may also be used. Because my use of these normal tests is as a descriptive statistic to indicate severe deviations from homogeneity, I am not extremely concerned whether the assumption of normality is met exactly or not.

The mathematical model for multiple linear regression can be expressed as follows:

$$y_i = \beta_0 + \beta_1 x_{i1} + \beta_2 x_{i2} + \cdots + \beta_p x_{ip} + e_i, \quad i = 1, 2, \ldots, n,$$

or in matrix formulation:

$$Y = 1\beta_0 + X\beta_1 + e,$$

where Y is an $(n \times 1)$ vector of observations, 1 is an $(n \times 1)$ vector of ones, β_0 is an unknown parameter, X is an $(n \times p)$ matrix of nonstochastic variables, β_1 is a $(p \times 1)$ vector of unknown parameters, and e is an $(n \times 1)$ vector of random errors. The traditional assumptions are $E(e) = 0$ and $\text{Cov}(e) = \sigma^2 I$. Thus $E(Y) = X\beta$ and $\text{Cov}(Y) = \sigma^2 I$. The least squares estimates of β_0 and β_1, $\hat{\beta}_0$ and $\hat{\beta}_1$, are obtained as follows:

$$\hat{\beta}_1 = (\tilde{X}'\tilde{X})^{-1}\tilde{X}'\tilde{Y}, \quad \hat{\beta}_0 = \bar{Y} - \bar{X}'\hat{\beta}_1, \tag{9.1}$$

where $\bar{X} = X'1/n$ is the vector of column means, $\tilde{X} = X - 1\bar{X}'$ is the centered X matrix so that $\tilde{x}_{ij} = x_{ij} - \bar{x}_j$, $\bar{x}_j = \frac{1}{n}\sum_{i=1}^{n} x_{ij}$, $\bar{Y} = Y'1/n$, and $\tilde{Y} = Y - 1\bar{Y}$. I will assume that $\tilde{X}'\tilde{X}$ matrix is nonsingular in each place where I require $(\tilde{X}'\tilde{X})^{-1}$. I may also obtain a sum of squared errors, $\text{SSE} = \tilde{Y}'\tilde{Y} - \tilde{Y}'\tilde{X}\hat{\beta}_1$.

To implement multiple linear regression in a parallel fashion, I partition the whole set of the observations and variables into k subsets of close to equal size. I then send each of these subsets to one of the k nodes. I denote data sent to, computed at, or received from node i by adding a subscript (i) to the item. Thus I send to node i: $Y_{(i)}$ and $X_{(i)}$ of n_i rows each. I compute at node i: $\bar{X}_{(i)}$, $\bar{Y}_{(i)}$, $(\tilde{X}'_{(i)}\tilde{X}_{(i)})$, $(\tilde{X}'_{(i)}\tilde{Y}_{(i)})$, $(\tilde{Y}'_{(i)}\tilde{Y}_{(i)})$, and $\text{SSE}_{(i)}$. I note two things at this point: (1) I center at each node. (I use a one pass recursive centering algorithm for speed and accuracy.) (2) I do not compute the slopes and intercepts at each node, although I could if I wished. I am not going to use the node estimates. I merely wish to use the information returned from the nodes to: (i) compute the least squares estimates for all the data and (ii) assess homogeneity of the results for fault checking.

To proceed with my homogeneity checks, I define three potential models for my data. Model 0 is the nominal model. For each node partition of the data, I assume that the slope vector β_1, and the intercept β_0 are the same for all nodes. For Model 1, I assume that the node partitions have the same slope vector, but different intercepts. For Model 2, I assume that the node partitions have both different slope vectors and different intercepts. In matrix terms, the three models are given by:

Model 0:
$$Y_{(i)} = 1\beta_0 + X_{(i)}\beta_1 + e_{(i)}.$$

Model 1:
$$Y_{(i)} = 1\beta_{0(i)} + X_{(i)}\beta_1 + e_{(i)}.$$

Model 2:
$$Y_{(i)} = 1\beta_{0(i)} + X_{(i)}\beta_{1(i)} + e_{(i)}.$$

After aggregating the computed node information at the host node, I proceed to compute a SSE for each of the three models as follows:

Model 2:
$$\text{SSE}_2 = \sum_{i=1}^{k} \text{SSE}_{(i)}.$$

Model 1:
$$(\tilde{Y}'\tilde{Y})_{(1)} = \sum_{i=1}^{k} (\tilde{Y}'_{(i)}\tilde{Y}_{(i)}),$$

$$(\tilde{X}'\tilde{Y})_{(1)} = \sum_{i=1}^{k} (\tilde{X}'_{(i)}\tilde{Y}_{(i)})$$

$$(\underset{\sim}{\tilde{X}}{}'\underset{\sim}{\tilde{X}})_{(1)} = \sum_{i=1}^{k}(\underset{\sim(i)}{\tilde{X}}{}'\underset{\sim(i)}{\tilde{X}})$$

$$SSE_1 = (\underset{\sim}{\tilde{Y}}{}'\underset{\sim}{\tilde{Y}})_{(1)} - (\underset{\sim}{\tilde{X}}{}'\underset{\sim}{\tilde{Y}})'_{(1)}(\underset{\sim}{\tilde{X}}{}'\underset{\sim}{\tilde{X}})_{(1)}^{-1}(\underset{\sim}{\tilde{X}}{}'\underset{\sim}{\tilde{Y}})_{(1)}$$

Model 0:
$$n = \sum_{i=1}^{k} n_i, \quad \bar{\bar{\underset{\sim}{X}}} = \sum_{i=1}^{k} n_i \bar{\underset{\sim(i)}{X}}/n, \quad \bar{\bar{Y}} = \sum_{i=1}^{k} n_i \bar{Y}_{(i)}/n,$$

$$(\underset{\sim}{\tilde{Y}}{}'\underset{\sim}{\tilde{Y}})_{(0)} = (\underset{\sim}{\tilde{Y}}{}'\underset{\sim}{\tilde{Y}})_{(1)} + \sum_{i=1}^{k} n_i(\bar{Y}_{(i)} - \bar{\bar{Y}})^2,$$

$$(\underset{\sim}{\tilde{X}}{}'\underset{\sim}{\tilde{Y}})_{(0)} = (\underset{\sim}{\tilde{X}}{}'\underset{\sim}{\tilde{Y}})_{(1)} + \sum_{i=1}^{k} n_i(\bar{\underset{\sim(i)}{X}} - \bar{\bar{\underset{\sim}{X}}})(\bar{Y}_{(i)} - \bar{\bar{Y}})$$

$$(\underset{\sim}{\tilde{X}}{}'\underset{\sim}{\tilde{X}})_{(0)} = (\underset{\sim}{\tilde{X}}{}'\underset{\sim}{\tilde{X}})_{(1)} + \sum_{i=1}^{k} n_i(\bar{\underset{\sim(i)}{X}} - \bar{\bar{\underset{\sim}{X}}})(\bar{\underset{\sim(i)}{X}} - \bar{\bar{\underset{\sim}{X}}})'$$

$$SSE_0 = (\underset{\sim}{\tilde{Y}}{}'\underset{\sim}{\tilde{Y}})_{(0)} - (\underset{\sim}{\tilde{X}}{}'\underset{\sim}{\tilde{Y}})'_{(0)}(\underset{\sim}{\tilde{X}}{}'\underset{\sim}{\tilde{X}})_{(0)}^{-1}(\underset{\sim}{\tilde{X}}{}'\underset{\sim}{\tilde{Y}})_{(0)}$$

I calculate the regression solutions using Equation (9.1) with the summary statistics computed for Model 0. The degrees of freedom associated with the error sums of squares are given by $df_0 = n - p - 1$, $df_1 = n - p - k$, $df_2 = n - kp - k$.

I may now calculate two test statistics. I may test for total homogeneity (which is the true redundancy test for fault checking) and for homogeneity of slopes only (which I have included simply because it is so easy to do and might provide some detail about what went wrong if something did). The test for total homogeneity uses the statistics: $SS_{(2,0)} = SSE_0 - SSE_2$, $df_{(2,0)} = df_0 - df_2 = (k-1)(p+1)$, $MS_{(2,0)} = SS_{(2,0)}/df_{(2,0)}$, $F_{(2,0)} = MS_{(2,0)}/MSE_2$, where $MSE_2 = SSE_2/df_2$.

The test for homogeneity of slopes only uses the statistics: $SS_{(2,1)} = SSE_1 - SSE_2$, $df_{(2,1)} = df_1 - df_2 = (k-1)(p+1)$, $MS_{(2,1)} = SS_{(2,1)}/df_{(2,1)}$, $F_{(2,1)} = MS_{(2,1)}/MSE_2$, where $MSE_2 = SSE_2/df_2$. If heterogeneity is detected, then further tests may be made to isolate the nodes with different and presumed faulty results. I note at this point that I set the significance level for my homogeneity test very small. This is because I want a very small false alarm rate. I only want to detect egregious deviations from homogeneity, as might be caused by a hardware or software failure.

The methodology described above is designed to isolate potentially catastrophic failures in the node hardware or software. However, it could also be used to effect a speedup of other kinds of homogeneity checks on data. For example, suppose that instead of parsing the data to allocate it to nodes in a manner which creates random samples, I allocated the data to correspond to some meaningful partition of the data such as orthants of the $\underset{\sim}{X}$ space. The parallel algorithm described above would then yield a speedup of this homogeneity check, but would no longer have any fault detection capability. However, a simple modification whereby I split each partition into two or more subpartitions via random sampling would still give a homogeneity check using straightforward extensions of the above methodology.

9.3.3.1 Timing Results

The timing study was designed to measure the effectiveness of the parallel scheme described above, and, in particular, to measure the effect of changing the size of the communications packets sent between the host and the nodes. I began by generating data files of similar data. I did this by taking an original data set with six independent variables and then generating data sets of arbitrary size. I then matched the covariance structure of the rows of the $\tilde{X}$ matrix with that of the original problem, made the regression coefficients the same as in the original problem, and matched the variability of the generated residuals with those of the original problem. Hence, regardless of the size of the test data set, I could be assured that it stochastically agreed with the original data set. In this sense, my test data sets were comparable. The sizes selected for this part of the study were $n = 8000$ and 16,000 observations. I also used various numbers of nodes, so that the speedup from parallelizing could be determined.

The study also measured the effect of differing sizes of communication packets sent from the host to the nodes. The sizes used in this study were 125, 250, and 500 observations per node. Because I used a "broadcast" transmission of the data for all nodes, with each node picking its data out of the message, the size of the packet transmitted also depends on the number of nodes. The size of the actual transmitted packet is number of nodes times package size. It might appear that the largest possible packet size would minimize the effect of communications start-up overhead. However, this choice could lead to nodes remaining idle while transmission is taking place. Hence, it may be more effective to use smaller packets, so that the nodes may continue doing productive work. I used several sizes to examine the effect of packet size on overall efficiency.

I measured at each node the overall time for the program, the time waiting for the host to read data, the computation time, and data transmission time (both sending and receiving). My measure of effective time for the computations is the maximum over nodes of overall time minus time waiting for the host to read the data. Hence, the computation time and the communications overhead time for the hypercube are included in my time measure, but the time for the host to initially read in the data is not included. The speedup for any given number of nodes for a particular configuration is given by the ratio of the time for one node divided by the time measure for that number of nodes. Times were measured by an internal clock subroutine on the Intel iPSC/2 hypercube using 80386 chips and are given in milliseconds. The results of my simulations are given in Table 9.1. Each number is the average for two runs.

I observe from Table 9.1 that the effective times indeed decrease as I add more nodes. I also note, as is well known, that the speedup is not linear in the number of nodes. The speedups achieved for the six rows of Table 9.1 (from 1 node to 16) are respectively: 12.65, 13.55, 11.79, 13.16, 10.43, and 12.31. The speedups are greater for the larger data set and are greater for package size 125

TABLE 9.1
Results of Part 1 of the Timing Study

Observations per Node per Package	Sample Size	Effective Time				
		Number of Nodes				
		1	2	4	8	16
125	8,000	15,337	7,722	3,931	2,058	1,212
	16,000	30,703	15,458	7,854	4,088	2,266
250	8,000	15,337	7,721	3,935	2,084	1,301
	16,000	30,699	15,448	7,852	4,106	2,332
500	8,000	15,358	7,733	3,957	2,155	1,472
	16,000	30,737	15,467	7,877	4,155	2,496

observations per node than for the larger packages. The reason that speedup is not perfectly linear is that communications overhead increases as the number of nodes increases. However, as might be expected for perfectly parallel computations as I have here, the computation time indeed decreases as the reciprocal of the number of nodes. In fact, a regression of computation time (again the maximum over nodes) versus sample size divided by number of nodes yields an R^2 of 0.999978. The communication overhead prevents achieving perfect linear speedup.

I also made some additional runs with larger sample sizes to explore the limiting behavior of the speedup. I wished to observe where the asymptote, if any, was with respect to increased speedup and sample size. Hence, I made additional runs with 1 and 16 nodes for each package size for sample sizes 32,000, 64,000, and 128,000. The results of these runs and some information from Table 9.1 are presented in Table 9.2.

As may be seen from Table 9.2, the speedup appears to have an asymptotic value of approximately 13.8 for 16 nodes, regardless of the packet size. Nevertheless, for any given sample size, the smaller packet size gives smaller effective times and larger speedups. Hence, I observe the phenomenon that the desired efficiency of the larger packet size (namely, the lesser number of times the communication start-up overhead is involved) is overcome by the fact that the nodes sit idle waiting for data to arrive with the larger packet sizes.

I make one final remark with regard to effective time. If the time at the host for reading in the data are included, the time to read in the data overwhelms the computations for this problem. I conceive of a situation in which the data are acquired in some automated mode which can bypass the reading step that I did here. This is reasonable, because the fault checking feature described above would be critical in a situation where the data arrived in huge amounts and was processed in an automated fashion. The effective time as I have measured it gives a fair reading of the speedup from the parallel implementation of regression. All communication overhead is included except the reading of the data. This is a commonly applied method for measuring speedup.

TABLE 9.2
Results of Part 2 of the Timing Study

Observations per Node per Package	Sample Size	Effective Time		
		Number of Nodes		
		1	2	16
125	8,000	15,337	1,212	12.65
	16,000	30,703	2,266	13.55
	32,000	61,471	4,460	13.78
	64,000	122,132	8,855	13.79
	128,000	243,648	17,667	13.79
250	8,000	15,337	1,301	11.79
	16,000	30,699	2,332	13.16
	32,000	61,445	4,521	13.59
	64,000	122,199	8,905	13.72
	128,000	243,714	17,685	13.78
500	8,000	15,358	1,472	10.43
	16,000	30,737	2,496	12.31
	32,000	61,515	4,661	13.20
	64,000	122,309	9,043	13.53
	128,000	244,085	17,815	13.70

Again I use the maximum of the node times as my measure of the node processing time. Time is in milliseconds and each time is for one run. The results for 1, 2, 4, 8, and 16 nodes were respectively: 8,292,879; 4,143,741; 2,082,504; 1,055,863; and 545,479. The speedup from 1 to 16 nodes was 15.20. The parallelism achieves close to perfect linear speedup in the number of nodes. The communication time is extremely small compared to the large amount of computation time for this application.

9.4   STOCHASTIC LOAD BALANCING

In the discussion of stochastic domain decomposition, I assumed that the data are divided into equal size pieces. This is done so that each node will have the same amount of data to process and hence finish the computation in approximately the same amount of time. For linear algorithms such as computation of means, variances, kernel density estimators, quadrature, and multiple linear regression as discussed above this is a reasonable strategy and I could expect to achieve an approximately balanced load. However, if the algorithms are adaptive and/or the nodes do not have the same capacity, then load balancing becomes a much more complex issue. A parallel computation is only as fast as the slowest node.

Thus for linear algorithms, a natural and obvious procedure in a k-node SIMD machine is to partition the data/algorithm into k equivalent pieces and send each of the k pieces to a node. In a perfectly deterministic machine, all nodes would complete their computations simultaneously and return their completed computations to the controller node. Actually, the simultaneous completion of the computations would be undesirable because the nodes would be simultaneously contending for communication channels or buses. Nonetheless, large discrepancies in completion time can significantly extend computation times. These discrepancies may arise because the execution time of an algorithm is data dependent and/or because the processing time at each node varies. As an example in the former case, consider a numerical integration routine whose speed of convergence depends on the gradient of the integrand. The natural partition might send to some nodes parts of the integrand where the gradient is large and to other nodes parts where the gradient is smaller. Completion times differing by a factor of two or more are not desirable.

Even when data differences are not a significant factor, computation times at each node may differ significantly. For example, in concurrent computers consisting of independent processors linked by communication channels, the impact of contention for communication channels may have significant effect. A processor node may complete a partial computation of about microseconds, but wait for about milliseconds for access to the communication linkage. Thus a difference of a few microseconds conceivably could cause one node to win the contention for a communication over another node and substantially alter the time to completion. Even when contention for a communications link is not an issue, there may be substantial differences among performances of nodes due to environmental and other factors. The fundamental theme is that for a number of reasons performance of nodes of a parallel computer with otherwise identical algorithms is subject to statistical fluctuation. This fluctuation causes an imbalance in the load on the various nodes and consequently will affect the total computation time adversely. In view of the stochastic character of the load imbalance, I construct load balancing algorithms designed to minimize overall computation times. The following discussion is based on Wegman [10].

9.4.1   A STOCHASTIC MODEL

To develop a stochastic algorithm for decomposing a computational task among a number of processors consider first basic computational task of size, W. The meaning of computational task is intentionally imprecise, but it could be, for example, segment of code or an algorithm and the associated data. Assume that I decompose task W into k subtasks of size W_j, $j = 1, \ldots, k$, for k different processors such as machines with $k = 16$ to $k = 128$. Let T_j be the computation time for the jth task. This model would also serve for a distributed computing environment consisting of a network of workstations (NOW). To construct the basic model, I make four simplifying assumptions:

 a. Any task can be partitioned arbitrarily finely.
 b. Computation time at any node is directly proportional to task size.
 c. Communication time is negligible.
 d. $T_1, \ldots, T_k$ are independent random variables with common density $f(t)$.

A very simple decomposable task is one or more nested DO loops. Obviously, the finest partition I can construct is to the level of the individual sequence of code in the innermost loop. The assumption (a) is obviously unrealistic taken to the limit. However, in any application demanding high-performance computing, the size of the loop will be very much bigger than the number of processors and the discretization approximation will be negligible. Assumption (b) is simply a linearity assumption and seems reasonable when applied to the nodes individually. Assumption (c) is the most problematic. Certainly, in architectures such as the hypercube or networks of workstations, which are communication intensive, this is hardly a realistic assumption for many types of tasks. Nonetheless, I have in mind applications where the tasks are computationally very intensive (say minutes or hours at each node) and the communication traffic is negligible when the nodes are operating. It should be clear that if I am dealing with tasks whose basic computation time at each node is about seconds or less, load balancing is something of a moot issue. Finally, assumption (d) is a statistical assumption most appropriate in the SIMD framework. The model discussed here may be extended to the case where nonidentical tasks are sent to distinct nodes (i.e., the MIMD framework).

To develop the model, I let $T_{(k)} = \max\{T_1, \ldots, T_k\}$. The complete task W will take time $T_{(k)}$ to compute on a k-node processor. Observe that $T^* = \sum_{j=1}^{k}(T_{(k)} - T_j)$ is the total idle time. The statistical distribution of $T_{(k)}$ is

$$g(t) = k[F(t)]^{k-1} f(t), \quad t > 0 \tag{9.2}$$

where $F(t)$ is the distribution corresponding to density, $f(t)$. Our approach to load balancing in general is to use the idle time, T^*, by creating a nondegenerate partition of W_j into pieces X_j and Y_j so that $T_j = U_j + V_j$ where U_j is the computation time for X_j, V_j is the computation time for Y_j. Here decomposition would be done so that the U_j would be iid and the V_j would also be iid. The procedure is to start processor j with task X_j, the Y_j's being held as load-balancing tasks. When a processor is finished with its current task, it is assigned one of the Y_j filler tasks.

At issue is to determine the balance between U_j and V_j. Several criteria can be proposed.

a. $$\sum_{j=1}^{k} EV_j = \sum_{j=1}^{k} E(U_{(k)} - U_j) = k \, (EU_{(k)} - EU_j)$$

In this case, all the filler tasks are expected to be computed in the expected idle time. Hence total expected compute time is $EU_{(k)}$ where $U_{(k)} = \max\{U_1, \ldots, U_k\}$. Since $V_j > 0$ with probability one, I have $U_j < T_j$ with probability one so that $U_{(k)} < T_{(k)}$ with probability one. It follows that $EU_{(k)} < ET_{(k)}$.

b. $$E \max V_j = EU_{(k)} - EU_{(1)}$$

In this case, the right-hand side is the longest expected idle time for any processor. The left-hand side corresponds to the expected computing time for the load balancing task Y_j requiring the longest computing time. Here $U_{(1)} = \min U_j$.

c.
$$EV_j = EU_{(k)} - EU_{(1)}.$$

Similar to criterion (b), but the left-hand side corresponding to the expected computing time for the average load balancing task Y_j rather than the worst case situation. Clearly this is a more conservative criterion.

These criteria may be implemented in the following way. Let $U_j = \theta T_j$ and $V_j = (1 - \theta)T_j$. Then criterion (a) becomes

a*.
$$n(1 - \theta) ET_j = n\theta(ET_{(k)} - ET_j)$$

or

$$\theta_a = ET_j / ET_k.$$

Similarly, criterion (b) becomes

b*.
$$(1 - \theta)ET_{(k)} = \theta ET_{(k)} - \theta ET_{(1)}$$

or

$$\theta_b = \frac{ET_{(k)}}{(2ET_{(k)} - ET_{(1)})}.$$

Finally, criterion (c) becomes

c*.
$$(1 - \theta) ET_j = \theta ET_{(k)} - \theta ET_{(1)}$$

or

$$\theta_c = \frac{ET_j}{(ET_j + ET_{(k)} - ET_{(1)})}.$$

Criteria (a*) through (c*) give explicit values for the load balancing parameter θ in terms of parameters, which may be estimated from knowing the probability density $f(t)$ associated with distribution of computation times T_j, in particular, in terms of moments of the random variables T_j or their order statistics. Notice that if μ and σ are, respectively, the mean and standard deviation associated with T_j, then as $\mu/\sigma \to \infty$,

$$E \max T_j = ET_{(k)} \to ET_j \quad \text{and} \quad E \min T_j = ET_{(1)} \to ET_j$$

so that θ_a, θ_b, and $\theta_c \to 1$. Thus, as I might hope for the deterministic case ($\sigma = 0$), $\theta = 1$ and no load balancing partitioning is necessary.

I close this section by observing that $\sum_i EU_i$ is the expected compute time for the X portion of the task whereas $\sum_i EV_i$ is the expected compute time for the Y portion of the compute task. However, $\sum_i (EU_{(k)} - EU_i)$ is the expected idle time that the Y tasks can use as filler. Thus the total expected computation time is $\sum_i EU_i + \sum_i EV_i - \sum_i (EU_{(k)} - EU_i) = k(ET_i + EU_i - EU_{(k)})$. Since kET_i is the expected computation time with no stochastic load balancing effort, $k(EU(n) - EU_i) = \theta ET^*$ is the expected reduction in computing time due to stochastic load balancing.

9.4.2   EXPECTATION CALCULATIONS

I calculate the moments for a number of simple cases. Perhaps the simplest reasonable model for the distribution $F(t)$ of computation times T_j is the exponential distribution. For numerical comparison, I also calculate the appropriate moments of the uniform.

9.4.2.1   Exponential Distribution

Consider the exponential density

$$f_\lambda(t) = \lambda^{-1} e^{-t/\lambda}, \quad t > 0$$

with distribution

$$F_\lambda(t) = 1 - e^{-t/\lambda}, \quad t > 0.$$

From (9.2), $T_{(k)} = \max\{T_1, \ldots, T_n\}$ has density

$$g_\lambda^{(k)}(t) = k\,[F_\lambda(t)]^{k-1} f_\lambda(t), \quad t > 0,$$

and cumulative distribution

$$G_\lambda^{(k)}(t) = [F_\lambda(t)]^k, \quad t > 0.$$

The expectation is then

$$ET_{(k)} = \int_0^\infty [1 - G_\lambda^{(k)}(t)]\mathrm{d}t = \int_0^\infty [1 - (1 - e^{-t/\lambda})^k]\mathrm{d}t.$$

By change of variables to $x = 1 - e^{-t/\lambda}$ so that $\mathrm{d}t = (\lambda\,\mathrm{d}x)/(1 - x)$. Thus

$$ET_{(k)} = \int_0^1 (1 - x^k)\frac{\lambda\,\mathrm{d}x}{(1 - x)}$$

which yields

$$ET_{(k)} = \int_0^1 \left(1 + x + x^2 + \cdots + x^{k-1}\right)\lambda\,\mathrm{d}x$$

or

$$ET_{(k)} = \lambda(1 + \frac{1}{2} + \frac{1}{3} + \cdots + \frac{1}{k}). \tag{9.3}$$

Similarly, the distribution of $T_{(1)} = \min T_j$ is given by

$$G_\lambda^{(1)}(t) = 1 - [1 - F_\lambda(t)]^k, \quad t > 0.$$

The expectation is

$$ET_{(1)} = \int_0^\infty [1 - G_\lambda^{(1)}(t)]dt = \int_0^\infty e^{-tk/\lambda} \, dt = \lambda/k. \tag{9.4}$$

Finally, the expected value of T_j is

$$ET_j = \lambda. \tag{9.5}$$

Formulae (9.3) through (9.5) allow me to compute the value of θ_i for $i = a, b, c$. In criteria (a) through (c), the solutions in the exponential case are

a[†].

$$\theta_a = \frac{1}{1 + \dfrac{1}{2} + \dfrac{1}{3} + \cdots + \dfrac{1}{k}}.$$

b[†].

$$\theta_b = \frac{1 + \frac{1}{2} + \dfrac{1}{3} + \cdots + \dfrac{1}{k}}{2(1 + \dfrac{1}{2} + \dfrac{1}{3} + \cdots + \dfrac{1}{k}) - \dfrac{1}{k}}.$$

c[†].

$$\theta_c = \frac{1}{2 + \dfrac{1}{2} + \dfrac{1}{3} + \cdots + \dfrac{1}{k-1}}.$$

Table 9.3 gives some values of the load balance parameter for various number of nodes. It is frequently used as a memoryless distribution, i.e., given that an item has not yet failed, it will have the same probability of failure as when it was new. Thus in the load balancing context, given that a computation is not yet finished, the probability distribution associated with the time to completion is the same as when it first began. The gamma family of distributions generalizes the exponential and

TABLE 9.3
Load Balance Parameters for the Exponential Distribution

k	θ_a	θ_b	θ_c
1	1.000	1.000	1.000
2	0.667	0.800	0.500
3	0.545	0.720	0.406
4	0.480	0.676	0.353
8	0.368	0.602	0.278
16	0.296	0.559	0.232
32	0.246	0.539	0.199
64	0.211	0.519	0.174
128	0.184	0.511	0.156
∞	0.000	0.500	0.000

perhaps provides some more realistic models for computation times. However, because of its more complicated form, computation of load balance parameters in closed form is not readily available.

9.4.2.2 Uniform Distribution, Case 1

A second illustrative example is the uniform density. The uniform density on $(0, m)$ is given by

$$f_m(t) = 1/m, \quad 0 < t < m$$

with distribution

$$F_m(t) = t/m, \quad 0 < t < m.$$

Again from (9.2) the distribution of $T_{(k)}$ has density

$$g_m^{(k)}(t) = \frac{k}{m^k} t^{k-1}, \quad 0 < t < m.$$

From this it follows that

$$ET_{(k)} = \int_0^m \frac{k}{m^k} t^k \, dt = \frac{km}{k+1}. \tag{9.6}$$

The density of $T_{(1)}$ is

$$g_m^{(1)}(t) = \frac{k}{m} \left(1 - \frac{t}{m}\right)^{k-1}, \quad 0 < t < m.$$

A simple calculation shows that

$$ET_{(1)} = \int_0^m \frac{k}{m} \left(1 - \frac{t}{m}\right)^{k-1} t \, dt = \frac{m}{k+1}. \tag{9.7}$$

Finally for a uniform density $ET_j = m/2$. Thus using (9.6) and (9.7) I may again calculate the load balance parameters for criteria (a) through (c).

a‡.
$$\theta_a = \frac{k+1}{2k}.$$

b‡.
$$\theta_b = \frac{k}{2k-1}.$$

c‡.
$$\theta_c = \frac{k+1}{3k-1}.$$

Table 9.4 gives some values of the load balance parameters for uniform densities. Notice again that the load balance parameter is independent of m. The reason for this is perhaps a bit subtle. It should be noted that in all of the load balance criteria (a) through (c), the left-hand side of the load balance equation is a location-dependent quantity whereas the right-hand side is a scale-dependent quantity. In both distribution examples, the location and scale parameters are both proportional to the parameters λ and m, respectively. Hence, the parameters themselves wash out. In particular, $\mu/\sigma = 1$ for the exponential case and $\mu/\sigma = \sqrt{3}$ for the uniform case. Both of these are comparatively small values and far from ∞. In the next example, I consider a situation where location and scale are not proportional.

TABLE 9.4
Load Balance Parameters for the Uniform Distribution (1)

k	θ_a	θ_b	θ_c
1	1.000	1.000	1.000
2	0.750	0.667	0.600
3	0.667	0.600	0.500
4	0.625	0.571	0.455
8	0.563	0.533	0.391
16	0.531	0.516	0.362
32	0.516	0.508	0.347
64	0.508	0.504	0.340
128	0.504	0.502	0.337
∞	0.500	0.500	0.333

9.4.2.3 Uniform Distribution, Case 2

In this third example I again consider the uniform case, but this time with density given by

$$f_m(t) = 1, \quad m - 1 < t < m.$$

In this case, it is straightforward to show that

$$ET_{(k)} = m - \frac{1}{k+1},$$

$$ET_{(1)} = m - \frac{k}{k+1},$$

and

$$ET_j = m - \frac{1}{2}.$$

Simple calculations show that

a′.
$$\theta_a = \frac{m - \dfrac{1}{2}}{m - \dfrac{1}{k+1}},$$

b′.
$$\theta_b = \frac{m - \dfrac{1}{k+1}}{m + \dfrac{k-2}{k+1}},$$

c′.
$$\theta_c = \frac{m - \dfrac{1}{2}}{m + \dfrac{k-1}{k+1} - \dfrac{1}{2}}.$$

Table 9.5 contains load balance parameters for this example with $m = 3$. In this case $\mu/\sigma = 5\sqrt{3}$. Table 9.6 to Table 9.8 contain calculations of the expected reduction in computing time for the three examples considered.

9.4.3 RANDOMIZED LOAD BALANCE PARAMETERS

In the previous section, I consider the case where θ was a fixed, but unknown number which could be calculated from the distribution associated with T_j. In this section, I consider θ_j to be a random variable with conditional density $g_v(\theta \mid t_j)$ where as before $U_j = \theta_j T_j$. The joint density of θ_j and T_j is given by $g_v(\theta \mid t) f(t)$. Letting $Z_j = T_j$, I have a Jacobian of 1 and it follows that the joint density of U_j and Z_j is

---

TABLE 9.5
Load Balance Parameters for the Uniform Distribution (2)

k	θ_a	θ_b	θ_c
1	1.000	1.000	1.000
2	0.938	0.889	0.882
3	0.909	0.846	0.833
4	0.893	0.824	0.806
8	0.865	0.788	0.763
16	0.850	0.769	0.739
32	0.842	0.760	0.727
64	0.838	0.755	0.720
128	0.835	0.752	0.717
∞	0.833	0.750	0.714

TABLE 9.6
Expected Reduction in Computing Time for the Exponential

k	a	b	c
1	0.000	0.000	0.000
2	0.667	0.800	0.500
3	1.364	1.800	1.000
4	2.080	2.930	1.529
8	5.056	8.277	3.825
16	11.267	21.296	8.821
32	24.115	52.249	19.468
64	50.509	124.416	41.829
128	104.441	289.873	88.313

TABLE 9.7
Expected Reduction in Computing Time for the Uniform (1)

k	a	b	c
1	0.000	0.000	0.000
2	0.250	0.222	0.200
3	0.500	0.450	0.375
4	0.750	0.686	0.545
8	1.750	1.659	1.217
16	3.750	3.643	2.553
32	7.750	7.634	5.721
64	15.750	15.630	10.555
128	31.750	31.627	21.222

TABLE 9.8
Expected Reduction in Computing Time for the Uniform (2)

k	a	b	c
1	0.000	0.000	0.000
2	0.313	0.296	0.294
3	0.682	0.625	0.625
4	1.071	0.988	0.968
8	2.692	2.451	2.373
16	6.000	5.430	5.217
32	12.650	11.418	10.925
64	25.979	23.412	22.350
128	52.642	47.409	45.206

$$h_v(u, z) = h_v\left(\frac{u}{z} \mid z\right) f(z)$$

and that the marginal density of U_j is

$$h_v(u) = \int h_v\left(\frac{u}{z} \mid z\right) f(z)\mathrm{d}z.$$

Using the formulae stated earlier, I calculate total expected computation time. Since T_j is a positive random variable a reasonably rich family of plausible distributions is the gamma family, i.e.,

$$f_{\gamma,\delta}(t) = \frac{1}{\Gamma(\gamma)\delta^\gamma}t^{\gamma-1}\mathrm{e}^{-t/\delta}, \quad t > 0.$$

Also since θ lies between 0 and 1, a reasonable family for θ is the beta family.

$$g_{\alpha,\beta}(\theta) = B(\alpha, \beta)\theta^{\alpha-1}(1 - \theta)^{\beta-1}, \quad 0 < \theta < 1.$$

Hence the joint density of U_j and Z_j is

$$h_{\alpha,\beta,\gamma,\delta}(u,z) = \frac{B(\alpha,\beta)}{\Gamma(\gamma)\delta^\gamma}\left(\frac{u}{z}\right)^{\alpha-1}\left(1-\frac{u}{z}\right)^{\beta-1}z^{\gamma-1}\mathrm{e}^{-z/\delta}, u, \quad z > 0.$$

The parameters α and β may depend on $Z_j = t_j$. Hence the marginal density of U_j may be written as

$$h_{\gamma,\delta}(u) = \frac{1}{\Gamma(\gamma)\delta^\gamma}\int_0^\infty B(\alpha,\beta)\, u^{\alpha-1}(z-u)^{\beta-1}z^{\gamma-\alpha-\beta+1}\mathrm{e}^{-z/\delta}\,\mathrm{d}z. \tag{9.8}$$

One possible choice for α and β is $\alpha = z/(1+z)$ and $\beta = 1/(1+z)$. In this case I have

$$h_{\gamma,\delta}(u) = \frac{1}{\Gamma(\gamma)\delta^\gamma}\int_0^\infty B\left(\frac{z}{1+z},\frac{1}{1+z}\right)u^{-1/(1+z)}(z-u)^{-z/(1+z)}\,z^\gamma\,e^{-z/\delta}\,\mathrm{d}z. \tag{9.9}$$

The density of V_j can be calculated in a similar manner by replacing θ with $1-\theta$, that is to say interchanging the roles of α and β. The procedure then is to choose a value of θ_j according to g_v. If I have some experience with the distribution of the T_j, I may make my choice dependent on the T_j's. I have in mind here potentially choosing θ_j larger (hence smaller filler tasks) to be used toward the end of the run. In any case, the conditional density of θ on t_j need not actually depend on t_j. It should be clear from the rather complicated formulae (9.8) and (9.9) that a closed-form expression for the densities of U_j and of V_j is not readily feasible. Nonetheless, an expression for EU_j can be found as

$$EU_j = \int_0^\infty uh_{\gamma,\delta}(u)\mathrm{d}u.$$

That is

$$EU_j = \int_0^\infty \int_0^\infty \frac{1}{\Gamma(\gamma)\delta^\gamma} B(\alpha,\beta)\, u^{\alpha+1-1}(z-u)^{\beta-1}z^{\gamma+1-(\alpha+1)-\beta+1}e^{-z/\delta}\mathrm{d}u\,\mathrm{d}z.$$

So that

$$EU_j = \frac{\Gamma(\gamma+1)B(\alpha,\beta)\delta}{\Gamma(\gamma)B(\alpha+1,\beta)}.$$

By symmetry

$$EV_j = \frac{\Gamma(\gamma+1)B(\beta,\alpha)\delta}{\Gamma(\gamma)B(\beta+1,\alpha)}.$$

Unfortunately, the computation of the expectation of $U_{(k)}$ is considerably more difficult so that numerical methods must be pursued.

9.5   SUMMARY AND CONCLUSIONS

High-performance computing has historically been the realm of those who have engaged in mathematical modeling using PDE. The recent focus on data mining of massive data sets has focused the interest of statisticians and related fields on the possibility, indeed, the necessity of focusing on high-performance parallel computation. The interaction between high-performance parallel computing and statistical methods is a two-way street. I have tried to illustrate this with the discussion of stochastic domain decomposition and stochastic load balancing. In the former, parallel computing offers the possibility of speedup of many different types of statistical algorithms whereas the stochastic component allows for an ability for fault detection, an ability not present in the usual domain decomposition methods. With the stochastic load balancing methodology, knowledge of statistical methods suggests an approach to improving performance of parallel computing. This two-way street should encourage more statisticians and computer scientists to interact.

ACKNOWLEDGMENTS

This work was completed under the sponsorship of the Office of Naval Research under contract DAAD19-99-1-0314 administered by the Army Research Office, by the Air Force Office of Scientific Research under contract F49620-01-1-0274 and contract DAAD19-01-1-0464, the latter also administered by the Army Research Office and finally by the Defense Advanced Research Projects Agency through cooperative agreement 8105-48267 with the Johns Hopkins University. The author would like to thank three anonymous referees and also Prof. Karen Kafadar for careful reading and helpful suggestions that substantially improved the presentation of this chapter.

REFERENCES

[1] E.J. Wegman. Huge data sets and the frontiers of computational feasibility. *Journal of Computational and Graphical Statistics* 4(4):281–295, 1995.

[2] W. Gropp and E. Lusk. Implementing MPI: the 1994 implementors' workshop. In *Proceedings of the 1994 Scalable Parallel Libraries Conference*, Mississippi, October, 1994.

[3] W. Gropp, E. Lusk, and A. Skjellum. *Using MPI — Portable Parallel Programming with the Message-Passing Interface*, 2nd ed. Cambridge, MA: The MIT Press, 1999.

[4] M. Snit, S. Otto, S. Huss-Lederman, D. Walker, and J. Dongarra. *MPI—The Complete Reference, the MPI Core*, Vol. I, 2nd ed. Cambridge, MA: The MIT Press, 1998.

[5] J. Dongarra, I. Duff, P. Gaffney, and J. McKee. *Vector and Parallel Computing: Issues in Applied Research and Development*. New York, NY: John Wiley & Sons, 1989.

[6] I. Foster. *Designing and Building Parallel Programs: Concepts and Tools for Parallel Software Engineering*. Reading, MA: Addison-Wesley, 1995.

[7] W. Gatu and E. Kontoghiorghes. Parallel algorithms for computing all possible subset regression models using the QR decomposition. *Parallel Computing* 29:505–521, 2003.

[8] E. Kontoghiorghes. *Parallel Algorithms for Linear Models: Numerical Methods and Estimation Problems*. Boston: Kluwer Academic Publishers, 2000.

[9] M. Xu, E.J. Wegman, and J.J. Miller. Parallelizing multiple linear regression for speed and redundancy: an empirical study. *Journal of Statistical Computation and Simulation* 39:205–214, 1991.

[10] E.J. Wegman. A stochastic approach to load balancing in coarse grain parallel computers. In A. Buja and P. Tukey, Eds. *Computing and Graphics in Statistics*. New York: Springer-Verlag, 1991, pp. 219–230.

Parallel Algorithms for Predictive Modeling

Markus Hegland

CONTENTS

ABSTRACT

Parallel computing enables the analysis of very large data sets using large collections of flexible models with many variables. The computational methods are based on ideas from computational linear algebra and can draw on extensive research in parallel algorithms. Many algorithms for the direct and iterative solution of penalized least squares problems and for updating can be applied. Both methods for dense and sparse problems are applicable. An important property of the algorithms is their scalability, i.e., their ability to solve larger problems in the same time using hardware which grows linearly with the problem size. Whereas in most cases large granularity parallelism is to be preferred, it turns out that smaller granularity parallelism can also be exploited effectively in the problems considered.

The development is illustrated by four examples of nonparametric regression techniques. In the first example, additive models are considered. While the backfitting method contains dependencies that inhibit parallel execution, it turns out that parallelization over the data leads to a viable method, akin to the bagging algorithm without replacement which is known to have superior statistical properties in many cases. The second example considers radial basis function fitting with thin plate splines. Here the direct approach turns out to be nonscalable but an approximation with finite elements is shown to be scalable and parallelizes well. One of the most popular algorithms in data mining is multivariate adaptive regression splines (MARS). This is discussed in the third example. MARS has been modified to use a multiscale approach and a parallel algorithm with a small granularity has been seen to give good results. The final example considers sparse grids. Sparse grids take up many ideas from the previous examples and, in fact, can be considered as a generalization of MARS and additive models. They are naturally parallel when the combination technique is used. We discuss limitations and improvements of the combination technique.

10.1 INTRODUCTION

10.1.1 PREDICTIVE MODELS AND FACTORS AFFECTING THE PERFORMANCE OF RELATED ALGORITHMS

The availability of increasing computational power has lead to new opportunities in data processing. This power can be used to process huge data collections and to fit increasingly complex models. The analysis of huge data collections, however, does pose new statistical and algorithmic challenges.

Larger data sets do not automatically imply better predictions. For estimating a fixed number of parameters, the law of large numbers implies estimation errors of the order $O(n^{-1/2})$ in terms of the data size n. For the estimation of functions and for density estimation larger data sets in principle allow fitting more complex data and the detection of more detail. However, large data does not automatically reduce the bias which may occur by the choice of the sampling frame. A famous example that illustrates this point is the Literary Digest Survey in 1936, which used a large sample of 2.4 million responses to predict that Roosevelt would get 43% of the votes where he actually received 62% in the US presidential elections. In contrast, a much smaller sample of 50,000 collected by Gallup predicted a 56% victory [1, 2]. Thus the larger sample did not automatically guarantee a better prediction. The bias due to the way the sample is chosen is typically not reduced by choosing a larger sample. These issues of bias and data size are comprehensively treated in Ref. [3]. Often a good model combined with a relatively small data set can provide superior results.

Larger data have a tendency to suggest rejection of a null hypothesis even if the effect in question is unimportant. One way out of this dilemma is to increase significance levels with the data size. One should also remember that statistical significance does not necessarily imply interestingness or substantive importance. For references and a further discussion of the implications of data size for testing see Ref. [4]. In a machine learning context, this problem may be viewed as an instance of spurious pattern detection. This is further discussed in Ref. [5], together with issues relating to data quality.

The second class of challenges mainly considered here relates to computational aspects. Special algorithms and techniques are required to harness the computational power available. In this review several recently developed methods and algorithms will be discussed. These developments have resulted from joint and independent efforts of several communities. These include statisticians with an interest in computational performance, numerical linear algebraists interested in data processing, the more recently emerging data mining community with interests in algorithms, the computational mathematics community interested in applications of approximation theory, the diverse parallel computing community, machine learners with a computational orientation interested in very large-scale problems, and many computational scientists, especially the ones interested in the computational solution of partial differential equations (PDE). Many other communities that I have not mentioned here have contributed to this exciting development as well. I hope the following discussion will be of interest to members of many of these communities. The computational techniques discussed in the following have originated at various places and the specific parallel algorithms covered have all been the topic of extensive research conducted at the Australian National University. For anyone interested in trying out these methods there are many data sets available that allow investigating the performance of computational techniques see Refs. [6–8], and also Matrix Market (http://math.nist.gov/MatrixMarket). While some of these data sets are not large by today's means, they are still useful for scalability studies. Software is available for many of the methods discussed, and while well-established companies have constantly been able to improve their standard software, maybe some of the most exciting developments in the area of algorithms and large-scale high-performance computing software has been happening in the open source community driven by publicly funded university research.

Except for the occasional brief comment, this review does not discuss statistical issues. However, statistical issues are extremely important, and they have been covered in many excellent statistical texts, e.g., Ref. [9], but also in the data mining literature, e.g., Ref. [5]. Here we will focus on computational aspects and the algorithms, their computational capability to handle very large data sets and complex models and the way in which they harness the performance of multiprocessor systems.

Predominantly, this review will consider computational techniques used in *predictive modeling*, and, in particular, algorithms based on linear algebra. Predictive modeling aims to induce functional patterns from data. More specifically, given two sequences

$$x^{(1)}, \ldots, x^{(N)} \quad \text{and} \quad y^{(1)}, \ldots, y^{(N)}$$

defining the data, predictive modeling aims to find functions

$$y = f(x)$$

such that

$$y^{(i)} \approx f(x^{(i)}).$$

Whereas the type of the predictor x can be fairly arbitrary, including real and binary vectors, text, images, and time series, we will focus in the following on the case where x is an array containing categorical (discrete) and real (continuous) components. The response y will be considered to be real, i.e., we will be discussing algorithms for *regression*. Many of the observations and algorithms can be generalized to other data types for both the predictors and the response. In particular, the algorithms discussed here are also used for classification, see, Ref. [9].

Practically the most important measure of the efficiency of a computational technique is the *time it takes to complete a particular task*. For a regression problem, this time depends firstly on the *data size*, modeled by N. The data available in data analysis and, in fact, the data which is actually used in individual studies has been growing exponentially. It has been suggested that this growth is akin to the growth in computational performance (according to Moore's law), especially for fast growing data areas like bioinformatics [10]. GByte-sized data sets are very common, TByte data sets are used in large studies and PByte data sets are emerging. Software and algorithm development, however, is done at a slower pace. Depending on the complexity of the computational methods in terms of the data size this discrepancy between speed of software development and simultaneous growth of data and computational speed may or may not be a problem. If for a particular application all the data need to be scanned, the complexity of any technique is at best linear in the data size. A technique with linear data complexity is called *scalable* in the data mining literature. A property of scalable techniques is that they will in the same time solve a problem twice the size using twice the computational resources. More generally, the concept of scalability considers the increase in computational power (e.g., processor speed, number of processors) required to keep constant the time needed to solve a problem when the data size increases. If this growth of computational resources is proportional to the growth in data size then the algorithm is called scalable. Note that this usage of the term scalable is quite different from the concept of scalability often used in parallel processing.

There are two fundamentally different ways to increase computational throughput. One way is to increase the speed (clock cycle, memory and communication bandwidth, latencies) of the hardware and the other way is to use multiple (often identical) hardware components that work simultaneously. To make full usage of this hardware parallelism one needs special algorithms and this is the focus of the following discussion. Ideally, the throughput of any computation grows linearly with the amount of hardware available. The amount of disk space required grows linearly with the data size and the work required to read the data grows linearly as well. Thus the computational resources to analyze a data set at the least have to grow linearly in the data size N. Consequently $O(N)$ is the optimal resource complexity and we will in the following assume that all the hardware resources available grow linearly with the data size.

The $O(N)$ hardware complexity has important implications for the feasible computational complexity of the algorithms. We can assume that the computations required are close to the limits of what is feasible. One would typically not allow for an increase in computational time even if the data sets grow. However, if we allow the number of processing units to grow linearly in the size of the data that means that in the ideal case the time to process the data stays constant — *if the algorithms used have $O(N)$ complexity*. Again this computational complexity is optimal as one at least needs to scan all the data once. Consequently, scalability is an essential requirement for the algorithms considered for large and growing data sets.

Besides the data size, another determining factor for performance is *model complexity*. This complexity relates to the number of components of the predictor variable vector x, and the dimension of the function space. Whereas in earlier parametric linear models there were just a handful of variables and parameters to determine, modern nonparametric models and, in particular, data mining applications consider thousands up to millions of parameters. Finally, the function is usually drawn from a large collection of possible candidate functions. While one was interested earlier in just "the best" of all models, in data mining applications one would like to find all models which are interesting and supported by the data. Thus the *size of the candidate set* is a third determining factor of performance. The algorithms considered here are characterized by the function class.

A simple way to generate spaces of functions with many variables is to start with spaces of functions of one variable and consider the tensor product of these spaces. In terms of the basis functions, for example, if V_i are spaces of functions of the form $f(x_i) = \sum_j u_{i,j} b_{i,j}(x_i)$, then the functions in the tensor product $\bigotimes_{i=1}^d V_i$ consists of functions of the form

$$f(x) = \sum_{j_1, \dots, j_d} u_{j_1, \dots, j_d} b_{1, j_1}(x_1) \cdots b_{d, j_d}(x_d).$$

If (for simplicity) we assume that each component space V_i has m basis functions $b_{i,j}(x_i)$, then the tensor product space has m^d basis functions. As the basis is local we expect that this expansion will contain essentially all basis functions for most functions of interest. For example, if tensor product hat functions are used for a basis, then one has for the constant function $f(x) = 1$ the expansion:

$$1 = \sum_{j_1,\ldots,j_d} b_{1,j_1}(x_1) \cdots b_{d,j_d}(x_d).$$

This problem has been called the *curse of dimensionality* and it occurs in many other situations as well. In the following we will consider function classes which, while often subsets of the tensor product space use a different basis and are able to deal effectively with the curse of dimensionality.

To find interesting functions one needs to measure success using a loss function, error criterion, or *cost function*. This cost function has also a tremendous influence on the development of the algorithm and the performance. For regression problems the simplest loss function used is the *sum of squared residuals*

$$\sum_{i=1}^{N}(f(x^{(i)}) - y^{(i)})^2.$$

This is the cost function we will mainly use here. Many other measures are used including penalized sum of squares and median absolute deviation. As always, one should be very careful when interpreting this function, in particular if the function has been determined by minimizing this sum. The sum of squares, or so-called resubstitution error, is in this case a very poor measure for the actual error of the fitted function as it will seriously underestimate the actual error. More reliable measures have been developed including holdout data sets, crossvalidation, and bootstrap.

The data is often thought of as drawn from an underlying theoretical data distribution. If we denote the expectation with respect to this distribution by $E(\cdot)$ the expected mean squared residual is

$$E((f(X) - Y)^2)$$

where X and Y are the random variables observed in the data. The best function f would be the one which minimizes this expectation. This theoretical best function is denoted as a conditional expectation:

$$f(x) = E(Y|X = x).$$

The algorithms we discuss here all aim to approximate this "best" function.

10.1.2 BASIC CONCEPTS OF PARALLEL COMPUTING

Parallelism is a major contributor to computational performance. Parallelism is exploited by splitting the instruction and data streams into several independent branches. Hardware supporting this parallelism ranges from different stages of arithmetic pipelines, multiple memory paths, different CPUs with a shared memory, multiple independent processors with their own memories each, collections of workstations and collections of supercomputers. In general, any computation will involve parallelism at many different levels. Slightly simplified, one has p streams of operations that run in parallel instead of having to run sequentially. A fundamental characteristic of parallel execution is *parallel speedup*, which is defined as the ratio of the time for sequential execution to parallel execution

$$S_p = \frac{T_{\text{sequential}}}{T_{\text{parallel}}}.$$

Ideally, if one has p streams running in parallel the speedup is p. Thus one introduces *parallel efficiency* as the ratio of actual speedup to the number of streams or parallel processes:

$$E_p = \frac{S_p}{p}.$$

Of course one cannot do everything in parallel, some tasks require that other ones are completed first. In some cases this *dependency* can be overcome. For example, the determination of a sum $\sum_{i=1}^{N} x^{(i)}$ can either be done with a recursion

$$s_0 = 0, \quad s_{i+1} = s_i + x^{(i)}$$

or by determining first the sums of each consecutive data points to get a new sequence then repeat this until finished:

$$z_{i-1,0} = x^{(i)}, \quad z_{i,k+1} = z_{2i,k} + z_{2i+1,k}.$$

This *recursive doubling* algorithm does increase the amount of parallelism considerably especially for large data sizes N. The number of operations has not increased, but the amount of parallelism is limited. Rather than completing at a time of approximately $(N-1)/p\tau$ this method would require

$$T_{\text{parallel}} = \left(\frac{N-p}{p} + \log_2(p) \right) \tau$$

where τ is the time for one addition. Thus the efficiency in this case is

$$E_p = \frac{1}{1 + \frac{p \log_2(p) - p + 1}{N - 1}}.$$

However, not all dependencies can be resolved in a way similar to the one discussed. Examples of this unresolvable case are many iterative methods where the next step depends in a complex way on the previous one.

A second example of dependency that can be overcome is the solution of a linear recursion:

$$z_0 = a, \quad z_{i+1} = b_i z_i + c_i.$$

This can be done by interpreting the recursion as a linear system of equations, which can then be solved with a parallel algorithm (we will talk about this more later). However, this comes at a cost and it turns out that in this case the time required is roughly doubled. These extra operations, which are required to make the algorithm parallel, are called *parallel redundancy*. This situation is again very common and for an algorithm where the amount of work is increased by a factor of r will at most be able to achieve a parallel efficiency of

$$E_p \leq \frac{1}{r}.$$

Note that while the efficiency is bounded at least it does not go to zero like in the previous case. If enough hardware parallelism p is available this may still be a very interesting option as there is no bound on the smallest amount of time required.

For both cases of overcoming dependencies the order of the operations and even the nature of the operations were changed. Whereas this would typically give the same results if exact arithmetic is used due mainly to the fact that the associative law of addition, this is not the case for floating point arithmetic. A careful analysis of the parallel algorithms is required to guarantee that the results of the computations are still valid. This is, in particular, the case for the second case where the parallel algorithm can even break down! There are, however, often stable parallel versions of the algorithms available (see Ref. [11] for some examples that generalize the second example).

An important general law that explains the effect of a sequential fraction of computation is *Amdahl's law*. It is seen that in the case where a proportion r of the total work has to be done sequentially the efficiency is at most

$$E_p = \frac{1}{1 + (p - 1)r}.$$

The speedup is limited to $1/r$ and so the efficiency drops to 0 for very large p. As every computation will have some proportion which cannot be shared among other processors this is a fundamental limitation to the amount of speedup available and thus to the computational time. This may be overcome by considering larger problems that have a smaller proportion r of sequential tasks.

Whereas in principle an efficiency higher than one is not possible there are cases where the parallel execution can run more than p times faster on p processors than on one processor. This happens, for example, when the data does not fit in the memory of one processor but does fit in the memory of p processors. In this case one requires the loading of the data from disk during the execution on one processor, say, in p chunks. If one counts the time that it takes to load the data from disk into memory in the parallel execution as well one can confirm that in this case $E_p \leq 1$ as well. However, this example points to an important aspect of parallel processing: the increase in available storage space. Not only are the computational processing capabilities increased in a parallel processor but also the memory, caches, and registers. This increase in storage can have a dramatic effect on processing times.

Maybe the main inhibitor of parallel efficiency is (small) problem size. Obviously one needs at least p tasks to get a speedup of p. However, smaller problem sizes do not require the application of parallel processing. Often one would like to consider various problems with different problem sizes and one is interested in maintaining parallel efficiency over all problems. The question is thus can we maintain parallel efficiency for larger numbers of processors if we at the same time increase the problem size. One chooses the problem size as a function of the amount of parallelism p. This is a case of "parallel scalability" and is referred to as *isoefficiency*. In the case of the determination of the sum of N numbers one gets scalability if $N(p) = O(p \log_2(p))$ and many other cases are scalable. Note that the case of Amdahl's law does give a hard limit to scalability, so in this case no problem size will help to maintain efficiency as speedup is limited to $1/r$, if r is the amount of sequential processing required.

When mapping a collection of tasks onto multiple resources a speedup of p can only be obtained if all the tasks take the same time — the time for the parallel execution is equal to the time required for the longest task. This is the problem of *load-balancing* the work. Thus one has to make sure that the size of the tasks are all about the same. One way to achieve this is to have very many tasks and none extremely large. The tasks can either be preallocated based on some estimate or may be allocated in "real-time" using a *master–worker* model. Note, however, that the master–worker approach can be inefficient if it requires a large data movement.

A typical parallel computing task will consist of periods where several independent subtasks are processed in parallel and other periods where information is exchanged between the processors. More generally, one defines granularity as the amount of work done between two points of synchronization. The following discussion applies to this definition as well where the communication overhead is now replaced by the synchronization overhead. The average volume of work that is done between the communication periods determines the *granularity* of the task. In the case of small granularity, the communication overhead or latency will strongly influence the time and so one would aim to do larger tasks between data interchange steps. In addition to the communication overhead, one finds that some parts of the communication are independent of the granularity, thus again larger granularity would show better overall performance due to a reduction in the communication to computation ratio. Finer granularity is used at the level of pipelines, vector instructions, and single instruction multiple data (SIMD) processing where larger granularity is usually preferred for collections of workstations and multiple instruction multiple data (MIMD) processing.

The effect of the hardware on the time to execute an algorithm is mainly determined by the access times of various storage devices, including registers, several levels of cache, memory, secondary storage, distributed storage in a cluster, and distributed storage across the Internet. These *memory hierarchies* are defined by access speeds. The fastest computations are the ones that rely for their data movements almost entirely on the faster memories and only occasionally have to access the slower memories. The reason for the existence of these hierarchies is because of the enormous price difference in the various types of memories. The amounts of slower memory available on any computing system will dwarf the amounts of fast memory available. Any algorithm that requires frequent disk access will be much slower than a similar algorithm that mostly accesses registers.

An example for slow data access is the data transfer between different processors without shared memory. Here the programmer directly has to deal with the data movements. The transfer of data between different processors is called *communication* and the most popular way to implement this is using *message passing*. Whereas the message passing routines do include various complex patterns of data interchange they all rely on the two primitives *send* — where one processor makes data available to another processor — and *receive* — where the processor accepts data from another processor. Thus any data transfer requires two procedure calls on two different processors. This provides a simple means of synchronization. In practice, software using message passing is even popular on shared memory processors due to portability and the simple synchronization mechanism. (Usually there is only a small time penalty through the application of message passing.) The time for message passing is modeled to be proportional to the data size plus some latency as:

$$t_{\text{communication}} = \tau + \beta n$$

where n is the amount of data transferred (typically in MByte), β is the *communication bandwidth*, and τ the *latency*. As mentioned earlier, it is important to communicate data in large batches to avoid the penalty occurred through the latency. Another attractive feature of message passing is that there are standard message passing libraries available, in particular message passing interface (MPI) [12].

Message passing is used for tasks that are distributed among processors in a multiprocessor system or a collection of workstations. In the case of a distributed system containing several multiprocessors even larger granularity is required. On this level data communication would be done using network data transfer commands (like "scp" and "rsync") and the tasks should have very large granularity. Running the jobs in batch mode on a collection of multiprocessors is appropriate for this large granularity; for an example of an open source implementation using an R language frontend, see http://datamining.anu.edu.au/software/qlapply/qlapply.html. There are several issues that have to be addressed when using this type of distributed computing. The latencies are enormous and thus only extremely large tasks are suitable for distributed processing. On the other hand, the data may be distributed and local processing of the data where it resides gives an advantage as the data does not need to be moved. In addition to the high latency one also has to deal with issues of access validation, privacy (some customers will not allow moving data to other sites), and different operating systems and software, and different data formats. Standards and protocols need to be developed to deal with these issues. Finally, performance assessment and optimal task distribution need to deal with heterogeneous environments with different computational and data access speeds.

10.1.3 CONCEPTS FROM NUMERICAL LINEAR ALGEBRA

We consider the problem of fitting data sets using functions from linear function spaces by (penalized) least squares methods. This problem has two computational components, the search for a good approximating subspace and the solution of a (penalized) least squares problem. We will review some of the developments and ideas from parallel numerical linear algebra which are relevant to the following discussion. Numerical linear algebra has for sometime been one of the main areas of parallel algorithm development. A cumulation of this work was the release of the LAPACK and ScaLAPACK software [13, 14] and a list of freely available software for the solution of linear algebra problems can be found at http://www.netlib.org/utk/people/JackDongarra/la-sw.html. The solution techniques can be grouped into two major classes: iterative methods and direct methods. These methods are applied either to the residual equations

$$Ax - b = r$$

or to the corresponding normal equations

$$A^{T}Ax = A^{T}b.$$

For the normal equations iterative methods require the determination of repeated matrix–vector products

$$y = Mx$$

with the same matrix $M = A^{T}A$ but with varying x. The most common parallel matrix–matrix multiplication algorithms start by partitioning the matrix into st blocks:

$$M = \begin{bmatrix} M_{1,1} & \cdots & M_{1,t} \\ \vdots & & \vdots \\ M_{s,1} & \cdots & M_{s,t} \end{bmatrix}$$

and with according blocking of the vector x the matrix–vector product has the form

$$y_k = \sum_{j=1}^{t} M_{k,j} x_j, \quad \text{for } k = 1, \ldots, s.$$

So now one uses st processors for the parallel multiplication step and successively less for the $\log_2(t)$ summation steps. Special variants that have been considered are the cases of $t = s$ and $t = 1$ and $s = 1$. The variants have slightly different characteristics. For example, in the case of $t = 1$ no reduction step and loss of parallelism occurs. However, the whole vector x needs to be broadcast at every step. There is no need for broadcast in the case of $s = 1$ but the reduction operation is more dominant in this case. An analysis shows that the communication costs are similar for all cases. The choice of st, which can be larger than p, influences the performance as well. See Refs. [15, 16] for a further discussion of these issues. The cases of $s = 1$ and $t = 1$ are especially of interest for sparse matrices as this provides better load balancing because these cases do not require even distribution of the nonzero elements over the whole matrix but only over the rows or columns.

The effect of memory hierarchies has been considered in-depth in the research leading to the ScaLAPACK and LAPACK libraries on direct solvers. The basic building blocks are collected in the BLAS libraries which are available for functions adapted to the multiple levels of the memory hierarchy [13] for the case of shared memories and PBLAS [14] for the case of distributed memories.

The underlying algorithms including Gaussian elimination, Cholesky factorization, and QR factorization all require simple vector operations like "SAXPY" which combines two vectors (e.g., rows of a matrix):

$$z = \alpha x + y,$$

where x and y are vectors and $\alpha \in \mathbb{R}$. Here all the products and additions can be done in parallel and the average amount of parallelism is $O(n)$. This fine-grain parallelism even in the case of vector processing suffers from the fact that there are three memory accesses per every two floating point operations. However, these operations have been fundamental in the further development; they form part of the level 1 BLAS subroutines. In terms of the memory hierarchy these operations are most suited if the access is on the level of registers.

Now one can group n subsequent updates into one matrix update. This is the basis for matrix factorizations and the update step is here a rank-one modification of a matrix:

$$A = A + \alpha x y^{\mathrm{T}}.$$

Here the total amount of parallelism is the same as in the case of a matrix–vector product and it turns out that the performance and granularity are basically the same as there. This is the case of a level 2 BLAS routine and it allows efficient usage of vector registers. This is considered to be of intermediate granularity in numerical linear algebra. Note, however, that the amount of data to be accessed is proportional to the number of floating point operations performed $O(n^2)$. In terms of the memory hierarchy these operations are most suited if the operands reside in registers or cache.

Highest granularity is obtained when matrix–matrix operations are used. In the case of the solution of linear systems of equations this means that operations of the type

$$A = A + BC \quad \text{and} \quad X = A^{-1}X$$

occur. Such operations allow the effective usage of various caches and efficient parallel execution. Here one sees the "volume to surface effect" of parallel linear algebra, the number of operations performed here are $O(n^3)$ whereas the amount of data accessed is $O(n^2)$. This is one aspect of the large granularity of the computation.

The key to the development of methods for complex computational systems including a hierarchy of memories from registers to distributed computers takes into account the different access times and the different sizes of memory in the different levels of the hierarchy.

These computational building blocks have been optimized for many computational platforms and architectures and implementations of methods which adapt to different architectures are available [17]. There are various approaches to the implementation on parallel architectures (see Refs. [14, 18]).

Granularity issues are important for iterative methods, and for direct methods both using the normal equations and the QR factorization of the design matrix A. The approach using normal equations has two computational steps:

1. Determine the normal matrix $M = A^T A$ and $A^T b$.
2. Solve the normal equations $A^T A x = A^T b$.

In our case these two steps have totally different characteristics. In the first step the data are accessed. This step is scalable with respect to the data size and parallelizes well if the data is equally distributed as

$$A = \begin{bmatrix} A_1 \\ A_2 \\ \vdots \\ A_p \end{bmatrix}.$$

The partial normal matrices $M_i = A_i^T A_i$ are determined on each processor and then assembled using a recursive doubling method to give:

$$M = \sum_{i=1}^{p} M_i.$$

This approach assumes that the complexity of the model is such that the matrix M can be held in memory of one processor. We will see examples where further structure of this matrix can be exploited to distribute it onto several processors. This scheme is very effective for very large data sets and large models. The solution step can now be parallelized as well to deal with the $O(n^3)$ complexity.

It is often pointed out that the QR factorization approach is more stable and will give better results than the normal equation approach. It turns out that this approach can also be performed in two steps:

1. Determine the QR factorization $A = QR$ and $Q^T b$.
2. Solve $Rx = Q^T b$.

Again only the first step is data dependent. It can be parallelized by partitioning the data as in the case above. The parallel QR factorization can be implemented using BLAS and PBLAS operations and most heavily relying on matrix–matrix operations.

So far we have considered the parallel solution of least squares problems. The algorithms considered here have another aspect as the actual model will be selected from many candidates. To compare several candidates one has to solve many different least squares problems. Often the algorithms used take the form of greedy algorithms where a certain model is increased by some extra degrees of freedom and various ways to increase a given model are compared. Thus one solves many problems with residual equations

$$A'x' - y = r$$

for fixed y and

$$A' = \begin{pmatrix} A & B \end{pmatrix}$$

for a given A and many different B. The normal matrix for this case is

$$\begin{bmatrix} A^{\mathrm{T}}A & A^{\mathrm{T}}B \\ B^{\mathrm{T}}A & B^{\mathrm{T}}B \end{bmatrix}.$$

Now the update requires a further scan of all the data to determine $A^{\mathrm{T}}B$ and $B^{\mathrm{T}}B$ which can be done in parallel like before. In addition one can do various updates for several possibilities B in parallel. The Cholesky factorization can also be updated efficiently see, Ref. [19]. Note, however, that the granularity of the update is substantially less than the granularity of the full factorization. We have so far focused on the case of dense linear algebra. This is an important case and, in fact, many of the matrices occurring in predictive modeling are dense, e.g., when linear combinations of radial basis functions are fitted or for additive models. However, for very large-scale computations the complexity of dense matrix computations is not feasible, and, in fact, we will see how such dense matrices can be avoided. So one has to ask the question if the principles discussed above are still applicable in the sparse matrix case. It turns out that this is indeed the case. The application of dense linear algebra for the direct solution of sparse least squares problems has been well established for the multifrontal methods [20]. These methods make use of the fact that during elimination procedure the original sparsity of the matrix tends to gradually disappear as more and more dense blocks appear. This is certainly the case for block bidiagonal matrices which have the form

$$\begin{pmatrix} A_1 & & & \\ B_2 & A_2 & & \\ & B_3 & A_3 & \\ & & \ddots & \ddots \end{pmatrix}.$$

The assembly stage is not fundamentally different from the dense case. In the factorization the amount of parallelism seems now confined by the inherent dependency of the factorization which proceeds in a frontal fashion along the diagonal. Each stage of the factorization will consider a new block row with B_{i+1} and A_{i+1} and will do a dense factorization on these blocks. Thus the parallelism in this factorization can be readily exploited. In addition, one can see that by an odd–even permutation (i.e., take first all the odd-numbered block rows and columns and then the even-numbered ones) one gets a block system with an extra degree of parallelism. This approach is very similar to the odd–even reduction approach to summation. However, it turns out that the amount of floating point operations in this approach is higher than in the original factorization. For more details on this, see Refs. [11,21,22].

In the case of more general matrices one can observe very similar effects. However, instead of having one front and dense factorizations with a constant size one now has multiple fronts which subsequently combine to form ever larger dense matrices. In addition to the dense parallelism one has also the parallelism across various fronts which is of the same nature as the parallelism for the block bidiagonal case. There is also the potential of trading of some parallelism with redundant computations. See Refs. [20, 23] for more details. These multifrontal methods have originated in the solution of finite element problems. For finite element problems one also has two stages very similar to the two stages (forming of normal equations and solution) for the least squares problems. It turns out that, like in the case of least squares problems, the assembly can be viewed as an elimination step for the augmented system of equations. The assembly and the factorization of the matrix can be combined to one stage like in the case of the QR factorization.

10.2   ADDITIVE MODELS

10.2.1   THE BACKFITTING ALGORITHM

Additive models are functions of the form

$$f(x_1, \ldots, x_d) = f_0 + \sum_{j=1}^{d} f_j(x_j).$$

For this subclass the curse of dimensionality is not present as only $O(d)$ storage space is required. The constant component f_0 is obtained as an average of the values of the response variable:

$$f_0 = \frac{1}{N} \sum_{i=1}^{N} y^{(i)}.$$

For the estimation of the f_j one-dimensional smoothers can be used. A smoothing parameter controls the trade-off between bias and variance. The choice of the value of the smoothing parameter is essential and has been extensively discussed in the literature, see Refs. [9, 24] for an introduction and further pointers.

Using a smoother S_j, the f_j are defined as the solution of the equations

$$f_j = S_j \left(\mathbf{y} - f_0 - \sum_{i \neq j} \mathbf{f}_i \mid \mathbf{x}_j \right) \tag{10.1}$$

where f_j are the vectors of values of f_j at the data points $x_j^{(i)}, i = 1, \ldots, N$ or $\mathbf{f}_j = f_j(\mathbf{x}_j)$. Uniqueness of the solution is obtained through orthogonality conditions on the f_j. This choice of the f_j is motivated by the fact that the functions f_j^e, which minimize expected squared residuals, satisfy

$$f_j^e(x_j) = E \left(Y - f_0^e - \sum_{i \neq j} f_i^e(X_i) \mid X_j = x_j \right)$$

and the equations are obtained by replacing the expectation by the smoother. Alternatively, the fixpoint equations are obtained from a smoothing problem for additive models [25, 26].

The simple form of these equations where in the jth equation the variable x_j only occurs explicitly on the left-hand side suggest the usage of a Gauss–Seidel or SOR process [27] for their solution. More specifically, for the Gauss–Seidel case a sequence of component functions f_j^k is produced with

$$f_j^0 = S_j(\mathbf{y} - f_0 \mid \mathbf{x}_j).$$

The iteration step is then

$$f_j^{k+1} = S_j \left(\mathbf{y} - f_0 - \sum_{i < j} \mathbf{f}_i^{k+1} - \sum_{i > j} \mathbf{f}_i^k \mid \mathbf{x}_j \right)$$

where $\mathbf{f}_j^k = f_j^k(\mathbf{x}_j)$. This is the *backfitting algorithm*, see Refs. [9, 25, 28] for more details and also a discussion of the local scoring algorithm that is used for generalized additive models.

In case where the smoother S_j is linear one introduces matrices $\mathbf{S}_j$ (which depend on $\mathbf{x}_j$) such that

$$\mathbf{f}_j = f_j(\mathbf{x}_j) = S_j(\mathbf{y} \mid \mathbf{x}_j)(\mathbf{x}_j) = \mathbf{S}_j \mathbf{y}.$$

As a consequence of the fixpoint equations (10.1) one obtains a linear system of equations for the function value vectors $\mathbf{f}_j$:

$$\begin{bmatrix} I & \mathbf{S}_1 & \cdots & \mathbf{S}_1 \\ \mathbf{S}_2 & I & \cdots & \mathbf{S}_2 \\ \vdots & \vdots & & \vdots \\ \mathbf{S}_d & \mathbf{S}_d & \cdots & I \end{bmatrix} \begin{bmatrix} \mathbf{f}_1 \\ \mathbf{f}_2 \\ \vdots \\ \mathbf{f}_d \end{bmatrix} = \begin{bmatrix} \mathbf{S}_1(\mathbf{y} - f_0) \\ \mathbf{S}_2(\mathbf{y} - f_0) \\ \vdots \\ \mathbf{S}_d(\mathbf{y} - f_0) \end{bmatrix}. \tag{10.2}$$

The backfitting procedure leads to an iterative characterization of the $\mathbf{f}_i^k$:

$$\mathbf{f}_j^0 = \mathbf{S}_j(\mathbf{y} - f_0)$$

and

$$\mathbf{f}_j^{k+1} = \mathbf{S}_j \left(\mathbf{y} - f_0 - \sum_{i<j} \mathbf{f}_i^{k+1} - \sum_{i>j} \mathbf{f}_i^{k} \right).$$

This is formally a block-Gauss–Seidel procedure for the component vectors and the established convergence theory for these methods can be applied. In particular, convergence is obtained if the matrices $\mathbf{S}_j$ have eigenvalues in $(0, 1)$. In fact, it turns out that in many practical applications backfitting converges in a couple of iterations.

The block matrix is singular for many important types of smoothers. Consider, for example, S_j to be the smoother of a smoothing spline [26]. In this case S_j will have an eigenvector $e = (1, \ldots, 1)^T$ with the eigenvalue one, as constant functions are preserved under smoothing with smoothing splines. It follows that any choice of $\mathbf{f}_j = \alpha_j e$ with $\sum_{j=1}^{d} \alpha_j = 0$ leads to an eigenvector to eigenvalue zero for the block matrix which is thus singular. The standard convergence theorems do not apply in this case but require a simple modification, see Ref. [28].

In general, the backfitting algorithm produces a sequence of functions

$$f_1^1, \ldots, f_d^1, f_1^2, \ldots, f_d^2, f_1^3, \ldots .$$

Each element of the sequence depends on d immediately preceding elements. For the determination of one element one smoother S_j is evaluated. Thus the complexity of the determination of one element in this sequence is $O(N)$. For the function d one requires d consecutive elements f_j. The complexity of the determination of $f_1^k, \ldots, f_d^k$ is thus $O(kdN)$. Due to the dependency of the elements in the sequence one cannot determine multiple elements of the sequence simultaneously. A parallel implementation of the backfitting procedure would thus have to exploit any parallelism which is inherent in the smoother S_j and in the evaluation of $\mathbf{y} - f_0 - \sum_{i<j} \mathbf{f}_i^{k+1} - \sum_{i>j} \mathbf{f}_i^{k}$. While this is a possible option, we will next discuss a different approach, which can be viewed as an alternative to backfitting.

There are other alternatives to the parallel algorithm discussed here. One simple alternative is based on a Jacobi or fixpoint iteration. For this method, the sequence of functions is defined by

$$f_j^{k+1} = S_j \left(\mathbf{y} - f_0 - \sum_{i \neq j} \mathbf{f}_i^{k} \mid \mathbf{x}_j \right).$$

This substantially decreases the convergence rate but allows the simultaneous determination of d functions f_j^{k+1}. One can improve convergence by accelerating this procedure with a conjugate gradient (CG) method. There are, however, additional problems with this approach. First the approach leads to a parallelism of at most degree d. But a speedup of d may not be possible due to unbalanced loads as it is unlikely that the time for all smoothers S_j is the same. This is true in particular if one includes binary and categorical predictor variables. Load balancing would also be typically lost if one includes interaction terms or functions $f_{ij}(x_i, x_j)$, inclusions that do not require major modifications of the backfitting method but that may improve predictive performance.

In the case of lower levels of parallelism p and large d one can get some load balancing using a master–worker approach where each processor can demand another component once it has completed earlier work. Such a dynamic load-balancing approach is well suited to a distributed computing environment. However, maybe the main problem with this approach of parallelizing over component functions f_j is that each processor will require reading a large proportion of all the data. Thus memory availability limits the feasibility of this approach. An out-of-core method may be considered but this can introduce substantial overheads as the data needs to be transferred between secondary store and memory as each iteration needs to visit all the data. Thus for the data-intensive iteration steps of backfitting one would prefer to have all the data points in memory. A difficulty of the parallel Jacobi method is also that at each step of the iteration data communication needs to be performed to move the new $\mathbf{f}_i$ (or at least their sum) to all the processors.

Further parallel iterative methods could be developed by applying known methods from linear algebra [29] to the block linear system of Equations (10.2). Little has been done in this area, however, see, Ref. [9] for some references. An issue one has to deal with is the singularity of the block matrix as most iterative methods are defined for nonsingular matrices.

10.2.2 A PARALLEL BACKFITTING ALGORITHM

A parallel backfitting algorithm is required for time-critical applications, when one deals with very large data sets and for the case of slow convergence. Here we discuss an algorithm which is based on data partitioning. The proposed method has applications outside of parallel processing as well as it can be used for the case where the data is distributed over multiple repositories and for the case where not all the data fits into main memory. Data mining applications commonly do require processing of very large data sets, which may be distributed, will not fit in the memory, and have many variables. Thus data mining predictive modeling problems are applications which can ideally profit from the parallel algorithms discussed here.

The algorithm is based on a partition of the data into p subsets. The elements of subset r are denoted by

$$\mathbf{x}_1^{[r]}, \ldots, \mathbf{x}_d^{[r]}, \mathbf{y}^{[r]}$$

where $r = 1, \ldots, p$. Using the same smoothers for each subset and the backfitting algorithm one obtains p partial models $f_j^{[r]}$ for $j = 0, \ldots, d$ and $r = 1, \ldots, p$. The final model is obtained as a weighted sum

$$f_j = \sum_{r=1}^{p} w_r f_j^{[r]}$$

for some weights w_r. The weights will characterize the influence of each partial component on the overall model. They should satisfy $w_r \geq 0$ and

$$\sum_{r=1}^{p} w_r = 1.$$

They might include some measure of reliability of each estimate, for example, they could just be N_r/N, where N_r is the number of data points in subset r. One can also iteratively refine this method, first fitting the data in this way, then the residuals etc.

If the subset r requires k_r iterations to find the partial model the time for this subset is $O(dk_r N_r)$, say $dk_r N_r \tau_1$. If all the subsets are processed in parallel the total time for this step is then

$$T_1 = d \max_r k_r N_r \tau_1.$$

The determination of the sum is a reduction operation. If we assume that the subsets are distributed, one can do this reduction in $\log_2(p)$ steps by first adding the even elements to the preceding ones to get an array of models of size $p/2$ then repeating this step until only one model is left. As all these models are of the same size the time to do this step is

$$T_2 = \log_2(p) \sum_{j=0}^{d} m_j \tau_2$$

for some constant τ_2 and where m_j is the complexity of the model f_j. Note that in the case of a shared memory one could make use of the higher parallelism by adding the components in parallel and one would get for this case (if enough processors are available) $T_2 = \log_2(p) \max_j m_j \tau_2$. The speedup obtained through this algorithm is

$$S_p = \frac{dkN\tau_1}{d \max_r k_r N_r \tau_1 + \log_2(p) \sum_{j=0}^{d} m_j \tau_2}.$$

Whereas the summing up part (T_2) is important for very large p and smaller N, in the cases of interest in data mining the first component is often dominating. However, data communication costs are included in the constant τ_2 and depending on the ratio of communication to computation time (basically τ_2/τ_1 in this case) the second component may be substantial. Where this is not the case, and if one partitions the data in a balanced way so that $N_r = N/p$ and introduces a parameter $\kappa = k/\max_r k_r$ then one gets a speedup of

$$S_p = \kappa_p p.$$

So to determine the speedup one now needs to find κ_p. If κ_p is bounded from below for fixed N/p and growing p then the algorithm is *scalable*. Clearly the value of κ_p depends on the actual partitioning of the data. We defer this discussion a bit and first consider another question: How does the resulting additive model compare to the additive model obtained from the backfitting algorithm? Again this would have to depend on the way the data is partitioned but also on the weights w_r.

In a very simple case it turns out that the iterates of the method presented above can also be generated by a smoothing procedure very similar to backfitting. This is the case of p times repeated observations (the same x values but arbitrary y values for all partitions) that are distributed over the

p processors. We formulate this simple observation as a proposition (the index j is always assumed to be $j = 1, \ldots, d$):

Proposition 10.1 *Let* $\mathbf{x}_j = [\mathbf{x}_j^{[1]}, \ldots, \mathbf{x}_j^{[p]}]$ *and* $\mathbf{y} = [\mathbf{y}^{[1]}, \ldots, \mathbf{y}^{[p]}]$. *Furthermore let* $f_0^{[r]}$ *and* $f_j^{[r]}$ *be the components of an additive model which have been fitted to the partial data* $\mathbf{x}_j^{[r]}$ *and* $\mathbf{y}$ *such that Equations (10.1) hold for these data points. Now let* $w_r \geq 0$ *with* $\sum_{r=1}^{p} w_r = 1$ *and set*

$$f_0 = \sum_{r=1}^{p} f_0^{[r]}, \quad and \quad f_j = \sum_{r=1}^{p} f_j^{[r]}.$$

If $\mathbf{x}_j^{[r]} = \mathbf{x}^*$ *for all* r, *and the smoothers for the partial components are linear with matrix* $\mathbf{S}_j^*$ *and* $\mathbf{f}_j = f_j(\mathbf{x})$ *then*

$$\mathbf{f}_j = \mathbf{S}_j \left(\mathbf{y} - f_0 - \sum_{i \neq j} \mathbf{f}_i \right)$$

with

$$\mathbf{S}_j = \begin{bmatrix} w_1 \mathbf{S}_j^* & \cdots & w_r \mathbf{S}_j^* \\ \vdots & & \vdots \\ w_1 \mathbf{S}_j^* & \cdots & w_r \mathbf{S}_j^* \end{bmatrix}.$$

Proof. The vector $\mathbf{f}_j$ consists of p identical blocks $f_j(\mathbf{x}^*)$. We need to show that

$$f_j(\mathbf{x}^*) = \sum_{r=1}^{p} w_r \mathbf{S}_j^* \left(\mathbf{y}^{[r]} - f_0 - \sum_{i \neq j} f_i(\mathbf{x}^*) \right).$$

By expanding the right-hand side using $\sum_{r=1}^{p} w_r = 1$ and the definitions of f_0 and f_j one gets:

$$\sum_{r=1}^{p} w_r \mathbf{S}_j^* \left(y^{[r]} - f_0 - \sum_{i \neq j} f_i(\mathbf{x}^*) \right) = \mathbf{S}_j^* \left(\sum_{r=1}^{p} w_r y^{[r]} - f_0 - \sum_{i \neq j} f_i(\mathbf{x}^*) \right)$$

$$= \mathbf{S}_j^* \left(\sum_{r=1}^{p} w_r y^{[r]} - \sum_{r=1}^{p} w_r f_0^{[r]} - \sum_{i \neq j} \sum_{r=1}^{p} w_r f_i^{[r]}(\mathbf{x}^*) \right)$$

$$= \mathbf{S}_j^* \left(\sum_{r=1}^{p} w_r (y^{[r]} - f_0^{[r]} - \sum_{i \neq j} f_i^{[r]}(\mathbf{x}^*) \right).$$

Now one can use property of Equation (10.2) for the components to see that this is equal to

$$\sum_{r=1}^{p} w_r f_j^{[r]}(\mathbf{x}^{[r]}) = f_j(\mathbf{x}^*).$$

$\square$

So the parallel algorithm satisfies the Equations (10.2) and thus the same methods for the analysis of convergence, etc. can be applied for this method. The underlying one-dimensional smoother is quite intuitive. Many actual smoothers will have the property that S_j determined as weighted average of partial smoothers with $w_r = 1/p$ is exactly the same as the smoother applied to the full data set. In this case the sequential and the parallel methods will give the same results. One can also see that the eigenvalues of the matrices S_j and the ones of partial matrices S_j^* are the same in this case and thus the $k_r = k$ and thus $\kappa_p = 1$ up to small variations which can depend on the data vectors $\mathbf{y}^{[r]}$. (The sequential backfitting turns out to be the backfitting for a problem with a matrix S_j^* and data $\sum_r w_r y^{[r]}$.)

In the more general case where the data partitions are all of the same size and have been randomly sampled (without replacement) from the original data the parallel algorithm is an instance of a *bagging* without replacement (see Refs. [30–32] for further analysis). Basically, the parallel algorithm gives results very close to the results of the sequential algorithm. There might even be some advantages of choosing the parallel backfitting algorithm over the sequential one due to higher variance reduction especially for nonlinear estimators. Parallel implementation of bagging has been suggested in Ref. [33] where it is also pointed out that the parallel algorithm of the kind discussed above may have superior statistical properties. We observe that the parallel backfitting procedure can be easily modified to provide a more general parallel bagging procedure. Besides bagging and parallel additive models other areas in data mining use the same idea of partitioning the data, applying the underlying method to all the parts and combining the results. An example where this is done in association rule mining can be found in Ref. [34].

10.2.3 IMPLEMENTATION — THE BLOCK AND MERGE METHOD

An implementation of a variant of the algorithm discussed above has been suggested and discussed in Ref. [35]. The method was tested on a SUN E4000 with eight SPARC processors with 167 MHz each and 4.75 GBytes of shared memory. The application considered was the estimation of risk for a large car insurance database. The implementation used a master–slave approach, was implemented in C++ using the MPI message passing package for data exchange between processors. Whereas this is not necessary for shared memory processors the overhead of using MPI is small and this makes the code much more portable. Tests showed that this approach is scalable in the data size, and in the number of processors and achieves good speedups for the small numbers of processors considered. However, for larger numbers of processors and distributed computing this approach has not been studied.

In this implementation the data is partitioned into m blocks where m is typically larger than the number of processors available. The partitions are random and all of equal size. The size of one block contains N/m records is chosen such that it fits into memory which provides a performance advantage for the backfitting algorithm as no swapping is required.

It was found that choosing a small smoothing parameter for the one-dimensional smoothers has an advantage. This reduces the bias of the method at the cost of increased variance. Merging the models from the m partial data sets may involve averaging but may actually also include some smoothing. Additional smoothing may be done particularly after the last merge to control the trade-off between bias and variance. The merging stages, however, were not very costly and consequently were done on one processor sequentially.

The smoothing parameter in the initial stages of the algorithm is now not only used to trade-off bias and variance. In addition it controls the number of iterations and resolves numerical difficulties due to uneven data distribution. Only at the final combination step the smoothing parameter is used exclusively to balance bias and variance.

For very large data sizes continuous data was discretized (binned) onto a regular grid. This allows effective data compression and leads to banded systems of equations if smoothing splines on equidistant grids are used. Note that binning can have an effect of substantially reducing the data size and thus increase computational performance, but it may also lead to load imbalance when the data sizes on the different processors are reduced differently.

10.3   THIN PLATE SPLINES

10.3.1   FITTING RADIAL BASIS FUNCTIONS

Radial basis function approximations are real functions $f(x)$ with $x \in \mathbb{R}^d$ of the form

$$f(x) = \sum_{i=1}^{N} c_i \, \rho(\|x - x^{(i)}\|) + p(x), \tag{10.3}$$

where p is a polynomial typically of low degree and is even zero in many cases. Radial basis functions have recently attracted much attention as they can efficiently approximate smooth functions of many variables. Examples for the function ρ include Gaussians $\rho(r) = \exp(-\alpha r^2)$, powers and thin plate splines $\rho(r) = r^\beta$ and $\rho(r) = r^\beta \ln(r)$ (for even integers β only), multiquadrics $\rho(r) = (r^2 + c^2)^{\beta/2}$, and others. Radial basis functions have been generalized to metric spaces where the argument of ρ is the distance $d(x, x^{(i)})$ (and the polynomial $p = 0$).

As always let $x^{(i)}$ be the data points of the predictor variables. The vector of coefficients of the radial basis functions $\mathbf{u} = (u_1, \ldots, u_N)$ and the vector of coefficients of the polynomial term $\mathbf{v} = (v_1, \ldots, v_m)$ are determined, for smoothing, by a linear system of equations of the form:

$$\begin{bmatrix} A + \alpha I & P \\ P^{\mathrm{T}} & 0 \end{bmatrix} \begin{bmatrix} \mathbf{u} \\ \mathbf{v} \end{bmatrix} = \begin{bmatrix} \mathbf{y} \\ 0 \end{bmatrix}. \tag{10.4}$$

The block A has elements $\rho(\|x^{(i)} - x^{(j)}\|)$, the elements of the matrix P are $p_j(x^{(i)})$ where the p_j are the basis functions for the polynomials, the matrix I is the identity, and α is the smoothing parameter. The vector $\mathbf{y}$ contains the observed values $y^{(i)}$ of the response variable. Existence, uniqueness, and approximation properties, in particular for the case of interpolation ($\alpha = 0$) have been well-studied. Reviews of radial basis function research can be found in Refs. [36–38].

The main computational task in the determination of a fit with radial basis functions is the solution of the system of equations (10.4). The degree of the polynomial term p is $m = O(d^s)$ for some small $s(m = d + 1$ for linear polynomials) and N is very large. Thus most elements of the matrix of the linear system of equations are in the block A. Whereas the radial basis functions ρ may be nonzero far from the origin, the effects of the combination of the functions cancels out, a consequence of the equation $P^{\mathrm{T}} \mathbf{c} = 0$. In spite of this locality, the matrix A is usually mostly dense. In high dimensions this may even be true when the radial basis function ρ have a small support $\{x \mid \rho(x) \neq 0\}$. The reason for this is the concentration effect that is observed in high dimensions. One finds that the distribution of pairwise distances becomes very narrow so that all randomly chosen pairs of points have close to the same distance. As the support of the radial basis function has to be chosen so that for every $x^{(i)}$ there are at least some $x^{(j)}$ for which it follows from the concentration and the smoothness of ρ that for any i one gets $\rho(\|x^{(i)} - x^{(j)}\|) \neq 0$ for most other j. It follows that A is fundamentally a dense matrix.

When solving the system (10.4) one could either use direct or iterative solvers. Due to the density of the matrix the direct solver will have complexity $O(n^3)$ and an iterative solver at least complexity $O(n^2)$ as it requires more than one matrix–vector product. Efficient parallel algorithms for both cases are available and have been widely discussed in the literature (see Ref. [14] for software and further references). The $O(n^2)$ or $O(n^3)$ complexity however implies nonscalability and thus very large problems are not feasible for this approach. Similar to the approach discussed in Section 10.2 one may use a bagging approach here as well and partition the data, fit a partial model to every part and average over the partial models. While this is a valid approach, we will now consider a different approach.

10.3.2   PARALLEL AND SCALABLE FINITE ELEMENT APPROXIMATION OF THIN PLATE SPLINES

The radial basis function smoother can also be characterized variationally. In the case of thin plate splines [39] the radial basis functions are Greens functions of a PDE and the radial basis smoother is the smoothing spline that minimizes

$$M_\alpha(f) = \frac{1}{n}\sum_{i=1}^{n}(f(\mathbf{x}^{(i)}) - y_i)^2 + \alpha \int_\Omega \left(\frac{\partial^2 f(\mathbf{x})}{\partial^2 x_1}\right)^2 + 2\left(\frac{\partial^2 f(\mathbf{x})}{\partial x_1 \partial x_2}\right)^2 + \left(\frac{\partial^2 f(\mathbf{x})}{\partial^2 x_2}\right)^2 d\mathbf{x} \quad (10.5)$$

for $d = 2$. The intuition behind this is that the parameter α allows to control the stiffness of the spline and thus the curvature of the fit. Small α will lead to a "closer fit" of the data where large α will lead to a "smoother fit." For our computations we will consider the case of a compact domain Ω rather than $\Omega = \mathbb{R}^2$. The domain Ω is either $\mathbb{R}^2$ or a compact subset.

Instead of using the radial basis functions (for the case $\Omega = \mathbb{R}^2$) and Equation (10.4) with linear polynomials one can develop a good approximation to this problem using numerical methods for the solution of elliptic PDEs, in particular finite element methods. This approach transforms the smoothing problem into a penalized finite element least squares fit. Finite element approximations have well-established approximation properties (see Refs. [40, 41]). The main results indicate that generic finite element approximations lead to good approximations for smooth enough functions and regular domains. We will only discuss the computational aspects further here.

Fundamentally, the finite element method approximates the function f with a function from a finite-dimensional linear space, in particular a space of piecewise polynomial functions. This approximation is written as a linear combination of basis functions which typically have a compact (small) support (the set of x where $f(x) \neq 0$). Thus we get a representation

$$f(x) = \sum_{j=1}^{M} u_j B_j(x)$$

with basis functions B_j and coefficients u_j. The degrees of freedom of the approximating space is denoted by m and is typically much smaller than N. We are now looking for a function f of this form which minimizes the functional M_α in (10.5). The vector of coefficients $u = [u_1, \ldots, u_m]$ which minimize the quadratic function

$$\mathbf{u}^T M \mathbf{u} - 2\mathbf{b}^T \mathbf{u} + \frac{1}{N}\mathbf{y}^T\mathbf{y} + \alpha \mathbf{u}^T S \mathbf{u}$$

for some matrices M and S and a vector $\mathbf{u}$ the determination of which will be discussed later. The minimum of this function satisfies the equations

$$(M + \alpha S)\mathbf{u} = \mathbf{b}.$$

Thus the fitting problem has now been reduced to two problems, the first one being the determination of M, S, and $\mathbf{b}$ and the second one the solution of this system. We will now discuss both steps and see how parallel processing can substantially improve performance but maybe most importantly how the scalability problem of the original radial basis function approach has suddenly disappeared!

These two steps can be found in any finite element solver. The first step is called the *assembly* step and the second is called *solution*. From the variational problem one sees that a general element of the mass matrix M is

$$[M]_{jk} = \frac{1}{N} \sum_{i=1}^{N} B_j(x^{(i)}) B_k(x^{(i)}),$$

the right-hand side b has the components

$$b_k = \frac{1}{N} \sum_{i=1}^{N} y^{(i)} B_k(x^{(i)})$$

and the elements of the stiffness matrix S are the integrals

$$[S]_{kj} = \int_{\Omega} \left(\frac{\partial^2 B_k(\mathbf{x})}{\partial^2 x_1} \right) \left(\frac{\partial^2 B_j(\mathbf{x})}{\partial^2 x_1} \right) + 2 \left(\frac{\partial^2 B_k(\mathbf{x})}{\partial x_1 \partial x_2} \right) \left(\frac{\partial^2 B_j(\mathbf{x})}{\partial x_1 \partial x_2} \right) + \left(\frac{\partial^2 B_k(\mathbf{x})}{\partial^2 x_2} \right) \left(\frac{\partial^2 B_j(\mathbf{x})}{\partial^2 x_2} \right) d\mathbf{x}.$$

The complexity of the assembly step depends very much on the support of the basis functions B_j. We will assume here that the support is such that for every point x at most K basis functions have $B_j(x) \neq 0$. Thus the complexity of the determination of b is $O(KN)$ and one can see that the complexity of the determination of M is $O(K^2N)$ as for every point x there are at most K^2 products $B_i(x)B_j(x) \neq 0$. The complexity of the determination of S is of the order $O(K^2m)$ in nondegenerate cases. For large data sets the time to generate the matrix will dominate. Note that this step is scalable in the data size. Moreover, it only requires one scan of the full data set. By partitioning the data into p equal parts this step can also be ideally parallelized up to a reduction step. The complexity for this stage is thus approximately

$$T_a = \frac{K^2 N \tau_1 + \log_2(p) v(M) \tau_2}{p}$$

where $v(M)$ is the number of nonzero elements of M. Typically the first part is dominating and a speedup of close to p will be obtained.

The complexity of the solution step depends on the method used. Both direct methods using, e.g., parallel multifrontal algorithms [20] and iterative method using, e.g., parallel multigrid algorithms [42] are applicable. The speedups obtained for these steps is not quite as good as for the assembly. This mostly has to do with increased communication requirements but also with the fact that (in particular for direct solvers) parallel algorithms require extra operations compared with the algorithms that have higher computational demands than the corresponding sequential algorithm. This extra work required for parallel execution is sometimes called "redundant."

10.3.3   A PARALLEL NONCONFORMING FINITE ELEMENT METHOD

The finite element method delivered a parallel algorithm which is scalable in the data size and thus capable to handle very large fitting problems. However, in our discussion we have glossed over one important point, namely the effect of the second derivatives in the penalty term. They lead to fourth-order partial derivatives in the Euler equations and can make the linear systems of equations ill-conditioned and hard to solve, which, e.g., leads to slow convergence of the iterative solvers. But maybe computationally even more of a problem is the fact that this requires the basis functions to be much more smooth which in turn leads to large K and matrices A and S which are more densely populated. The higher smoothness requires also that one uses higher order polynomials which are more expensive to estimate. Of course many of this problems are not unique for the fitting problem, in fact, they are known to occur for general finite element problems for plates.

This contrasts very much with the situation where the penalty is the norm of the gradient of f in which case the Euler equations require the Laplacian, and one can use a triangulation with piecewise linear functions (or rectangular grids with piecewise bilinear functions) with the computational advantages that one only needs to evaluate linear functions and that the matrices have a very sparse structure. Whereas such a penalty is sometimes used for penalized finite element fitting this is not applicable for two or more dimensions as the corresponding Greens function is unbounded at zero and thus does not lead to an ordinary radial basis function expansion.

To handle these situations so-called nonconforming finite element methods have been introduced and we will now discuss such a nonconforming method which has been used to approximate a thin plate spline in Refs. [43–47]. The smoothing function f is approximated by

$$f(x) = u_0(x) + c_0 + c_1 x_1 + c_2 x_2$$

where

$$\int_\Omega u_0 \, \mathbf{dx} = 0. \tag{10.6}$$

The key idea is the introduction of the gradient $\nabla u_0 = (u_1 + c_1, u_2 + c_2)$ where the functions u_i have integral zero and satisfy

$$(\nabla u_0, \nabla v) = (\mathbf{u}, \nabla v), \quad \text{for all } v \in H^1(\Omega) \tag{10.7}$$

where $H^1(\Omega)$ denotes the Sobolev space of functions with square integrable first derivatives. With these functions the functional M_α can now be rewritten as a functional

$$J_\alpha(u_0, u_1, u_2) = \frac{1}{n} \sum_{i=1}^{n} (u_0(\mathbf{x}^{(i)}) + c_0 + c_1 x_1^{(i)} + c_2 x_2^{(i)} - y^{(i)})^2$$

$$+ \alpha \int_\Omega \left(\frac{\partial u_1(\mathbf{x})}{\partial x_1} \right)^2 + \left(\frac{\partial u_1(\mathbf{x})}{\partial x_2} \right)^2 + \left(\frac{\partial u_2(\mathbf{x})}{\partial x_1} \right)^2 + \left(\frac{\partial u_2(\mathbf{x})}{\partial x_2} \right)^2 \mathbf{dx}. \tag{10.8}$$

Whereas the number of unknowns has tripled, only first-order derivatives appear in the functional and in the constraints.

Let the functional values of u_0 at the observation points be denoted by

$$Pu_0 = [u_0(\mathbf{x}^{(1)}) \cdots u_0(\mathbf{x}^{(n)})]^\mathrm{T}.$$

Furthermore let

$$X = \begin{bmatrix} 1 & \cdots & 1 \\ \mathbf{x}^{(1)} & \cdots & \mathbf{x}^{(n)} \end{bmatrix}^\mathrm{T} \in \mathbb{R}^{n,3} \quad \text{and} \quad \mathbf{y} = [y^{(1)}, \dots, y^{(n)}]^\mathrm{T}.$$

Then the functional J_α is more compactly written as

$$J_\alpha(u_0, u_1, u_2, \mathbf{c}) = \frac{1}{n}(Pu_0 + X\mathbf{c} - \mathbf{y})^\mathrm{T}(Pu_0 + X\mathbf{c} - \mathbf{y}) + \alpha(|u_1|_1^2 + |u_2|_1^2)$$

and the computation will find approximations of the minimum of J_α for $u_0, u_1, u_2 \in H^1(\Omega)$ with zero mean and constants $\mathbf{c} = [c_0 \ c_1 \ c_2]^T \in \mathbb{R}^3$ subject to constraint (10.7).

The finite element approximation is obtained if the minimum is chosen over a finite-dimensional space V instead of $H^1(\Omega)$ and the constraint (10.7) has to hold for $v \in V$ only. The authors now consider a rectangular domain Ω and introduce a regular grid. As space V the space of continuous piecewise (on the grid cells) bilinear functions is chosen and the basis functions are the ones which are one on a grid point and zero everywhere else. In the coefficients the problem is a quadratic optimization problem with linear constraints that leads to a saddle point problem with the following linear system of equations:

$$
\begin{bmatrix}
\alpha A & 0 & 0 & 0 & -B_1^T \\
0 & \alpha A & 0 & 0 & -B_2^T \\
0 & 0 & M & F & A \\
0 & 0 & F^T & E & 0 \\
-B_1 & -B_2 & A & 0 & 0
\end{bmatrix}
\begin{bmatrix}
\mathbf{u}_1 \\
\mathbf{u}_2 \\
\mathbf{u}_0 \\
\mathbf{c} \\
\mathbf{w}
\end{bmatrix}
=
\begin{bmatrix}
0 \\
0 \\
n^{-1} N^T \mathbf{y} \\
n^{-1} X^T \mathbf{y} \\
0
\end{bmatrix}
\tag{10.9}
$$

Here $\mathbf{w}$ is the vector of Lagrangian multipliers introduced to satisfy the constraints and A is a symmetric positive matrix that approximates the Laplacian, $E = X^T X / n$, $F = N^T X$, and $M = N^T N$, etc. The convergence theory of this approximation is further discussed in Ref. [46]. Practical tests confirmed good approximation. The same method can also be applied to approximate kernel density estimation (see Ref. [48]).

The parallel implementation is very similar to the case of the conforming method of the previous section. Due to the nature of the grid one has here $K = 4$ which is much smaller than what one expects in the earlier case. The most costly part of the assembly is the determination of the matrix M. However, due to the regular grid and the piecewise bilinear functions this stage has been made very efficient. In fact, in some examples this stage was faster than the later solution stage! After the assembly the matrices are communicated to all the processors, which does introduce a little overhead. The assembly of A and B_i is cheap and does not require any data scanning.

For the solution of the system it is first reordered and scaled to get

$$
\begin{bmatrix}
\alpha A & 0 & -\alpha B_1^T & 0 & 0 \\
0 & \alpha A & -\alpha B_2^T & 0 & 0 \\
-\alpha B_1 & -\alpha B_2 & 0 & 0 & \alpha A \\
0 & 0 & 0 & E & F^T \\
0 & 0 & \alpha A & F & M
\end{bmatrix}
\begin{bmatrix}
\mathbf{u}_1 \\
\mathbf{u}_2 \\
\mathbf{w} \\
\mathbf{c} \\
\mathbf{u}_0
\end{bmatrix}
=
\begin{bmatrix}
0 \\
0 \\
0 \\
n^{-1} X^T \mathbf{y} \\
n^{-1} N^T \mathbf{y}
\end{bmatrix}
\tag{10.10}
$$

The upper left 2×2 block is regular, so the elimination of the two unknown vectors $\mathbf{u}_1$ and $\mathbf{u}_2$ results in the reduced linear system

$$
\begin{bmatrix}
-\alpha G & 0 & \alpha A \\
0 & E & F^T \\
\alpha A & F & M
\end{bmatrix}
\begin{bmatrix}
\mathbf{w} \\
\mathbf{c} \\
\mathbf{u}_0
\end{bmatrix}
=
\begin{bmatrix}
0 \\
n^{-1} X^T \mathbf{y} \\
n^{-1} N^T \mathbf{y}
\end{bmatrix}
\tag{10.11}
$$

where $G = B_1 A^{-1} B_1^T + B_2 A^{-1} B_2^T$. Elimination of $\mathbf{c}$ and $\mathbf{w}$ from Equation (10.11) leads to the equation

$$
\left(M + \alpha A G^{-1} A - F E^{-1} F^T \right) \mathbf{u}_0 = n^{-1} (N^T \mathbf{y} - F E^{-1} X^T \mathbf{y})
\tag{10.12}
$$

for $\mathbf{u}_0$. Recalling that $E = X^T X$, $F = N^T X$, and $M = N^T N$, one can rewrite (10.12) as

$$(N^T(I - XE^{-1}X^T)N + \alpha AG^{-1}A)\mathbf{u}_0 = n^{-1}(N^T\mathbf{y} - FE^{-1}X^T\mathbf{y}) \qquad (10.13)$$

where the matrix is symmetric positive definite because the first part is symmetric nonnegative and the second part is symmetric positive definite. This equation is very similar to the equation in the previous section in that it is a positive definite system for most coefficients of the function f. The system can thus be solved with the CG algorithm. Note that for the iteration one does not need to assemble the matrix of the system but only requires to perform the matrix–vector product

$$\mathbf{u} \mapsto (M - FE^{-1}F^T + \alpha AG^{-1}A)\mathbf{u}$$

efficiently. For example, the second term is $FE^{-1}F^T\mathbf{u}$ and this is done as three consecutive multi-plications. The third term is $AG^{-1}A\mathbf{u}$ and here one requires a first multiplication with A then the multiplication with G^{-1}, which is done as a solution of a linear system of equations and finally another multiplication with A. As the matrix $G = B_1 A^{-1} B_1^T + B_2 A^{-1} B_2^T$ is positive definite one may use again the method to find the solution of this system. In this inner iteration one needs to apply A^{-1} which is done using a fast Laplace solver or again a CG method which gives three nested iterations. One problem with nested iterations is that the number of iterations appear as factors in the time complexity of the solver.

The solution step was parallelized by partitioning the domain into equal strips and solving the systems on the strips independently. The strips were chosen to overlap and due to the locality of the thin plate spline fit if one observes that choosing a large enough overlap did not require any further outer iterations like they are often necessary for other PDEs partial differential equations. Thus the parallel redundancy of this method was limited to the redundant computations introduced through the overlapping subdomains and good speedup was obtained. Care has to be taken when partitioning the domain as the first and the last partition only have overlap on one side. This has the potential to lead to load imbalance if one chooses the domains to be "the same with some overlap on both sides." Thin plate splines with large numbers of data points have many applica-tions. The ones considered in the study were elevation data and aeromagnetic data. The imple-mentation was tested on a SUN Enterprise 4000 with 10 Ultra-Sparc processors, 475 GBytes of main memory, and 256 GBytes of disk space. An overall parallel efficiency of around 80% was reported.

10.4   MULTIVARIATE ADAPTIVE REGRESSION SPLINES

10.4.1   THE MARS ALGORITHM

Multivariate adaptive regression splines (MARS) [9, 49] are piecewise multilinear functions $f(x)$ for $x \in \mathbb{R}^d$. The space of multilinear functions using a rectangular grid with m grid points per dimension is computationally not feasible for large d as the grid has m^d grid points and so the space has m^d degrees of freedom. This is an instance of the curse of dimensionality [50]. The MARS procedure addresses this challenge by using only selected subspaces of the full space. In particular, it generates a sequence of subspaces

$$V_0 \to V_1 \to V_2 \to \cdots$$

with $V_k \subset V_{k+1}$ and where V_0 is the space of constant functions. The ingenuity of the method is in the way how these spaces are selected and how the elements are represented. The construction proceeds in a greedy fashion, i.e., given V_k a set of candidate spaces for V_{k+1} are considered and out of those the space which minimizes the residual is selected. More concisely, given the functional

$$J(V) = \min_{f \in V} \sum_{i=1}^{N} (f(x^{(i)}) - y^{(i)})^2$$

one selects the V_{k+1} such that

$$J(V_{k+1}) \le J(V)$$

for all candidate spaces V. The approximating function $f^k \in V_k$ is then

$$f^k = \operatorname*{argmin}_{f \in V_k} \sum_{i=1}^{N} (f(x^{(i)}) - y^{(i)})^2.$$

In particular it follows that the value of the constant function f^0 is the average of $y^{(i)}$.

To understand how the candidate spaces are constructed we will first consider the one-dimensional case ($d = 1$). In this case V_j is a space of piecewise linear functions with j knots and the candidates for V_{j+1} are the spaces of functions that contain all the knots of V_j (as $V_j \subset V_{j+1}$) but in addition, they contain one new knot that remains to be chosen. As possible knot location any data point $x_1^{(i)}$ (which has not been chosen before) is eligible. Thus one gets $N - j$ candidates for V_{j+1}. To find the best candidate one needs to solve a least squares problem. As the complexity of the solution of one least squares problem is $O(N)$ it now appears that the complexity of this whole procedure is $O(N(N - j))$ and the method is thus not scalable. However, the least squares problems for different knots are not independent and we will see how one may choose the generating functions such that the algorithm is $O(N)$.

Interestingly, in the MARS procedure the elements in the set V_j are not represented by (linear independent) basis functions but by more general generating functions (which in the literature are referred to as basis functions). These generating functions are introduced in pairs; in particular, V_{j+1} contains the following new generating functions, which are not in V_j:

$$(x_1 - x_1^{(t)})_+ \quad \text{and} \quad (x_1^{(t)} - x_1)_+$$

for some data point $x_1^{(t)}$ where for any $x \in \mathbb{R}$:

$$(x)_+ = \begin{cases} x & \text{if } x > 0 \\ 0 & \text{else.} \end{cases}$$

It follows that the space V_j requires $2j + 1$ generating functions. However, from the definition of $(\cdot)_+$ it follows that

$$(x_1 - x_1^{(k)})_+ - (x_1^{(k)} - x_1)_+ = x_1 - x^{(k)}.$$

And so only in the construction of V_1 are the degrees of freedom increased by two, for later V_j the degrees of freedom are only increased by one. Thus V_0 has one degree of freedom and all the other V_j have $j + 2$ degrees of freedom.

Now consider the vector of function values $f^j(\mathbf{x})$ on the data points. In terms of the generating functions this can be written as a matrix–vector product:

$$f^j(\mathbf{x}) = A_j \mathbf{u}_j$$

where $\mathbf{u}_j$ is the vector of coefficients. The matrix A_j is defined recursively with $A_0 = \mathbf{e}$ where $\mathbf{e} = (1, 1, \ldots, 1)^T \in \mathbb{R}^N$ and

$$A_{j+1} = \left[A_j, \; (\mathbf{x}_1 - x_1^{(t)})_+, \; (x_1^{(t)} - \mathbf{x}_1)_+ \right] \tag{10.14}$$

where $\mathbf{x} = (x_1^{(1)}, \ldots, x_1^{(N)})^T$. The $\mathbf{u}_j$ are determined by a least squares fit, and they are the solution of the normal equations

$$A_j^T A_j \mathbf{u}_j = A_j^T \mathbf{y}.$$

The new knot in Equation (10.14) is chosen from the set of unused knots such that the sum of squared residuals $\|A_j \mathbf{c}_j - \mathbf{y}\|^2$ is minimized. The way this is done is by considering all admissible knots according to their order. It turns out that the assembly of the normal equations only has to visit the data once for each j and after that can be updated without having to read the full data set. Assume that some A_{j+1} has been found with some knot $x_1^{(t)}$ and consider the knot $x_1^{(s)}$ with $s > t$. The normal equations are then updated, the new system solved, and the knot is accepted if it leads to a reduction of the residual. The matrix with the new knot is

$$A'_{j+1} = \left[A_j, \; (\mathbf{x}_1 - x_1^{(t+1)})_+, \; (x_1^{(t+1)} - \mathbf{x}_1)_+ \right]$$

and the difference between the two matrices becomes

$$C = A'_{j+1} - A_{j+1} = \Delta \, [0, \; \mathbf{f}_t, \; (\mathbf{f}_t - \mathbf{e})]$$

where $\Delta = x_1^{(t+1)} - x_1^{(t)}$ and $\mathbf{f}_t = \delta(x_1^{(t+1)} > \mathbf{x})$ consists of the vector with ones where the corresponding component of $\mathbf{x} < x_1^{(t+1)}$ and zero else. Note the special structure of C that contains elements zero and Δ only.

The normal matrix for the new knot is now

$$M'_{j+1} = \left(A'_{j+1} \right)^T A'_{j+1} = A_{j+1}^T A_{j+1} + C^T A_{j+1} + A_{j+1}^T C + C^T C$$

and the right-hand side is

$$b' = A_{j+1}^T \mathbf{y} + C^T \mathbf{y}.$$

We now need to show that we can determine the new matrix and right-hand side with having to read the full data set. We will need a vector $\mathbf{z}$ with components $z^{(t)} = \mathbf{f}_t^T \mathbf{y}$. This is just the cumulative sum of the components of $\mathbf{y}$ ordered according to the data points $x_1^{(j)}$. The determination of the vector $\mathbf{z}$ thus can be done at the outset of the algorithm and requires $O(n)$ operations. We do the same for the data vector $\mathbf{x}_1$ and, after A_{j+1} has been determined for the new columns $(\mathbf{x}_1 - x_1^{(t)})_+$ and $(x_1^{(t)} - \mathbf{x}_1)_+$.

For the right-hand side one needs to update b by $C^T\mathbf{y} = \varDelta[0, \mathbf{f}_t^T\mathbf{y}, \mathbf{f}_T^T\mathbf{y} - \mathbf{e}^T\mathbf{y}]^T$. As the terms $\mathbf{f}_t^T\mathbf{y}$ have been precomputed this does not require any further access of the data. For the matrix one needs to update M_{j+1} by $A_{j+1}^TA_{j+1} + C^TA_{j+1} + A_{j+1}^TC + C^TC$. For this one needs to determine

$$C^TC = \varDelta^2 \begin{pmatrix} 0 & 0 & 0 \\ 0 & t & 0 \\ 0 & 0 & N-t \end{pmatrix}$$

as $f_t^Tf_t = t$ and one gets

$$C^TA_{j+1} = \varDelta \begin{pmatrix} 0 & 0 & 0 \\ f_t^TA_j & \mathbf{e}^T(\mathbf{x}_1 - x_1^{(t)})_+ & 0 \\ (\mathbf{e} - f_t)^TA_j & 0 & -\mathbf{e}^T(x_1^{(t)} - \mathbf{x}_1)_+ \end{pmatrix}.$$

The components in the first block column have been determined earlier as they involve exclusively columns of A_j. The other components like $\mathbf{e}^T(\mathbf{x}_1 - x_1^{(t)})_+$ have actually been computed earlier and appear as components of M_{j+1}. As a consequence one does not need to scan the data again and can simply add up the updates to the matrix and solve the normal equations.

In the d dimensional case again two new generating functions are added at each step. Let $\{B_0, B_1, B_2, \ldots, B_{2j}\}$ be the generators of the space V_j. Then the new generators for V_{j+1} have the form

$$B_{2j+1}(x) = B_k(x)(x_s - x_s^{(t)})_+ \quad \text{and} \quad B_{2j+2}(x) = B_k(x)(x_s^{(t)} - x_s)_+.$$

The indices k, s, and t are chosen such that $B_k(x)$ is independent of x_s. They are chosen such that V_{j+1} gives minimal least sum of squared residuals. The generator B_k is called a parent of the new generator functions and the spaces V_j can be represented by trees where the nodes are the generator functions and the edges are labeled with the factors $(x_s - x_s^{(t)})_+$ and $(x_s^{(t)} - x_s)_+$. The independence condition makes that any generator can have at most d factors and the resulting functions are piecewise multilinear, i.e., piecewise linear in each variable. The length of any path starting from the root B_0 to any leave is less than d, thus d is the maximal depth of the tree. For $d = 1$ any node $B_j \neq B_0$ is a leave. Any node other than the ones at depth d can potentially have N offsprings. A limit to this is given by the fact that one can generate the same basis functions in different sequences. In Figure 10.1 the possible growth of the tree of generating functions associated with spaces $V_0, \ldots, V_5$ for the case of $d \leq 3$ is given. Typically the depth of the tree is limited, frequently used bounds include 2 and 5. In the case where only trees of depth 1 are admitted one gets additive models. In a final step, after a model has been generated that fits the data well, pruning of any unneeded and overfitting generating functions is done using crossvalidation. This pruning also removes redundant generating functions.

The MARS procedure was developed by Friedman [49] (see also Ref. [9] for a more recent discussion). A variant is the Polymars [51] procedure that limits the depth of the tree to two. The algorithm determines the spaces V_j in sequential and due to the dependence of V_{j+1} on the earlier

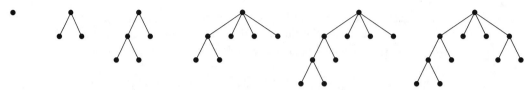

FIGURE 10.1 Possible trees for the MARS spaces $V_0, \ldots, V_5$.

spaces a detection of more than two generating functions in parallel would require a fundamental algorithm redesign. At each stage many different possible candidates for V_{j+1} are investigated. These investigations could be done in parallel. There is, however, a dependency as to overcome $O(N^2)$ complexity some of the matrices are generated sequentially. However, this is a small part and the solutions of the systems could still be done in parallel. As yet a lower granularity of the algorithm requires the determination of a large number of sums and scalar products, which can have $O(N)$ complexity and are readily parallelized. The search, while greedy, is very comprehensive and considers a large number of candidate basis functions. In a way the algorithm is the ultimate in adaptivity as every data point is considered for the generation of the generating functions. This may well contribute to the success of MARS; however, as a consequence, the complexity of the algorithm is, even with the clever choice of generating functions, typically much higher than the complexity of the previously discussed finite element scheme (and for higher dimensions the sparse grid scheme discussed next) where the data processing and the choice of the function space is separated.

10.4.2 MARS WITH B-SPLINES

Although the approach with generating functions is very effective and has proved very successful in data mining applications, there is the question if one cannot replace the generating functions by more standard hat basis functions or B-splines. The advantage of the hat functions is based on their locality. To model local effects with the truncated linear functions one needs a linear combination of three generator functions. This may lead to ill-conditioned linear systems of equations and also to instabilities if the coefficients of the three components are not determined exactly so that the components cancel out for faraway points.

Again, consider the case of $d = 1$ first. Hat functions are a special case of B-splines [52] and are defined on a grid. They are linear between the grid points and zero on all grid points except for one. For example, if the data points are ordered such that $x_1^{(i+1)} > x_1^{(i)}$, then the hat functions defined by the data points are

$$
h_i(x_1) = \begin{cases} \dfrac{x - x_1^{(i-1)}}{x^{(i)} - x_1^{(i-1)}} & x \in [x_1^{(i-1)}, x^{(i)}] \\[2ex] \dfrac{x_1^{(i+1)} - x_1}{x^{(i+1)} - x_1^{(i)}} & x \in [x_1^{(i)}, x^{(i+1)}] \\[2ex] 0 & \text{else.} \end{cases}
$$

(with slight modifications at the boundaries.) With these hat functions one gets a piecewise linear interpolant $s(x_1) = \sum_{i=1}^{N} y(i) h_i(x)$. The direct application of the h_i in adaptive regression, whereas computationally possible is inefficient as the functions are "too local." In particular they do not have any variance reducing capabilities and the omission of any of the basis results usually in a large local error. This is why one introduces hat functions $h_{i,j}$ with a slightly larger support.

They are defined by

$$
h_i(x_1) = \begin{cases}
\dfrac{x - x_1^{(i-k)}}{x^{(i)} - x_1^{(i-k)}} & x \in [x_1^{(i-k)}, x^{(i)}] \\[3ex]
\dfrac{x_1^{(i+k)} - x_1}{x^{(i+k)} - x_1^{(i)}} & x \in [x_1^{(i)}, x^{(i+k)}] \\[3ex]
0 & \text{else.}
\end{cases}
$$

for some $k \geq 1$. (Again, one has to modify this at the boundary.) With k large enough one gets similar benefits as for the truncated linear functions and one could now develop an algorithm that adaptively introduces a basis function $h_{i,k}$ to get a new space V_{j+1}. However, the search space has now suddenly increased dramatically as one now has to search over two parameters i and k and such an algorithm would certainly not be scalable any more as k can take $O(N)$ values.

This challenge is addressed by only considering one value of k at a time, starting with relatively large k and successively reducing the k as soon as a larger scale cannot provide any fitting improvement. For these functions one needs to again address the computational issue of nonscalability, because $O(N)$ candidates for new knots are considered. This, however, turns out to be more difficult than in the original case except for the case of equidistant data points. This is a consequence of the fact that in general the difference $h_{j+1,k} - h_{j,k}$ is not piecewise constant. The locality (for small k) may provide some saving as in the update of the linear systems of equations only $2k + 1$ data points would be involved.

More specifically, one now constructs the spaces V_j by setting V_0 to the space of constant functions and $V_{j+1} = V_j + \langle h_{t,k} \rangle$ for some t and where $\langle h \rangle$ denotes the space of all multiples of h and where for two spaces V and W we denote by $V + W$ the linear span of these two spaces. Of course only $h_{t,k}$ are admissible which generate a space $V_{j+1} \neq V_j$. Similar to the original MARS algorithm one gets a sequence of matrices A_j with $A_0 = \mathbf{e}$ and

$$
A_{j+1} = \begin{bmatrix} A_j & h_{s,k}(\mathbf{x}) \end{bmatrix}.
$$

Thus the updated normal matrix is

$$
M_{j+1} = A_{j+1}^{\mathrm{T}} A_{j+1} = \begin{bmatrix} M_j & A_j^{\mathrm{T}} h_{s,k}(\mathbf{x}) \\ h_{s,k}(\mathbf{x})^{\mathrm{T}} A_j & h_{s,k}(\mathbf{x})^{\mathrm{T}} h_{s,k}(\mathbf{x}) \end{bmatrix}.
$$

An efficient implementation of this requires $O(k)$ operations. Now in the case of large k, say, if $k = N/10$ this would lead to a nonscalable algorithm. In this case one may choose not to search over all possible data points but only over a subset. A reasonable compromise is to take the points $x_1^{(1)}, x_1^{(k+1)}, x_1^{(2k+1)}, \dots$. Thus one gets N/k grid points, e.g., in the case of $k = N/10$ one gets the 10% quantiles. With this choice the underlying function space in one dimension is now the space of piecewise linear functions defined on the grid where the grid points are quantiles. This again leads to a scalable algorithm.

The parameter k can be adaptively reduced. In the example that started with the 10% quantiles, one could try $k = N/100$ such that one obtains the 1% quantiles as the new 100 grid points of this finer grid. This step could be repeated and in the end one would get the same approximation space as for the original MARS just searched slightly differently. In principle the approximating space obtained using hat functions could be different from the one obtained by MARS, but this needs to be studied further.

In the case of higher dimensions one proceeds with the generation of basis functions like in the original MARS procedure except that only one new function (which is always a basis function) is

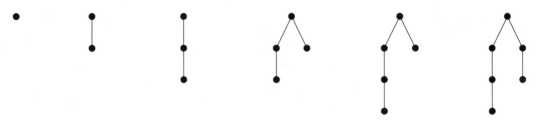

FIGURE 10.2 Possible trees for the BMARS spaces $V_0, \ldots, V_5$.

added to any V_j to get V_{j+1}. So in this case the dimension of the space V_j is $j + 1$. The new basis function is now of the form

$$B_{j+1}(x) = B_r(x)h^t_{sk+1,k}(x_t)$$

and admissible functions $h^t_{sk+1,k}(x_t)$ are the ones where B_r is independent of x_t, and where B_{j+1} is not an element of V_j. A possible tree relating to the first spaces $V_0, \ldots, V_5$ in this case is displayed in Figure 10.2. Note that one only has one new branch for every space V_j and one of the reasons for the simpler tree is that the usage of basis functions removes some of the redundancy present in the earlier MARS method.

An implementation of this method, which has been called BMARS, is discussed in Refs. [47,53, 54], see also Ref. [55], where this method was first suggested, for a more comprehensive discussion. The parallelism in this algorithm is very similar to the previous case. Instead of solving the normal equations this method uses a Gram–Schmidt method to solve the least squares problem. The scalar products required for Gram–Schmidt have been parallelized using a data partitioning approach. The parallelism used is fine-grain. The parallel implementation of BMARS does message passing with PVM and has been tested on shared memory systems (Sun Enterprise with 10 processors) as well as on distributed memory systems (Beowulf). On the shared memory system a speedup very close to the optimum was obtained for a data set with 10^6 data records from motor insurance [47] and taxation data [55]. Again no effect of the overhead for using PVM in the shared memory algorithm was observed. The benefit of using message passing, in this case PVM, is that the same code can be used for both the distributed and shared memory implementation.

10.5 SPARSE GRIDS

10.5.1 SPARSE GRID SPACES AND PENALIZED LEAST SQUARES

Like MARS, sparse grids define a space of multilinear functions. Compared to the adaptively chosen grid points of the MARS algorithms the sparse grid points are more regular. As will be seen, a consequence of this regularity is that methods for lower dimensional rectangular grids can be utilized. These solvers are combined to generate high-dimensional approximations. As this combination technique has a natural parallel implementation it will be discussed in a subsequent subsection in more detail. Sparse grid functions have been used successfully for the solution of PDEs with up to around 10 variables, and for quadrature [56]. Like MARS, sparse grids generalize the concepts of additive models by adding interactions terms of various complexities.

Ultimately, sparse grids are constructed as superpositions of regular grids. In the case of $d = 1$ a sparse grid is equal to regular grid with grid points $\xi_j = x_1^{\min} + j2^{-\alpha+1}(x_1^{\max} - x_1^{\min})$ for $j = 0, \ldots, 2^{\alpha-1}$ for some level $\alpha > 0$. In the case of $\alpha = 0$ there is only one grid point $x_1^{\min}$. In more general cases, functions for $d = 1$ are linear between the grid points and hat functions are the canonical basis functions in this case. (Unlike MARS which uses truncated powers.) We will denote the space of piecewise linear functions of level α by V_α. Note that $V_\alpha \subset V_{\alpha+1}$.

A regular (rectangular) grid is defined as the Kronecker product of one-dimensional regular grids and thus the corresponding function spaces are tensor products of spaces of functions of one variable. We denote these (rectangular) spaces by

$$V_{\boldsymbol{\alpha}} = V_{\alpha_1} \times \cdots \times V_{\alpha_d}.$$

(with the index or level vector $\boldsymbol{\alpha}$ denoted by bold type font.) Recall that $\alpha_i = 0$ means that the functions are constant with respect to variable x_i. The basis functions are products of hat functions of the type $h^1_{j_1}(x_1) \cdots h^d_{j_d}(x_d)$. If $\boldsymbol{\alpha} \leq \boldsymbol{\beta}$ in the sense of the partial (componentwise) order then $V_{\boldsymbol{\alpha}} \subset V_{\boldsymbol{\beta}}$.

Any set A of level vectors defines a sparse grid space by

$$V = \sum_{\boldsymbol{\alpha} \in A} V_{\boldsymbol{\alpha}}.$$

An element f of this sparse grid function space can be represented as the sum of regular grid functions:

$$f = \sum_{\boldsymbol{\alpha} \in A} f_{\boldsymbol{\alpha}}$$

where $f_{\boldsymbol{\alpha}} \in V_{\boldsymbol{\alpha}}$.

Whereas for higher dimensions adaptive methods (similar to MARS) appear to be feasible [57] this point needs to be further investigated. Most research so far has concentrated on algorithms for the efficient determination of penalized least squares fits of sparse grid functions once the space is given. For the original sparse grid methods the component spaces $V_{\boldsymbol{\alpha}}$ all have the same overall level k, i.e., $|\boldsymbol{\alpha}| = \sum_{j=1}^d \alpha_i = k$ and furthermore the values of the spaces were restricted to $\alpha_i > 0$. A consequence of this last constraint is that the dimension of the smallest nontrivial sparse grid is 2^d. This limitation has been lifted in the generalization of sparse grids considered here.

The penalized least squares problem for sparse grids consists of the determination of a function $f \in \sum_{\boldsymbol{\alpha} \in A} V_{\boldsymbol{\alpha}}$ that minimizes

$$J(f) = \frac{1}{n} \sum_{i=1}^n \left(f(\mathbf{x}^{(i)}) - y^{(i)} \right)^2 + \beta \, \|Lf\|^2 \tag{10.15}$$

for some (differential) operator L. Whereas this problem can be recast as a problem of orthogonal projections, the projections onto the subspaces $V_{\boldsymbol{\alpha}}$ generally do not commute and the spaces $V_{\boldsymbol{\alpha}}$ are generally not pairwise orthogonal.

Like in the case of additive models, the matrix of the normal equations for the sparse grid space is mostly dense, except for sparse diagonal blocks, which are associated with the spaces $V_{\boldsymbol{\alpha}}$.

Let $A_{\boldsymbol{\alpha}}$ be the matrix of the function values of the basis functions of $V_{\boldsymbol{\alpha}}$ and let $\mathbf{u}_{\boldsymbol{\alpha}}$ be the vector of coefficients of $f_{\boldsymbol{\alpha}}$ with respect to the basis. Furthermore, let the penalty term be additive in the components such that $D_{\boldsymbol{\alpha}}$ is the matrix of the component of $\|Lf\|^2$ corresponding $V_{\boldsymbol{\alpha}}$ in this basis. Then one gets

$$J(f) = \frac{1}{n} \left\| \sum_{\boldsymbol{\alpha} \in A} A_{\boldsymbol{\alpha}} \mathbf{u}_{\boldsymbol{\alpha}} - \mathbf{y} \right\|^2 + \alpha \sum_{\boldsymbol{\alpha} \in A} \mathbf{u}_{\boldsymbol{\alpha}}^{\mathrm{T}} D_{\boldsymbol{\alpha}} \mathbf{u}_{\boldsymbol{\alpha}}.$$

Now let $S_{\boldsymbol{\alpha}} = A_{\boldsymbol{\alpha}}(A_{\boldsymbol{\alpha}}^{\mathrm{T}} A_{\boldsymbol{\alpha}} + \alpha D_{\boldsymbol{\alpha}})^{-1} A_{\boldsymbol{\alpha}}^{\mathrm{T}}$ be the "smoothing matrix" for space $V_{\boldsymbol{\alpha}}$ and $\mathbf{f}_{\boldsymbol{\alpha}} = A_{\boldsymbol{\alpha}} \mathbf{u}_{\boldsymbol{\alpha}}$ be the vector of function values of the component in $V_{\boldsymbol{\alpha}}$ then one has the fix point equations

$$\mathbf{f}_{\boldsymbol{\alpha}} + S_{\boldsymbol{\alpha}} \sum_{\beta \neq \alpha} \mathbf{f}_{\beta} = S_{\boldsymbol{\alpha}} \mathbf{y}. \tag{10.16}$$

This follows directly from the normal equations

$$A_\alpha^{\mathrm{T}} \sum_{\beta \in A} A_\beta \mathbf{u}_\beta + \alpha D_\alpha \mathbf{u}_\alpha = A_\alpha^{\mathrm{T}} \mathbf{y}$$

which, by premultiplication with $(A_\alpha^{\mathrm{T}} A_\alpha + \alpha D_\alpha)^{-1}$ gives

$$\mathbf{u}_\alpha + (A_\alpha^{\mathrm{T}} A_\alpha + \alpha D_\alpha)^{-1} A_\alpha^{\mathrm{T}} \sum_{\beta \neq \alpha} A_\beta \mathbf{u}_\beta = (A_\alpha^{\mathrm{T}} A_\alpha + \alpha D_\beta)^{-1} A_\alpha^{\mathrm{T}} \mathbf{y}.$$

Multiplication of this equation with A_α and inserting the definitions of f_α and S_α gives the desired result. As in the case of the additive models one could now use the backfitting algorithm for these equations. Here we will, however, discuss a different approach based on the combination method.

10.5.2 THE COMBINATION METHOD

The solution of a penalized least squares problem is the orthogonal projection of the smoothing spline (with respect to an appropriately chosen norm) into the sparse grid space [57, 58]. Let P_α denote the orthogonal projection into the subspace V_α. If it is known that these projections commute pairwise, i.e.,

$$P_\alpha P_\beta = P_\beta P_\alpha, \quad \alpha, \beta \in A,$$

and if the set of component spaces V_α is complete under intersection, i.e., if for every $\alpha, \alpha' \in A$ there is a $\beta \in A$ such that

$$V_\alpha \cap V_{\alpha'} = V_\beta$$

then there are integer coefficients $\gamma_\alpha \in \mathbb{Z}$ such that the projection into the sum of the spaces V_α is

$$P = \sum_{\alpha \in A} \gamma_\alpha P_\alpha,$$

see Refs. [57,59–61]. In this case one can determine the solution of the penalized least squares problem on the sparse grid space by solving the corresponding problems on the component spaces and then combining these solutions linearly. Moreover, all the projections into the component spaces can be determined in parallel. This approach, which is not dissimilar to the additive Schwarz methods, is called the *combination method*. Before discussing actual implementations of the combination method, we will consider conditions under which the combination method works.

The original combination method gives exact results for the interpolation problem, i.e., if P_α are the interpolants onto the regular grid defining the spaces V_α. Furthermore, the combination method is known to act like an extrapolation technique in the case of simple elliptic PDEs [59], and it is thought that this extrapolation behavior occurs more generally. However, the combination technique was also seen to provide good results for classification problems [62, 63] that do not satisfy the requirements of the extrapolation theory nor the sparse grid formula.

When two spaces V_α and V_β are orthogonal, the projection onto the (direct) sum of these spaces is the sum of the projections onto the individual spaces. This is a simple case of the combination formula. It turns out that in the general case a similar geometric characterization holds:

Lemma 10.1 *Two projection operators P_α and P_β commute if and only if the two spaces V_α and $V_\beta \cap (V_\alpha^\perp + V_\beta^\perp)$ are orthogonal.*

Proof. As any f which is orthogonal to V_α is also orthogonal to $V_\alpha \cap V_\beta$ (and similar for V_β) one has

$$(V_\alpha \cap V_\beta)^\perp \supset V_\alpha^\perp + V_\beta^\perp.$$

Let $f \in (V_\alpha \cap V_\beta)^\perp$, set $r_n := (P_\beta P_\alpha)^n f$ and recursively $f_0 := 0$ and $f_{n+1} = f_n + ((I - P_\alpha) + (I - P_\beta) P_\alpha) r_n$ then $f = f_n + r_n$. Furthermore, for the operator $P_\beta P_\alpha$ one has $(f, P_\beta P_\alpha f) = (P_\beta f, P_\alpha f) = \cos(\theta) \| P_\beta f \| \| P_\alpha f \|$ where $\theta \in [0, \pi]$ is the angle between the projections. It follows that the only eigenvalues of magnitude one correspond to the case where $f = P_\alpha f = P_\beta f$, i.e., $f \in V_\alpha \cap V_\beta$. However, as r_0 is orthogonal to this space it follows that all r_n are orthogonal as well and $P_\beta P_\alpha$ is a strictly decreasing operator on $(V_\alpha \cap V_\beta)^\perp$ so that $r_n \to 0$. By construction one has $f_n \in V_\alpha^\perp + V_\beta^\perp$. As this set is closed it follows that f is in $V_\alpha^\perp + V_\beta^\perp$ and so

$$(V_\alpha \cap V_\beta)^\perp = V_\alpha^\perp + V_\beta^\perp.$$

If P_α and P_β commute, the product $P_\beta P_\alpha$ is the orthogonal projection onto the intersection $V_\alpha \cap V_\beta$. Moreover, $P_\alpha P_\beta (I - P_\beta P_\alpha) = 0$, $V_\alpha = \text{range}(P_\alpha)$ and $V_\beta \cap (V_\alpha \cap V_\beta)^\perp = \text{range} P_\beta (I - P_\alpha P_\beta)$ and it follows that V_α is orthogonal to $V_\beta \cap (V_\alpha \cap V_\beta)^\perp$, which was shown to be the same as $V_\beta \cap \left(V_\alpha^\perp + V_\beta^\perp \right)$.

Conversely, if V_α and $V_\beta \cap (V_\alpha^\perp + V_\beta^\perp)$ are orthogonal, then one gets the decomposition:

$$V_\alpha + V_\beta = V_\alpha \cap (V_\alpha \cap V_\beta)^\perp \oplus V_\beta \cap (V_\alpha \cap V_\beta)^\perp \oplus (V_\alpha \cap V_\beta)$$

and the three components are pairwise orthogonal, $V_\alpha = V_\alpha \cap (V_\alpha \cap V_\beta)^\perp \oplus (V_\alpha \cap V_\beta)$ and similar for V_β. If the projections onto the three components of the above sum are denoted by P_1, P_2, and P_3 then $P_1 + P_2 + P_3$ is the projection onto the sum and $P_\alpha = P_1 + P_3$ and $P_\beta = P_2 + P_3$. It follows that $P_\alpha P_\beta = P_3 = P_\beta P_\alpha$ and one has established the commutativity. $\qquad \square$

Consider now the case of random variables (vectors) $\mathbf{X}$ where the scalar product is given by

$$(f, g) = E(f(\mathbf{X}) g(\mathbf{X}))$$

and where $E(\cdot)$ is the expectation. If $\mathbf{X} = (\mathbf{X}_1, \mathbf{X}_2)$ partitions such that the two components are independent and $f_1 \in V_1$ only depend on the first component, i.e., $f_1(\mathbf{X}) = g_1(\mathbf{X}_1)$ and $f_2 \in V_2$ only depends on the second, i.e., $f_2(\mathbf{X}) = g_2(\mathbf{X}_2)$ for some g_1 and g_2. Then one gets for $f_1 \in V_1$ and $f_2 \in V_2$:

$$(f_1, f_2) = E(f_1(\mathbf{X})) E(f_2(\mathbf{X}))$$

and it follows that $f_1 - E(f_1)$ and $f_2 - E(f_2)$ are orthogonal so that the orthogonal projections are additive and the orthogonal projection onto $V_1 + V_2$ is thus $Pf = P_1 f + P_2 f - E(f)$. This is again a very simple case of the combination formula.

More generally, let $\mathbf{X} = (\mathbf{X}_1, \mathbf{X}_2, \mathbf{X}_3)$ and V_i be the space of functions that only depends on $\mathbf{X}_i$ and $\mathbf{X}_3$ for $i = 1, 2$. Then $V_1 \cap V_2$ is the space of functions that only depends on $\mathbf{X}_3$. The scalar product can be decomposed as

$$(f_1, f_2) = E(E(f_1(\mathbf{X}) f_2(\mathbf{X}) | \mathbf{X}_3)).$$

Assume now that the $\mathbf{X}_1, \mathbf{X}_2$ are independent conditionally on $\mathbf{X}_3$. It follows that for $f_i \in V_i$ one has

$$E(f_1(\mathbf{X}) f_2(\mathbf{X}) | \mathbf{X}_3) = E(f_1(\mathbf{X}) | \mathbf{X}_3) E(f_2(\mathbf{X}) | \mathbf{X}_3).$$

A direct consequence is that g_1 and g_2 with $g_i(\mathbf{X}) = E(f_i(\mathbf{X}) | \mathbf{X}_3)$ are orthogonal and one gets for the projection P into $V_1 + V_2$:

$$Pf(\mathbf{X}) = P_1 f(\mathbf{X}) + P_2 f(\mathbf{X}) - E(f | \mathbf{X}_3),$$

which is another application of the combination formula. Note that $E(f|\mathbf{X}_3)$ is the orthogonal projection into $V_1 \cap V_2$ and that a consequence of the conditional independence is that the projections P_1 and P_2 commute.

The problem is that the commutation property of the projectors does not hold in general. In fact, the combination method can fail quite spectacularly in very simple cases. Consider, for example, the case of two spaces V_{α_1} and V_{α_2} with intersection 0 and with a very small (but nonzero) angle between them, such that

$$\max_{f \in V_{\alpha_1}, g \in V_{\alpha_2}} \frac{(f, g)}{\|f\| \|g\|} \approx 1.$$

Then the two projections $P_1 f$ and $P_2 f$ can be very close. If in particular Pf is also close to $P_1 f$ then we have

$$Pf \approx P_1 f \approx P_2 f.$$

The combination technique for this case, however, would yield $P_1 f + P_2 f$ and one thus gets

$$P_1 f + P_2 f \approx 2Pf$$

so that the error introduced by the combination method is maximal. Note, however, that in this case approximations like $P_1 f$ or $(P_1 f + P_2 f)/2$ give good results. This suggests that modifying the coefficients γ_i of the combination method may provide substantially better performance at little extra cost [58].

An effective way to determine adapted combination coefficients uses the values of the functions $f_j(\mathbf{x})$ as new variables and considers a linear model with $f_j(\mathbf{x})$ as predictor variables for the response $\mathbf{y}$. The coefficients γ_j are then solved by minimizing the sum of the squared residuals

$$\sum_{i=1}^{N} \left(\sum_{j=1}^{m} \gamma_j f_j(\mathbf{x}^{(i)}) - y^{(i)} \right)^2.$$

The method can be further improved by using the approximation as a component for an iterative method where at each step the residual of the previous iteration is used as a new response and fitted using a combination of projections into the component spaces. This method has been implemented and shown to give good performance in Ref. [58].

10.5.3 IMPLEMENTATION OF THE COMBINATION TECHNIQUE

The smoothing technique with sparse grids using the combination method has been implemented for a distributed memory system and the code is available from http://datamining.anu.edu.au. However, the code is still in beta stage and under development. The code is developed in the Python scripting language, uses some C code for the time-critical assembly stage, and uses a Python binding of MPI (PyPar, which can also be obtained from the same website).

Basically, the method has the following main components:

1. Assembly of the normal equations for all the subspaces V_{α_j}
2. Solution of the equations on the subspaces
3. Combination of the solutions on the subspaces

The approximation that is obtained in this way is enhanced by iterative refinement and the subspaces can be determined adaptively. So far, several important components for this method have been implemented and tested on shared and distributed memory processors (see Ref. [64] and the references therein). Here we review some of the issues covered and methods used.

The "outermost loop" in the software is for the subspace determination. This search procedure requires the solution of many candidate problems. The solutions are compared using crossvalidation. This search is naturally parallelized over the candidates. A greedy algorithm is used here to determine a good model in a stepwise fashion.

For each solution procedure one has at the outer level an iterative refinement accelerated with a CG method. This procedure is not distributed as it implements a sequential process.

At the next level, the sparse grid fitting is done. Here parallelism is available both over the individual component grids but also for the solution on one component grid. These two types of parallelism do have different memory access patterns and thus are suited to different levels in the memory hierarchy. Early results for parallel sparse grid solvers can be found in Refs. [61, 65]. Consider now the sparse grid solution step in a bit more detail.

Assume that there are m subgrids and p processors each with local memory and disks. Especially for high-dimensional data the number of subgrids can be large and typically larger than the number of p processors available. It is assumed that the data is distributed by records (rows). In a first parallel step all the processors assemble the components of the normal equations relating to their local data for all subgrids. Then the components need to be added to form the normal equations for the full data set for each model. A more scalable variant, at the cost of some synchronization overhead, is obtained when the models are distributed and as soon as a processor has assembled a part for a particular model this part is then communicated to the processor which will do the solution. This method does require an *a priori* assignment of models to processors and some later communication might be needed if tasks need to be redistributed. The matrices that are assembled in this way only need to be assembled in the first steps whereas the right-hand sides need to be reassembled during the process. Thus the heaviest part of the assembly is before the actual solution step and only some minor assembly is required during the iterative refinement steps.

In the actual solution step the component grids are distributed evenly to the processors. As approximate estimates of the computation costs are known an effective static load-balancing strategy is available [66]. The only communication required is for a short setup phase initially and gather all the results at the end. The solution for each component is done with a sparse iterative solver. This does require more frequent exchange of data between the parallel components. It has been implemented using threading on an SMP architecture. Parallelism of the sparse matrix–vector product is achieved by partitioning the matrix into blocks of columns. The amount of speedup obtained in this way is typically less than what can be achieved on higher level discussed above. However, this is still an important source of parallelism and a study [64] showed that a combination of the two types of parallelism may yield much improved speedups by smoothing out the effect of the phases of low processor usage, which occurs if only the between-grid parallelism is used.

10.6 CONCLUSION

One of the main challenges of predictive modeling relates to the curse of dimensionality. We have discussed four methods that effectively deal with this curse and for which parallel algorithms have been developed. These algorithms are based on ideas from parallel numerical linear algebra. We have considered scalability, granularity, and the implementation of the algorithms.

There are, however, many other methods that can deal with the curse of dimensionality and which have not been discussed above. Maybe the two most glaring omissions include tree-based methods and artificial neural nets. Whereas these methods have also been parallelized they rely less on parallel numerical linear algebra. Tree-based methods provide piecewise constant functions that are used for classification and regression [67, 68]. One motivation for the MARS algorithm was to provide a smooth version of the regression trees. The bagging idea we discussed for additive models is directly applicable to classification trees [33]. Trees also have a natural parallelism that is based on the partitioning of the domain but that only gradually emerges. Thus one has a similar situation as in the case of sparse linear systems with two levels of parallelism, where one level is emerging whereas the other is decaying. For a discussion of parallel classification trees see Ref. [69].

Another very popular method for predictive modeling uses artificial neural nets [70]. Again one can consider parallelism over the data and parallelism within the model. Various approaches to parallel implementation, including hardware implementations, have been discussed in Ref. [71].

I hope I have convinced the reader that it is worthwhile to revisit modern statistical techniques with a view towards computation and parallel implementation. The exploitation of parallelism in the algorithms and the development of parallel variants is the main motivation for this. However, for larger-scale problems, issues like stability become more important, and it is important that these issues are considered as well. The examples discussed here have been chosen to illustrate that high throughput and reliable computation can be achieved simultaneously.

ACKNOWLEDGMENTS

The developments reviewed here are the results of research done over the last 20 years including several research communities. Many of the specific parallel algorithms have been developed at the ANU in the *CRC for Advanced Computational Systems* and in the *Australian Partnership for Advanced Computing Expertise Program in Data Mining*. Contributors to this research included several students, postdoctoral researchers, and collaborators. In particular I would like to acknowledge contributions from Australia by I. Altas, S. Bakin, P. Christen, W. Clarke, K. Ding, G. Hooker, S. Laffan, J. Maindonald, I. McIntosh, O. Nielsen, M. Osborne, N. Potter, S. Roberts, H. Silcock, P. Strazdins, G. Williams, A. Welsh and from overseas P. Arbenz, J. Garcke, G. Golub, M. Gutknecht, W. Light, and Z. Shen. These researchers have all contributed to the research discussed and helped me form the ideas that I discussed in this paper. I would also like to thank the referees for their helpful comments, and, in particular, the editor Prof. E. Kontoghiorghes for his assistance and patience.

REFERENCES

[1] Maindonald, J.H. (in press). Statistical computing. In *The Encyclopedia of Life Support Systems*. EOLSS Publishers, Oxford, UK.

[2] Gallup, G. (1976). *The Sophisticated Poll Watcher's Guide*. Princeton Opinion Press, Princeton, NJ.

[3] Rosenbaum, P.R. (2002). *Observational Studies*. 2nd ed. Springer-Verlag, Heidelberg.

[4] Welsh, A.H. (1996). *Aspects of Statistical Inference*. Wiley Series in Probability and Statistics: Probability and Statistics. John Wiley & Sons, New York.

[5] Hill, C.M., Malone, L.C., and Trocine, L. (2004). Data mining and traditional regression. In Hamparsum Bozdogan, Ed., *Statistical Data Mining and Knowledge Discovery*. Chapman & Hall/CRC, London, pp. 233–249.

[6] Bay, S.B., Kibler, D.F., Pazzani, M.F., and Smyth, P. (2000). The UCI KDD archive of large data sets for data mining research and experimentation. *SIGKDD Explorations*, 2(2):81–85

[7] Duff, I.S., Grimes, R.G., and Lewis, J.G. (1992). Users' guide for the Harwell–Boeing Sparse Matrix Collection (Release I). Technical report RAL 92-086, Rutherford Appleton Labs, Chilton, Oxon, England.

[8] Ihaka, R. and Gentleman, R. (1996). R: A language for data analysis and graphics. *Journal of Computational and Graphical Statistics*, 5(3):299–314.

[9] Hastie, T., Tibshirani, R., and Friedman, J. (2001). *The Elements of Statistical Learning*. Springer Series in Statistics. Springer-Verlag, New York (Data mining, inference, and prediction).

[10] Hey, T. and Trefethen, A. (2003). The data deluge: an e-science perspective. In Berman, F., Fox, G.C., and Hey, A.J.G., Eds., *Grid Computing: Making the Global Infrastructure a Reality*, Wiley Series in Communications, Networking and Distributed Systems, Wiley, New York, chap. 36, pp. 809–824.

[11] Hegland, M. and Osborne, M.R. (1998). Wrap-around partitioning for block bidiagonal linear systems. *IMA Journal of Numerical Analysis*, 18(3):373–383.

[12] Gropp, W., Lusk, E., and Skjellum, A. (1994). *Using MPI: Portable Parallel Programming with the Message Passing Interface*. MIT Press, Cambridge, MA.

[13] Anderson, E., Bai, Z., Bischof, C., Demmel, J., Dongarra, J., Du Croz, J., Greenbaum, A., Hammarling, S., McKenney, A., Ostrouchov, S., and Sorensen, D. (1992). *LAPACK's User's Guide*. SIAM, Philadelphia, PA.

[14] Blackford, L.S., Choi, J., Cleary, A., D'Azevedo, E., Demmel, J., Dhillon, I., Dongarra, J., Hammarling, S., Henry, G., Petitet, A., Stanley, K., Walker, D., and Whaley, R.C. (1997). *ScaLAPACK Users' Guide*. SIAM, Philadelphia, PA.

[15] Ortega, J. (1988). *Introduction to Parallel and Vector Solution of Linear Systems*. Plenum Press, New York.

[16] Van de Velde, E.F. (1994). *Concurrent Scientific Computing*, Vol. 16 of Texts in Applied Mathematics. Springer-Verlag, New York.

[17] Whaley, R.C. and Dongarra, J. (1998). Automatically tuned linear algebra software (ATLAS). In *SC'98: High Performance Networking and Computing: Proceedings of the 1998 ACM/IEEE SC98 Conference: Orange County Convention Center, Orlando, FL, USA, November 7–13, 1998* (Best Paper Award for Systems).

[18] Strazdins, P. (2001). A comparison of look ahead and algorithmic blocking techniques for parallel matrix factorization. *International Journal of Parallel and Distributed Systems and Networks*, 4(1):26–35.

[19] Golub, G.H. and Van Loan, C.F. (1996). *Matrix Computations*, 3rd ed. Johns Hopkins Studies in the Mathematical Sciences. Johns Hopkins University Press, Baltimore, MD.

[20] Amestoy, P.R., Duff, I.S., L'Excellent, J.Y., and Koster, J. (2001). A fully asynchronous multifrontal solver using distributed dynamic scheduling. *SIAM Journal of Matrix Analysis and Application*, 23(1):15–41 (electronic).

[21] Arbenz, P. and Hegland, H. (1998). On the stable parallel solution of general narrow banded linear systems. In *High Performance Algorithms for Structured Matrix Problems*, Advanced Theory in Computing Mathematics, II. Nova Scientific Publishing, Commack, NY, pp. 47–73.

[22] Hegland, M., Osborne, M.R., and Sun, J. (2002). Parallel interior point schemes for solving multistage convex programming. *Annals of Operation Research*, 108:75–85 (Operations research and constraint programming in the Asia Pacific region).

[23] Duff, I.S., Erisman, A.M., and Reid, J.K. (1989). *Direct Methods for Sparse Matrices, 2nd ed., Monographs on Numerical Analysis*. Oxford University Press, New York.

[24] Härdle, W. (1991). *Smoothing Techniques*. Springer Series in Statistics. Springer-Verlag, New York (with implementation in S).

[25] Hastie, T. and Tibshirani, R. (1986). Generalized additive models. *Statistical Science*, 1(3):297–318 (with discussion).

[26] Wahba, G. (1990). *Spline Models for Observational Data*, vol. 59 of *CBMS-NSF Regional Conference Series in Applied Mathematics*. SIAM, Philadelphia, PA.

[27] Ortega, J.M. and Rheinboldt, W.C. (2000). *Iterative Solution of Nonlinear Equations in Several Variables*, vol. 30 of Classics in Applied Mathematics. SIAM, Philadelphia, PA (Reprint of the 1970 original).

[28] Hastie, T.J. and Tibshirani, R.J. (1990). *Generalized Additive Models*, vol. 43 of Monographs on Statistics and Applied Probability. Chapman & Hall, London.

[29] Saad, Y. (2001). Parallel iterative methods for sparse linear systems. In *Inherently Parallel Algorithms in Feasibility and Optimization and their Applications (Haifa, 2000)*, vol. 8 of Studies on Computational Mathematics, North-Holland, Amsterdam, pp. 423–440.

[30] Breiman, L. (1996). Bagging predictors. *Machine Learning*. 24(2):123–140.

[31] Friedman, J. and Hall, P. (2000). On bagging and nonlinear estimation, unpublished manuscript.

[32] Buja, A. and Stuetzle, W. (2000). The effect of bagging on variance, bias and mean squared error.

[33] Skillikorn, D.B. (2000). Parallel predictor generation. In Zaki, M.J. and Ho, C.T., Eds., *Large-Scale Parallel Data Mining*, Vol.1759 of Lecture Notes in Artificial Intelligence. Springer-Verlag, Heidelberg, pp. 190–196.

[34] Savasere, A., Omieski, E, and Navanthe, S. (1995). An efficient algorithm for mining association rules in large databases. In *Proceedings of 21st International Conference. Very Large Data Bases*. Morgan Kaufmann, San Francisco. pp. 432–444.

[35] Hegland, M., McIntosh, I., and Turlach, B.A. (1999). A parallel solver for generalised additive models. *Computational Statistics and Data Analysis*, 31(4):377–396.

[36] Dyn, N. (1989). *Interpolation and Approximation by Radial and Related Functions*, vol. 1. Academic Press, New York. pp. 211–234.

[37] Powell, M.J.D. (1992). The theory of radial basis function approximation in 1990. In Light, W., Ed., *Advances in Numerical Analysis*, Vol. II (Lancaster, 1990), Oxford University Press, New York, pp. 105–210.

[38] Buhmann, M.D. (1993). New developments in the theory of radial basis function interpolation. In *Multivariate Approximation: From CAGD to Wavelets (Santiago, 1992)*, Vol. 3 Series Approximate Decomposition., World Scientific Publishing, River Edge, NJ, pp. 35–75.

[39] Duchon, J. (1977). Splines minimizing rotation-invariant semi-norms in Sobolev spaces. In *Constructive Theory of Functions of Several Variables (Proceedings Conference of Mathematical Research Institute, Oberwolfach, 1976)*, Lecture Notes in Mathematics, Vol. 571. Springer, Berlin. pp. 85–100.

[40] Ciarlet, P.G. (2002). *The Finite Element Method for Elliptic Problems*, vol. 40 of *Classics in Applied Mathematics*. SIAM, Philadelphia, PA. Reprint of the 1978 original (North-Holland, Amsterdam; MR **58** #25001).

[41] Braess, D. (2001). *Finite elements, 2nd ed.* Cambridge University Press, Cambridge, MA. (Theory, fast solvers, and applications in solid mechanics, Translated from the 1992 German edition by Larry L. Schumaker).

[42] Trottenberg, U., Oosterlee, C.W., and Schüller, A.W. (2001). *Multigrid*. Academic Press, San Diego, CA (With contributions by A. Brandt, P. Oswald, and K. Stüben).

[43] Hegland, M., Roberts, S., and Altas, I. (1998). Finite element thin plate splines for data mining applications. In *Mathematical Methods for Curves and Surfaces, II (Lillehammer, 1997)*, Innovations of Applied Mathematics, Vanderbilt University Press, Nashville, TN, pp. 245–252.

[44] Hegland, M., Roberts, S., and Altas, I. (1998). Finite element thin plate splines for surface fitting. In *Computational Techniques and Applications: CTAC97 (Adelaide)*. World Scientific Publishing, River Edge, NJ, pp. 289–296.

[45] Christen, P., Altas, I., Hegland, M., Roberts, S., Burrage, K., and Sidje, R. (2000). Parallelization of a finite element surface fitting algorithm for data mining. In *Proceedings of the 1999 International Conference on Computational Techniques and Applications (Canberra)*, Vol. 42. pp. C385–C399.

[46] Roberts, S., Hegland, M., and Altas, I. (2003). Approximation of a thin plate spline smoother using continuous piecewise polynomial functions. *SIAM Journal on Numerical Analysis*, 41(1):208–234.

[47] Williams, G., Altas, I., Bakin, S., Christen, P., Hegland, M., Marquez, A., Milne, P., Nagappan, R., and Roberts, S. (2000). The integrated delivery of large-scale data mining: the ACSys data mining project. In Zaki, M.J. and Ho, C.T., Eds., *Large-Scale Parallel Data Mining*, Vol. 1759 of Lecture Notes in Artificial Intelligence. Springer-Verlag, Heidelberg, pp. 24–55.

[48] Hegland, M., Hooker, G., and Roberts, S. (2000). Finite element thin plate splines in density estimation. In *Proceedings of the 1999 International Conference on Computational Techniques and Applications (Canberra)*, Vol. 42, pp. C712–C734.

[49] Friedman, J.H. (1991). Multivariate adaptive regression splines. *The Annals of Statistics*, 19:1–141.

[50] Bellman, R. (1961). *Adaptive Control Processes: A Guided Tour*. Princeton University Press, Princeton, NJ.

[51] Stone, C.J., Hansen, M.H., Kooperberg, C., and Truong, Y.K. (1997). Polynomial splines and their tensor products in extended linear modeling. *The Annals Statistics*, 25(4):1371–1470 (With discussion and a rejoinder by the authors and Jianhua Z. Huang).

[52] De Boor, C. (2001). *A Practical Guide to Splines*, Vol. 27 of Applied Mathematical Sciences. Springer-Verlag, New York, revised edition.

[53] Bakin, S., Hegland, M., and Osborne, M. (1998). Can MARS be improved with *B*-splines? In *Computational Techniques and Applications: CTAC97 (Adelaide)*. World Scientific Publishing, River Edge, NJ, pp. 75–82.

[54] Bakin, S., Hegland, M., and Osborne, M.R. (2000). Parallel MARS algorithm based on *B*-splines. *Computing Statististics*, 15(4):463–484.

[55] Bakin, S. (1999). *Adaptive Regression and Model Selection in Data Mining Problems*. Ph.D. thesis, Australian National University, Canberra, Australia.

[56] Zenger, C. (1991). Sparse grids. In *Parallel Algorithms for Partial Differential Equations (Kiel, 1990)*, Vieweg, Braunschweig, pp. 241–251.

[57] Hegland, M. (2003). Adaptive sparse grids. In Burrage, K. and Sidje, R.B., Ed., *Proceedings of 10th Computational Techniques and Applications Conference CTAC-2001*, vol. 44, pp. C335–C353. [Online] http://anziamj.austms.org.au/V44/CTAC2001/Hegl (April 1, 2003).

[58] Hegland, M. (2003). Additive sparse grid fitting. In *Curve and Surface Fitting: Saint-Malo 2002*, Nashboro Press, Brentwood, pp. 209-219.

[59] Bungartz, H.J., Griebel, M., Röschke, D., and Zenger, C. (1994). Two proofs of convergence for the combination technique for the efficient solution of sparse grid problems. In *Domain Decomposition Methods in Scientific and Engineering Computing (University Park, PA, 1993)*, American Mathematical Society, Providence, RI, pp. 15–20.

[60] Griebel, M., Schneider, M., and Zenger, C. (1992). A combination technique for the solution of sparse grid problems. In *Iterative Methods in Linear Algebra (Brussels, 1991)*. North-Holland, Amsterdam, pp. 263–281.

[61] Griebel, M. (1992). The combination technique for the sparse grid solution of PDEs on multiprocessor machines. *Parallel Processing Letters*, 2(1):61–70.

[62] Garcke, J., Griebel, M., and Thess, M. (2001). Data mining with sparse grids. *Computing*, 67:225–253.

[63] Garcke, J., Griebel, M., and Thess, M. (2001). Data mining with sparse grids. In *Proceedings of KDD2001*.

[64] Garcke, J., Hegland, M., Nielsen, O. (2003). Parallelisation of sparse grids for large scale data analysis. In *Proceedings of the ICCS, Lecture Notes in Computer Science*. Springer-Verlag, Heidelberg (to appear).

[65] Garcke J. and Griebel, M. (2001). On the parallelization of the sparse grid approach for data mining. In Margenov, S., Wasniewski, J., and Yalamov, P. Ed., *Large-Scale Scientific Computations, Third International Conference*, Sozopol, Bulgaria, Vol. 2179 of Lecture Notes in Computer Science, pp. 22–32.

[66] Griebel, M., Huber W., Störtkuhl, T., and Zenger, C. (1993). On the parallel solution of 3d pdes on a network of workstations and on vector computers. In Bode, A. and Dal Cin, M. Ed., *Parallel Computer Architectures: Theory, Hardware, Software, Applications*, Vol. 732 of Lecture Notes in Computer Science. Springer-Verlag, Heidelburg, pp. 276–291.

[67] Quinlan, J.R. (1993). *C4.5: Programs for Machine Learning*. Morgan Kaufmann, San Francisco.

[68] Breiman, L., Friedman, J.F., Olshen, R.J., and Stone, C.J. (1984). *Classification and Regression Trees*. Wadsworth International Group, Blemont, California.

[69] Srivastava, A., Han, E.H., Kumar, V., and Singh, V. (1999). Parallel formulations of decision-tree classification algorithms. *Data Mining and Knowledge Discovery*, 3(3):237–261.

[70] Ripley, B.D. (1996). *Pattern Recognition and Neural Networks*. Cambridge University Press, Cambridge.

[71] Sundararajan, N. and Saratchandran, P., Ed. (1998). *Parallel Architectures for Artificial Neural Networks – Paradigms and Implementations*. IEEE Computer Society, Washington DC, USA.

11 Parallel Programs for Adaptive Designs

Quentin F. Stout and Janis Hardwick

CONTENTS

ABSTRACT

Parallel computing can provide important assistance in the design and analysis of adaptive sampling procedures, and some efficient parallel programs have been developed to allow one to analyze useful sample sizes. Response adaptive designs are an important class of learning algorithms for a stochastic environment and can be applied in numerous situations. As an illustrative example, the problem of optimally assigning patients to treatments in clinical trials is examined. Although response adaptive designs have significant ethical and cost advantages, they are rarely utilized because of the complexity of optimizing and analyzing them. Computational challenges include massive memory requirements, few calculations per memory access, and multiply-nested loops with dynamic indices. The effects of various parallelization options are analyzed, showing that, while standard approaches do not necessarily work well, efficient, highly scalable programs can be developed. This allows one to solve problems thousands of times more complex than those solved previously, which helps to make adaptive designs practical.

11.1   INTRODUCTION

In situations where data are collected over time, adaptive sampling methods often lead to more efficient results than do fixed sampling techniques. When sampling or "allocating" adaptively, sampling decisions are based on accruing data. In contrast, when using fixed sampling procedures, the sample sizes taken from different populations are specified in advance and are not subject to change. Using adaptive techniques can reduce cost and time, or improve the precision of the results for a given sample size. For example, in many financial situations one tries to optimize rewards by constantly adjusting decisions as information is collected. In a pharmaceutical setting one may initially sample several different compounds to estimate their efficacy, and then concentrate a second round of testing on those compounds deemed most promising. When buying a present one may look at many options and stop when one has found something that seems sufficiently nice. In this latter case, which is an optional stopping problem, we assume some cost for looking time and a reward for appropriateness of the gift.

Fully sequential adaptive procedures, in which one adjusts after each observation, are the most efficient. For example, OECD TG 425 [21] contains guidelines for determining acute toxicity of potentially hazardous compounds. Because experiments are conducted on rats, there has been a strong motivation to develop experimental designs that expose as few animals as possible. Whereas in the past test guidelines had recommended predetermined sampling rules calling for approximately 30 rats per compound, the standard has changed, and TG 425 recommends adaptive procedures requiring far fewer, typically 8 to 12. Unfortunately, fully sequential procedures are rarely used due to difficulties related to generating and implementing good procedures as well as to complications associated with analyzing the resulting data. For example, they assume immediate responses and the ability to switch rapidly between alternatives, and they can involve designs, which greatly reduce average sample size but increase the maximal sample size.

A long-term goal of this research program has been to increase access to adaptive designs by creating a collection of algorithms that optimize and analyze a variety of sequential procedures. In particular we have focused on developing serial and parallel algorithms that allow researchers greater flexibility to incorporate diverse statistical objectives and operational considerations. Some of these techniques and applications are detailed in Refs. [12–14, 23].

We differentiate between the design and the analysis of sampling procedures. The design phase might generate optimal or nearly optimal sampling procedures, whereas the analysis phase may be applied to an arbitrary sampling procedure and may involve a wide range of operational characteristics. The analysis itself may be either exact or approximate. In some situations, optimal procedures may not be used because they are complex or difficult to employ and explain. Still, they provide a basis of comparison to establish the efficiency of suboptimal designs. If one can show that the relative efficiency of a sampling procedure is high compared with the optimal one, then investigators may be justified in implementing a simpler and, typically, more intuitive suboptimal option. Because this collection of algorithms also allows for the optimization of strategies that are constrained to have desirable operational characteristics, the likelihood that investigators can incorporate such goals and still achieve statistical efficiency is increased.

Historically, it has been not only analytically, but also computationally, infeasible to attain exact solutions to most adaptive allocation problems. As an example, in Ref. [5] the authors argue that if, for a specific problem, the optimal sequential procedure were "practically obtainable, the interest in any other design criteria which have some justification although not optimal is reduced to pure curiosity." They immediately add, however, that obtaining optimal procedures is not practicable. Then, as an illustration of the "intrinsically complicated structure" of optimal procedures, the authors detail the first step of the optimal solution to a simple sequential design problem involving only three Bernoulli observations. During this same year, 1956, it was pointed out in Ref. [4] that problems of this nature could, in principle, be solved via dynamic programming. However, such solutions are still typically viewed to be infeasible. Thirty-five years later, in Ref. [32], when addressing a variation of the problem in Ref. [5], the author reiterates that "in theory the optimal strategies can always be found by dynamic programming but the computation required is prohibitive."

This situation motivated us to work on greatly extending the range of problems that could be analyzed and optimized computationally. While some of the gains can be attributed to the ever increasing power of computers, much is due to algorithms and implementation, as will be shown. To state that a problem can be "solved via dynamic programming" is as vague as saying that one need only "do the math." Careful implementations of complex dynamic programming variations, along with new algorithmic techniques such as path induction, have been necessary to achieve the results reviewed here. While the models for which optimal solutions can be computed are often, albeit arguably, deemed to be "too simplistic," it is nevertheless the case that the insight one garners from evaluating these models is likely to lead to better heuristics that apply as well to more complex scenarios.

The remainder of this chapter is organized as follows. In Section 11.2 there is a discussion of the basic types of parallel computers available and some useful computing paradigms for working with adaptive designs. In Section 11.3 we introduce the multiarm bandit model, which will be used to illustrate the steps used to obtain an efficient parallel algorithm. Section 11.4 gives a naive serial algorithm for the three-arm bandit, and then an improved version. Sections 11.5 and 11.6 show increasingly more efficient (and complex) parallel algorithms for the same problem, and Section 11.7 transfers this to a different type of parallel computer. Section 11.8 gives some illustrative results showing how the optimal three-arm bandit design is superior to simpler alternatives. Section 11.9 explains the parallelization of a related but more complex problem, namely two-arm bandits with delayed responses. Section 11.10 has some concluding remarks.

11.2   PARALLEL COMPUTING MODELS AND PARADIGMS

In this section we discuss parallel computing platforms and their basic characteristics, along with some computing paradigms that are used in conjunction with adaptive designs.

11.2.1   PARALLEL COMPUTING PLATFORMS

Most of our parallel algorithms are developed for *distributed memory*, or *message-passing*, computers, in which data is stored with processors and all communication and access to data is via the exchange of messages between processors. Conceptually these machines are similar to a standard network of computers, where all communication is via basic read and write operations. Distributed memory machines are the most common form of parallel computer, especially because of the widespread introduction of *clusters*. These relatively inexpensive systems consist of commodity processor boards interconnected by commodity communication systems, typically utilizing open source software.

In Section 11.7, the distributed memory algorithm is modified into a form written explicitly for a shared memory parallel computer. Such a computer has its memory organized so that any processor can directly access any data, without messages. In general, shared memory machines are preferable because they simplify the process of converting a serial program into a parallel one. Whereas small-shared memory machines are increasingly becoming commodities for departmental computing, larger machines require specialized, more expensive, components, and hence are far less common than distributed memory clusters.

By far the most common way to create a parallel program for distributed memory systems is to use message passing interface (MPI). MPI has an extensive collection of operations for exchanging messages and collecting information. Furthermore, because it is a well-developed standard available for most platforms, it helps one to develop programs that can be run on a variety of systems. MPI is available for shared memory machines as well, but for them there is an additional standard, OpenMP. This provides compiler directives for automatically parallelizing many loops, which simplifies the parallelization process.

Various timing results are presented throughout to illustrate the performance improvements achieved through various means. Note that the absolute values of the numbers are of little interest because processor performance rapidly improves over time. Relative performance, however, is a useful measure. Further, the basic techniques remain applicable no matter how fast the system is. The distributed memory results presented were obtained using MPI on an IBM SP2, where each processor was an 160 MHz POWER2 Super Chip (P2SC) processor with 1 GB of RAM and 1 GB additional virtual memory. The shared memory results were obtained on a 16-processor SGI Origin with 12 GB RAM, where each processor was a 250 MHz MIPS R10000. Throughout, all times are elapsed wall-clock time measured in seconds. Rerunning the same problem showed very little timing variation, so we merely report average time.

11.2.2 FARMING

There is a simple form of parallelization for adaptive designs that is widely used. It is sometimes called *farming*, and the resultant algorithms are often referred to as *embarrassingly parallel*. As an example, suppose we have an adaptive design and wish to determine certain of its operating characteristics. It may be infeasible to do this exactly, even with the aid of parallel computers, and hence it is done via simulation. One way to accomplish such simulations, especially if each is itself rather lengthy, is to have several different processors do their own collection of simulations, and then to combine them at the end. As long as one takes care to insure that the random number generators on the processors are independent, this method is quite simple and extremely efficient. There is no communication among the processors except at start-up and in the final collection of data. Thus, even a low-cost distributed memory system with slow communication channels imposes very little overhead and can achieve high efficiency.

A variation on this theme is to do parameter sweeps to tune performance. Many adaptive designs have a variety of adjustable components, such as start-up, stopping, and decision rules. For example, with staged sampling, one collects k observations deterministically before any adaptation is employed. However, the most suitable value of k may be unclear, and hence a suite of evaluations, using different values of k, may be required to determine the optimal one. Here too, the parallelization is trivial, because different processors can work on different values of k. Note that this approach is applicable whether the evaluations for a single k are exact or are obtained via simulation.

When farming is possible, it is almost always the most efficient form of parallelization. That is, one may be able to parallelize the evaluation of a single simulation or exact evaluation, but it is usually more efficient to run different ones in parallel and then combine results rather than run each in parallel. This is because the parallelization of a single task typically adds communication and other overhead, and thus while a single task will be completed most quickly if it is run in parallel, the total set of tasks will be completed quickest if the tasks are run serially.

In a few cases of, say, parameter sweeps, the optimal performance is obtained by a mixture of the embarrassingly parallel and standard parallelization. This occurs when a single task, such as exact evaluation for a specific value of k, runs most efficiently on a small number of processors, rather than on a single one. Examples of this, shown in subsequent sections, can easily occur if the memory requirements exceed the memory available on a single processor. In this case, it is best to find the number of processors that runs a single evaluation at the highest efficiency. If this value is, say p_e, and there are p total processors, then the total time to complete all evaluations is optimized by running p/p_e evaluations simultaneously.

11.2.3 EXACT OPTIMIZATION

At the opposite end of the spectrum, in terms of the programming effort required, lies the problem of determining optimal adaptive sampling designs. While there are a variety of techniques needed for different problem types, we concentrate here on the problem of optimizing an objective function. Suppose the objective function $\mathcal{O}$ is defined on the terminal states of the experiment, and the goal is to maximize the expected value of $\mathcal{O}$. We assume that the sampling options available, and responses obtained, are discrete.

During the experiment, suppose we are at some state σ and can sample from populations $P_1, \ldots, P_k$. For population P_i, suppose that at state σ there are $r(i)$ possible outcomes, $o_0^i, \ldots, o_{r(i)-1}^i$, and that these occur with probability $\pi_0^i(\sigma), \ldots, \pi_{r(i)-1}^i(\sigma)$, respectively. Let $\sigma + o_j^i$ denote the state where o_j^i has been observed by sampling P_i. Let $\mathcal{E}_{\text{opt}}(\sigma)$ denote the expected value of the objective function attained by starting at state σ and sampling optimally, and let $\mathcal{E}_{\text{opt}}^i(\sigma)$ denote the expected value of the objective function attained by starting at state σ, observing P_i, and then proceeding optimally.

The important recursive relationship, sometimes called the principle of optimality, is that

$$\mathcal{E}_{\text{opt}}^i(\sigma) = \sum_{j=0}^{r(i)-1} \pi_j^i(\sigma)\, \mathcal{E}_{\text{opt}}(\sigma + o_j^i). \tag{11.1}$$

Because the only actions available are either to stop with value $\mathcal{O}(\sigma)$ or sample one of the populations, we thus have

$$\mathcal{E}_{\text{opt}}(\sigma) = \max\left\{\mathcal{O}(\sigma), \max\{\mathcal{E}_{\text{opt}}^i(\sigma) : i = 1, \ldots, k\}\right\}$$

where the maximum is restricted to those options that are permissible at σ.

Note that not all adaptive designs are for problems with objective functions satisfying such recursive equations. For example, many mini–max objectives cannot be presented this way because they are not defined in terms of expectations with respect to a distribution on the populations, but rather a maximum or minimum over the populations. Thus they do not have the additivity property used above. As an example, in Section 11.8, we compute a criterion known as min $P(\text{CS})$, which is the minimum probability of correctly identifying the best arm at the end of the experiment. To do this, we use path induction [13], which is described in the next section.

When recurrences such as (11.1) do hold, then there is a very powerful technique, *dynamic programming*, for obtaining the optimal design. One starts at the terminal states, and then for each of their predecessor states, determines the population to sample that will optimize the expected value. The optimal action, and the resulting optimal expected value, are recorded for each of these states. Then the optimal actions at predecessors of these states are determined and so on until the initial state is reached.

One important limiting factor of dynamic programming is the need to determine the value and optimal action of every state that can be reached. As will be shown in Section 11.3, the state space can be exceeding large. This fact is one of the reasons for utilizing parallel computing, because the computational demands of dynamic programming can be more than feasible with a single processor.

Note that to be able to employ dynamic programming, not only does one need for the recursive equations to hold, but one also needs the transition probabilities, $\pi_j^i(\sigma)$, at each state σ. Thus, dynamic programming requires a Bayesian statistical framework in which the π_j^i are random variables whose distributions are updated as data are observed. Technically, this means that one begins with a joint *prior* distribution, Γ, on the π_j^i and proceeds to calculate a *posterior* distribution, which is simply the prior conditioned on the outcomes observed so far. In the calculations, one then uses the posterior mean $E^\Gamma(\pi_j^i \mid \sigma)$ as the value of $\pi_j^i(\sigma)$.

11.2.4 EXACT EVALUATION

Since exact analytical evaluations of design operating characteristics are rarely accessible for complex adaptive designs, they are typically obtained computationally. These characteristics can be estimated via simulation or they can be calculated exactly. Often a serial program can be used to generate simulations, or farming can be used to exploit multiple processors. On the other hand, exact evaluations are typically far more complex and may require more sophisticated parallel algorithms.

One well-known technique is *backward induction*, in which the calculations are performed just as in dynamic programming, moving from the end of the experiment towards the beginning. However, to determine the expected value of a state, one uses the possibly random choice the design would make at that state, rather than considering all choices and choosing the best. Thus, backward induction can be as computationally challenging as dynamic programming. In some cases, however, it may be considerably simpler, such as when it is known that most of the states can never be reached by the given design. Note that such evaluations can be carried out regardless of how the design was created. They may involve either an evaluation for a specific Bayesian distribution, or a collection of evaluations for robustness studies or a frequentist overview.

If many evaluations are needed, then it may be more efficient to use *path induction*. With path induction, there is a preliminary pass from the beginning of the experiment towards the end, and then repeated evaluations are performed on the terminal states. A detailed explanation of this approach appears in Ref. [13]. For the purpose of this work, however, the most important feature is that the calculations for the preliminary pass proceed in the opposite order of those for dynamic programming. Hence the same parallelization techniques can be applied to path induction as for dynamic programming, and similar efficiencies can be obtained. Calculations for the evaluations are typically expected values summed over terminal states, and hence these are similar to farming, in that each processor sums over its terminal states and then a global sum is computed. Thus it is quite easy for a parallel computer to perform each evaluation efficiently.

Because backward induction and path induction are so similar to dynamic programming, only the parallelizations for dynamic programming will be discussed in detail.

11.3 EXAMPLE: BANDIT MODELS

In clinical trials, there are multiple goals that must be considered when designing an experiment. One of these goals is to treat all patients as well as possible, but there are differing viewpoints as to the relevant patient population. For example, if you were a patient, you'd like to be given the treatment currently deemed the best. Physicians sometimes use this viewpoint as well. This is known as *myopic* allocation, because there is no attempt to allocate you with the hope of gaining information from your result to help treat patients in the future. A second viewpoint is that it is all the patients in the clinical trial that are important, in which case one tries to maximize the total number of successful outcomes by the end of the trial. A third viewpoint emphasizes "future" patients who will be treated as a result of the decision made at the trial's termination. Addressing the optimal treatment of this last group has long been considered the classical goal of clinical trials. Still, the need to optimize the well-being of the subjects in the trial itself, be they humans or animals, has drawn increasing attention. Ultimately, the best designs will balance the needs of trial subjects and future patients, although, unfortunately, there is no design that can optimize these goals simultaneously. One way to approach this problem is to attempt to find a design that is optimal from each viewpoint (as is tackled here), and then to develop methods that utilize heuristics from each optimal design. For discussion of the latter approach see Ref. [9].

In this section, we describe the design for allocating patients to treatment options such that, on average, the maximal number of positive outcomes is obtained for trial patients (see Section 11.8 for discussion of the immediate and future patient criteria). This objective can be modeled as a *bandit problem* [3]. Such models are important in stochastic optimization as well as in decision and learning theory. In a k-arm bandit problem one can sample from any of k independent arms (populations) at each stage of the experiment. (Here, "arm" = "treatment option.") Statistically speaking, bandit problems are usually presented within a Bayesian decision theoretic framework. Thus, associated with each arm is an initial or prior distribution on the unknown outcome or "reward" function. After sampling from an arm (e.g., allocating a patient to a treatment) one observes the outcome and updates the information to get the posterior for that arm. The goal is to *determine how best to utilize accruing information to optimize the total outcome for the experiment*.

For our example, the outcome functions are Bernoulli random variables such that, from state σ and using the notation of Section 11.2.3, $o_0 = 0$ represents a treatment failure that occurs with probability $1 - p_i(\sigma) = \pi_0^i(\sigma)$, and $o_1 = 1$ represents a success that occurs with probability $p_i(\sigma) = \pi_1^i(\sigma)$, $i = 1, \ldots, k$. Our goal is to maximize the number of successes in n observations. At each stage, $m = 0, \ldots, n - 1$, of an experiment of length n, an arm is selected and the response is observed. At stage m, let (s_i, f_i) represent the number of successes and failures from arm i. Then the state $(s_1, f_1, \ldots, s_k, f_k)$, is a vector of sufficient statistics.

Figure 11.1 illustrates a simple two-arm bandit design with $n = 4$. The rate of success of each arm has a uniform or "flat," uninformative, prior distribution. Using a nonadaptive design, one would expect to achieve two successes. With the optimal adaptive design one expects 2.27 successes. The advantages of adaptation become more pronounced the longer the trial is and the more arms there are. For example, with $n = 100$ and uniform priors on each arm, nonadaptive allocation will average 50 successes no matter how many arms there are. However, the optimal two-arm bandit will average 65 successes, and the three-arm bandit averages 72.

Optimal designs for the bandit problem can be obtained via dynamic programming, but the number of states, and hence the time and space required to evaluate them all, have the formidable growth rate of $\Theta\left(n^{2k}/(2k)!\right)$. We concentrate on the three-arm version, which has $\Theta(n^6)$ complexity. Note that the state space grows exponentially in the number of arms. This "curse of dimensionality" often makes exact solutions infeasible. As a result, approximations may be used and the quality of the solution reduced.

Even when parallel computing is employed, major difficulties include:

- Time and space grow rapidly with the input size, so intensive efforts are needed to obtain a useful increase in problem size.
- The time–space ratio is low, making RAM the limiting factor.

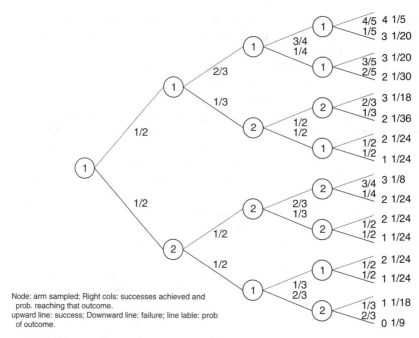

FIGURE 11.1 A two-arm bandit, with $n = 4$ and uniform priors on each arm.

- There are few calculations per memory access.
- The nested loops have dynamic index dependencies (see Algorithm 1).

Performance is further exacerbated by the interaction of these aspects. Table 11.10 illustrates, for example, the dramatic limitations imposed by space constraints and imperfect load balance caused by the loop structure.

11.3.1 PREVIOUS WORK

The three-arm problem had never previously been solved exactly because it was considered infeasible. As noted earlier, researchers have long indicated frustration with the far simpler *two-arm* bandit problem. In particular, in Ref. [17], the authors remark that "the space and time requirements for this computation grow at a rate proportional to n^4 making it impractical to compute the decision even for moderate values of say $n \geq 50$." Previously, the largest exact two-arm bandit solution utilized a IBM 3090 supercomputer with six processors to solve $n = 320$ [8]. Here, we solve a problem more than 300 times harder, namely the three-arm bandit with $n = 200$. Further, because to obtain operating characteristics we must evaluate this design many times, the total work is at least 10,000 times harder than that done earlier; since, had it been feasible, researchers would have used methods available at the time rather than the path induction exploited here. Whereas there has been scant previous work on the parallel solution of bandit problems, in the computer science community there has been more work on the parallel solution of similar recurrences. Most of this concentrates on theoretical algorithms where the number of processors scales far faster than the input size [15, 20, 25, 26, 30], or where special purpose systems are created [18, 27]. Others [19, 29] look at dynamic programming difficulties when the subproblems are not as well understood.

Algorithm 1 Serial Algorithm for Determining Optimal Adaptive Three-Arm Allocation

---

$\{\widehat{s_i}, \widehat{f_i}$: one success, failure on arm i }
{si, fi: number of successes, failures arm i }
{m: number of observations so far }
{n: total number of observations }
$\{|\sigma|$: number of observations at state σ }
{V: the value of the optimal design starting at state σ, i.e., $\mathcal{E}_{\text{opt}}(\sigma)$.
      V(**0**) is the value of the optimal design starting at the beginning.}
{pi(si,fi): prob of success on arm i, if si successes and fi failures have been observed }

for all states σ with $|\sigma|$=n **do** {i.e., for all terminal states}
   V(σ)=number of successes in σ

for m=n−1 downto 0 **do** {compute for all states of size m}
  for s3=0 to m **do**
    for f3=0 to m−s3 **do**
      for s2=0 to m−s3−f3 **do**
        for f2=0 to m−s3−f3−s2 **do**
          for s1=0 to m−s3−f3−s2−f2 **do**
            f1 = m−s3−f3−s2−f2−s1
            $\sigma = \langle$s1,f1,s2,f2,s3,f3$\rangle$
            V(σ) = max{(p1(s1,f1) · V($\sigma + \widehat{s_1}$) + (1−p1(s1,f1)) · V($\sigma + \widehat{f_1}$)) ,
                    (p2(s2,f2) · V($\sigma + \widehat{s_2}$) + (1−p2(s2,f2)) · V($\sigma + \widehat{f_2}$)) ,
                    (p3(s3,f3) · V($\sigma + \widehat{s_3}$) + (1−p3(s3,f3)) · V($\sigma + \widehat{f_3}$)) }

11.4 SERIAL IMPLEMENTATION

The goal of a bandit problem with dichotomous responses is to determine, at each state, which arm should be selected so as to maximize the expected number of successes over the course of the experiment. To solve this via standard dynamic programming (Algorithm 1), first the values of each terminal state (those with n observations) are computed. Then, the optimal solution is found for all states with m observations based on the optimal solutions for all states with $m + 1$ observations, for $m = n - 1$ down to 0.

The recurrence at the heart of this dynamic programming algorithm is in the center of the loops. At state $\sigma = (s_1, f_1, s_2, f_2, s_3, f_3)$, one may sample any of the three arms. If arm i is sampled, then the resulting state will be either $\sigma + \widehat{s_i}$ or $\sigma + \widehat{f_i}$, where $\widehat{s_i}$ and $\widehat{f_i}$ denote a single additional success or failure, respectively, on arm i. Given the prior distribution on the success rate for arm i, along with the observations in σ on arm i, one can then compute the probability of these two outcomes. Let the probability of observing a success be denoted by $p_i(s_i, f_i)$. In the previous stage of the dynamic programming, the expected values of these states, assuming one proceeds optimally to the end of the experiment, have been determined. Thus the expected value of sampling from arm i and proceeding optimally, i.e., $V^i(\sigma) = \mathcal{E}^i_{\text{opt}}(\sigma)$, is given by

$$V^i(\sigma) = p_i(s_i, f_i) \cdot V(\sigma + \widehat{s_i}) + [1 - p_i(s_i, f_i)] \cdot V(\sigma + \widehat{f_i})$$

Choosing the arm that yields the highest expected value is the optimal decision at σ.

For the purposes of efficient implementation and parallelization, the specific recurrence used to combine values is less important than the indices of the values being referenced, because they determine the memory accesses and communication required. We have a stencil of dependencies, whereby the value at state σ depends only on the neighbor values at $\sigma + \widehat{s_1}$, $\sigma + \widehat{f_1}$, $\sigma + \widehat{s_2}$, $\sigma + \widehat{f_2}$, $\sigma + \widehat{s_3}$, and $\sigma + \widehat{f_3}$, along with the priors and σ itself. With minor changes to this equation (and no change in the dependencies), the same program can also perform backward induction (see Section 11.2.4) to evaluate the expected number of successes for an arbitrary three-arm design. This allows one to evaluate suboptimal designs which may be desirable for reasons such as simplicity and intuitiveness. The same stencil of dependencies can also be used to optimize and evaluate designs for two Bernoulli arms with randomly censored observations [23].

Note that the recurrences involve extensive memory accesses, with little computation per access. There are $\binom{n+6}{6} = \Theta(n^6)$ states, and the time and space complexities are also $\Theta(n^6)$.

11.4.1 SPACE OPTIMIZATIONS

Given the vast space requirements needed to solve these problems, good algorithms must incorporate several space reduction techniques. The first of these results from the observation that values of V for a given m depend only on the values for $m + 1$, so only the states corresponding to these two stages need to be kept simultaneously. This reduces the working space to $\Theta(n^5)$, and by properly arranging the order of the calculations, the space can be further reduced to only that required for one stage's worth of states, i.e., we gain another factor of 2. This corresponds to the *collapsed* column in Table 11.1. In this table, max n shows the maximum problem solvable by a 1 GB RAM machine with a time limit of 18 h, *limitation* shows which limit was reached, and *prog len* is the size of the version in lines of source code. Note that the collapsed version allows us to solve problems substantially larger, and also results in a slight speedup.

TABLE 11.1
Serial Versions, Time (sec) to Solve Problem of Size *n*
(From Ref. [24])

n	First	Collapsed	Naive comp	Best
10	0.009	0.004	0.082	0.004
20	0.18	0.1	3.2	0.092
30		1.4	35	0.71
40		4.1	186	3.9
50		15	689	13
60			2,024	35
70			5,047	86
80			11,242	185
90			22,756	362
100			42,900	659
110				1,225
120				1,922
130				34,961
Max n	27	54	100	135
Limitation	memory	memory	time	time
Prog len	193	193	282	419

max n: Maximum problem solvable with 1 GB and time $\leq 64,400$ sec (18 h).

Another significant space reduction results from the fact that, due to the constraint

$$s_3 + f_3 + s_2 + f_2 + s_1 + f_1 \leq n,$$

only a corner of the five-dimensional V array is used (approximately $1/5! = 1/120$ of the total). To take advantage of this, the five-dimensional V array is mapped 1-1 onto a linear array V_ℓ. Unfortunately, this mapping also requires that all array references be translated from the original five indices into their position in the linear array. From a software engineering viewpoint, the best way to implement this translation is to use a function which takes as input the five indices and yields their position in the array, i.e., a mapping of the form

$$V(\mathtt{s1,s2,f2,s3,f3}) \mapsto V_\ell\bigl(T(\mathtt{s1,s2,f2,s3,f3})\bigr).$$

Unfortunately, this is extremely costly as the translation function T is a complicated fifth degree polynomial that must be evaluated for every array access. This version, the *naive comp* in Table 11.1, can solve larger problems but is significantly slower than the *collapsed* version. For the *best* version, we broke the translation function into a series of offset functions, where each offset function corresponds to a given nested loop level, i.e.,

$$\begin{aligned}
T(s_1, s_2, f_2, s_3, f_3) &= T_{s_3}(m) + T_{f_3}(m-s_3) + T_{s_2}(m-s_3-f_3) + T_{f_2}(m-s_3-f_3-s_2) \\
&\quad + T_{s_1}(m-s_3-f_3-s_2-f_2) + s_1.
\end{aligned}$$

An offset function only needs to be recalculated before its corresponding loop is entered, and the more expensive offset functions correspond to the outermost loops.

This method dramatically reduces the translation cost down to a usable level but greatly increases program complexity, as is shown by the increase in *prog len*.

The simplified Algorithm 1 ignores the fact that to utilize the design, one needs to record the arm selected at each state. Unfortunately, these values cannot be overwritten and the storage required is $\Theta(n^6)$. Fortunately, this too involves only values in one corner, allowing a reduction by a factor of $1/6! = 1/720$. These values are stored on disk and do not reduce the amount of memory available for calculation. Using run-length encoding or other compression techniques would likely reduce this to $\Theta(n^5)$, but so far this has not been necessary. Note that if one only needs the value of the optimal design, but not the design itself, then this storage is not needed. Such a situation arises, for example, when the optimal design is only used to gauge the performance of simpler designs.

11.5 INITIAL PARALLEL ALGORITHM

To parallelize the recurrence, we first address load balancing. In the initial parallelization the natural step of dividing the work among the processors was taken. The outermost m loop behaves very much like "time" in many simulations and cannot be parallelized, so instead one parallelizes the second outermost loop, s3. At stage m, processor $\mathcal{P}_j$ is assigned the task of computing all values where s3 is in the range start_s3(j,m) ... end_s3(j,m).

Because the number of states corresponding to a given value of s3 grows as $(\mathtt{m-s3})^4$, determining the range of s3 values assigned to each processor is nontrivial. Thus, simply assigning all processors an equal number of s3 values would result in massive load imbalance and poor scaling. We evaluated two solutions to this problem. Optimal s3 partitioning is itself a small dynamic programming problem that takes time and space $\Theta(mp)$. However, it was easy to develop a fast $\Theta(m)$ greedy heuristic that was nearly optimal, and it is this heuristic that was used in the initial program.

11.5.1 COMMUNICATION

The communication needed can be divided into *array redistribution* and *external neighbor acquisition*. Array redistribution occurs because, as the calculation proceeds, the number of states shrinks. To maintain load-balance, the s3 range owned by a processor changes over time. At stage m, processor P_j needs the states with s3 values in the range start_s3(j,m)...start_s3(j,m+1)-1 from P_{j-1}. Redistribution includes the cost of moving the states currently on the processor to create space for these new states.

External neighbor acquisition occurs because the calculations for a state may depend on its neighbors in other processors. To calculate states with s3=end_s3(j,m) during stage m, P_j needs to obtain a copy of the states with s3=end_s3(j,m)+1 from P_{j+1}. Note that external neighbor acquisition negates round-robin or self-scheduling approaches to load-balancing the s3 loops, as this would result in a dramatic increase in the communication requirements. This does not necessarily hold for shared memory systems, however, as can be seen from the OpenMP version in Section 11.7. Shared memory computers are able to utilize these approaches because their much faster communication systems reduce the latency to a tolerable level.

11.6 SCALABLE PARALLEL ALGORITHM

The initial load-balancing approach is simple to implement and debug because it makes minimal changes to the serial version. Unfortunately, it has imperfect load and working space balancing and this severely limits scalability (see Table 11.2) and solvable problem size (see Table 11.3).

For a more scalable version (Algorithm 2), instead of partitioning the states using the coarse granularity of the s3 values, we partition them as finely as possible, i.e., by individual states. The assigned states are specified by start_σ and end_σ. However, this leads to numerous difficulties. The first is that a processor's V array can now start or end at arbitrary values of s3, f3, s2, f2, s1, and f1, so one can no longer use a simple set of nested loops to iterate between the start and end value. Our first attempt to solve this problem had nested if-statements within the innermost loop, where the execution rarely went deep within the nest. While logically efficient, this turned out to be quite slow because it was too complex for the compiler to optimize. A solution that the compiler was able to cope with was to use a set of nested loops with if-statements in front of each loop so that it starts and stops appropriately. This solution was almost as fast as the original serial nested loops.

TABLE 11.2
Scaling Results, $n = 100$
(From Ref. [24])

p	Efficiency $e(p)$	
	Initial	Scalable
1	1.00	1.00
2	0.96	0.96
4	0.93	0.94
8	0.81	0.91
16	0.64	0.86
32	0.48	0.81

TABLE 11.3
Stepwise Improvements in Scalable Version,
$n = 100$, 1, and 8 Processors (From Ref. [24])

Version	$t(1)$	$t(8)$	$e(8)$
First scalable	1044	178	0.734
Improved loops	775	143	0.678
Offsets in array	766	134	0.715
Scalable comm	762	106	0.903
Nonblocking comm	760	104	0.913

Another difficulty was that the offset calculations are not uniformly distributed along the range of the V array, and this leads to a noticeable load imbalance. Storing the results of the offset equations in arrays significantly decreases the cost of each offset calculation and reduces the load imbalance to a more acceptable level. However, there is still some slight load imbalance that could be addressed by including the cost of these array lookups in the load balancing.

11.6.1 COMMUNICATION

The move to perfect division of the V array also caused complications in the communication portion of the program. The main complication was that data needed for either external or redistribution

---

Algorithm 2 Scalable Parallel Algorithm

{$\mathcal{P}_j$: processor j}
{start_σ(j,m), end_σ(j,m): range of σ values assigned to $\mathcal{P}_j$ for this m value,
  with start_σ(j+1,m)=end_σ(j,m)+1 }

{For all processors $\mathcal{P}_j$ simultaneously, do}

for σ=start_σ(j,n) to end_σ(j,n) **do** {i.e., for all terminal states}
  V(σ)=number of failures in σ

for m=n−1 downto 0 **do** {compute for all states of size m}
  for σ=start_σ(j,m) to end_σ(j,m) **do**
    determine s1, f1, s2, f2, s3, f3 from σ
    compute V as before

{Array redistribution}
Send needed V values to other processors
Receive V values from other processors

{External data acquisition}
Send needed V values to other processors
Receive V values from other processors

---

aspects was no longer necessarily located on adjacent processors. This resulted in a considerable increase in the complexity of the communication portions of the program.

Our initial version of the communication functions used a natural strategy when space is a concern: each processor sent the data it needed to send, shifted its remaining internal data, and then received the data sent to it. Blocking sends were used to insure that there was space to receive the messages. Unfortunately, this serialized the communication, because the only processor initially ready to receive was the one holding the end of the array, i.e., the only processor which does not redistribute to any other processor. The next processor able to receive was the second from the end, because it sent only to the end processor, and so on. The performance of this version was unacceptable. The next version removed the interaction and performed adequately, but synchronization costs became more of a problem. To remove these, we switched to nonblocking communication wherever possible. This made communication fairly efficient, although there may still be room for some slight additional improvement. Unfortunately, nonblocking communication requires additional buffers to accommodate incomplete sends and receives. In general there is a serious conflict between extensive user space requirements and minimizing communication delays. The communication buffers needed to overlap communication and calculation, and to accommodate nonblocking operations, can be large.

11.6.2 SCALABLE TIMING RESULTS

Table 11.2 shows the efficiency, $e(p)$, of the initial and scalable parallel versions as the number of processors p increases. Table 11.3 shows the effect on timing and scaling of each of the major changes detailed in Section 11.6, contrasting one processor and eight processor versions, where $t(p)$ is the time. Note that the improvements reduced the serial time and increased the parallel efficiency relative to the reduced serial time.

Table 11.4 contains the percentage of the total running time taken by different parts of the scalable program as the number of processors increases. *Calc* is the percentage of time taken by the dynamic programming calculations, *file* is the cost of writing the decisions to disk, and *misc* is the part of the time not attributed elsewhere. Under *array redist*, we show the cost of shifting data among the processors to maintain load-balance, where *comm* is the cost of calculating the redistribution and communicating the data between the processors, and *shift* is the cost of moving the data on the processor. Below *external comm* is the cost of getting neighbor states from other processors, including the cost of determining which processor has the data, where to put it on the current processor, and the cost of communicating the data.

TABLE 11.4
Percentage Distribution of Time within Scalable Version, $n = 100$
(From Ref. [24])

p	Calc.	File	Misc.	Array redist. Comm.	Shift	External Comm.
1	98	1.9	0.1	0.0	0.0	0.0
2	94	1.6	0.9	1.9	1.2	0.4
4	88	1.6	0.1	4.5	2.0	3.8
8	84	1.4	0.2	6.5	2.0	5.9
16	73	1.2	0.7	11.0	2.1	12.0
32	57	1.1	0.0	16.1	1.7	24.1

TABLE 11.5
Timing Results, $n = 200$,
Scalable Version.
(From Ref. [24])

p	$t(p)$
16	10,463
32	1,965

Table 11.5 presents the running times of the scalable program for $n = 200$ for 16 and 32 processors. Note that the speedup is more than a factor of two. This occurred because on 16 processors the program must make extensive use of disk-based virtual memory. A similar effect can be seen in Table 11.1 as n increases from 120 to 130. This illustrates an often overlooked advantage of parallel computers, a bonus increase in speed simply because dividing a problem among more processors allows it to run in RAM instead of in virtual memory. However, this can be successful only if the parallelization load-balances the memory and computation requirements.

11.7   SHARED MEMORY IMPLEMENTATIONS

To measure the performance of the three-arm bandit code on a shared memory machine we implemented four separate versions.

The first version, which we call MPI, uses the shared memory implementation of the MPI libraries. Aside from a few changes due to differences in the versions of Fortran on the two machines, this version is identical to the scalable version of the code previously described.

The next version, OpenMP, uses OpenMP directives to implement a shared memory version of the code. This version is very similar to that in Algorithm 1, except for the addition of a second copy of the V array. This second copy is necessary because, while using a shared memory implementation the same V array is shared among all the processors, which may be acting on different sections of it at arbitrary times. This means there is no longer a guarantee that every calculation that uses a state will have read the state's value before it is overwritten. Thus, we need to have a second array to hold the current stage's inputs whereas the current stage's outputs are being stored. After a stage is completed its output array is copied into the input array for the next stage. To convert the code, OpenMP parallel-do directives were used around the outermost loop, s3, of the dynamic programming setup, and the s3 loop in the dynamic programming. Both of these loops use OpenMP dynamic scheduling, where each processor grabs a user-defined chunk size number of iterations, performs them, and then, when completed, grabs another set. This process continues until all the iterations of the loop have been completed.

To compute the chunk size for each stage, we first determine the average amount of work per processor at that stage. The chunk size is then the maximum number of initial iterations whose combined work is not greater than the average work. Note that this will not be the number of iterations divided by the number of processors because the work per iteration varies dramatically. This will create many chunks with diminishing amounts of work which will be taken by underloaded processors as they complete their tasks, helping to insure approximately even load balance. Such a dynamic scheduling approach is not useful for distributed memory systems because of the increase in complexity that would result from tracking the location of the states and synchronizing access to them, and the cost of moving so much state information among processors.

TABLE 11.6
Efficiency of Shared Memory Implementations, $n = 100$ (From Ref. [24])

	MPI		OpenMP		Auto		Auto+Copy	
p	$t(p)$	$e(p)$	$t(p)$	$e(p)$	$t(p)$	$e(p)$	$t(p)$	$e(p)$
1	439	1.00	406	1.00	471	1.00	454	1.00
2	290	0.76	209	0.97	473	0.49	419	0.54
4	155	0.70	113	0.90	465	0.25	404	0.28
8	90	0.61	72	0.70	473	0.13	403	0.14
16	73	0.38	59	0.43	470	0.06	397	0.07

The third version of shared memory code, Auto, was generated by using the SGI Fortran auto-parallelizer on the serial version of our code. Unfortunately, due to the dependencies inside the V array described above, the auto-parallelizer was only able to parallelize the innermost, s1, loop of the dynamic programming setup.

The final version of shared memory code, Auto+Copy, again used the auto-parallelizer, but this time on the double V array code described above for OpenMP. The reduction in dependencies allowed it to do slightly better. It parallelized the innermost, s1, loops of both the setup and the main body of the dynamic programming.

Table 11.6 shows the results of our measurements on these four versions. As can be seen, the hand-parallelized versions perform far better than those done automatically. In fact, Auto has almost no discernible increase in speed as the number of processors increases. Auto+Copy does slightly better but is still far inferior to the others. The winner clearly is OpenMP, which was to be expected as it has far less overhead than MPI. Note, however, that OpenMP's scalability will degrade as the number of processors increases because it cannot allocate less than one s3 loop per processor. (Because we had only 16 nodes on our SGI Origin, we cannot provide numbers for more processors.) Implementing a fully scalable code using OpenMP would be difficult, and in the end would probably result in something quite similar to the MPI version.

11.8   ILLUSTRATIVE RESULTS FOR THREE-ARM BANDIT

To illustrate the use of the parallel algorithm in Algorithm 2, it was applied to the problem of comparing three sequential allocation procedures involving three arms. We continue with the example of designing a clinical trial to address the ethical obligation to optimize patient treatment. In Section 11.3, three interpretations of treating patients optimally were offered. We examine one design for each interpretation and then look to see which of these designs seems to address all three interpretations the best. Computationally, the intent is to show that the parallel program provides heretofore unattainable exact evaluations of these procedures and their operating characteristics for practical sample sizes. The procedures are:

Bandit: The fully sequential design that maximizes the expected number of successes within the experiment. It is determined via dynamic programming.

Myopic: A fully sequential design that chooses, at each state, the arm that has the highest probability of producing a success. For the current patient, this is the desirable "personal physician" approach.

Equal Allocation (EA): A commonly used fixed sampling approach, in which each arm receives $n/3$ pulls. This is the classical allocation procedure that is expected to perform well with respect to choosing the best treatment to apply to future patients once the trial has terminated.

As noted, to optimize the bandit procedure, a Bayesian approach is taken in the design phase. Myopic allocation also utilizes a Bayesian approach. Recall, however, that the procedures can be analyzed from either a Bayesian or frequentist perspective. To illustrate this, the allocation schemes were compared (analyzed) according to two criteria — one Bayesian and the other frequentist.

The first is the Bayesian criterion, *expected number of failures*, "$E^\Gamma(F)$." For this example, the prior distribution, Γ, on the treatment means was the product of independent uniform, Beta(1,1), distributions. Because we use the same Γ throughout the example, we drop the superscript to simplify notation. Recall that $E(F)$ is the criterion minimized by the bandit procedure. Myopic allocation, on the other hand, assumes that the next observation is the last one, and as such it calls for allocation to the treatment that presently looks the best. For clinical trials with small sample sizes, this goal is virtually the same as trying to minimize $E(F)$ among the n trial patients. Note that while dynamic programming is needed to determine $E(F)$ for bandit allocation, backward induction is used to determine the $E(F)$ for myopic allocation. For EA, $E(F)$ is simply the sum of $n/3$ times the prior probabilities of failure for each arm. In the uniform case, this sum is simply $n/2$.

To address the third interpretation, which is to optimize future patient well-being, we focus on correctly identifying the best treatment arm at the end of the trial. The decision rule is to select the arm with the highest observed rate of successes, with the intent to treat all future patients with the selected therapy. In case of ties, the winner is selected randomly, as is standard. We wish to do this with high probability, so we examine the *probability of correct selection, P(CS)=P*(CS $\mid p_1, p_2, p_3$), where $(p_1, p_2, p_3) \in \Omega = [0, 1]^3$.

There exists no allocation procedure that maximizes P(CS) for all combinations of $(p_1, p_2, p_3) \in \Omega$. While one can carry out a pointwise comparison of two designs, assessing P(CS) over Ω for each, in general it is more tractable to utilize a summary measure to assess overall performance. As with the $E(F)$ criterion, one could work with a Bayesian version of P(CS), and integrate with respect to a prior distribution on the treatment success rates. Optimizing this measure can be done via dynamic programming. A different approach is needed, however, if a frequentist measure is desired. First, let $(p_{(1)}, p_{(2)}, p_{(3)})$ be the order statistics for (p_1, p_2, p_3). In other words, $p_{(1)}$ is the smallest success rate, $p_{(2)}$ the second smallest, and $p_{(3)}$ the largest. Fix a $\delta > 0$, and say that a selection of arm i is *correct to within* δ, denoted CS_δ, if $p_i > p_{(3)} - \delta$. (In the examples, $\delta = 0.1$.) Let Ψ be the class of all allocation designs of length n (notation for length suppressed). For $\psi \in \Psi$ define

$$P_\delta(\psi) = \min_{(p_1, p_2, p_3) \in \Omega} P(CS_\delta \mid \psi; p_1, p_2, p_3)$$

Then, a popular optimization goal is to locate $\psi^* \in \Psi$ such that

$$P_\delta(\psi^*) = \max_{\psi \in \Psi} P_\delta(\psi)$$

Unfortunately, ψ^* is unknown when there are three or more arms. Standard dynamic programming approaches cannot be used to solve this problem because of the nonlinear nature of the minimum operation in the definition of $P_\delta(\psi)$. However, when $k = 2$, the optimal procedure is to allocate equally to each arm. Thus, while for three or more arms there exist adaptive procedures that are better than EA on this measure, EA has the potential to be a very good suboptimal procedure.

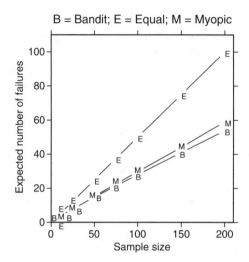

FIGURE 11.2 Sample size vs. E(failures), uniform priors (From Ref. [10]).

For an arbitrary allocation algorithm, it is not known which values of (p_1, p_2, p_3) yield the minimum over Ω, and it is not possible to determine this exactly through backward induction. This indicates that a search throughout the parameter space is needed to determine $P_\delta(\psi)$. $P_\delta(\psi)$ is an example of a criterion for which an allocation design needs to be evaluated multiple times. Because of these multiple evaluations, path induction was used to search for P_δ for the bandit and myopic designs. For two arms it can be shown that the minimum always occurs when $p_{(1)} = p_{(2)} - \delta$, reducing the dimension of the relevant search space. Here the search was over arm probabilities such that $p_{(1)} = p_{(2)} = p_{(3)} - \delta$. whereas for three or more arms there are contrived designs where P_δ is not attained in this region, for the designs considered here it seems to be a reasonable assumption.

In Figure 11.2, $E(F)$ for each procedure is plotted as a function of the sample size. Similarly, P_δ versus sample size is presented in Figure 11.3. As noted, uniform priors have been used throughout.

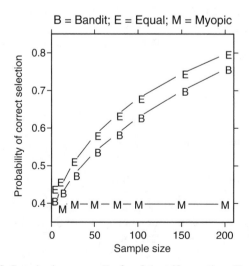

FIGURE 11.3 Sample size versus P_δ, $\delta = 0.1$, uniform priors (From Ref. [10]).

These were used mainly because they are a common basis for comparison across the literature. Naturally, had other priors been used then the results would be somewhat different. The program can easily handle a wide range of prior distributions.

Note that the bandit allocation comes very close to achieving the high P_δ of EA, while incurring far fewer failures. Myopic allocation also incurs few failures but has a very poor ability to correctly locate the best arm. Thus, we find that the bandit design seems to be the preferred design overall because it performs optimally on $E(F)$ and also attains high values of P_δ. For the indifference region of $\delta = 0.1$, the minimum $P(CS_\delta)$ for myopic allocation occurs when one arm has a success probability of 1 and the others have probability 0.9. In this situation, there is greater than a 60% chance that the trial will never even try the superior arm. This is largely due to the prior assumption that the average success rates for the different arms are $1/2$. The myopic rule randomizes in the first stage and if a success is obtained, the parameter estimate for the arm selected is updated to $2/3$, while the other arm estimates remain at $1/2$. This procedure selects the next observation from the arm with the expected success rate of $2/3$. With the true parameter having a value ≥ 0.9, the outcome is again likely to be a success. This result inclines the rule even further in favor of the arm already sampled. There are simple ways to alter myopic allocation so that P_δ significantly improves with very little increase in failures; however, a discussion of this is beyond the scope of this work.

11.9 DELAYED RESPONSE PROBLEM

An interesting dilemma that arises with adaptive designs is that information may not accrue at the rate of allocation. In this case, new questions that arise are (a) how to optimize an experimental design knowing that responses will be delayed and (b) how to model the response delay. *Delayed responses* are a significant practical problem in clinical trials, and are often cited as a difficulty when trying to apply adaptive designs [1, 28]. We have applied our scalable parallelization approach to a version of this problem in which there are two Bernoulli arms. Moreover, like the three-arm problem, no nontrivial delayed response problem has been fully optimized previously, either analytically or computationally. Again, determining the optimal design has been seen as being intractable, although some special cases have been analyzed. These include two-stage designs where the first stage is EA [6], and designs where one arm has a known success rate and the problem is to decide when to stop trying the unknown arm [7, 31].

There are many different models of the delay, appropriate for varying circumstances. Here we assume that the response times for each arm are exponentially distributed, and that patients arrive according to a Poisson process. We call the optimal design for this problem the *delayed two-armed bandit*, D2AB. In this setting, the natural states are of the form $(s_1, f_1, u_1, s_2, f_2, u_2)$, where u_i is the number of patients assigned to treatment i but whose outcomes are unknown.

As before, we have the condition that $s_1 + f_1 + u_1 + s_2 + f_2 + u_2 \leq n$, which allows compression, and all nonnegative values of s_1, f_1, s_2, f_2, s_3, f_3 satisfying this constraint are valid, just as they were for the three-arm bandit. However, a critical difference is that the recurrence for $V(\sigma)$ depends upon $V(\sigma + \widehat{u_1})$, $V(\sigma + \widehat{s_1} - \widehat{u_1})$, $V(\sigma + \widehat{f_1} - \widehat{u_1})$, $V(\sigma + \widehat{u_2})$, $V(\sigma + \widehat{s_2} - \widehat{u_2})$, and $V(\sigma + \widehat{f_2} - \widehat{u_2})$. That is, either a patient is assigned a treatment and the outcome is initially unknown, or we have just observed the outcome of a treatment. See Ref. [11] for the detailed form of the recurrence and its derivation.

While the recurrences for the delayed response model again have a stencil of neighbor dependencies, they are much more complicated. To go through the calculations systematically, one needs the appropriate notion of "stage," corresponding to m in the three-arm program. In general, the stage of a state σ should be the maximum path length to the state from the initial state **0**. In the

TABLE 11.7

Analysis of Delay Program on New System, $n = 100$ (From Ref. [24])

p	$e(p)$	Calc.	Misc.	Array Redist. Comm.	Array Redist. Shift	External Comm.
1	1.00	95.8	0.0	0.0	4.2	0.0
2	0.93	89.5	0.0	3.7	3.8	3.0
4	0.79	75.7	0.0	12.4	3.6	8.3
8	0.67	61.9	0.1	18.4	2.8	16.8
16	0.41	41.5	0.2	28.2	2.0	28.1
32	0.27	25.8	0.2	31.8	1.2	41.0

three-arm problem, all paths to σ from $\mathbf{0}$ took the same number of steps, which was the sum of the entries. Here again all paths have the same length, but it is $2(s_1 + f_1 + s_2 + f_2) + u_1 + u_2$, i.e., the components do not contribute uniformly.

Because all the paths from σ from $\mathbf{0}$ are the same length, states at stage k (i.e., at distance k) depend only on states at stage $k + 1$, which allows one to store only two stages at a time. Further, as in the three-arm problem, by carefully analyzing the dependencies and going through the loops in the correct order, this can be reduced down to one stage. However, there are now $2n$ stages for the outermost loop, as opposed to the n used previously. This has the negative effect of doubling the number of rounds of communication, which significantly reduces the parallel efficiency. It does have a positive effect, however, of slightly reducing the memory requirements because the same number of states are spread over more stages. The nonuniform roles of the indices make the array compression calculations somewhat more complex, and make it harder to determine the indices of the states depended on.

An additional complication comes from the fact that for the three-arm problem, any combination of nonnegative entries having a sum of m was a valid state at stage $m \leq n$. Now, however, there can be a valid stage $m \leq 2n$, and a combination of nonnegative entries having that weighted sum, but the combination does not correspond to a state. For example, if $n = 100$, then $(0, 0, 75, 0, 0, 75)$ is not a valid state, even though it is at stage 150. The reason is that it violates the constraint that $s_1 + f_1 + u_1 + s_2 + f_2 + u_2 \leq n$. Previously this constraint was automatically satisfied, but this is no longer true. This situation complicates the compressed indexing and access processes. Details can be found in Ref. [22].

Table 11.7 contains the timing and scaling analysis of the program, which incorporates all of the features of the most scalable three-arm program. This was run on a newer version of the computer system where policies had been adjusted to improve disk usage but which had the unintended effect of reducing scalability. Hence we would expect the performance to be degraded somewhat, but the drop in efficiency is rather significant, caused by the complex indexing and extra rounds of communication. Perhaps further tuning would have improved this, but it was sufficient for our purposes. This is an important aspect of parallel computing, in that improving parallel performance can be a never-ending process, and hence one needs to assess the trade-offs between programmer effort and time versus computer time.

11.9.1 RANDOMIZED PLAY-THE-WINNER

One popular *ad hoc* sampling rule is known as the randomized play-the-winner (RPW) rule, which first appeared in Ref. [33]. The RPW is an urn model containing "initial" balls that represent the

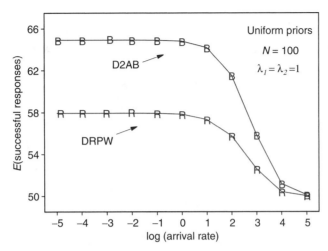

FIGURE 11.4 Expected successes for D2AB and DRPW, $\lambda_1 = \lambda_2 = 1$ (from Ref. [11]).

treatment options. Patients are assigned to arms according to the type of ball drawn at random from the urn. Sampling is with replacement, and balls are added to the urn according to the last patient's response.

An advantage of urn models like RPW is the natural way in which delayed observations can be incorporated into the allocation process. When a delayed response eventually comes in, balls of the appropriate type are added to the urn. Because sampling is with replacement, any delay pattern can be accommodated. We call this design the *delayed RPW* (DRPW) *rule*. A DRPW strategy, in which responses occur with a fixed delay, is mentioned in Ref. [16]. In Ref. [2] the authors consider a slightly altered version of this rule for a related best selection problem. However, only asymptotic results have been obtained for these cases. These results are consistent with ours when the delay is not large compared to the arrival rate, but they do not correctly predict behavior when the delay is comparable to the sample size times the arrival rate (see Figure 11.4).

11.9.2 SAMPLE RESULTS

We carried out exact analyses of the exponential delay model for both the D2AB and DRPW. Here we present results for $n = 100$. For the DRPW the urn is initialized with one ball for each treatment. This particular initial urn may be thought of as having roughly the effect of the uniform priors used in the bandit design. If a success is observed on treatment i then another ball of type i is added to the urn, while if a failure is observed then another ball of type $3 - i$ is added, $i = 1, 2$. For comparative purposes, we look at base and best case scenarios. The best fixed-in-advance allocation procedure is the base case, i.e., the optimal solution when no responses will be available until after all n patients have been allocated. To maximize successes, one should allocate all patients to the treatment with the higher expected success rate. We denote the expected number of successes in the base fixed case by $E_{\mathrm{bf}}[S]$. Here, we consider only uniform priors on the treatment success rates p_1 and p_2, in which case any fixed allocation rule works equally well, yielding an expected return of $E_{\mathrm{bf}}[S] = n/2$.

The best possible case arises when all responses are observed immediately (full information). In this situation, DRPW is simply the regular RPW and the D2AB is the regular two-armed bandit. Recall that the regular two-armed bandit optimizes the problem of allocating to maximize total successes. Letting $E_{\mathrm{opt}}[S]$ represent expected successes in the best case, $E_{\mathrm{opt}}[S] = 64.9$ for this

TABLE 11.8
Bandit: $E[S]$ as (λ_1, λ_2) vary, $n = 100$, $\lambda_s = 1$, uniform priors (From Ref. [11])

λ_1	λ_2						
$\downarrow$	10^{-5}	10^{-4}	10^{-3}	10^{-2}	10^{-1}	10^{0}	10^{1}
10^{-5}	50.1						
10^{-4}	51.2	51.2					
10^{-3}	55.4	55.4	55.8				
10^{-2}	59.3	59.4	59.9	61.5			
10^{-1}	60.9	61.0	61.6	63.1	64.1		
10^{0}	61.3	61.3	61.9	63.5	64.5	64.8	
10^{1}	61.3	61.3	62.0	63.5	64.6	64.8	64.9

example. Using the difference $E_{opt}[S] - E_{bf}[S]$ as a scale for improvement, one can think of the values on this scale, $(0, 14.9)$, as representing the "extra" successes over the best fixed allocation of 100 observations. For an allocation rule ψ define

$$R_\psi = \frac{E_\psi[S] - E_{bf}[S]}{E_{opt}[S] - E_{bf}[S]}$$

to be the *relative improvement* over the base case. While R_ψ also depends on n, the prior parameters, and the response and arrival rates, these are omitted from the notation.

Note that, for fixed arrival and delay rates, $R_{D2AB} \to 1$ as $n \to \infty$. However, this is not true for R_{DRPW}, because asymptotically the urn contains a nonzero fraction of balls corresponding to the inferior arm. If the arm probabilities are p_1, p_2, let $q_{(i)} = 1 - p_{(i)}$, $i = 1, 2$ (using the order statistic notation introduced in Section 11.8). Then $R_{DRPW}(p_1, p_2) \to (q_{(1)} - q_{(2)})/(q_{(1)} + q_{(2)})$. For uniform priors, $R_{DRPW} \to 0.545$. However, this asymptotic behavior gives little information about the values for practical sample sizes, and exact solutions for fixed values of n are not known. Hence their performance must be determined computationally.

Tables 11.8 and 11.9 contain the expected successes for the D2AB and the DRPW rules, respectively. Patient response rates, λ_1 and λ_2, vary over a grid of values between 10^{-5} and 10^{1}, and

TABLE 11.9
RPW: $E[S]$ as (λ_1, λ_2) vary, $n = 100$, $\lambda_s = 1$, uniform priors (From Ref. [11])

λ_1	λ_2						
	10^{-5}	10^{-4}	10^{-3}	10^{-2}	10^{-1}	10^{0}	10^{1}
10^{-5}	50.0						
10^{-4}	50.2	50.4					
10^{-3}	51.6	51.7	52.6				
10^{-2}	54.8	54.8	54.9	55.7			
10^{-1}	56.5	56.5	56.5	56.7	57.3		
10^{0}	56.9	56.9	56.9	57.1	57.6	57.8	
10^{1}	57.0	57.0	57.0	57.2	57.6	57.8	57.9

the patient arrival rate is fixed at 1. Note that, for both rules, when $\lambda_1 = \lambda_2 = 10^{-5}$, $E[S] \approx 50$. When $\lambda_1 = \lambda_2 = 10$, the delayed bandit rule gives $E[S]=64.9$ as one would expect. Note that in the best case scenario for the DRPW, $E[S] = 57.9$, which gives an R of 0.53. With the RPW, we can expect to gain only 7.9 successes as compared to the 14.9 for the optimal bandit.

Moving away from the extreme points, consider the case when λ_1, λ_2, and λ_s are all the same order of magnitude. The D2AB rule is virtually unaffected, with an R value of 0.99. This is true because, on average, there is only one allocated patient whose outcome has not been observed throughout the trial. For the DRPW, $R_{DRPW} = 0.52$, which is only slightly smaller than the nondelayed case. Both rules seem quite robust to mild to moderate delays in adaptation. It is only when *both* response rates are at least three orders of magnitude below the arrival rate that results begin to degrade seriously. When $\lambda_1 = \lambda_2 = 10^{-3}$, for example, R_{D2AB} is only 0.40, and R_{DRPW} is a dismal 0.17. It is also interesting to note that even when the response rate is only 1/100th the arrival rate, the D2AB does better than the RPW with immediate responses. Figure 11.4 illustrates the expected successes for DRPW and D2AB when the response rates are both one but the arrival rate varies between 10^{-5} and 10^5.

When we consider scenarios in which only one treatment arm supplies information to the system, we see an interesting result. For example, using uniform priors, when $\lambda_1 = \lambda_s = 1$ but $\lambda_2 = 10^{-5}$, the relative improvement is 0.76 for the D2AB and 0.47 for the DRPW. This is an intriguing result for the DRPW because its R-value is 89% of the best possible RPW value. Still, one clearly prefers the D2AB because there is only a 24% loss over the optimal solution whereas excluding half the information.

One way to view this problem independently from the allocation rules is to examine the expected number of allocated but unobserved patients when a new patient allocation decision must be made. As noted, when the response delay rate is 1, at any point in time one expects only a single observation to be delayed, and the impact on performance is minimal. When $\lambda_1 = \lambda_2 = 0.1$, once approximately 20 patients have been allocated there is a consistent lag of about 10 patients. Connecting this value to the results in Tables 11.8 and 11.9, one finds that a loss of roughly 10% of the total information at the time of allocation of the last patient (and a significantly higher loss rate for earlier decisions), corresponds to a loss of only about 5% in terms of the improvement available from D2AB, and about 8% from DRPW.

When the response rate is about 100 times slower that the arrival rate, asymptotically there will be approximately 100 unobserved patients at any point in time. Fortunately, for a sample size of 100, one is quite far from this asymptotic behavior, and approximately 37% of the responses have been observed by the time the last allocation decision must be made. This allows the D2AB to achieve 77% of the relative improvement possible, whereas the DRPW rule attains only 38%. Note that this is an example where asymptotic results would be quite misleading, and thus a computational approach is required to determine the true behavior.

While for space reasons this work has only analyzed problems in which both treatments have uniform priors, similar results hold for more general priors.

11.10 CONCLUSIONS

There is considerable interest in using adaptive designs in various experiments because they can save lives (human or animal), time, cost, or other resources. For example, for a representative delayed response problem with $n = 100$, uniform priors, and response delay rates 10 times patient arrival rates, simple EA averages 50 successes. The most commonly suggested adaptive technique, RPW, achieves only a 14.7% improvement, while the newly obtained optimal solution (D2AB) achieves a 28.4% improvement (see Figure 11.4). In fact, the optimal solution is nearly as good as the optimal solution for the case where there are no delays. Note that this is also the first exact evaluation of RPW in this setting, accomplished via backward induction. Its improvement over EA, as well as its degradation relative to the results obtained by the optimal design, were not known. The former could have been estimated via simulation, while the latter could not have been.

Note that a Bayesian approach was needed for dynamic programming to create the optimal designs. However, analysis phases, such as the evaluation of RPW or the myopic rule in Section 11.8, can be done on any design, whether it is *ad hoc*, Bayesian, or frequentist. The analyses may evaluate a mix of Bayesian or frequentist criteria, independent of the design. This point is pursued further in Ref. [12].

However, overall the complexity of adaptive designs has proven to be a major hurdle impeding their use. Our goal is to reduce computational concerns to the point where they are not a key issue in the selection of appropriate designs. This chapter has concentrated on the parallel computational aspects of this work, while other papers analyze the statistical and application impact of new serial algorithms [12].

Unfortunately, the recurrences involved have attributes that make it difficult to achieve high performance and scalability. Memory requirements tend to be the limiting factor, and trying to ameliorate this causes overhead and a significant increase in program complexity. As noted in Section 11.6, increases in program complexity can cause severe performance problems when the compiler is unable to optimize the innermost loops, and hence one must select alternatives with the compiler's limitations in mind. Space constraints, and low calculation to communication ratios, also complicate the ability to reduce the effects of communication latencies and overhead. However, by working diligently, it is possible to achieve significant speedups and scalable parallelizations, although this comes at a cost of increased program length and more complex program maintenance.

In Table 11.10, the effects of memory limitations on the three-arm problem, using 1 GB per processor, are illustrated. *Uncompressed* refers to a parallel program using load balancing as in the initial parallel version, but without compressing to a one-dimensional array. Note how the scalable version needs fewer processors to solve large problems, and that it can solve arbitrarily large problems, while the other versions cannot go beyond a fixed problem size no matter how many processors are available. This is due to the imperfect load balancing in the earlier versions which were unable to allocate less than a single s3 loop per processor.

Besides being able to compare alternative parallelizations, we can also compare to the work of others. Using only 16 processors of an IBM SP2 we solved the three-arm, $n = 200$ problem. This is approximately 500,000 times harder than the problem called "impractical" in Ref. [17], and more than 300 times harder than that solved in Ref. [8] on a parallel IBM 3090 of approximately the same computational power. Further, more than 100 evaluations used for determining P_δ would have taken

TABLE 11.10

Minimum Processors (*p*) Needed to Solve Problem of Size *n*, Using 1 GB per Processor (From Ref. [10])

n	Uncompressed	Initial	Scalable
100	100	1	1
200	∞	21	16
300	∞	∞	173

Max. problem solvable: uncompressed: 105; initial: 231; scalable: ∞.

these authors a 100-fold increase in time, whereas by using path induction it only roughly tripled the time required to find the design.

Similar parallelization steps can be used to solve problems involving four or more arms, arms with more than two outcomes, designs with staged allocation, and so forth. However, since the computational requirements of these problems grow more rapidly than those of the problems considered here, the largest problems solvable with the same resources will be smaller. Note that the parallelization process described applies much more broadly than adaptive designs for clinical and preclinical trials, although this in itself is an important application. The bandit model is widely used in areas such as operations research, artificial intelligence, economics, and game theory. Further, our work generally applies to neighbor recurrences using stencils. This common class of recurrences includes many dynamic programming problems, such as the generalized string matching used in some data mining and bioinformatics applications.

Despite some successes, it is important to realize the limitations of parallel computing. Parallel computing can only do a little to overcome the curse of dimensionality that plagues many uses of dynamic programming (and relatives such as backward induction and path induction) for adaptive designs. When computational time increases as $\Theta(n^6)$, as is true for determining the optimal design for the three-arm bandit and two-arm bandit with delayed response problems, then doubling the sample size results in a 64-fold increase in computational time. Thus to double the size of the largest problem solvable on a serial computer, yet solve it in the same amount of time, would require 64 processors with perfect efficiency, or even more processors with more realistic efficiency. One may have access to such a computer and program, and the doubled problem size may be what is needed to solve a specific problem. However, if another doubling is needed then it is unlikely the researcher will have access to suitable parallel resources and hence other methods will have to be employed.

Finally, merely throwing a parallel computer at a problem is unlikely to be of much help. For farming-like applications this is a relatively simple process and likely to attain the desired improvements, but for many other problems the process is far more complicated. As was shown, extensive work was needed to achieve useful problem sizes and reasonable efficiency. Often it is easiest to utilize shared memory systems, but typically only modest performance will be achieved without significant work. One important aspect of the parallelization process that should be kept in mind is that to make the most of the programming effort, one should emphasize the use of software standards. MPI is the dominant message-passing system, and is widely and freely available. Similarly, OpenMP is the dominant parallelization method for shared-memory machines. By using these, porting code to new, typically more powerful, platforms is greatly simplified. Because significant effort may have been put into the parallelization process, one would like to be able to use the resulting program for an extended period of time.

ACKNOWLEDGMENTS

The parallel programming was done by Robert Oehmke, and further details of algorithm improvements and parallelization aspects can be found in his thesis [22]. This research was partially supported by NSF grant DMS-0072910. Parallel computing support was provided by the University of Michigan's Center for Advanced Computing. Some of the material presented here first appeared in Refs. [10, 11, 24].

REFERENCES

[1] Armitage, P. (1985). The search for optimality in clinical trials. *Int. Stat. Rev.* 53:1–13.

[2] Bandyopadhyay, U. and Biwas, A. (1996). Delayed response in randomized play-the-winner rule: a decision theoretic outlook. *Calcutta Stat. Assoc. Bull.* 46:69–88.

[3] Berry, D.A. and Fristedt, B. (1985). *Bandit Problems: Sequential Allocation of Experiments*. Chapman & Hall, London.

[4] Bellman, R. (1956). A problem in the sequential design of experiments. *Sankhya A* 16:221–229.

[5] Bradt, R. and Karlin, S. (1956). On the design and comparison of certain dichotomous experiments. *Ann. Math. Stat.* 27:390–409.

[6] Douke, H. (1994). On sequential design based on Markov chains for selecting one of two treatments in clinical trials with delayed observations. *J. Japanese Soc. Comput. Stat.* 7:89–103.

[7] Eick, S. (1998). The two-armed bandit with delayed responses. *Ann. Stat.* 16:254–264.

[8] Hardwick, J. (1989). Computational problems associated with minimizing the risk in a simple clinical trial. In *Contemporary Mathematics: Statistical Multiple Integration*, Flournoy, N. and Tsutakawa, R. (Eds) American Mathematics Association, 115:239–257.

[9] Hardwick, J. (1995). A modified bandit as an approach to ethical allocation in clinical trials. In *Adaptive Designs*, Vol. 25, Flournoy, N. and Rosenberger, W., Eds. IMS Lecture Notes — Monograph Series, pp.65–87.

[10] Hardwick, J., Oehmke, R., and Stout, Q.F. (1999). A program for sequential allocation of three Bernoulli populations. *Comp. Stat. Data Anal.* 31:397–416.

[11] Hardwick, J., Oehmke, R., and Stout, Q.F. (2001). Optimal adaptive designs for delayed response models: exponential case. In *MODA6: Model Oriented Data Analysis*, Atkinson, A., and Hackl, P., and Müller, W., Eds. Physica Verlag, Heidelberg, pp. 127–134.

[12] Hardwick, J. and Stout, Q.F. (1998). Flexible algorithms for creating and analyzing adaptive sampling procedures. In *New Developments and Applications in Experimental Design*, Vol. 34 IMS Lecture Notes — Monograph Series, pp. 91–105.

[13] Hardwick, J. and Stout, Q.F. (1999). Using path induction to evaluate sequential allocation procedures. *SIAM J. Sci. Comput.* 21:67–87.

[14] Hardwick, J. and Stout, Q.F. (2001). Optimal few-stage designs. *J. Stat. Plan. Inf.* 104:121–145.

[15] Ibarra, O.H., Wang, W., and Jiang, T. (1993). On efficient parallel algorithms for solving set recurrence equations. *J. Algorithms* 14:244–257.

[16] Ivanova, A. and Rosenberger, W. (2000). A comparison of urn designs for randomized clinical trials of $k > 2$ treatments. *J. Biopharm. Stat.* 10:93–107.

[17] Kulkarni, R. and Kulkarni, V. (1987). Optimal Bayes procedures for selecting the better of two Bernoulli populations. *J. Stat. Plan. Inf.* 15:311–330.

[18] Lew, A. and Halverson, A., Jr. (1993). Dynamic programming, decision tables, and the Hawaii parallel computer. *Comput. Math. Appl.* 27:121–127.

[19] Lewandowski, G., Condon, A., and Bach, E. (1996). Asynchronous analysis of parallel dynamic programming algorithms. *IEEE Trans. Parallel Distrib. Syst.* 7:425–438.

[20] Lokuta, B. and Tchuente, M. (1988). Dynamic programming on two dimensional systolic arrays. *Inf. Proc. Lett.* 29:97–104.

[21] Test Guideline 425: Acute Oral Toxicity — Up-and-Down Procedure, Organization for Economic Cooperation and Development (OECD), www.epa.gov/oppfead1/harmonization/docs/E425guideline.pdf, 2001.

[22] Oehmke, R. (2004). High-Performance Dynamic Array Structures on Parallel Computers. Ph.D. dissertation, University of Michigan, Ann Arbor, MI.

[23] Oehmke, R., Hardwick, J., and Stout, Q.F. (1998). Adaptive allocation in the presence of censoring. *Comput. Sci. Stat.* 30:219–223.

[24] Oehmke, R., Hardwick, and J., Stout, Q.F. (2000). Scalable algorithms for adaptive statistical designs. *Sci. Program.* 8:183–193.

[25] Ranka, S. and Sahni, S. (1990). String editing on a SIMD hypercube multicomputer. *J. Parallel Distrib. Comput.* 9:411–418.

[26] Rytter, W. (1988). On efficient parallel computations for some dynamic programming problems. *Theoret. Comput. Sci.* 59:297–307.

[27] Sastry, R. and Ranganathan, N. (1993). A systolic array for approximate string matching. *Proceedings IEEE International Conference on Computer Design*, pp. 402–405.

[28] Simon, R. (1977). Adaptive treatment assignment methods and clinical trials. *Biometrics* 33:743–744.

[29] Strate, S.A., Wainwright, R.L., Deaton, E., George, K.M., Bergel, H., and Hedrick, G. (1993). Load balancing techniques for dynamic programming algorithms on hypercube multicomputers. *Applied Computing: States of the Art and Practice*, 562–569.

[30] Tang, E. (1995). An efficient parallel dynamic programming algorithm. *Comput. Math. Appl.* 30:65–74.

[31] Wang, X. (2000). A bandit process with delayed responses. *Stat. Prob. Lett.* 48:303–307.

[32] Wang, Y.-G. (1991). Sequential allocation in clinical trials. *Comm. Stat.: Theory Meth.* 20:791–805.

[33] Wei, L.J. and Durham, S. (1978). The randomized play the winner rule in medical trials. *J. Am. Stat. Assoc.* 73:840–843.

12 A Modular VLSI Architecture for the Real-Time Estimation of Higher Order Moments and Cumulants

Elias S. Manolakos*

CONTENTS

* This work has been partially supported by a grant from the National Science Foundation, MIP - 9309319.

ABSTRACT

Higher order statistics (HOS), such as the higher order moments (HOM) and cumulants, have been recognized as important tools in modern signal processing and data analysis, because they overcome well-known limitations of the autocorrelation/power spectrum based second-order methods. We present here the systematic synthesis of a modular VLSI processor array architecture that can compute in real-time HOM and cumulants up to any specified maximal order. We describe first the steps of a novel design methodology that can be followed to explore the large space of available solutions to identify and characterize those architectures that meet certain designer-imposed constraints. Then we use it to (i) systematically formulate locally recursive algorithms (LRA) and linear space–time operators that can transform them into planar parallel array architectures with optimal characteristics for the real-time estimation of *all* moment lags, up to any desirable order higher than three and (ii) determine how the moments generating VLSI architecture can be minimally extended to provide in real-time estimates of the fourth-order cumulants, whereas still retaining its optimal characteristics.

Keywords: Higher Order Statistics; VLSI Architectures; Pipelining; Parallel Processing; Array Synthesis Methodologies

12.1 INTRODUCTION

Methods based on second-order statistics, such as the autocorrelation or the power spectrum, are very commonly used in modern data analysis and signal processing. However, there are many situations in practice that require utilizing higher (than the second) order statistics (HOS) [1, 2], such as the HOM, the *cumulants*, or their frequency counterparts, also known as *polyspectra* [3]. The main reasons for using HOS are among others [2]: (a) to suppress additive Gaussian noise and improve detection and estimation performance, (b) to identify nonminimum phase systems, (c) to exploit information related to deviations from Gaussianity for classification purposes, and (d) to perform nonlinear systems identification. Among the diverse scientific domains where HOS have been utilized, we mention here: nonlinear system identification [4, 5], blind source separation [6], econometrics [7, 8], portfolio theory [9], stock market modeling [10], etc. For an excellent review on HOS and their applications in science and engineering, the interested reader is referred to Ref. [3]. A comprehensive bibliography on the subject has appeared in Ref. [11].

Although several HOS estimators have been proposed and their properties analyzed (see for example Refs. [12, 13]), the efficient real-time computation of HOS has remained a challenge. The computational complexity, memory, and data input requirements of applications that rely on HOS estimates grow rapidly with the number of samples in the data sequence to be processed. Therefore, appropriately designed multiprocessor architectures should be employed when high performance is demanded. However, the formulation parallelizable HOS estimation algorithms is an interesting and challenging problem. Such algorithms could be translated into application specific array processors (ASAP) implementable directly in hardware (using VLSI devices such as FPGAs or ASICs), or realized as message passing parallel programs (using clusters of workstations).

To the best of our knowledge the first attempt to use parallel processing and pipelining for the real-time estimation of the third order statistics was reported in Ref. [14]. Alternative solutions were published more recently in Refs. [15, 16]. For the much more complex problem of fourth-order statistics estimation, earlier work from our group includes: (i) The parallelization of a realization of the "indirect" estimation algorithm [3], leading to the design of a triangular array architecture with acceptable performance for computing the fourth order moments [17]. (ii) A novel sliding window time- and order-recursive moments estimation approach and its processor array implementation

[18]. In this contribution we generalize earlier results and discuss in detail the systematic synthesis of a fine granularity, high throughput, VLSI architecture that computes in the same hardware more than one type of HOS. In particular, we (i) formulate locally recursive HOS estimation algorithms that compute different statistics up to *any desirable higher order k* (certainly larger than $k = 3$), (ii) and find the space–time mapping linear operators that map these algorithms to a modular and universal moments and cumulants computing VLSI architecture with optimal characteristics.

The dimensionality (number of nested loops) of HOS estimation algorithms increases with the statistics order k. However, as we are targeting a planar VLSI implementation, we have to limit the number of processor array dimensions to two. So as k increases, we are in need of a systematic methodology that would allow us to identify *analytically* any available options (algorithmic formulations and corresponding array architectures), that can best meet implementation technology specific restrictions and/or our design preferences. To carry the synthesis effectively and improve upon previous suboptimal architectures derived using *ad hoc* techniques, we had to modify the traditional mapping methodology used to transform a serial nested-loop algorithm to parallel computational structures [19]. In particular, as we discuss in Section 12.2, we introduced a new design flow which, given a nested-loops specification of a problem, allows the step-by-step *construction* of optimal locally recursive algorithms (LRA). In essence, an optimal LRA is constructed by finding the most appropriate set of vectors (localized data dependencies) that span the algorithm's multidimensional iteration space and distribute the input data to the computations that should use them to produce results. The novelty in the proposed modification is that the selection of these dependence vectors is driven by the desirable characteristics of the target parallel machine (which will execute the LRA). These characteristics are captured as constraints and are allowed to drive the LRAs formulation, so the new design approach is called *constraint-driven* localization.

In Section 12.3 we apply the new methodology to synthesize a family of minimum latency, parallel array architectures for the efficient computation of HOMs. A unique property of this array is that it may provide estimates of *all moment lags*, up to any desirable maximal order $k > 3$ specified by the designer. In practice, $k = 4$ is usually sufficient, but it is interesting that the same architecture can deal with any larger k, should that be of interest. The range of applications that can benefit from the availability of such a generic *moments generating engine* is very broad because HOMs are used in many higher level interpretation tasks [1]. Furthermore, the computation of HOM estimates, using a block of M data samples, has time complexity in $O(M^k)$ and is the most expensive step toward the calculation of several other statistics of interest, including the fourth-order cumulants and the polyspectra (obtained via the "indirect" estimation algorithm [20]).

The proposed array synthesis methodology can be used to derive closed-form analytical expressions for the permissible scheduling functions in terms of problem size parameters for *all* valid architectural solutions to a given problem. This characterization is especially useful when intermediate results produced by a processor array are consumed by another array stage, and therefore it is important to construct systematically pairs of pipelined architectures and scheduling functions that achieve optimal overall *space–time matching*. In Section 12.4, we show how this capability can be exploited as we attempt to expand the HOMs architecture, so that the first set of statistics (moments) can be reused to produce a second set (the fourth-order cumulants), whereas preserving the desirable properties of overall minimum latency and memory.

It is worth noting that although applied to a particular class of problems here, the developed unified array synthesis methodology can be employed in general to derive computation structures with desirable properties, as long as the iteration space and indexing functions of the variables involved in the original nested-loop algorithm targeted for parallel implementation are known at compile-time (as usually the case in scientific computing). The same methodology, but with a different set of constraints, has been used in Ref. [21] to derive highly efficient single instruction multiple data (SIMD) parallel algorithms for HOMs estimation.

Before we proceed, let us clarify the notation to be used throughout this manuscript: the set of vectors with n integer elements will be denoted by $\mathcal{Z}^n$, the matrices/column vectors by bold capital/lowercase letters respectively, the matrix transposition sign by superscript "T", index points in $\mathcal{Z}^n$ as $\vec{v}$, and scalar quantities by lowercase letters. In addition, we let $\mathbf{e}_i^{(n)} \in \mathcal{Z}^n$ be the ith canonical basis vector (ith column of identity matrix $\mathbf{I}_n$). Finally, for simplicity by $\mathbf{e}_i$ and $\mathbf{u}_i$ (without superscript) we denote the canonical basis vectors in the algorithm's and processors' space, respectively.

12.2 SYNTHESIS OF VLSI PROCESSOR ARRAY ARCHITECTURES

12.2.1 TRADITIONAL SPACE–TIME MAPPING METHODOLOGIES

Developing methodologies for the systematic synthesis of ASAP has been a very active research area. The main objective is to transform systematically the nested-loop specification of a sequential algorithm targeted for parallelization to a multiprocessor parallel architecture model. Following the seminal paper of Karp et al. [22], various approaches have been proposed (a detailed discussion can be found in Refs. [23, 24]). In this context, an algorithm is represented by a set of computations "attached" to the points of an *index space*, $\mathcal{I}$ (also called *iteration* space) that forms a multidimensional lattice of integer points, subset of $\mathcal{Z}^{n_a}$, where n_a is the number of indices used in the algorithm's nested-loop description. Each computation may also be though as a node in a graph that is embedded to the lattice, known as the Dependencies graph (DG) [19]. The "work" (computational cost) of the algorithm is directly related to its dimensionality n_a and the number of nodes in the DG.

Array synthesis involves several *behavior-preserving* transformations. First, the original nested-loop algorithmic specification is translated into an equivalent LRA. Then an array architecture emerges as the result of applying linear *space and time* mapping to the index space of the constructed LRA [19, 24]. Traditionally array synthesis has been treated as a two-phase procedure that starts with the *localization of the dependencies* of the algorithm (construction of an LRA) followed by the *mapping* of the constructed LRA to a parallel architecture. It is important to note that in both phases, starting from a specific point (nested-loop algorithm/LRA) we can derive numerous valid solutions (LRAs/ASAPs) with very different characteristics. Therefore methodologies that facilitate the efficient exploration of the large design space are very desirable.

An LRA may be uniquely represented by an acyclic directed DG whose nodes correspond to computations in the algorithm's index space and arcs capture the precedence relationships among computations. That is, given two DG nodes $\vec{v}_1, \vec{v}_2 \in \mathcal{I}$ with the former dependent on the latter, the associated dependence is defined as the vector difference, $\mathbf{d} = \vec{v}_1 - \vec{v}_2$. If all index points in $\mathcal{I}$ have an identical dependence structure we say that the LRA corresponds to a regular iterative algorithm (RIA) [25] comprised of a set of uniform recurrent equations (UREs) [26].

The localization of the data dependencies of a nested-loop algorithm has been studied by several researchers [27–30]. During this phase, new variables may be introduced so that in the resulting LRA formulation: (a) each variable is assigned only once and (b) all dependencies among variables can be represented by vectors of length that is independent of the problem size. To perform localization, all variables with values that need to be broadcasted should be identified. A variable is said to be a *broadcast* if its value is required by several computations taking place at different index points in $\mathcal{I}$ (one-to-many distribution). A broadcast variable is "localized" by distributing its value in a pipelined fashion along a propagation contour to the computations (index points) that need it. It is possible to identify if a variable, say w, is a broadcast and characterize all valid propagation contours that can be used for its localization by examining its *indexing matrix* $\mathbf{A}_w$ [27, 29]. Let us assume that variable w has an indexed *instance* $w(f[\vec{v}])$, where the indexing function $f[\vec{v}]$ is affine, i.e., $f[\vec{v}] \equiv \mathbf{A}_w \cdot \vec{v} + \mathbf{b}_w$ and $\vec{v}$ is the vector of indices spanning $\mathcal{I}$. If $\mathbf{A}_w$ is a row rank deficient matrix, then variable w is a broadcast. Moreover, the contours that can be used to distribute the value of w should form a subspace of the right null space of $\mathbf{A}_w$, namely $\mathcal{N}(\mathbf{A}_w)$. Therefore, any dependence vector $\mathbf{d}_w$ introduced for this localization should be of the following integer linear combination form

$$\mathbf{d}_w = \sum_{\forall q} a_{w,q} \mathbf{v}_{w,q}, \quad \text{where } \mathbf{v}_{w,q} \text{ is a basis vector of } \mathcal{N}(\mathbf{A}_w) \quad \text{and } a_{w,q} \in \mathcal{Z}. \quad (12.1)$$

That is, depending on the choice of the integers $\{a_{w,q}\}$, there may exist numerous valid propagation contours for the localization of a broadcast variable. Considering also that an algorithm may involve several broadcast variables a very large number of valid LRA formulations may be possible. On the other hand, if w is not a broadcast (i.e., $\mathbf{A}_w$ is full rank), the dependence associated with propagating the value of w is already specified in the original nested-loop algorithm and this is the reason why it is often called a "true" dependence.[1] When the dependence vectors for *all* broadcast variables have been selected (after fixing the constants $\{a_{w,q}\}$, $\forall w, q$) an LRA equivalent of the original algorithm has been constructed that can be represented uniquely by its index space $\mathcal{I}$ and the *dependencies matrix* $\mathbf{D}$ having as columns the vectors $\mathbf{d}_w$ for all variables w.

During the second phase of the synthesis, also known as *space–time mapping*, the constructed LRA is transformed onto an ASAP. Several systematic mapping methodologies have been proposed including those in Refs. [19, 24–26, 31–33]. Two linear (or affine) *space* and *time* transformations are chosen to map the DG of the selected LRA to a graphical model for the array architecture, known as the *signal flow graph* (SFG) [19]. The linear space transformation allocates (places) the algorithm's computations to SFG nodes, or processing elements (PEs). Analytically, it is represented as a matrix $\mathbf{S} \in \mathcal{Z}^{n_p \times n_a}$ with full row rank [24] that maps the n_a-dimensional algorithm's index space onto the n_p-dimensional processor space, i.e., $\mathbf{S}: \mathcal{Z}^{n_a} \supset \mathcal{I} \longmapsto \mathcal{P} \subset \mathcal{Z}^{n_p}, n_p < n_a$. Hence, an algorithm index point (DG node) $\vec{v} \in \mathcal{I}$ will be transformed to the PE (SFG node) $\mathbf{S} \cdot \vec{v} \in \mathcal{P}$. Furthermore, the linear time transformation (schedule) $\mathbf{T} \in \mathcal{Z}^{1 \times n_a}$ assigns to every computation, $\vec{v} \in \mathcal{I}$, an integer $\mathbf{T} \cdot \vec{v}$ representing the schedule time instant of its execution, i.e., $\mathbf{T}: \mathcal{I} \longmapsto \mathcal{Z}$.

A chosen schedule has to be permissible, i.e., satisfy the *precedence* (causality) as well as the *space–time compatibility* constraints [24, 33]. The former ensures that if there exists a dependence from a DG node $\vec{v}_1$ to node $\vec{v}_2$, then $\vec{v}_2$ will be executed at least one schedule time instant after $\vec{v}_1$. The latter constraints guarantee that there will be no data *collisions* in the array, i.e., two distinct computations in the algorithm's index space allocated to the same PE will not be scheduled for execution at the same time period. For linear transformations $\mathbf{S}$ and $\mathbf{T}$, both sets of constraints can be expressed analytically as:

Precedence constraints (PCs): For every data dependence $\mathbf{d}_w$ it should be true that $\mathbf{T} \cdot \mathbf{d}_w \geq 1$.

Space–time compatibility constraints STCs: Let $\vec{v}_1, \vec{v}_2$ be two arbitrary and distinct index points (computations) allocated to processors $\mathbf{S} \cdot \vec{v}_1, \mathbf{S} \cdot \vec{v}_2$, and scheduled for time instants $\mathbf{T} \cdot \vec{v}_1$, $\mathbf{T} \cdot \vec{v}_2$, respectively. Let $\vec{\delta} \equiv \vec{v}_2 - \vec{v}_1$; it should be true that $\begin{bmatrix} \mathbf{S} \\ \mathbf{T} \end{bmatrix} \cdot \vec{\delta} \neq \mathbf{0}_{n_p+1}, \forall \vec{\delta}$.

To find a permissible schedule that achieves minimum *latency* (number of schedule steps) we should solve an integer nonlinear programming problem [19, 22, 25] or use heuristic techniques [34]. These methods do not provide analytical expressions for the elements of the minimum latency schedule vectors in terms of the problem size parameters, a feature that is very desirable in practice as we will demonstrate later. To overcome this limitation we have introduced, and discussed in Section 12.2.2, a methodology which, given the space transformation $\mathbf{S}$ and the algorithm's index space $\mathcal{I}$, can be used to characterize analytically the elements of *all* permissible schedule vectors and the latency that each schedule vector achieves.

---

[1] We may be able to reverse the dependence vector direction, if the DG node operation is associative, but this implies a change to the original nested-loop algorithm specification.

The SFG resulting from the space–time mapping is a graph embedded into the processors' space $\mathcal{P} \in \mathcal{Z}^{n_p}$. Its vertices denote the PEs and edges communication links. Every communication edge in $\mathcal{P}$ is the outcome of a dependence vector transformation. That is, the edge due to $\mathbf{d}_w$ has orientation $\mathbf{c}_w = \mathbf{S} \cdot \mathbf{d}_w$ and a number of delay buffers $\mathbf{T} \cdot \mathbf{d}_w$ associated with it. As with the localization, the mapping is not unique, meaning that many SFGs may arise from the same LRA by applying different linear space–time transformation pairs $(\mathbf{S}, \mathbf{T})$. Consequently, given an LRA, the optimal mapping problem is to find those $\mathbf{S}$ and $\mathbf{T}$ that yield ASAPs with desirable characteristics. Optimality may be defined in terms of minimum latency or minimum *block-pipelining period* (β), which corresponds to the time interval between initiations of two consecutive instances of the same problem in the array architecture [19]. Minimizing β maximizes the array's throughput.

The transformations $\mathbf{S}$ and $\mathbf{T}$ may be selected so that the resulting architecture also meets constraints related to the implementation technology. With VLSI architectures in mind, the following constraints are usually considered in practice:

C1: The architecture model (SFG) should correspond to an at most two-dimensional graph.

C2: The input of data and output of results (I/O operations) should take place at PEs in the periphery of the array architecture.

C3: Data broadcasting should be avoided and PEs should communicate via edges of short and fixed length (that is independent of the problem size) for the architecture to support very high data rates.

Traditionally, the two phases of array synthesis, localization of data dependencies and space–time mapping, have been treated as two separate problems. Following a vertical two-phase synthesis approach, it is possible that no architecture that meets all design objectives is found, or that a good mapping results to a suboptimal architecture due to a suboptimal LRA selection in the first phase. A flexible array synthesis approach that addresses these limitations is outlined in Section 12.2.2.

12.2.2 A CONSTRAINT-DRIVEN UNIFIED ARRAY SYNTHESIS METHODOLOGY

We highlight here the steps of a *unified* methodology that allows the localization and mapping phases to closely interact and affect each other. Its main difference from other traditional approaches is that it aims at *constructing gradually* an LRA (for the problem to be solved) in a way that leads (after the space–time mapping) to an optimized array architecture, i.e., one that meets specified architectural constraints imposed by the target implementation technology or reflecting the designer's objectives and preferences. The steps of the methodology are depicted schematically in Figure 12.1 and described briefly below. In Sections 12.3 and 12.4 we will demonstrate the flexibility of this general methodology by using it to derive systematically optimal array structures for problems such as the real-time estimation of the HOM and the fourth-order cumulants under different design objectives and constraints.

12.2.2.1 Data Dependence Analysis of the Algorithm

It is assumed that the index space of the original nested-loop algorithm is known at compile-time. First, we analyze all indexed instances of variables, find the broadcasts and characterize the allowable directions of propagation of each one (see Equation (12.1)). We seek, in other words, to identify all degrees of freedom in selecting the dependence vector (direction of propagation) associated with each variable instance present in the nested-loop algorithm.

12.2.2.2 Determining the Space Transformation

Among all possible choices for linear space transformations $\mathbf{S}$, we consider only those that satisfy implementation technology-related constraints. It is assumed that implementations violating these

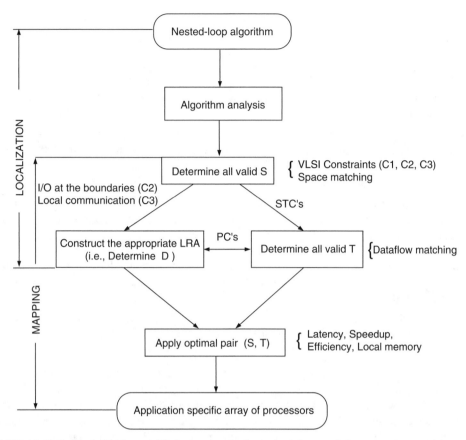

FIGURE 12.1 Constraints-driven unified array synthesis methodology and the parameters each stage may depend upon. Reprinted from [38] with permission from Taylor & Francis.

constraints will suffer from a significant cost overhead and are not acceptable. For instance, if we demand I/O operations occurring only at boundary PEs of the architecture (constraint C2), we may engineer the space transformation **S** such that the resulting processor space is parallel to the normal vectors of computation hyperplanes (in the DG) that include I/O operations [25, 35]. In addition, to further reduce the number of valid **S** candidates we may also consider only those leading to arrays with certain desirable architectural properties, such as minimal number of PEs or minimal block pipelining period β. It is noteworthy that even though numerous choices for **S** may exist, some of them will yield isomorphic array topologies and therefore the number of distinct array structures is in practice limited and depends on the dimensionality of the nested-loop algorithm, n_a.

12.2.2.3   Constraint-Driven Localization

An LRA is uniquely specified by its index space $\mathcal{I}$ and dependence matrix **D**. The transformation of the original nested-loop algorithm to an appropriate (optimized) LRA for the given problem is achieved by eliminating data broadcasting in a way that (i) satisfies the VLSI technology-related constraints and (ii) can yield *minimum latency*, when scheduled by a linear time transformation **T**. To facilitate the search for optimal schedules, we conjecture that a good LRA is designed if its matrix **D** is of minimum rank, i.e., **D** has the smallest possible number of linearly independent columns. In addition, we require all other columns of **D** to be linear combinations of the independent columns with nonnegative integer coefficients. The rational behind this conjecture is threefold:

- The smaller the number of linearly independent dependence vectors, the less the PCs that should be satisfied by permissible schedules **T**, i.e., the larger the freedom when searching for optimal schedules.

- The smaller the number of distinct dependence vectors, the less the number of distinct SFG edges. Because parallel SFG edges (resulting from the mapping of parallel dependence vectors) can potentially be implemented by a *shared* communication link, the interconnection area in the hardware architecture is minimized.

- The smaller the number of linearly independent dependence vectors, the larger the number of components the DG may be decomposed to. It has been established [36] that the rank of matrix **D** is inversely proportional to the *partitionability* of the algorithm.

It is possible that one or more of the architectural constraints and an LRA has been constructed (fully specified) when all the integer coefficients $\{a_{w,q}\}$ in expression (12.1) for every dependence $\mathbf{d}_w$ have been fixed.

12.2.2.4 Determining All Permissible Linear Schedules

Next, we seek to determine *all* valid time transformations **T** that are compatible: (i) with the chosen **S** (satisfy the STCs), (ii) with the already specified dependences (satisfy the PCs), and (iii) meet the data-flow constraints for a multistage architectural design. Furthermore, we are aiming at deriving expressions that characterize the schedule elements in terms of the problem size parameters, as already explained in Section 12.1. However, when arriving at this step, there may still exist multiple localization choices for some of the variables. Thus, depending on the particular application, one may find it necessary to determine first the appropriate dependencies (construct the LRA) and then all valid **T**s, or vice versa, as suggested by the double-headed horizontal arrow in Figure 12.1.

12.2.2.5 Applying the Space–Time Mapping

Having identified the pairs (**S**, **T**) that yield desirable characteristics, it is straightforward to apply the space and time transformations to the DG of the LRA and derive SFG models of parallel architectures as discussed in Section 12.2. (For more details see Ref. [19].)

The main advantages of the proposed unified array synthesis methodology are the following: First, it maintains the search space for optimal designs as large as possible by not assuming that the data dependencies are fully specified from the very beginning, in an attempt to maximize the architectural solutions that have the chance to meet the constraints and desired optimality metrics. Furthermore, it advocates deriving closed-form expressions for the elements of all permissible schedule vectors as functions of the problem-size parameters. This is especially useful when trying to find optimal ways to map multidimensional (deeply nested) algorithms onto linear or two-dimensional mesh architectures, because in this case the elements of the schedule vectors (and consequently the resulting latency) will necessarily be related to the problem size (number of nodes in the DG). The availability of parameterized synthesis quality-related metrics is also very useful when designing *multi stage* tightly pipelined architectures. By fully characterizing families of optimal solutions for each stage we can maintain the largest possible design exploration space and select only at the very end, the pair of solutions that can be optimally space–time matched to each other, as we demonstrate in Section 12.4 and discuss in more detail in article [37].

12.3 HOM ESTIMATION

12.3.1 THE PROBLEM

Assuming a realization of a real process $\mathcal{X}$ to be given by M samples $\{x_0, x_1, \ldots, x_{M-1}\}$, estimates of the kth order joint moments (kOMs) of $\mathcal{X}$ can be computed over the lags $i_j, \ j = 1, \ldots, k-1$ as follows [1, 20]:

$$\widehat{m}_k(i_1, i_2, \ldots, i_{k-1}) \equiv \frac{1}{M} \sum_{i=s_1}^{s_2} x_i \, x_{i+i_1} \, \cdots \, x_{i+i_{k-1}} \tag{12.2}$$

where $s_1 = \max\{0, -i_1, \ldots, -i_{k-1}\}$ and $s_2 = \min\{M-1, M-1-i_1, \ldots, M-1-i_{k-1}\}$. Due to definition (12.2), the kOMs exhibit symmetries [1, 20], thus the domain of support $\mathcal{R}_0 \equiv \{(i_1, i_2, \ldots, i_{k-1}) : \ 0 \le i_{k-1} \le i_{k-2} \le \cdots \le i_2 \le i_1 \le M-1\}$ is sufficient for their estimation. Let us define the region $\mathcal{I}_p \equiv \{[i_1, \ldots, i_{k-1}, i]^T : [i_1, \ldots, i_{k-1}]^T \in \mathcal{R}_0 \text{ and } 0 \le i \le M-1-i_1\}$, and the kth order product term $p_k(i_1, \ldots, i_{k-1}, i), (i_1, i_2, \ldots, i_{k-1}, i) \in \mathcal{I}_p$ as:

$$\begin{aligned} p_k(i_1, i_2, \ldots, i_{k-1}, i) &\equiv \; x_i \prod_{l=1}^{k-1} x_{i+i_l} \\ &= \; p_{k-1}(i_1, i_2, \ldots, i_{k-2}, i) \, x_{i+i_{k-1}}; \quad p_1 = x_i, \; p_0 = 1. \end{aligned}$$

Clearly the products $\{p_k(i_1, i_2, \ldots, i_{k-1}, i)\}$ with $(i_1, i_2, \ldots, i_{k-1}, i) \in \mathcal{I}_p$ form a minimal size set of terms, thus estimates of the kOMs in the *nonredundant* region $\mathcal{R}_0$ can be equivalently computed using the following formula:

$$\widehat{m}_k(i_1, i_2, \ldots, i_{k-1}) = \frac{1}{M} \sum_{i=0}^{M-i_1-1} p_{k-1}(i_1, i_2, \ldots, i_{k-2}, i) \, x_{i+i_{k-1}}. \tag{12.3}$$

Equation (12.3) demonstrates that every kth order moment term $\widehat{m}_k$ can be computed using product terms $\{p_{k-1}\}$ (participating in the computation of the terms $\{\widehat{m}_{k-1}\}$) in at most M steps; we call this a "forward" order recursion.

In Figure 12.2 we provide a nested-loop algorithm that computes *all* moment terms, up to a desired order k, in a forward-order recursive fashion based on (12.3). The indexes $i_1, i_2, \ldots, i_{k-1}$ correspond to the moment lags, and i, l to the iteration and order index, respectively. The motivation behind the reformulation of (12.2) as in (12.3) is that we want to be able to compute moments of any order as it is necessary for the computation of higher order *cumulants* [37]. It is well known that the lth order cumulants can be obtained from lth and lower order moments for every l [1, 20].

12.3.2 LRA SYNTHESIS

We first apply the unified synthesis methodology and construct systematically an appropriate LRA that can lead to the synthesis of a family of optimal triangular arrays for moments estimation, up to an arbitrary order k, within $\mathcal{R}_0$. The starting point is the nested-loop algorithm of Figure 12.2 and the results obtained here are valid *for any maximal order k*.

12.3.2.1 Data Dependence Analysis of the Algorithm

The algorithm's index space is $\mathcal{I}_m \equiv \{\vec{v} = [i_1, \ldots, i_{k-1}, i, l]^T : \ 0 \le i_{k-1} \le \cdots \le i_2 \le i_1 \le M-1, \ 0 \le i \le M-1-i_1, \text{ and } 1 \le l \le k\} \subset \mathcal{Z}^{n_a}$, where $n_a = k+1$ denotes the dimensionality of the algorithm, and to simplify the notation we let $\mathbf{e}_i \equiv \mathbf{e}_i^{(n_a)}$ denote the ith column of the identity

Let: $\mathcal{R}_0 = \{(i_1, ..., i_{k-1}) : 0 \le i_{k-1} \le ... \le i_2 \le i_1 \le M - 1\}$ and
$\mathcal{I}_m = \{(i_1, ..., i_{k-1}, i, l) : (i_1, ..., i_{k-1}) \in \mathcal{R}_0, 0 \le i \le M - 1 - i_1, l = 1, ..., k\}, i_0 \equiv 0$
Initially: $p(i_1, ..., i_{k-1}, i, 0) = 1, m(i_1, ..., i_{k-1}, l) = 0, \forall (i_1, ..., i_{k-1}) \in \mathcal{R}_0, 0 \le i \le M - 1 - i_1, 1 \le l \le k$.

for $i_1 = 0$ to $M - 1$
   for $i_2 = 0$ to i_1
      ...
      for $i_{k-1} = 0$ to i_{k-2}
         for $i = 0$ to $M - i_1 - 1$
            for $l = 1$ to k
               $p(i_1, ..., i_{k-1}, i, l) \leftarrow p(i_1, ..., i_{k-1}, i, l - 1) \, x_{i+i_{l-1}}$
               $m(i_1, ..., i_{k-1}, l) \leftarrow m(i_1, ..., i_{k-1}, l) + p(i_1, ..., i_{k-1}, i, l)$
            endfor $\{l\}$
         endfor $\{i\}$
      endfor $\{i_{k-1}\}$
      ...
   endfor $\{i_2\}$
endfor $\{i_1\}$

At the end: $\widehat{m}_1 \leftarrow \frac{1}{M} m(0, ..., 0, M - 1, 1)$, $\widehat{m}_2(i_1) \leftarrow \frac{1}{M} m(i_1, 0, ..., 0, M - 1 - i_1, 2), ...,$
$\widehat{m}_{k-1}(i_1, ..., i_{k-2}) \leftarrow \frac{1}{M} m(i_1, i_2, ..., i_{k-2}, 0, M - 1 - i_1, k - 1),$
$\widehat{m}_k(i_1, ..., i_{k-1}) \leftarrow \frac{1}{M} m(i_1, i_2, ..., i_{k-1}, M - 1 - i_1, k)$

FIGURE 12.2 A nested-loop, forward order recursive algorithm for the estimation of all the joint moments, up to a desired order k, at their nonredundant region of support $\mathcal{R}_0$. Reprinted from [38] with permission from Taylor & Francis.

matrix $\mathbf{I}_{n_a}$. The variables involved in the algorithm are x, p, and m with *instances* at the right-hand side of the assignment statements $x_{i+i_{l-1}}$, $p_{l-1}(i_1, \ldots, i_{k-1}, i)$, and $m_l(i_1, \ldots, i_{k-1})$, $l = 1, 2, \ldots, k$, respectively.

Because l is the $k + 1$'st index in $\mathcal{I}_m$, $p_{l-1}(i_1, \ldots, i_{k-1}, i)$ is rewritten as $p(f[\vec{v}]) = p(i_1, \ldots, i_{k-1}, i, l - 1)$. In addition, $(i_1, \ldots, i_{k-1}, i, l - 1)^{\mathrm{T}} = \mathbf{I}_{k+1} \cdot \vec{v} + (0, ..., 0, 0, -1)^{\mathrm{T}} = \mathbf{A}_p \cdot \vec{v} + \mathbf{b}_p$, its indexing matrix is $\mathbf{A}_p = \mathbf{I}_{k+1}$ and $\mathbf{b}_p = -\mathbf{e}_{k+1}$. Note that $\mathbf{A}_p$ is full rank (dim $\mathcal{N}(\mathbf{A}_p) = 0$) and therefore variable p *is not* a broadcast. Moreover, $p(i_1, \ldots, i_{k-1}, i, l)$ depends on $p(i_1, \ldots, i_{k-1}, i, l - 1)$, so there is a *true* dependence for variable p that can be described by a *constant* dependence vector $\mathbf{d}_p = -\mathbf{b}_p = \mathbf{e}_{k+1}$ (pointing towards increasing l-index values). On the other hand variable m, whose instance $m(i_1, \ldots, i_{k-1}, l)$ has indexing matrix $\mathbf{A}_m = [\mathbf{e}_1 \mid \cdots \mid \mathbf{e}_{k-1} \mid \mathbf{e}_{k+1}]^{\mathrm{T}}$, is a broadcast. Since $\mathcal{N}(\mathbf{A}_m)$ is the space spanned by $\mathbf{e}_k$ (note that the "missing" index in $m(i_1, \ldots, i_{k-1}, l)$ is i which is the kth index in $\mathcal{I}_m$) and dim $\mathcal{N}(\mathbf{A}_m) = 1$, the vector $\mathbf{d}_m \equiv a_m \mathbf{e}_k$, $a_m \in \mathcal{Z}$ characterizes all *valid* directions of propagation for the localization of variable m.

Determining the indexing matrix associated with variable x is more complicated because the indexing function of the instance $x_{i+i_{l-1}}$ does not correspond to an affine transformation of the index point $\vec{v} \equiv [i_1, \ldots, i_{k-1}, i, l]^{\mathrm{T}}$. For each value of $l = 1, 2, \ldots, k$ we get a different instance $x_{i+i_{l-1}}$ of variable x that will be renamed as w_j, $j = 0, \ldots, k - 1$. Each one of these variables is defined only within the $n_a - 1 = k$-dimensional subspace of $\mathcal{I}_m$, namely the hyperplanes with $l = 1, \ldots, l = k$ respectively. So we define $w_0(i) \equiv x_i$, $w_1(i_1, i) \equiv x_{i+i_1}, \ldots, w_{k-1}(i_{k-1}, i) \equiv x_{i+i_{k-1}}$ which now have indexing functions that are affine transformations of the index point $\vec{v}$. It is not hard to verify that the corresponding indexing matrices are $\mathbf{A}_{w_0} = \mathbf{e}_k^{(k)^{\mathrm{T}}}$, and $\mathbf{A}_{w_j} = [\mathbf{e}_j^{(k)} + \mathbf{e}_k^{(k)}]^{\mathrm{T}}$, for $j = 1, ..., k - 1$, respectively. Since dim $\mathcal{N}(\mathbf{A}_{w_0}) = $ dim $\mathcal{N}(\mathbf{A}_{w_j}) = k - 1$, all the variables w_j, $j = 0, ..., k - 1$ are broadcasts and their propagation contours can be selected within the null spaces of their respective indexing matrices. Therefore the valid directions of propagation for the

w_j (possible data dependencies $\mathbf{d}_{w_j}$) may be expressed as:

$$\mathbf{d}_{w_0} = a_{0,1}\mathbf{e}_1 + \cdots + a_{0,k-1}\mathbf{e}_{k-1}$$
$$\mathbf{d}_{w_1} = a_{1,2}\mathbf{e}_2 + \cdots + a_{1,k-1}\mathbf{e}_{k-1} + b_1(\mathbf{e}_1 - \mathbf{e}_k)$$
$$\cdots$$
$$\mathbf{d}_{w_j} = a_{j,1}\mathbf{e}_1 + \cdots + a_{j,j-1}\mathbf{e}_{j-1} + a_{j,j+1}\mathbf{e}_{j+1} + \cdots + a_{j,k-1}\mathbf{e}_{k-1} + b_j(\mathbf{e}_j - \mathbf{e}_k)$$
$$\cdots$$
$$\mathbf{d}_{w_{k-1}} = a_{k-1,1}\mathbf{e}_1 + \cdots + a_{k-1,k-2}\mathbf{e}_{k-2} + b_{k-1}(\mathbf{e}_{k-1} - \mathbf{e}_k) \tag{12.4}$$

where $a_{i,j}$ and $b_i \in \mathcal{Z}$.

The above analysis shows that only variable p has a true dependence with a direction specified by the original nested-loop algorithm. This dependence *is fixed* in contrast with the rest of the dependences for variables w_j, which may assume different vectors in a $(k-1)$-dimensional subspace. Therefore by appropriately selecting these dependences we may be able to *construct* a "suitable" LRA, i.e., one that meets our design objectives and target implementation constraints, and avoid adopting a valid, yet "inappropriate," LRA that will lead to a suboptimal architectural design after the mapping.

12.3.2.2 Determining the Space Transformation

To ensure I/O operations from the VLSI architecture's boundaries (constraint C2), the moment terms m should be produced at boundary PEs. Given the nested-loop algorithm of Figure 12.2, the final values for m become available at the algorithm's output hyperplane $i + i_1 = M - 1$ (with normal vector $\mathbf{e}_1 + \mathbf{e}_k$) which should be mapped to the periphery of the array architecture. Similarly, the input data for m (possibly corresponding to the estimates produced by the previous data block), are fed at the input hyperplane $i = 0$ (with normal vector $\mathbf{e}_k$). To satisfy (C2) the normal vectors to I/O hyperplanes, $\mathbf{e}_1 + \mathbf{e}_k$ and $\mathbf{e}_k$, must be parallel to the physical processor space, or equivalently the vectors $\mathbf{e}_1$, $\mathbf{e}_k$ should constitute the two-column vectors of space mapping matrix $\mathbf{S}$.

There is some freedom in selecting a third basis vector. Because there is irregularity in the algorithm (variables w_j) for different values of l, we should avoid projecting the DG along dimension l. So $\mathbf{e}_{k+1}$ is chosen as the third basis vector and $\mathbf{S}$ is given by:

$$\mathbf{S} = [\mathbf{e}_1 \mid \mathbf{e}_k \mid \mathbf{e}_{k+1}]^T, \quad \mathbf{S} \in \mathcal{Z}^{3 \times (k+1)}. \tag{12.5}$$

As a result of this space mapping, the algorithm's index space $\mathcal{I}_m \subset \mathcal{Z}^{n_a}$, $n_a = k + 1$, is transformed onto the processor space $\mathcal{P} \equiv \{\vec{v}_p = [i_1, i, l]^T : 0 \leq i_1 \leq M-1, 0 \leq i \leq M-1-i_1$, and $1 \leq l \leq k\} \subset \mathcal{Z}^{n_p}$, $n_p = 3$. In other words, the resulting physical architecture is a triangular array in the (i_1, i) plane, with sides of size M. Furthermore, every physical PE consists of a pipeline of k multiply accumulate (MAC) units organized along the third (internal) direction l ($1 \leq l \leq k$). For the rest of the paper and to simplify the notation we will denote as $\mathbf{u}_i \equiv \mathbf{e}_i^{(n_p)}$, the ith column vector of the identity matrix $\mathbf{I}_{n_p}$.

12.3.2.3 Constructing a Suitable LRA

Now it remains to determine the dependence matrix $\mathbf{D} \equiv [\mathbf{d}_{w_0} \mid \mathbf{d}_{w_1} \mid \mathbf{d}_{w_2} \mid \cdots \mid \mathbf{d}_{w_{k-1}} \mid \mathbf{d}_m \mid \mathbf{d}_p]$ to fully specify the LRA. Therefore, we have to decide which one of the valid directions of propagation will be used for every broadcast type variable. In Ref. [38], we have examined every variable and constructed an LRA that under the space mapping of Equation (12.5) it meets all the VLSI-imposed constraints C1–C3 and in addition minimizes the number of PCs. This "suitable"

Let: $\mathcal{R}_0 = \{(i_1, ..., i_{k-1}) : 0 \leq i_{k-1} \leq ... \leq i_2 \leq i_1 \leq M - 1\}$
Initially $\forall (i_1, ..., i_{k-1}) \in \mathcal{R}_0, 0 \leq i \leq M - 1 - i_1, 1 \leq l \leq k$: $p(i_1, ..., i_{k-1}, i, 0) = 1$,
$m(i_1, ..., i_{k-1}, -1, l) = 0$.

for $i_1 = 0$ to $M - 1$
  for $i_2 = 0$ to i_1
    $\cdots$
    for $i_{k-1} = 0$ to i_{k-2}
      for $i = 0$ to $M - 1 - i_1$
        for $l = 1$ to k

$$w(i_1, ..., i_{k-1}, i, l - 1) \leftarrow \begin{cases} w(i_1 + 1, i_2, ..., i_{k-1}, i, l - 1), & \text{if } l \neq 2, i_1 < M\text{-}1\text{-}i \\ (\mathbf{d}_{w_{l-1}} = -\mathbf{e}_1)x_{i+i_{l-1}}, & \text{if } l \neq 2, i_1 = M\text{-}1\text{-}i \\ \\ w(i_1 + 1, i_2, ..., i_{k-1}, i - 1, l - 1), & \text{if } l = 2, i > 0 \\ (\mathbf{d}_{w_1} = -\mathbf{e}_1 + \mathbf{e}_k)x_{i+i_1}, & \text{if } l = 2, i = 0 \end{cases}$$

          $p(i_1, ..., i_{k-1}, i, l) \leftarrow p(i_1, ..., i_{k-1}, i, l - 1) \, w(i_1, ..., i_{k-1}, i, l - 1)$ $(\mathbf{d}_p = \mathbf{e}_{k+1})$
          $m(i_1, ..., i_{k-1}, i, l) \leftarrow m(i_1, ..., i_{k-1}, i - 1, l) + p(i_1, ..., i_{k-1}, i, l)$ $(\mathbf{d}_m = \mathbf{e}_k)$
        endfor $\{l\}$
      endfor $\{i\}$
    endfor $\{i_{k-1}\}$
    $\cdots$
  endfor $\{i_2\}$
endfor $\{i_1\}$
At the end: $\widehat{m}_1 \leftarrow \frac{1}{M}m(0, ..., 0, M - 1, 1)$, $\widehat{m}_2(i_1) \leftarrow \frac{1}{M}m(i_1, 0, ..., 0, M - 1 - i_1, 2)$, $...$,
$\widehat{m}_{k-1}(i_1, ..., i_{k-2}) \leftarrow \frac{1}{M}m(i_1, i_2, ..., i_{k-2}, 0, M - 1 - i_1, k - 1)$,
$\widehat{m}_k(i_1, ..., i_{k-1}) \leftarrow \frac{1}{M}m(i_1, i_2, ..., i_{k-1}, M - 1 - i_1, k)$

FIGURE 12.3 The systematically constructed locally recursive algorithm for the estimation of all the joint moments, up to a desired order k. Reprinted from [38] with permission from Taylor & Francis.

LRA is represented by the following dependencies matrix

$$\begin{aligned} \mathbf{D} &\equiv \left[\mathbf{d}_{w_0} \mid \mathbf{d}_{w_1} \mid \mathbf{d}_{w_2} \mid \cdots \mid \mathbf{d}_{w_{k-1}} \mid \mathbf{d}_m \mid \mathbf{d}_p\right] \\ &= \left[-\mathbf{e}_1 \mid -\mathbf{e}_1 + \mathbf{e}_k \mid -\mathbf{e}_1 \mid \cdots \mid -\mathbf{e}_1 \mid \mathbf{e}_k \mid \mathbf{e}_{k+1}\right]. \end{aligned} \tag{12.6}$$

Observe that rank$(\mathbf{D}) = 3$ *independent of the maximum order parameter k* ($\mathbf{d}_{w_0}$, $\mathbf{d}_m$, and $\mathbf{d}_p$ are linearly independent and $\mathbf{d}_{w_1} = \mathbf{d}_{w_0} + \mathbf{d}_m$, $\mathbf{d}_{w_2} = \mathbf{d}_{w_3} = \cdots = \mathbf{d}_{w_{k-1}} = \mathbf{d}_{w_0}$). The constructed LRA, that is equivalent to the algorithm of Figure 12.2, is given in conditional uniform recurrent equations form in Figure 12.3. Variables w_j, $j = 0, ..., k - 1$ are all embedded into variable w in the $k + 1$-dimensional index space. The constructed LRA is not a RIA because the dependence vectors $\mathbf{d}_{w_{l-1}}$, $l = 1, ..., k$ may be different depending on l. The dependence that introduces this irregularity is $\mathbf{d}_{w_1}$ ($l = 2$). However, vector $\mathbf{d}_{w_1}$ has been chosen to be a linear combination of vectors $\mathbf{d}_{w_0}$ and $\mathbf{d}_m$ with nonnegative integer coefficients, it does not introduce any additional constraints when searching for minimum latency schedules, i.e., if a schedule is compatible with $\mathbf{d}_{w_0}$ and $\mathbf{d}_m$, it will not violate $\mathbf{d}_{w_1}$.

12.3.3 MOMENTS GENERATING TRIANGULAR ARRAY ARCHITECTURES

In this section, we present the systematic design of a family of minimum latency, minimum local memory, triangular systolic arrays for the real-time estimation of *all moments lags up to the fourth order*. From a practical standpoint, this is the most interesting special case ($k = 4$) of the general problem, so all results derived in Section 12.3 are applicable here.

The algorithm's index space becomes now $\mathcal{I}_m = \{[i_1, i_2, i_3, i, l]^T : 0 \le i_3 \le i_2 \le i_1 \le M - 1, \ 0 \le i \le M - 1 - i_1, \ 1 \le l \le 4\} \subset \mathcal{Z}^5$. The linear space transformation $\mathbf{S} \in \mathcal{Z}^{3 \times 5}$ that guarantees data/results I/O from the architecture boundaries becomes (see Equation (12.5)):

$$\mathbf{S} = [\mathbf{e}_1 \mid \mathbf{e}_4 \mid \mathbf{e}_5]^T = \begin{bmatrix} 1 & 0 & 0 & 0 & 0 \\ 0 & 0 & 0 & 1 & 0 \\ 0 & 0 & 0 & 0 & 1 \end{bmatrix}. \tag{12.7}$$

The above choice makes the processor space to be $\mathcal{P} \equiv \{\vec{i}_p = [i_1, i, l]^T : 0 \le i_1 \le M - 1, \ 0 \le i \le M - 1 - i_1, \text{ and } 1 \le l \le 4\} \subset \mathcal{Z}^3$ with the physical array realized in the two-dimensional (i_1, i) plane. The same processor space would result from *multiprojection* [19] if the successive projection vectors applied were first $\mathbf{p}_1 = [0, 1, 0, 0, 0]^T \in \mathcal{Z}^5$ and then $\mathbf{p}_2 = [0, 0, 1, 0]^T \in \mathcal{Z}^4$. When $k = 4$, the dependencies matrix of the LRA constructed in Section 12.3.2.3 reduces to:

$$\mathbf{D} \equiv [\mathbf{d}_{w_0} \mid \mathbf{d}_{w_1} \mid \mathbf{d}_{w_2} \mid \mathbf{d}_{w_3} \mid \mathbf{d}_m \mid \mathbf{d}_p] = \begin{bmatrix} -1 & -1 & -1 & -1 & 0 & 0 \\ 0 & 0 & 0 & 0 & 0 & 0 \\ 0 & 0 & 0 & 0 & 0 & 0 \\ 0 & 1 & 0 & 0 & 1 & 0 \\ 0 & 0 & 0 & 0 & 0 & 1 \end{bmatrix}. \tag{12.8}$$

The resulting LRA is shown in Figure 12.4.

One can easily verify that $\mathbf{d}_{w_1} = \mathbf{d}_{w_0} + \mathbf{d}_m$. So we have managed to construct an LRA (and DG) where all data dependences are characterized by only three linearly independent vectors $(\mathbf{d}_{w_0}, \mathbf{d}_m, \mathbf{d}_p)$. This construction facilitated optimal scheduling of the algorithm and allowed

Let: $\mathcal{R}_0 = \{(i_1, i_2, i_3) : 0 \le i_3 \le i_2 \le i_1 \le M - 1\}$
Initially: $p(i_1, i_2, i_3, i, 0) = 1$, $m(i_1, i_2, i_3, -1, l) = 0$, $\forall (i_1, i_2, i_3) \in \mathcal{R}_0, 0 \le i \le M - 1 - i_1, 1 \le l \le 4$.

for $i_1 = 0$ to $M - 1$
  for $i_2 = 0$ to i_1
    for $i_3 = 0$ to i_2
      for $i = 0$ to $M - 1 - i_1$
        for $l = 1$ to 4

$$w(i_1, i_2, i_3, i, l - 1) \leftarrow \begin{cases} w(i_1 + 1, i_2, i_3, i, l - 1), & \text{if } l = 1, 3, 4, i_1 < M - 1 - i \quad (\mathbf{d}_{w_0} = \mathbf{d}_{w_2} = \mathbf{d}_{w_3} \\ = [-1, 0, 0, 0, 0]^T) x_{i + i_{l-1}}, & \text{if } l = 1, 3, 4, i_1 = M - 1 - i \\ \\ w(i_1 + 1, i_2, i_3, i - 1, l - 1), & \text{if } l = 2, i > 0 \quad (\mathbf{d}_{w_1} = [-1, 0, 0, 1, 0]^T) \\ x_{i + i_1}, & \text{if } l = 2, i = 0 \end{cases}$$

$p(i_1, i_2, i_3, i, l) \leftarrow p(i_1, i_2, i_3, i, l - 1) \, w(i_1, i_2, i_3, i, l - 1), \quad (\mathbf{d}_p = [0, 0, 0, 0, 1]^T)$
$m(i_1, i_2, i_3, i, l) \leftarrow m(i_1, i_2, i_3, i - 1, l) + p(i_1, i_2, i_3, i, l), \quad (\mathbf{d}_m = [0, 0, 0, 1, 0]^T)$
        endfor $\{l\}$
      endfor $\{i\}$
    endfor $\{i_3\}$
  endfor $\{i_2\}$
endfor $\{i_1\}$

At the end: $\widehat{m}_1 \leftarrow \frac{1}{M} m(0, 0, 0, M - 1, 1)$, $\widehat{m}_2(i_1) \leftarrow \frac{1}{M} m(i_1, 0, 0, M - 1 - i_1, 2)$
$\widehat{m}_3(i_1, i_2) \leftarrow \frac{1}{M} m(i_1, i_2, 0, M - 1 - i_1, 3)$, $\widehat{m}_4(i_1, i_2, i_3) \leftarrow \frac{1}{M} m(i_1, i_2, i_3, M - 1 - i_1, 4)$

FIGURE 12.4 The constructed locally recursive algorithm for the estimation of all the joint moments, up to the fourth order, $k = 4$. Reprinted from [38] with permission from Taylor & Francis.

us (see Ref. [38] for details) to (a) determine and characterize all permissible schedule vectors of the format $\mathbf{T} = [t_1, t_2, t_3, t_4, t_5]$, and (b) find among them those yielding minimum latency.

For example, using the space transformation $\mathbf{S}$ of Equation (12.7) and the minimum latency schedule $\mathbf{T}_{m_2} = [-1, 1, M - 1, 1, 1]$ (derived in Ref. [38]) yields the triangular parallel VLSI architecture realized in the (i_1, i) plane, as shown in Figure 12.5 (for $M = 4$). Every square node corresponds to a PE of the array consisting of four pipelined stages (and are organized along the third direction l of the processor space). Each PE stage is a MAC unit. The delay D between two successive schedule time instants models the time required for a MAC computation and the communication between near-neighboring stages. Due to the appropriate choice of $\mathbf{S}$ and the dependence matrix $\mathbf{D}$, all I/O operations occur from the boundary PEs. In particular, the input data for variable

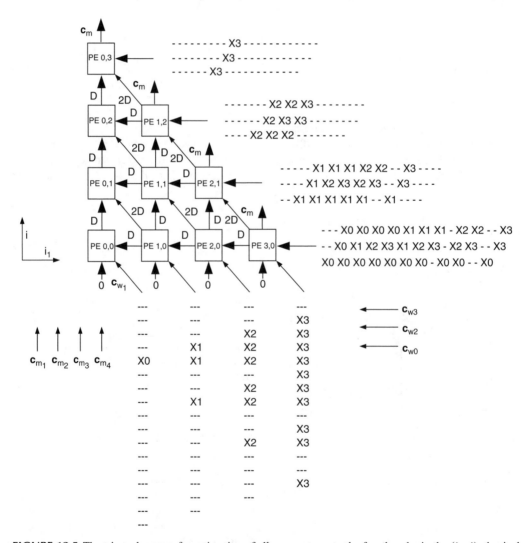

FIGURE 12.5 The triangular array for estimation of all moments up to the fourth order in the (i_1, i) physical processor space, when $M = 4$. The minimum latency schedule used is $\mathbf{T}_{m_2} = [-1, 1, M - 1, 1, 1]$. Reprinted from [38] with permission from Taylor & Francis.

w_1 ($l = 2$) are fed from the boundary line $i = 0$. The other input data $\{w_0, w_2, w_3\}$ ($l = 1, 3, 4$, respectively) are fed at the line $i_1 + i = M - 1$, the hypotenuse of the triangular array where also all the results $\{m_l\}$ become available.

The data dependencies of the DG (or LRA) are transformed to SFG edges by:

$$[\mathbf{c}_{w_0} \mid \mathbf{c}_{w_1} \mid \mathbf{c}_{w_2} \mid \mathbf{c}_{w_3} \mid \mathbf{c}_m \mid \mathbf{c}_p] \equiv \mathbf{S} \cdot \mathbf{D} = \begin{bmatrix} -1 & -1 & -1 & -1 & 0 & 0 \\ 0 & 1 & 0 & 0 & 1 & 0 \\ 0 & 0 & 0 & 0 & 0 & 1 \end{bmatrix}.$$

For example, dependence $\mathbf{d}_{w_1} = [-1, 0, 0, 1, 0]^{\mathrm{T}}$ is mapped into the SFG edge $\mathbf{c}_{w_1} = [-1, 1, 0]^{\mathrm{T}}$ pointing toward "decreasing i_1 and increasing i." It becomes apparent now why we have chosen the same dependence vector for the three different variable instances x_i, x_{i+i_2}, x_{i+i_3}: *they will all propagate through parallel SFG edges*. These three SFG edges ($\mathbf{c}_{w_0}$, $\mathbf{c}_{w_2}$, and $\mathbf{c}_{w_3}$) may be realized as a single communication link when the array is implemented in VLSI hardware.

The number of delays per SFG edge are determined by the time transformation $\mathbf{T}_{m_2}$ as follows:

$$\mathbf{T}_{m_2} \cdot \mathbf{D} = [1, 2, 1, 1, 1, 1].$$

Notice that these delays do not depend on the data block size M, leading to $O(1)$ local memory requirements per PE. This was made possible by selecting *the* pair of schedule and $\mathbf{D}$, among numerous permissible choices, that would yield a vector $\mathbf{T} \cdot \mathbf{D}$ with elements independent of M. The main advantage of the unified synthesis methodology is that it reveals such options, should they be available.

Each PE requires minimum local memory. In fact, the number of edge delays does not depend on the problem size M, and only seven storage elements (buffers) per PE are needed for any problem size. This is in contrast with an earlier design [39] where the *arbitrary selection* of a different but valid LRA resulted in $O(M)$ local memory requirements per processor. Furthermore a simple FIFO or LIFO external memory organization is sufficient to support the array's operation.

However, processors utilization is not uniform but decays as the i coordinate of the PE increases. This is unavoidable when the shape of the index space is pyramidal and only linear space–time transformations are considered. Techniques that seek to maximize the PE utilization by interleaving the computation of successive data blocks are presented in Ref. [14]. It can be easily verified that the block pipelining period [19] is $\beta = M(M - 1) + 1$. Because the sequential complexity of the estimation algorithm for $k = 4$ is in $O(M^4)$ and $O(M^2)$ processors are used, β is asymptotycally optimal.

12.4　FOURTH-ORDER CUMULANTS ESTIMATION

12.4.1　THE PROBLEM

Given a real, zero-mean, one-dimensional process, the set of its fourth-order cumulants $\{C_4\}$ can be estimated by using

$$\begin{aligned} C_4(i_1, i_2, i_3) &= \widehat{m}_4(i_1, i_2, i_3) - \widehat{m}_2(i_1)\widehat{m}_2(i_2 - i_3) - \widehat{m}_2(i_2)\widehat{m}_2(i_1 - i_3) \\ &\quad - \widehat{m}_2(i_3)\widehat{m}_2(i_1 - i_2), \end{aligned} \tag{12.9}$$

where i_1, i_2, i_3 are the lag indices and $\{\widehat{m}_l\}, l = 2, 4$, are the second and fourth-order moment estimates.

Our main objective is to design an integrated, fully pipelined (HOS) VLSI architecture that can compute in real-time estimates of *all the nonredundant moment and cumulant terms up to the fourth order*. Here, we will show how the constraint-driven unified array synthesis methodology discussed in Section 12.2.2 can be applied to construct an ASAP architecture that (i) computes efficiently all terms of the three-dimensional cumulants array $\{C_4\}$ and (ii) is optimally space–time matched (interfaced) with the moments generating triangular array, presented in Section 12.3, which accepts as input the samples of the incoming data sequence and provides estimates of *all* the $\{\widehat{m}_2\}$ and $\{\widehat{m}_4\}$ moment terms, as needed to compute the cumulant terms $\{C_4\}$.

12.4.2 LRA SYNTHESIS

12.4.2.1 Data Dependence Analysis of the Algorithm

A nested-loop algorithm providing the $\{C_4\}$ in the primary domain of support $\mathcal{R}_0 = \{(i_1, i_2, i_3) : 0 \le i_3 \le i_2 \le i_1 \le M-1\}$ is given in Figure 12.6. The algorithm's index space is four-dimensional, $\mathcal{I}_c \equiv \{(i_1, i_2, i_3, j)^{\mathrm{T}} : 0 \le i_3 \le i_2 \le i_1 \le M-1, 1 \le j \le 3\} \subset \mathcal{Z}^4$. We determine first all the valid directions of propagation for every instance of *broadcast* variables encountered in the right-hand side of the assignment statement. The variables to be considered here are C_4, w_{lj} and w_{rj}, where the subscripts l and r simply stand for "left" and "right" term in the assignment statement. The indexed instance of C_4 is $C_4(i_1, i_2, i_3)$ with indexing matrix $\mathbf{A}_c = [\mathbf{e}_1 \mid \mathbf{e}_2 \mid \mathbf{e}_3]^{\mathrm{T}}$. For the other two variable instances their indexing matrix depends on the value of j. In particular, $\mathbf{A}_{w_{lj}} = [\mathbf{e}_j]^{\mathrm{T}}$, $1 \le j \le 3$ and

$$
\mathbf{A}_{w_{rj}} = \begin{cases} [\mathbf{e}_2 - \mathbf{e}_3]^{\mathrm{T}}, \text{ for } j = 1 \\ [\mathbf{e}_1 - \mathbf{e}_3]^{\mathrm{T}}, \text{ for } j = 2 \\ [\mathbf{e}_1 - \mathbf{e}_2]^{\mathrm{T}}, \text{ for } j = 3 \end{cases}.
$$

Primary Domain of Support: $\mathcal{R}_0 = \{(i_1, i_2, i_3) : 0 \le i_3 \le i_2 \le i_1 \le M-1\}$
Index space: $\mathcal{I}_c \equiv \{(i_1, i_2, i_3, j) : 0 \le i_3 \le i_2 \le i_1 \le M-1, 1 \le j \le 3\}$

Let: $w_{lj}(i_1, i_2, i_3) \equiv \widehat{m}_2(i_j), j = 1, 2, 3,$
$w_{r1}(i_1, i_2, i_3) \equiv \widehat{m}_2(i_2 - i_3),$
$w_{r2}(i_1, i_2, i_3) \equiv \widehat{m}_2(i_1 - i_3),$
and $w_{r3}(i_1, i_2, i_3) \equiv \widehat{m}_2(i_1 - i_2).$
Initially: $C_4(i_1, i_2, i_3) = \widehat{m}_4(i_1, i_2, i_3).$
for $i_1 = 0$ to $M - 1$
    for $i_2 = 0$ to i_1
        for $i_3 = 0$ to i_2
            for $j = 1$ to 3
                $C_4(i_1, i_2, i_3) = C_4(i_1, i_2, i_3) - w_{lj}(i_1, i_2, i_3) \cdot w_{rj}(i_1, i_2, i_3)$
            endfor $\{j\}$
        endfor $\{i_3\}$
    endfor $\{i_2\}$
endfor $\{i_1\}$

FIGURE 12.6 Nested-loop algorithm for estimating the $\{C_4\}$ in the nonredundant domain of support $\mathcal{R}_0$. Reprinted from [37] with permission from Elsevier Science.

Therefore the dependencies introduced for the propagation of variables C_4, w_{lj}, and w_{rj} should have the general form

$$\mathbf{d}_{w_{lj}} = \begin{cases} \mathbf{d}_{l1} \equiv a_{l1}\mathbf{e}_2 + a_{l2}\mathbf{e}_3, & j = 1 \\ \mathbf{d}_{l2} \equiv b_{l1}\mathbf{e}_1 + b_{l2}\mathbf{e}_3, & j = 2 \\ \mathbf{d}_{l3} \equiv c_{l1}\mathbf{e}_1 + c_{l2}\mathbf{e}_2, & j = 3 \end{cases}, \qquad \mathbf{d}_c = a_c\mathbf{e}_4 \qquad (12.10)$$

$$\mathbf{d}_{w_{rj}} = \begin{cases} \mathbf{d}_{r1} \equiv a_{r1}\mathbf{e}_1 + a_{r2}(\mathbf{e}_2 + \mathbf{e}_3), & j = 1 \\ \mathbf{d}_{r2} \equiv b_{r1}\mathbf{e}_2 + b_{r2}(\mathbf{e}_1 + \mathbf{e}_3), & j = 2 \\ \mathbf{d}_{r3} \equiv c_{r1}\mathbf{e}_3 + c_{r2}(\mathbf{e}_1 + \mathbf{e}_2), & j = 3 \end{cases}$$

where the coefficients $\{a, b, c\}$ are integers to be fixed after determining the valid space and time mapping operators, so that certain architectural constraints and performance goals are met (constraint-driven localization of dependencies).

12.4.2.2 Determining the Space Transformation

To determine the space transformation $\mathbf{S}_c$, we consider the *space matching* between the two stages of the pipelined HOS architecture, the first one producing the moments and the second one computing the fourth-order cumulants. An optimal matching should result to minimal data migration between the two stages. However, it may not be possible to minimize the data migration and delays with respect to both the $\{\widehat{m}_2\}$ and $\{\widehat{m}_4\}$ streams of tokens that flow from the first to the second stage. Notice that the algorithm estimating the $\{C_4\}$ (see Equation (12.9)) consumes $O(M^3)$ tokens $\{\widehat{m}_4\}$ but only M tokens $\{\widehat{m}_2\}$, where M is the length of the data sequence. Therefore, the space matching of the two stages will be attempted only with respect to the stream $\{\widehat{m}_4\}$.

The space transformation $\mathbf{S}$ determined in Section 12.3.2.2 yields a moments generating triangular array architecture on the (i_1, i) plane with sides equal to M (see Figure 12.5). The set of fourth-order moment lags $\{\widehat{m}_4(k, i_2, i_3)\}$ become available at processing element $PE_{i_1=k, i=M-1-i_1}$ located at the hypotenuse of the triangular array. These results are needed as input at points $\{[k, i_2, i_3, 1]^T\} \in \mathcal{I}_c$ in the index space of the algorithm computing the $\{C_4\}$. To achieve a space matching that minimizes the complexity of the in-between stages interconnection hardware, these index points should also be allocated to a PE with index $i_1 = k$. The following linear transformation achieves this space matching objective

$$\mathbf{S}_c = [\mathbf{e}_1 \mid \mathbf{e}_4]^T = \begin{bmatrix} 1 & 0 & 0 & 0 \\ 0 & 0 & 0 & 1 \end{bmatrix}. \qquad (12.11)$$

The index space $\mathcal{I}_c$ is mapped under $\mathbf{S}_c$ onto the two-dimensional processors space $\mathcal{P} \equiv \{(i_1, j): 0 \leq i_1 \leq M - 1, \text{ and } 1 \leq j \leq 3\} \subset \mathcal{Z}^2$. This may be thought of as a size M one-dimensional array of physical processors organized along i_1. The second "pseudo dimension" j (whose range does not grow with the problem size M) models the internal structure of every physical PE, which consists of three pipelined MAC stages along j. Note that since the physical architecture is an one-dimensional array space-matched to the whole hypotenuse of the moment generating triangular array, we can relax constraint C_2 (see Section 12.2) demanding I/O operations from the array boundaries.

12.4.2.3 Construction of the LRA

Constructing the LRA, is equivalent to fixing all the dependence vectors. It amounts to assigning a direction of propagation to every localized variable, or equivalently to fixing the values of all coefficients $\{a, b, c\}$ in expressions (12.10). So far, following our unified synthesis approach, we have identified *all* valid directions of propagation for localizing each variable. Thus, we are now in the position to assess whether specific design objectives can be met by properly using the available degrees of freedom (coefficient values). Furthermore, we can consider design trade-offs and evaluate candidate solutions before committing to any specific one. We discuss next the step-by-step construction of an LRA that enjoys certain desirable properties.

1. Let us try to keep the number of *linearly independent* dependence vectors as small as possible and select these few basis vectors so that all other dependencies can be expressed as linear combinations of them with nonnegative integer coefficients. By doing so, we impose the smallest possible number of precedence constraints when searching for permissible schedules, thus exploiting all the inherently available parallelism, as explained in Section 12.2. According to Equation (12.10) three such sets of basis vectors can be found: $\{\mathbf{e}_1, \mathbf{e}_2, \mathbf{e}_4\}$, $\{\mathbf{e}_1, \mathbf{e}_3, \mathbf{e}_4\}$, and $\{\mathbf{e}_2, \mathbf{e}_3, \mathbf{e}_4\}$.

2. Let us try to partition the processors space into a set of disjoint PEs and eliminate interprocessor communication. Given that $\mathbf{S_c} = [\mathbf{e}_1|\mathbf{e}_4]^\mathrm{T}$ is fixed (selected to achieve optimal space matching), we should avoid using a basis vector depending on $\mathbf{e}_1$, because $\mathbf{S_c} \cdot \mathbf{e}_1 = \mathbf{u}_1$ introducing inter-PE communications along the first (i_1) dimension. Therefore, with $\mathbf{S}_c$ given, we can achieve a partitioned design if we choose as basis vectors the set $\{\mathbf{e}_2, \mathbf{e}_3, \mathbf{e}_4\}$. Using Equation (12.10), the dependence vectors then become:

$$
\begin{aligned}
\mathbf{d}_c &= a_c\mathbf{e}_4 \\
\mathbf{d}_{l1} &= a_{l1}\mathbf{e}_2 \quad or \quad \mathbf{d}_{l1} = a_{l2}\mathbf{e}_3, \quad \mathbf{d}_{l2} = b_{l2}\mathbf{e}_3, \quad \mathbf{d}_{l3} = c_{l2}\mathbf{e}_2, \quad (12.12) \\
\mathbf{d}_{r1} &= a_{r2}(\mathbf{e}_2 + \mathbf{e}_3), \qquad\qquad\qquad \mathbf{d}_{r2} = b_{r1}\mathbf{e}_2, \quad \mathbf{d}_{r3} = c_{r1}\mathbf{e}_3
\end{aligned}
$$

3. Among all permissible and minimum latency schedules we prefer those that provide the cumulants in increasing lag-index order, so that the most significant terms, which are close to the origin, are produced first. Note that the terms $\{C_4(i_1, i_2, i_3)\}$ become available at index points $\{[i_1, i_2, i_3, 3]^\mathrm{T} \in \mathcal{I}_c\}$. They will be produced in an increasing lag-index order (with respect to i_2 and i_3) only if $\tau_2 = \mathbf{T}_c \cdot \mathbf{e}_2 > 0$ and $\tau_3 = \mathbf{T}_c \cdot \mathbf{e}_3 > 0$. In Ref. [37] we have derived and characterized all permissible schedules that are compatible with the moments generating array schedule and achieve an overall minimum latency. Among those we can select schedules that also provide the cumulants in increasing lag-index order.

4. The range of coefficients $\{a, b, c\}$ affects the length and/or number of delays in the SFG links resulting after the mapping [19]. For example, the SFG link corresponding to the dependence $\mathbf{d}_c$ is $\mathbf{c}_c \equiv \mathbf{S} \cdot \mathbf{d}_c = a_c\mathbf{u}_2$, i.e., it points along the second "pseudo dimension" j, modeling the internal structure of each PE consisting of a pipeline of MAC stages. Its length is $|\mathbf{c}_c| = |a_c|$, and given a schedule $\mathbf{T}_c = [\tau_1, \tau_2, \tau_3, \tau_4]$ the number of delays associated with $\mathbf{c}_c$ will be equal to $|\mathbf{T}_c \cdot \mathbf{d}_c| = |\tau_4||a_c|$. Hence, if we want to minimize the number of delay buffers we should choose coefficient a_c to be ± 1. For the same reason, we use the same set of values for the rest of the coefficients. A set of dependence vectors that are of the form given in (12.12) and meet the precedence constraints by fixing the integer coefficients $\{a, b, c\}$ to unity is given below:

$$
\mathbf{D} \equiv [\mathbf{d}_{l1} \mid \mathbf{d}_{l2} \mid \mathbf{d}_{l3} \mid \mathbf{d}_{r1} \mid \mathbf{d}_{r2} \mid \mathbf{d}_{r3} \mid \mathbf{d}_c] = \begin{bmatrix} 0 & 0 & 0 & 0 & 0 & 0 & 0 \\ 1 & 0 & 1 & 1 & 1 & 0 & 0 \\ 0 & 1 & 0 & 1 & 0 & 1 & 0 \\ 0 & 0 & 0 & 0 & 0 & 0 & 1 \end{bmatrix}. \quad (12.13)
$$

Clearly, we have introduced the smallest possible number of linearly independent dependence vectors ($\mathbf{d}_{l1}$, $\mathbf{d}_{l2}$, and $\mathbf{d}_c$) whereas all the other vectors can be expressed as linear combinations of those with nonnegative integer coefficients. The constructed LRA, corresponding to the dependence matrix $\mathbf{D}$ in Equation (12.13), is shown in Figure 12.7.

12.4.2.4   A Farm of Processors Cumulants Generating Architecture

Among the available minimum latency schedules we have selected $\mathbf{T}_{c_2} = [0, 1, M - 1, 1]$ which achieves data flow matching with the minimum latency schedule $\mathbf{T}_{m_2} = [-1, 1, M - 1, 1, 1]$ used for the moments generating tri-array (for a detailed discussion on the schedule vector selection, see Ref. [37]).

By applying the space–time transformation pair $(\mathbf{S}_c, \mathbf{T}_{c_2})$ to the LRA of Figure 12.7, we obtain the SFG shown in the Figure 12.8 (for $M = 4$). This is a *farm* architecture with M independent non-communicating PEs organized along dimension i_1. Estimates of the cumulants slice $\{C_4(k, i_2, i_3)\}$ are produced in increasing lag-indexes (i_2, i_3) order at $PE_{i_1=k}$, $k = 0, 1, ..., M - 1$. A node corresponds to a PE composed of three-pipelined MAC stages (organized along the *internal* pseudo-dimension j).

The data dependencies of the LRA are transformed to SFG edges that are "self-loops," except from the dependence $\mathbf{d}_c$ that translates to the internal PE link directly connecting two MAC stages

Algorithm index space: $\mathcal{I}_c \equiv \{(i_1, i_2, i_3, j) : 0 \leq i_3 \leq i_2 \leq i_1 \leq M - 1, 1 \leq j \leq 3\}$
Initially: $C_4(i_1, i_2, i_3, 0) = \widehat{m}_4(i_1, i_2, i_3)$
for $i_1 = 0$ to $M - 1$
   for $i_2 = 0$ to i_1
      for $i_3 = 0$ to i_2
         for $j = 1$ to 3

$$w_l(i_1, i_2, i_3, j) \leftarrow \begin{cases} w_l(i_1, i_2 - 1, i_3, j), & \text{if } j = 1,\ i_2 \neq i_3 \quad (\mathbf{d}_{l1} = \mathbf{e}_2) \\ \widehat{m}_2(i_1), & \text{if } j = 1,\ i_2 = i_3 \\ w_l(i_1, i_2, i_3 - 1, j), & \text{if } j = 2,\ i_3 \neq 0 \quad (\mathbf{d}_{l2} = \mathbf{e}_3) \\ \widehat{m}_2(i_2), & \text{if } j = 2,\ i_3 = 0 \\ w_l(i_1, i_2 - 1, i_3, j), & \text{if } j = 3,\ i_2 \neq i_3 \quad (\mathbf{d}_{l3} = \mathbf{e}_2) \\ \widehat{m}_2(i_3), & \text{if } j = 3,\ i_2 = i_3 \end{cases}$$

$$w_r(i_1, i_2, i_3, j) \leftarrow \begin{cases} w_r(i_1, i_2 - 1, i_3 - 1, j), & \text{if } j = 1,\ i_3 \neq 0 \quad (\mathbf{d}_{r1} = \mathbf{e}_2 + \mathbf{e}_3) \\ \widehat{m}_2(i_2), & \text{if } j = 1,\ i_3 = 0 \\ w_r(i_1, i_2 - 1, i_3, j), & \text{if } j = 2,\ i_2 \neq i_3 \quad (\mathbf{d}_{r2} = \mathbf{e}_2) \\ \widehat{m}_2(i_1 - i_3), & \text{if } j = 2,\ i_2 = i_3 \\ w_r(i_1, i_2, i_3 - 1, j), & \text{if } j = 3,\ i_3 \neq 0 \quad (\mathbf{d}_{r3} = \mathbf{e}_3) \\ \widehat{m}_2(i_1 - i_2), & \text{if } j = 3,\ i_3 = 0 \end{cases}$$

$C_4(i_1, i_2, i_3, j) = C_4(i_1, i_2, i_3, j - 1) - w_l(i_1, i_2, i_3, j)w_r(i_1, i_2, i_3, j) \quad (\mathbf{d}_c = \mathbf{e}_4)$
         endfor $\{j\}$
      endfor $\{i_3\}$
   endfor $\{i_2\}$
endfor $\{i_1\}$

FIGURE 12.7 The systematically constructed locally recursive algorithm for the estimation of the $\{C_4\}$, in conditional uniform recurrence equations (CURE) format. Reprinted from [37] with permission from Elsevier Science.

as shown by:

$$[\mathbf{c}_{l1} \mid \mathbf{c}_{l2} \mid \mathbf{c}_{l3} \mid \mathbf{c}_{r1} \mid \mathbf{c}_{r2} \mid \mathbf{c}_{r3} \mid \mathbf{c}_c] \equiv \mathbf{S} \cdot \mathbf{D} = \begin{bmatrix} 0 & 0 & 0 & 0 & 0 & 0 & 0 \\ 0 & 0 & 0 & 0 & 0 & 0 & 1 \end{bmatrix}.$$

Clearly, no SFG edge exists with direction along i_1. Thus, the appropriate selection of the space transformation $\mathbf{S}_c$ and data dependence vectors for the localization of broadcast variables leads to the discovery of a farm architecture with internally pipelined processors and no interprocessor communication. The number of delays per SFG edge is given by:

$$\mathbf{T}_{c_2} \cdot \mathbf{D} = [0, 1, M-1, 1] \cdot \begin{bmatrix} 0 & 0 & 0 & 0 & 0 & 0 & 0 \\ 1 & 0 & 1 & 1 & 1 & 0 & 0 \\ 0 & 1 & 0 & 1 & 0 & 1 & 0 \\ 0 & 0 & 0 & 0 & 0 & 0 & 1 \end{bmatrix} = [1, M-1, 1, M, 1, M-1, 1].$$

For the dependencies that are transformed to self-loops, an edge delay of M units models a circular queue with M buffers. The control scheme of these queues is very simple: At every schedule time instant, the contents of the queue are shifted down in a cyclical manner. If the incoming token is an autocorrelation $\widehat{m}_2(k) = r_k$ term then it overwrites the first position of the queue; else if it is a dummy ("–") it is discarded. After this downshift operation, the datum residing in the first position of the queue is used in the multiplication.

At every schedule time step, in MAC stage $j = k$ of a PE the two autocorrelation tokens received through links $\mathbf{c}_{lk}$ and $\mathbf{c}_{rk}$ (associated with dependencies $\mathbf{d}_{lk}$ and $\mathbf{d}_{rk}$) are multiplied and the product from the result of the previous MAC stage in the pipeline is then subtracted. This partial

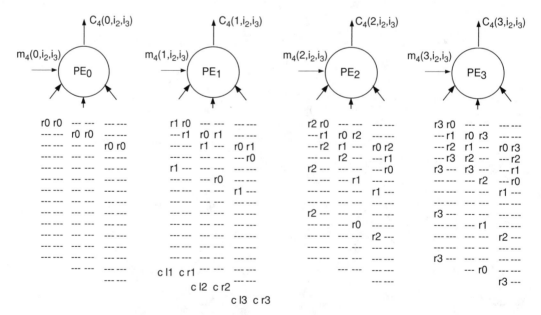

FIGURE 12.8 The farm of processors for the computation of the $\{C_4\}$ terms, $(M = 4)$. Each thick input line corresponds to the pair of links $\{c_{lj}, c_{rj}\}$ feeding with autocorrelation tokens the corresponding internal MAC stage $j = 1, 2, 3$ (not shown). Reprinted from [37] with permission from Elsevier Science.

result is initialized with the appropriate $\widehat{m}_4$ token supplied to the first ($j = 1$) MAC stage by the tri-array, and takes its final value corresponding to a $\{C_4\}$ term at the last ($j = 3$) MAC stage, as shown in Figure 12.9. Notice that due to the space–time matching, the terms $\{\widehat{m}_4\}$ are supplied to each PE_{i_1} of the farm in the order that they are produced by the corresponding $PE_{i_1, M-1-i_1}$ in the hypotenuse of the tri-array designed in section 12.3

As it can be seen in Figure 12.8 the autocorrelation term r_k is needed to all MAC stages of the PEs with index $i_1 \geq k$. Because the farm PEs are disjoint, r_k should be broadcasted via a bus. A total of $3M$ multiplications are performed in the farm at every schedule time step. Some of these computations are redundant, because the same product may be calculated multiple times at different PEs that need it. This computational redundancy, however, does not affect the array's latency.

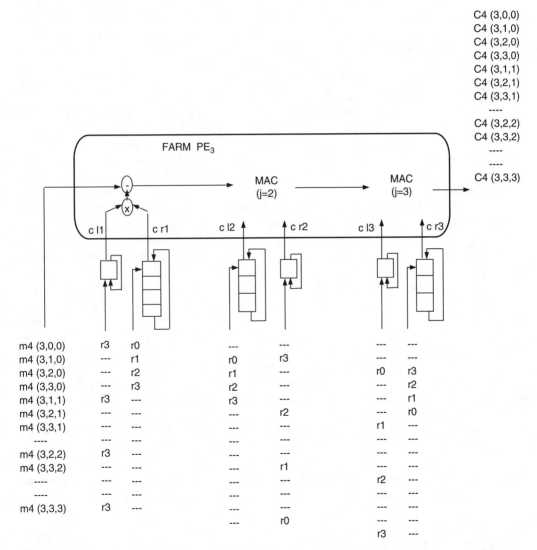

FIGURE 12.9 The pipelined structure and the input/output data skews for PE_{M-1} of the farm, computing the terms $\{C_4(M - 1, i_2, i_3), i_3 \leq i_2 \leq M - 1\}$ ($M = 4$). Reprinted from [37] with permission from Elsevier Science.

In Ref. [37] we have presented an alternative one-dimensional array solution whose operation does not require any data broadcasting. This array implements a modified version of the fourth-order cumulants estimation algorithm with minimal computational redundancy. Both one-dimensional array solutions use the same number of PEs and achieve the same latency, which is the smallest possible under linear scheduling. The linear array solution saves one multiplier per PE (relatively to the farm) but requires near-neighbor PE communications and a moderate amount of additional control logic.

12.5 CONCLUSIONS

A highly pipelined two-dimensional VLSI array architecture has been designed for the real-time estimation of HOM and cumulants. By introducing a unified, constraint-driven array synthesis methodology, it became possible to discover a new LRA that (i) remains applicable up to any maximal order k and (ii) can be transformed, after suitable linear space–time mapping, to a modular two-dimensional array processor architecture that meets the VLSI-imposed constraints as well as all user-specified design objectives. In particular, the synthesized two-dimensional triangular VLSI array architecture achieves minimum latency, minimum local memory, and consists of highly pipelined PEs that can provide estimates of *all* nonredundant moment lags.

The proposed design characterization methodology provides analytical expressions for architecture performance metrics in terms of problem size parameters that are very useful when exploring architectural options for realizing algorithms with consecutive nested-loop stages in hardware. For instance, when results of the first algorithmic stage are consumed by subsequent stages, analytical performance characterization allows us to find if there exist $(\mathbf{S}, \mathbf{T})$ mapping operator pairs that achieve optimal *space–time matching* between the two stages, so that buffering is minimized and the in-between architectural stages interconnection network is simplified, whereas the overall design remains a minimum latency solution. Following the proposed constraint-driven design methodology we have synthesized such a two-stage VLSI architecture, which first produces one set of statistics (the HOM) that are then used to compute another set (the fourth order cumulants). In general, the proposed constraint-driven array synthesis methodology can be employed to transform a high-dimensional nested-loop algorithmic specification (with known index space and indexing functions at compile-time) to optimal planar parallel array architectures with several pipelined stages.

All dependence graphs and space–time mapping transformation pairs derived here can be supplied as input to the design tool suite DG2VHDL developed in our group [40, 41] to generate automatically: (i) high-quality synthesizable VHDL models for the SFG processor array and its PEs, (ii) VHDL testbenches used to perform behavioral simulation of the derived array architectures for validation purposes, and (iii) scripts for behavioral and RTL level synthesis. Given a DG, the user of the DG2VHDL tool suite can search for optimal space–time operators and obtain with no effort synthesizable, optimized VHDL code for the different array models that meet the search criteria. This feature can facilitate the exploration and comparison of candidate processor array solutions before synthesis. As a last step, an industrial strength behavioral compiler (e.g., by Synopsys) may be used to translate the optimized VHDL description of the selected SFG array architecture to a netlist suitable for hardware implementation using a variety of application-specific device technologies, such as the increasingly popular field programmable gate arrays (FPGAs) [42].

REFERENCES

[1] J.M. Mendel. Tutorial on higher-order statistics (spectra) in signal processing and system theory: theoretical results and some applications. *Proc. IEEE*, 79(3):278–305, 1991.

[2] C.L. Nikias and A.P. Petropulu. *Higher-Order Spectral Analysis: A Nonlinear Signal Processing Framework*. Prentice Hall, New York, 1993.

[3] C.L. Nikias and J.M. Mendel. Signal processing with higher-order spectra. *IEEE Signal Process.*, 10(3):10–37, 1993.

[4] J.K. Tugnait. Identification of linear stochastic systems via second- and fourth-order cumulant matching. *IEEE Trans. Inform. Theory*, 33(3):393–407, 1987.

[5] S.W. Nam and E.J. Powers. Application of higher order spectral analysis to cubically non-linear system identification. *IEEE Trans. Signal Process.*, 42(7):1746–1765, 1994.

[6] J.F. Cardoso. Super-symmetric decomposition of the fourth-order cumulant tensor, blind identification of more sources than sensors. In *Proc. ICASSP*, 1991, pp. 3109–3112.

[7] M. Binder and H. Pesaran. Stochastic growth models and their econometric implications. *J. of Econ. Growth*, 4(2):139–183, 1999.

[8] R.C. Mlttelhammer. *Mathematical Statistics for Economics and Business*. Springer-Verlag, New York, 1996.

[9] D. Sornette, P. Simonetti, and J.V. Andersen. Nonlinear covariance matrix and portofolio theory for non-gaussian multivariate distributions. In *Res. Papers Econ. (RePEc)*, 1999.

[10] M.J. Hinich and D.M. Patterson. Evidence of nonlinearity in the trade-by-trade stock market return generating process. In *Chaos, Sunspots, Bubbles, and Nonlinearity*, Cambridge University Press, London, 1989, pp. 383–409.

[11] A. Swami, G.B. Giannakis, and G. Zhou. Bibliography on higher order statistics. *Signal Process.*, 60:65–126, 1997.

[12] G.C.W Leung and D. Hatzinakos. Implementation aspects of various higher-order statistics estimators. *J. Franklin Inst. - Eng. Appl. Math.*, 333B:349–367, 1996.

[13] D. Hatzinakos. Analysis of floating point roundoff errors in estimation of higher order statistics. *IEEE Proc.-F Radar Signal Process.*, 140:371–379, 1993.

[14] E.S. Manolakos, H.M. Stellakis, and D.H. Brooks. Parallel processing for biomedical signal processing: higher order spectral analysis — an application. *IEEE Comp.*, 24(3):33–43, 1991.

[15] S. Aloqeely, M. Al-Turaigi, and S. Alshebeili. A new approach for the design of linear systolic arrays for computing third order cumulants. *Integration — VLSI J.*, 24(1):1–17, 1997.

[16] M. Al-Turaigi and S. Alshebeili. A high speed systolic array for computing third-order cumulants. *Integration — VLSI J.*, 22(1):19–23, 1997.

[17] H.M. Stellakis and E.S. Manolakos. A high speed systolic array for computing third-order cumulants. *Signal Process.*, 36(3):341–354, 1994.

[18] H.M. Stellakis and E.S. Manolakos. Adaptive computation of higher order moments and its systolic realization. *Int. J. of Adapt. Control Signal Process.*, 10(2/3):283–302, 1996.

[19] S.Y. Kung. *VLSI Array Processors*. Prentice Hall, New York, 1989.

[20] C.L. Nikias and M.R. Raghuveer. Bispectrum estimation: a digital signal processing framework. *IEEE Proc.*, 75(7):869–891, 1987.

[21] J. Kalamatianos and E.S. Manolakos. Parallel computation of higher order moments on the MasPar-1 machine. In *IEEE Int. Conf. Acoust. Speech Signal Process. (ICASSP'95)*, 1995, pp. 1832–1835.

[22] R. Karp, R. Miller, and S. Winograd. The organization of computations for uniform recurrence equations. *J. ACM*, 14(3):563–590, 1967.

[23] M.T. O'Keefe, J.A.B. Fortes, and B.W. Wah. On the relationship between two systolic array design methodologies. *IEEE Trans. Comp.*, 41(12):1589–1593, 1992.

[24] W. Shang and J.A.B. Fortes. On time mapping of uniform dependence algorithms into lower dimensional processor arrays. *IEEE Trans. Parallel Distributed Process.*, 3(3):350–363, 1992.

[25] S.K. Rao. Architecture design for regular iterative algorithms. In Earl. E. Swartzlander, Jr. Ed., *Systolic Signal Processing Systems*. Marcel Dekker, New York, Basel, 1987, pp. 209–257.

[26] P. Quinton and Y. Robert. *Systolic Algorithms and Architectures*. Prentice Hall, Masson, 1991.

[27] J. Bu, E.F. Deprettere, and L. Thiele. Systolic array implemenation of nested loop programs. In *Proc. IEEE Int. Conf. ASAP*, 1990, pp. 31–42.

[28] V. Van Dongen and P. Quinton. Uniformization of linear recurrence equations: a step towards the automatic synthesis of systolic arrays. In *Proc. IEEE Int. Conf. Syst. Arrays*, 1988, pp. 473–482.

[29] J.A.B. Fortes and D.I. Moldovan. Data broadcasting in linearly scheduled array processors. In *Proc. IEEE Ann. Int. Symp. Comp. Arch.*, 1984.

[30] Y. Wong and J.-M. Delosme. Transformation of broadcasts into propagations in systolic algorithms. *Parallel Distributed Comput.*, 12(2):121–145, 1992.

[31] G.-J. Li and B.W. Wah. The design of optimal systolic arrays. *IEEE Trans. Comput.*, c-34(1):66–77, 1985.

[32] D. Moldovan and J. Fortes. Partitioning and mapping algorithms into fixed size, systolic arrays. *IEEE Trans. Comput.*, 35(1):1–12, 1986.

[33] P. Lee and Z.M. Kedem. Mapping nested loop algorithms into multidimensional systolic arrays. *IEEE Trans. Parallel Distributed Syst.*, 1(1):64–76, 1990.

[34] W. Shang and J.A.B. Fortes. Time optimal linear schedules for algorithms with uniform dependencies. *IEEE Trans. Comput.*, 40(6):723–742, 1991.

[35] F. Lorenzelli and K. Yao. A systematic partitioning approach for LS and SVD problems to fixed size arrays with constraints. In *Proc. IEEE ICASSP*, 1992, pp. V-585–V-588.

[36] W. Shang and J.A.B. Fortes. Independent partitioning of algorithms with uniform dependencies. *IEEE Trans. Comput.*, 41(2):190–206, 1992.

[37] E.S. Manolakos and H.M. Stellakis. Systematic synthesis of parallel architectures for the computation of higher order cumulants. *Parallel Comput.*, 26:655–676, 2000.

[38] E.S. Manolakos and H.M. Stellakis. Systematic synthesis of parallel architectures for the real-time estimation of the higher order statistical moments. *Parallel Algorithms Appl.*, 15:77–111, 2000.

[39] H.M. Stellakis and E.S. Manolakos. A tri-array for the real-time computation of higher order moments estimates. In *VLSI Signal Process. V*. IEEE Press, New Jersey, 1992, pp. 510–519.

[40] A. Stone and E.S. Manolakos. DG2VHDL: a tool to facilitate the synthesis of parallel array architectures. *J. VLSI Signal Process. Syst.*, 24:99–120, 2000.

[41] DG2VHDL: A tool to facilitate the synthesis of parallel VLSI architectures, http://www.cdsp.neu.edu/info/faculty/manolakos/dg2vhdl_root.html.

[42] A. Stone and E.S. Manolakos. Using DG2VHDL to synthesize an FPGA implementation of the 1-D discrete wavelet transform. In *Proc. IEEE Signal Process. Syst. (SiPS)*, October 1998, pp. 489–498.

13 Principal Component Analysis for Information Retrieval

Michael W. Berry and Dian I. Martin

CONTENTS

ABSTRACT

Advances in computational statistics and high-performance computing have significantly reduced the computational burden of converting raw text to encoded vectors for the purpose of information retrieval (IR) modeling and testing. Using principal component analysis (or PCA) to encode large text collections into low-dimensional vector spaces within a few minutes is an important advancement in the study and development of IR models such as latent semantic indexing (LSI). The development and tuning of parallel Lanczos algorithms for computing the singular value decomposition (SVD) of sparse term-by-document matrices has greatly facilitated the use of LSI-based models for large-scale text mining and indexing applications. However, future software development (in languages such as C++) is still needed for efficient up- and down-dating of low-rank singular subspaces. Time-sensitive or dynamic text collections require regular updates and simply computing the principal components of an updated (even larger) term-by-document matrix may be unwieldy with limited RAM. Exploiting distributed storage (Internet-based) may well be the next advancement for vector space IR models. Storing (and replicating) the principal components or singular subspaces themselves on distributed (remote) servers could significantly increase query response time. Through parallel processing, the *fusion* of separate information sources into one *global* index can be accomplished.

13.1 INTRODUCTION

From traditional multivariate statistics literature [1, 7, 12], PCA attempts to explain the variance–covariance structure of data using relatively few linear combinations of the corresponding variables. Ideally, k principal components would be used to account for the total system variability involving the n original variables. Data reduction is achieved by replacing the original data set of m

measurements on n variables to one involving m measurements on $k \ll n$ principal components. PCA is commonly used to reveal underlying or *latent* relationships among the variables that would normally go unnoticed.

For a mean-centered[1] $m \times n$ data matrix A, the first step of PCA involves the construction of the *covariance* matrix

$$C = \frac{1}{m-1} A^T A . \tag{13.1}$$

The diagonal and off-diagonal elements of the matrix C can be used to explain the variance of each variable (associated with each column of matrix A) and the covariance of variable pairs. More specifically, the variance for variable i is given by the diagonal element

$$c_{ii} = \frac{1}{m-1} \sum_{k=1}^{m} a_{ki}^2,$$

and the covariance of variables i and j are given by the off-diagonal element

$$c_{ij} = c_{ji} = \frac{1}{m-1} \sum_{k=1}^{m} a_{ki} a_{kj} .$$

In the context of IR, these variables (or columns of the matrix A), might represent any report, article, paragraph, etc. and their measures (rows of the matrix A) would constitute the frequency of any *token*, word, or word phrase used in the document or document fragment. As a simple illustration, consider the classic Mother Goose Rhyme *A Flea and a Fly* (see below) where each *document* is just a line of the nursery rhyme.

A Flea and a Fly

> (L1) A flea and a fly,
> (L2) Flew up in a flue.
> (L3) Said the flea, "Let us fly!"
> (L4) Said the fly, "Let us flee!"
> (L5) So they flew through a flap in the flue.

Ignoring unimportant words or *stop words* as they are commonly referred to in the IR literature [2, 3], we obtain the following 12×5 term-by-document matrix A:

$$A = \begin{pmatrix} & L_1 & L_2 & L_3 & L_4 & L_5 \\ flap & 0 & 0 & 0 & 0 & 1 \\ [F, f]lea & 1 & 0 & 1 & 0 & 0 \\ flee & 0 & 0 & 0 & 1 & 0 \\ [F, f]lew & 0 & 1 & 0 & 0 & 1 \\ flue & 0 & 1 & 0 & 0 & 1 \\ [F, f]ly & 1 & 0 & 1 & 1 & 0 \\ Let & 0 & 0 & 1 & 1 & 0 \\ Said & 0 & 0 & 1 & 1 & 0 \\ they & 0 & 0 & 0 & 0 & 1 \\ through & 0 & 0 & 0 & 0 & 1 \\ up & 0 & 1 & 0 & 0 & 0 \\ us & 0 & 0 & 1 & 1 & 0 \end{pmatrix} \tag{13.2}$$

---

[1] That is, the columns of A, A_j, are replaced by $A_j - \bar{x}u$, where u is a $m \times 1$ vector of all ones, and $\bar{x}$ is the mean of A_j.

The mean-centered version (say $\tilde{A}$) of the 12×5 term-by-document (or term-by-line) matrix in (13.2) is given by

$$\tilde{A} = \begin{pmatrix}
-0.1667 & -0.2500 & -0.4167 & -0.4167 & 0.5833 \\
0.8333 & -0.2500 & 0.5833 & -0.4167 & -0.4167 \\
-0.1667 & -0.2500 & -0.4167 & 0.5833 & -0.4167 \\
-0.1667 & 0.7500 & -0.4167 & -0.4167 & 0.5833 \\
-0.1667 & 0.7500 & -0.4167 & -0.4167 & 0.5833 \\
0.8333 & -0.2500 & 0.5833 & 0.5833 & -0.4167 \\
-0.1667 & -0.2500 & 0.5833 & 0.5833 & -0.4167 \\
-0.1667 & -0.2500 & 0.5833 & 0.5833 & -0.4167 \\
-0.1667 & -0.2500 & -0.4167 & -0.4167 & 0.5833 \\
-0.1667 & -0.2500 & -0.4167 & -0.4167 & 0.5833 \\
-0.1667 & 0.7500 & -0.4167 & -0.4167 & -0.4167 \\
-0.1667 & -0.2500 & 0.5833 & 0.5833 & -0.4167
\end{pmatrix}, \tag{13.3}$$

and the 5×5 covariance matrix $\tilde{C}$ corresponding to $\tilde{A}$ is given by

$$\tilde{C} = \begin{pmatrix}
0.1515 & -0.0455 & 0.1061 & 0.0152 & -0.0758 \\
-0.0455 & 0.2045 & -0.1136 & -0.1136 & 0.0682 \\
0.1061 & -0.1136 & 0.2652 & \mathbf{0.1742} & -0.1894 \\
0.0152 & -0.1136 & \mathbf{0.1742} & 0.2652 & -0.1894 \\
-0.0758 & 0.0682 & -0.1894 & -0.1894 & 0.2652
\end{pmatrix}. \tag{13.4}$$

The largest diagonal elements of the covariance matrix $\tilde{C} = [\tilde{c}_{ij}]$ are $\tilde{c}_{33}$, $\tilde{c}_{44}$, and $\tilde{c}_{55}$ which indicate the last three lines of the rhyme have the largest variance of term usage. Also, the largest (positive) off-diagonal element $\tilde{c}_{34} = \tilde{c}_{43} = 0.1742$ suggests that the strongest correlation in term use lies between lines three and four of the rhyme which one would expect. The covariance matrix can be difficult to interpret as the number of variables (columns of the matrix A or $\tilde{A}$) increases. Fortunately, its eigenvalues and eigenvectors can provide valuable information as the problem size (number of documents) increases.

Before discussing how the eigensolution of the covariance matrix C (or $\tilde{C}$) can account for the variance of the term usage in documents, it is important to note that in practical IR applications the original term-by-document matrix A may be used for statistical modeling rather than the mean-centered modified matrix $\tilde{A}$. As illustrated in (13.2), such matrices will typically be sparse[1] so that compressed storage formats [3] can be used to store and retrieve only the nonzero elements. Computing the covariance[2] of the matrix A in (13.2) produces

$$C = \begin{pmatrix}
0.1818 & 0 & 0.1818 & 0.0909 & 0 \\
0 & 0.2727 & 0 & 0 & 0.1818 \\
0.1818 & 0 & 0.4545 & \mathbf{0.3636} & 0 \\
0.0909 & 0 & \mathbf{0.3636} & 0.4545 & 0 \\
0 & 0.1818 & 0 & 0 & 0.4545
\end{pmatrix}. \tag{13.5}$$

Notice that the matrix element $c_{34} = c_{43} = 0.3636$ is still the largest off-diagonal element for the nursery rhyme example. Avoiding mean subtraction for data sets may not be advisable in other PCA settings, of course. In IR, the original matrix A is essentially an *incidence* matrix of nonnegative

---

[1] That is, have relatively few nonzeros compared to zeros.
[2] In a deviation from standard notation, this covariance matrix (as defined in (13.1)) is constructed for a data matrix A which is not mean-centered.

term frequencies. As will be discussed in Section 13.3, these frequencies are considered as *weights* for quantifying the importance of a term in describing the semantic content of a document [2, 3, 5]. Hence, PCA in this particular context is applied to strictly nonnegative sparse matrices.

13.2 RELATED EIGENSYSTEM

Suppose $\Lambda = \mathrm{diag}\,(\lambda_1, \lambda_2, \ldots, \lambda_n)$, where $\lambda_1 \geq \lambda_2 \geq \cdots \lambda_n$, is the diagonal matrix of eigenvalues of the covariance matrix C in (13.1), and $Q = (q_1, q_2, \ldots, q_n)$ is the matrix of corresponding eigenvectors. Then,

$$CQ = Q\Lambda,$$

and all the eigenvalues (λ_i) will be real and nonnegative since $A^\mathrm{T}A$ is symmetric and positive semidefinite. Suppose we define another $m \times n$ matrix B such that $B = [b_{ij}] = AQ$ or

$$b_{ij} = \sum_{k=1}^{n} a_{ik} q_{kj}.$$

Since the ith row of the matrix A defines the frequencies of occurrence for term i in all documents of the collection, b_{ij} is the linear combination of all term i occurrences as represented by the jth eigenvector of the covariance matrix C. Hence, the matrix B defines an alternative data matrix for transformed term-to-document associations. The covariance matrix for B is then given by

$$\frac{1}{m-1}(AQ)^\mathrm{T}(AQ) = \frac{1}{m-1}Q^\mathrm{T}A^\mathrm{T}AQ = \Lambda.$$

This implies that the transformed data has no correlation between the variables.

The *principal components* for such term frequency data are simply the leading[1] eigenvectors (q_i) of the covariance matrix C. It can be shown [1, 7, 12] that the total variance of the original variables (documents) as given by the trace of the covariance matrix C is equal to the total variance of the transformed variables. This total variance is represented by the sum of the eigenvalues, i.e.,

$$\frac{1}{m-1}\sum_{k=1}^{n}\sum_{i=1}^{m} a_{ik}^2 = \sum_{k=1}^{n} \lambda_k = \frac{1}{m-1}\sum_{k=1}^{n}\sum_{i=1}^{m} b_{ik}^2.$$

The first or primary principal component (q_1) is the linear combination of the variables (documents) for which the variance of term frequency is maximum. The second principal component (q_2) is the linear combination of the variables with the second-largest variance and is uncorrelated[2] with the first principal component q_1. The third, fourth, and higher order principal components are defined in similar fashion. Hence, data reduction is achieved by the projection of the covariance matrix C onto its k leading (or dominant) eigenvectors [4].

The eigensystem of the covariance matrix C in (13.5) is given by

$$\mathrm{diag}(\Lambda) = (0.8724 \quad 0.5669 \quad 0.1739 \quad 0.1604 \quad 0.0446),$$

---

[1] The term *leading* refers to the eigenvectors corresponding to the largest eigenvalues of the covariance matrix C.
[2] That is, orthogonal to q_1 from a linear algebra perspective.

$$Q = \begin{pmatrix} 0.2708 & 0 & 0.7693 & 0 & -0.5786 \\ 0 & 0.5257 & 0 & 0.8507 & 0 \\ 0.6963 & 0 & 0.2586 & 0 & 0.6696 \\ 0.6648 & 0 & -0.5842 & 0 & -0.4656 \\ 0 & 0.8507 & 0 & -0.5257 & 0 \end{pmatrix}.$$

Because the trace of the matrix Λ is 1.8182, the proportion of the total variance that can be attributed to each of the transformed variables (documents) is obtained by dividing the eigenvalues (λ_i's) by the value 1.8182. The first principal component or q_1 having all of its nonzero elements positive and of similar magnitude (i.e., not spanning different orders of magnitude) reflects similar term usage patterns in lines 1, 3, and 4 of the nursery rhyme *A Flea and a Fly*. This trend/behavior accounts for about half (near 48%) of the total variance. The second principal component or q_2 depicts a different pattern by indicating that lines 2 and 5 would tend to have the same term usage. This pattern accounts for 31% of the total variance so that the remaining principal components (collectively) account for only 21% of the total variance.

PCA is commonly used to distinguish correlations between the variables from the effects of random variations or *noise*. In many applications, closely spaced eigenvalues of the covariance matrix identify variance attributed to random variations (noise). On the other hand, if the leading (dominant) eigenvalues are sufficiently well-separated from the others, the corresponding eigenvectors (of the covariance matrix) may account for significant correlations between variables. As first discussed in Refs. [4, 6] and later in Ref. [5], the spectrum of $A^T A$ for most term-by-document matrices A does not necessarily provide these types of interpretations. Although the separation of eigenvalues of the covariance matrix greatly diminishes for most term-by-document matrices after a few hundred eigenpairs,[1] the optimal number of principal components to use for clustering documents according to semantic content is still an open question. Noise in this context is due to word usage variability — synonymy and polysemy to be more exact. Synonymy or the use of synonyms refers to many ways that a single concept can be expressed (using words).

Using our nursery rhyme example, a *flea* could be referred to as an *insect*, *bug*, *parasite*, or *pest* depending on the context. The word *flap* from the rhyme is polysemous in that it could refer to a motion (slap), a piece of garment, a piece of tissue for surgical grafting, part of a carton or box, the motion of a sail, or even a state of excitement. The proper meaning of the term can only be revealed from interpreting its use with other co-occurring terms. This mismatch in word use between authors of documents and those searching for relevant information from text-based collections (especially the World Wide Web [WWW]) is precisely what vector space models such as LSI [3–5, 9] attempt to correct via the use of PCA to project users' queries into high-dimensional vector spaces in which nearby document clusters (of vectors) reflect correlated information.

Producing term and document representations encoded in a k-dimensional vector space is the most computationally demanding task of LSI-based IR models. The approach used to produce such encodings is the SVD [10]. As will be discussed in the next section, computing the truncated SVD of a sparse term-by-document matrix A amounts to computing the k-leading principal components of the (unscaled) covariance matrix $C = A^T A$. As digital libraries and repositories of textual information continue to grow at staggering rates, the need for automated approaches to gauge document relatedness (without having to manually read or index each document) is becoming ever so critical. The remaining sections of this chapter focus on the computational aspects of PCA as defined by the sparse SVD problem arising in IR applications.

---

[1] *Eigenpair* refers to an eigenvalue and its corresponding eigenvector.

13.3   SINGULAR VALUE DECOMPOSITION

To encode terms and documents into k-dimensional subspaces, vector space IR models such as LSI [3–5, 9] require the factorization of large sparse term-by-document matrices (A) into orthogonal component matrices (or factors). The SVD [10] is one approach for computing the required factors. For an $m \times n$ matrix A (assuming $m \gg n$) of rank r, the SVD of A is defined as

$$A = U \Sigma V^{\mathrm{T}}, \tag{13.6}$$

where $U = (u_1, \ldots, u_m)$ and $V = (v_1, \ldots, v_n)$ are orthogonal matrices and $\Sigma = \mathrm{diag}(\sigma_1, \sigma_2, \ldots, \sigma_n)$, $\sigma_1 \geq \sigma_2 \geq \cdots \geq \sigma_r > \sigma_{r+1} = \cdots = \sigma_n = 0$. The σ_i's are the singular values of A, and the u_i's and v_i's are, respectively, the left and right singular vectors associated with σ_i, $i = 1, \ldots, r$. The three-tuple $\{u_i, \sigma_i, v_i\}$ is generally referred to as the ith singular triplet of the matrix A.

If iterative methods such as Lanczos or Arnoldi [3] are used to compute eigensystems of $A^{\mathrm{T}}A$, where A is an $m \times n$ sparse term-by-document matrix, substantial memory and CPU time can be required to compute the required k-dominant singular subspace. Although typical values of k tend to range from 100 to 300 [4, 9], further investigations into the effects of using more of the spectrum with singular subspaces on the order of $k = 1000$ or larger will be needed.

Investigating the properties of term-by-document matrices arising from IR applications facilitates better construction of vector space models for automated indexing and searching. Selecting the appropriate term weighting scheme [3, 5] to stress the importance of a term (keyword) within a document (local weighting) and across the text collection (global weighting) requires immediate knowledge of the incidence matrix or *graph* of the term frequencies or occurrences within the parsed documents. As discussed in Ref. [3], each element of the term-by-document matrix $A = [a_{ij}]$ can be defined by

$$a_{ij} = l_{ij} \times g_i \times n_j, \tag{13.7}$$

where l_{ij} is the local weight for term i occurring in document j, g_i is the global weight for term i in the collection, and n_j is a document normalization factor which specifies whether or not the columns of A (i.e., the documents) are normalized.[1]

Such matrices can be quite large with hundreds of thousands of rows (number of terms) and perhaps millions of columns (number of documents) for evolving document collections. As mentioned at the beginning of this chapter, the number of nonzeros per row is determined by how many different documents use a given term. As term usage can be widely varying, these matrices tend to be quite sparse (most of the a_{ij}'s are zero) with little or no exploitable nonzero structure, unlike matrices from finite element applications in structural mechanics or computational fluid dynamics. Visualization of these matrices is not practical and so extraction of the statistical properties of the term–document associations (as reflected by the nonzeros) is the only viable means of gaining insight. As these matrices are typically factored into orthogonal components via the SVD, SDD, ULV, or similar decompositions [3, 4], it is desirable to partition them for (parallelized) sparse matrix multiplication kernels used by iterative methods such as Lanczos and Arnoldi.

Current research in vector space models such as LSI [4, 5] is focused on determining the significance/contribution of any given singular triplet (or principal component) toward the effective representations of terms and documents in low-dimensional subspaces. Fast yet robust techniques for generating spectral[2] and rank information (or simple bounds) on term-by-document matrices will greatly facilitate the removal or inclusion of orthogonal factors for subsequent query processing.

---

[1]   That is, have unit-length or Euclidean norm of 1.

[2]   That is, singular triplets.

A recent study by Martin and Berry [13] quantified how much speed improvement can be expected for the parallel and distributed computation of the SVD of matrices arising in IR models. The target-computing platform used was that of the most commonly available parallel environment — a network of workstations (NOW). In the following sections, we review the basics of the Lanczos algorithm for computing the sparse SVD followed by parallelization issues and some of the recent performance results obtained for benchmark text collections [13].

13.4 LANCZOS ALGORITHM

The Lanczos algorithm (ca. 1950) is certainly one of the most popular methods for solving sparse symmetric eigenproblems [6, 10, 16]. The algorithm involves the generation of a sequence of $i \times i$ tridiagonal matrices T_i whose extremal eigenvalues approximate the corresponding eigenvalues of the original matrix A. For computing term and document vectors for IR models, we require the SVD of the (typically rectangular) term-by-document matrix A and so the Lanczos algorithm is used to approximate eigenpairs of the matrix $B = A^T A$. It is certainly possible for the matrix A to be $n \times n$ but the number of terms (m) parsed from a collection of (n) documents is usually much larger. For document collections spanning many contexts (e.g., newsfeeds, general reference, or the World Wide Web), the ratio of terms to documents can be 10 to 1 or even higher.

Suppose A is a (sparse) term-by-document matrix whose largest singular values and corresponding singular vectors are sought, and let q_1 be a randomly generated starting $n \times 1$ vector such that $\|q_1\|_2 = 1$. For $i = 1, 2, ..., k$, the Lanczos (tridiagonal) matrices T_i and Lanczos vectors q_i are defined by the following recursion [15]:

$$\beta_{i+1} q_{i+1} = B q_i - \alpha_i q_i - \beta_i q_{i-1}, \text{ and} \tag{13.8}$$

$$\alpha_i = q_i^T (B q_i - \beta_i q_{i-1})$$

$$|\beta_{i+1}| = \|B q_i - \alpha_i q_i - \beta_i q_{i-1}\|_2,$$

where $B = A^T A$, $\beta_1 \equiv 0$, and $q_0 \equiv 0$. For each Lanczos step i, the matrix T_i is a real symmetric tridiagonal matrix having diagonal entries α_j ($1 \le j \le i$), and subdiagonal (superdiagonal) entries β_{j+1} ($1 \le j \le i - 1$), i.e.,

$$T_i \equiv \begin{pmatrix} \alpha_1 & \beta_2 & & & & \\ \beta_2 & \alpha_2 & \beta_3 & & & \\ & \beta_3 & \cdot & \cdot & & \\ & & \cdot & \cdot & \cdot & \\ & & & \cdot & \cdot & \beta_i \\ & & & & \beta_i & \alpha_i \end{pmatrix}. \tag{13.9}$$

By construction, the vectors $\alpha_i q_i$ and $\beta_i q_{i-1}$ in (13.8) are respectively, the orthogonal projections of $B q_i$ onto the most recent Lanczos vectors q_i and q_{i-1}. Hence, for each i, the next Lanczos vector q_{i+1} is obtained by orthogonalizing $B q_i$ with respect to q_i and q_{i-1}.

In matrix notation, we can cast (13.8) as

$$B Q_i = Q_i T_i + \beta_{i+1} q_{i+1} e_i^T, \tag{13.10}$$

where $Q_i \equiv [q_1, q_2, ..., q_i]$ is the $n \times i$ matrix whose kth column is the kth Lanczos vector, and e_i^T is the ith column of the $n \times n$ identity matrix. This Lanczos recursion (13.10) produces a family of real symmetric tridiagonal matrices generated by both B and q_1. Table 13.1 outlines the basic steps

---

TABLE 13.1
Lanczos Algorithm for Computing the Sparse SVD

(1) Use the Lanczos recursion (13.8) to generate a family of real symmetric tridiagonal matrices T_i $(i = 1, 2, ..., s)$

(2) For some $k \leq s$, compute relevant eigen-values of T_k

(3) Select some or all of these eigenvalues as approximations to the eigenvalues of the matrix $A^T A$ (square roots are the singular values of A)

(4) For each eigenvalue λ compute a corresponding unit eigenvector z such that $T_k z = \lambda z$. Map such vectors into corresponding Ritz vectors $y \equiv Q_s z$, which are then used as approximations to the desired eigenvectors of $A^T A$ or right singular vectors of the matrix A

(5) For any accepted eigenpair (λ, y), compute the corresponding singular triplet of A, i.e., $\{u, \sigma, v\}$, via $\sigma = \sqrt{\lambda}$, $v = y$, and $u = (1/\sigma) A v$

---

of the Lanczos algorithm for computing the singular values and corresponding singular vectors of the sparse matrix A.

One can view the Lanczos recursion in (13.8) as a Gram–Schmidt orthogonalization of the set of Krylov vectors $\{q_1, Bq_1, ..., B^{k-1}q_1\}$ for $B = A^T A$ [10]. Moreover, since the span$\{Q_i\}$ is a Krylov subspace for $A^T A$, the Lanczos recursion is simply a method for computing the orthogonal projection of $A^T A$ on the subspace. Computing the eigenvalues of the T_i's produces approximations to the singular values of A (and corresponding singular vectors) restricted to these Krylov subspaces.

In finite-precision arithmetic, the Lanczos procedure suffers from the loss in the orthogonality of the Lanczos vectors, q_i. This phenomenon leads to the approximation of numerically multiple eigenvalues of T_i (for large i) for simple eigenvalues of $A^T A$, and the appearance of spurious (or ghost) eigenvalues among the computed eigenvalues for some T_i. Reorthogonalization of the Lanczos vectors (q_i) is one remedy to these problems, but how much does it cost? The *total* reorthogonalization of every Lanczos vector with respect to every previously generated Lanczos vector is one extreme [10]. A different approach accepts the inevitable loss in orthogonality and then tries to account for the spurious eigenvalue approximations [8]. As pointed out in Ref. [6], total reorthogonalization is certainly one way of maintaining orthogonality, however, it may complicate the Lanczos recursion with regard to storage requirements and arithmetic operations.

Lanczos implementations in recent studies [6, 13] have been based on the recursion in (13.8) with a selective reorthogonalization strategy, LANSO (ca 1989) [17]. In this case, the Lanczos vectors (q_i) are periodically reorthogonalized whenever a threshold for mutual orthogonality is exceeded. A recent parallel/distributed version (pLANSO) of this technique in Fortran-77 with message passing interface (MPI) constructs [18] has been developed for solving large symmetric generalized eigenvalue problems on massively parallel processors (MPPs) [19]. In the next section, we discuss a redesign of pLANSO in C++ for computing the SVD of large sparse term-by-document matrices on a NOW.

13.5 PARALLELIZATION OF THE LANCZOS ALGORITHM

In parallelizing the Lanczos algorithm with selective reorthogonalization, the Lanczos vectors can be uniformly and conformally mapped onto all the processors with no element in the vector residing on two processors. If $A^T A$ is $n \times n$ and there are p processors on a NOW, the same n/p elements of each Lanczos vector can reside on a single processor [13]. In other words, for every Lanczos vector computed, the first n/p elements of each vector will reside on the first processor, the second n/p elements of each vector will reside on the second processor, and so forth with the last n/p elements of each vector will reside on the pth processor. Each processor then independently solves the same eigenvalue problem (size dependent on the number of Lanczos steps taken) for the matrix $A^T A$, and thereby generates its own copy of the singular values of A. The Lanczos iterations and reorthogonalization steps require vector-based operations such as the dot product, saxpy ($\vec{y} = a\vec{x} + \vec{y}$), and the Euclidean or two-norm. By uniformly and conformally mapping the vectors, dense vector–vector operations required by the Lanczos algorithm are easily parallelized.

The saxpy operation can be perfectly parallelized with no communication among processors being necessary, and the computation of the eigenvectors, or Ritz vectors, can be *perfectly* parallelized after solving the small eigenvalue problem. The dot product is the key to successful parallelization of the algorithm because it is involved in the reorthogonalization steps, vector norm computations, and construction of the tridiagonal matrix T_i in (13.9). Successful parallelization of the dot product allows the processors to run in parallel while keeping computed Ritz values up-to-date. Each processor can compute the local dot product of the distributed vectors in parallel followed by a global sum operation [13]. While each processor solves the same small eigenvalue problem in parallel, each processor can also trigger the reorthogonalization of the distributed Lanczos vectors at the same time. Judging the loss of orthogonality, however, must be based on the global problem size, i.e., $\epsilon \sqrt{n}$, where n is the order of $A^T A$ and ϵ is machine unit-roundoff [19].

The other major operation required in parallelizing the Lanczos algorithm (with selective reorthogonalization) is sparse matrix–vector multiplication. The sparse $m \times n$ term-by-document matrix A can be partitioned across p processors, where each processor i stores m/p of the terms. The matrix–vector multiplications $y = A^T A x$ and $y = A x$ are then parallelized using localized dot products followed by a global sum, distributing vectors across processors, and sending portions of vectors back to the appropriate processors.

13.6 BENCHMARK COLLECTIONS

A recent performance assessment [13] of pLANSO for IR applications was based on benchmark text collections acquired from the Fifth Text Retrieval Evaluation Conference (TREC-5) [11]. Each collection consisted of text corpora and relevance judgments relating documents in each corpus to the given set of queries. The first benchmark collection was a subset derived from Foreign Broadcast Information Service (FBIS1). The corpus was constructed using the relevant judgments so that each document in the corpus was relevant to at least one query. The second benchmark collection was a text corpus taken from a collection of Los Angeles Times articles (LATIMES) and included relevance judgments. The third benchmark collection was the entire Foreign Broadcast Information Service corpus with relevance judgments included (FBIS2). Table 13.2 lists the TREC-5 text collections and the properties of the associated term-by-document matrices generated. In parsing each collection, log-entropy term weighting [3] (13.7) was applied to each term (keyword) appearing more than once globally in the collection.

13.7 TESTING ENVIRONMENTS

The first computing environment used in the pLANSO assessment was a network of 12 Sun Ultra-2 Model 2170 SPARCstations (each with dual 167 MHz Ultra-SPARC-1 processors and 256 MB RAM) connected via asynchronous transfer mode (ATM) switches at 155 mbps. Parallel task scheduling was accomplished via MPICH version 1.2.0 and serial results were obtained on one of the nodes (Sun Ultra-2 Model SPARCstation) powered by a 167 MHz Ultra-SPARC-1 processor

TABLE 13.2
TREC-5 Document Collections [11] Used for pLANSO Testing

	FBIS1	FBIS2	LATIMES
Source	Foreign Broadcast Information Service Data (1996)	Foreign Broadcast Information Service Data (1996)	Los Angeles Times Articles (1989–1990)
Terms	54,204	270,835	156,413
Documents	4,974	130,471	131,896
Nonzeros	1,585,012	21,367,169	23,527,979
Density(%)	0.59	0.06	0.11

with 2 2.1 GB internal disk drives and 11 GB of swap space (via an external drive). We refer to this network of workstations as NOW1.

As an alternative to the local area network (LAN) configuration of distributed nodes (processors and memories), the second computing environment used was a Sun Enterprise 220R symmetric multiprocessor (SMP) cluster with distributed shared memory. This cluster was comprised of 16 dual (450 MHz) UltraSPARC II 64-bit processors each with a 4 MB L2 cache, 512 MB SDRAM, and 27.3 GB hard drive. High-bandwidth communication between cluster nodes was accomplished using a gigabit network with parallel tasks scheduled via MPICH version 1.2. All nodes on this cluster were running the Sun OS (Solaris) 5.7 operating system where as the nodes of the previous (12-node) network were running Sun OS (Solaris) 5.5.1. Compilations on this cluster (NOW2) were based on g++ version 2.95.2 as compared to version 2.8.1 for NOW1.

13.8 RESULTS

Figures 13.1, 13.3, and 13.5 illustrate the elapsed (wall-clock) time for pLANSO (parallel Lanczos) when approximating the 100 largest singular triplets of the three test collections on the NOW1 test environment. Not more than 300 Lanczos iterations were completed in any experiment. Similarly, Figures 13.2, 13.4, and 13.6 demonstrate the performance on the higher-bandwidth NOW2 environment.

For each of the three collections, pLANSO was applied to the symmetric-positive definite matrix $A^{T}A$, where A is the corresponding (sparse) term-by-document matrix. The term vectors (or left singular vectors of A) were obtained via multiplication of A with the corresponding document vectors (right singular vectors) as defined in Step (5) of the Lanczos algorithm in Table 13.1. The elapsed wall-clock time for these two tasks, Lanczos execution and matrix multiplication for term vector generation, were measured separately. The retrieval performance of LSI (in terms of *recall* and *precision*) for these TREC-5 subcollections is documented elsewhere [14].

For the smaller of the three collections, FBIS1, the best speedup for pLANSO over a single processor sequential implementation (LANSO) was 3.78 on NOW1 for $p = 11$ and 3.5 on NOW2 for $p = 12$ processors (workstations). However, the improvement beyond $p = 5$ processors was not that significant. The percentage of the total elapsed wall-clock time due to the generation of term vectors (via sparse matrix–vector multiplication) ranged between 13% and 30% on both computing environments.

As illustrated in Figures 13.3 and 13.4, pLANSO computes the 100-largest singular triplets of the FBIS2 collection much faster (almost a factor of 3) on NOW2 compared to NOW1 (as expected).

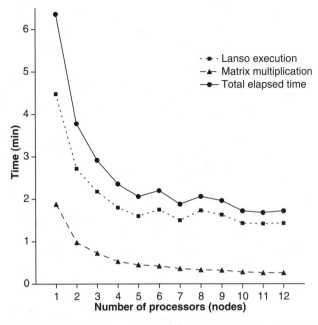

FIGURE 13.1 Performance of pLANSO with FBIS1 data set on NOW1.

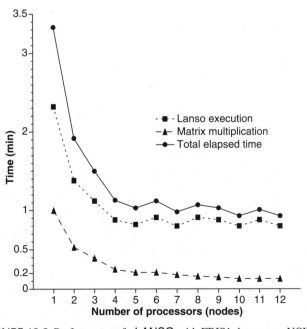

FIGURE 13.2 Performance of pLANSO with FBIS1 data set on NOW2.

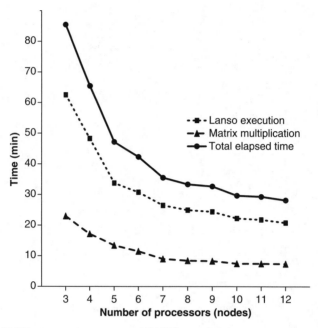

FIGURE 13.3 Performance of pLANSO with FBIS2 data set on NOW1.

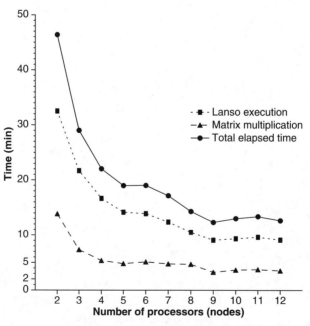

FIGURE 13.4 Performance of pLANSO with FBIS2 data set on NOW2.

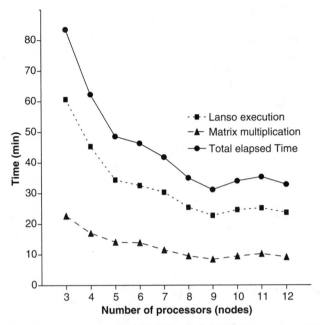

FIGURE 13.5 Performance of pLANSO with LATIMES data set on NOW1.

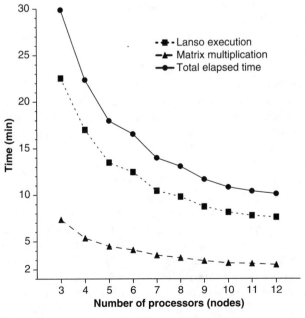

FIGURE 13.6 Performance of pLANSO with LATIMES data set on NOW2.

For this collection, no viable result for two processors was obtained (due to limited RAM) on NOW1. In that environment, a speed *improvement* factor of 2.7 over the three-processor run was obtained using nine processors. On NOW2, the speed improvement over an initial two-processor run of pLANSO was nearly four when nine processors were used. Due to limited RAM, no serial Lanczos implementation (LANSO) on NOW2 was possible. Term vector processing via sparse matrix multiplication was very consistent at about 28% of the total elapsed wall-clock time for both NOW1 and NOW2.

For the LATIMES collection, a tremendous reduction in elapsed computation time was obtained on NOW2 (using gigabit network connections). Although not shown to maintain linear scale in the y-axis in Figure 13.6, a two-processor run of pLANSO required over 3 h of elapsed wall-clock time to compute the 100 largest singular triplets of the corresponding term-by-document matrix. In contrast, using all 12 processors on NOW2 required just over 10 min of elapsed time. The best speed improvement factor over three-processor runs (on both NOW1 and NOW2, Figure 13.5 and Figure 13.6) was nearly 3 when all 12 workstations were exploited. Similar to the results obtained for FBIS2, the final sparse matrix multiplication to produce term vectors (i.e., the left singular vectors) consumed about 25–27% of the total elapsed time.

13.9 SUMMARY AND FUTURE DIRECTIONS

The pLANSO implementation of the parallel Lanczos algorithm for computing the SVD of sparse term-by-document matrices has obtained speedups ranging from 3 to 4 on two different workstation clusters of 12 processors (one configured with an ATM network and the other with a gigabit network). This implementation was a complete rewrite of the original Fortran and MPI version described in Ref. [19] using the C++ language (and MPI constructs). A high-performance SVD module in C++ was needed for integration with existing text parsing and indexing software. Portability requirements associated with Web-based indexing applications may soon require Java implementations of pLANSO and related SVD software.

Advances in computational statistics and high-performance computing have significantly reduced the computational burden of converting raw text to encoded vectors for the purpose of IR modeling and testing. The ability to encode benchmark collections such as those from TREC-5 [11] into as many as 100 dimensions in no more than 10 min is an important advancement in the study and development of IR models such as LSI [3–5, 9]. Future software development (in C++ and perhaps Java) is still needed for efficient up- and downdating of low-rank singular subspaces generated by codes such as pLANSO. Time-sensitive or dynamic text collections will require regular updates and simply computing the principal components of an updated (even larger) term-by-document matrix may be unwieldy with limited RAM.

The vary nature of indexing and text mining is likely to change over the next few years. Downloading distributed repositories of textual information on the WWW to a single server or cluster of machines for parsing, encoding, and query-matching presents a *static* view or snapshot of the WWW. Capturing immediate changes in the information still poses a computational challenge for current IR models and search engines. A paradigm shift from centralized indexing to one in which indices are stored (remotely) and automatically updated with document changes might require the deployment of agent-based computational algorithms/software for computing the principal components of term-by-document matrices assembled on remote sites. Exploiting distributed storage on the WWW may well be the next advancement for vector space IR models. Storing (and replicating) the principal components or singular subspaces themselves on distributed (remote) servers can significantly increase query response time and facilitate the *fusion* of information sources into one *global* index via parallel processing.

REFERENCES

[1] T. Anderson, *An Introduction to Multivariate Statistical Analysis*, 2nd ed. John Wiley, New York, 1984.

[2] R. Baeza-Yates and B. Ribeiro-Neto, *Modern Information Retrieval*. Addison-Wesley, Boston, MA, 1999.

[3] M. Berry and M. Browne, *Understanding Search Engines: Mathematical Modeling and Text Retrieval*. SIAM, Philadelphia, PA, 1999.

[4] M. Berry, Z. Drmač, and E. Jessup, Matrices, vector spaces, and information retrieval. *SIAM Review* 41 (1999), 335–362.

[5] M. Berry, S. Dumais, and G. O'Brien, Using linear algebra for intelligent information retrieval. *SIAM Review*, 37 (1995), 573–595.

[6] M.W. Berry, Large scale singular value computations. *International Journal of Supercomputer Applications* 6 (1992), 13–49.

[7] G. Bhattacharyya and R. Johnson, *Statistical Concepts and Methods*. John Wiley, New York, 1977.

[8] J. Cullum and R. Willoughby, *Lanczos Algorithm for Large Symmetric Eigenvalue Computations. Volume 1: Theory*. Birkhäuser, Boston, MA, 1985.

[9] S. Deerwester, S. Dumais, G. Furnas, T. Landauer, and R. Harshman, Indexing by latent semantic analysis. *Journal of the American Society for Information Science* 41 (1990), 391–407.

[10] G. Golub and C.V. Loan, *Matrix Computations*, 3rd ed. Johns-Hopkins, Baltimore, 1996.

[11] D. Harman and E. Voorhees, Eds., *Proceedings of the Fifth Text Retrieval Conference (TREC-5)*, Gaithersburg, MD. Department of Commerce, National Institute of Standards and Technology. NIST Special Publication 500-238, 1997.

[12] R. Johnson and D. Wichern, *Applied Multivariate Statistical Analysis*, 5th ed. Prentice Hall, Upper Saddle River, NJ, 2002.

[13] D. Martin and M. Berry, Parallel SVD for scalable information retrieval. In *Proceedings of the International Workshop on Parallel Matrix Algorithms and Applications*, Neuchâtel, Switzerland, 2000.

[14] D. Martin, M. Berry, and B. Thompson, Latent semantic indexing using term proximity. In *Proceedings of the Text Mining Workshop, First SIAM International Conference on Data Minining*, Chicago, IL, 2001.

[15] C. Paige, Error analysis of the lanczos algorithm for tridiagonalizing a symmetric matrix. *Journal of the Institute of Mathematics and its Applications* 18 (1976), 341–349.

[16] B. Parlett, *The Symmetric Eigenvalue Problem*. Prentice Hall, Englewood Cliffs, NJ, 1980.

[17] H. Simon, Analysis of the symmetric Lanczos algorithm with reorthogonalization methods. *Linear Algebra and its Applications* 61 (1984), 101–131.

[18] M. Snir, S. Otto, S. Huss-Lederman, D. Walker, and J. Dongarra, *MPI: The Complete Reference*. MIT Press, Cambridge, MA, 1995.

[19] K. Wu and H. Simon, A Parallel Lanczos Method for Symmetric Generalized Eigenvalue Problem. Technical Report LBNL-41284, Lawrence Berkeley Laboratory, NERSC, Berkeley, CA, 1997.

14 Matrix Rank Reduction for Data Analysis and Feature Extraction

Haesun Park[1] *and Lars Eldén*

CONTENTS

ABSTRACT

Numerical techniques for data analysis and feature extraction are discussed using the framework of matrix rank reduction. The singular value decomposition (SVD) and its properties are reviewed, and the relation to latent semantic indexing (LSI) and principal component analysis (PCA) is described. Methods that approximate the SVD are reviewed. A few basic methods for linear regression, in particular the partial least squares (PLS) method, are presented, and analyzed as rank reduction methods. Methods for feature extraction, based on centroids and the classical linear discriminant analysis (LDA), as well as an improved LDA based on the generalized singular value decomposition (LDA/GSVD) are described. The effectiveness of these methods are illustrated using examples from information retrieval and two-dimensional representation of clustered data.

---

[1] The work of this author was supported in part by the National Science Foundation grants CCR-0204109 and ACI-0305543. Any opinions, findings, and conclusions or recommendations expressed in this material are those of the authors and do not necessarily reflect the views of the National Science Foundation (NSF).

14.1 INTRODUCTION

The concepts of *rank reduction*, *matrix factorization*, *data compression*, *dimension reduction*, and *feature selection/extraction* are very closely related. The key link between the concepts is the *Wedderburn rank reduction theorem* from 1934 (comprehensive bibliographies are given in Refs. [1, 2]).

Theorem 14.1 (Wedderburn [3]). *Suppose $A \in \mathbb{R}^{m \times n}$, $f \in \mathbb{R}^{n \times 1}$, and $g \in \mathbb{R}^{m \times 1}$. Then*

$$rank(A - \omega^{-1} A f g^{\mathrm{T}} A) = rank(A) - 1$$

if and only if $\omega = g^{\mathrm{T}} A f \neq 0$. ∎

Let $A^{(1)} = A$, and define a sequence of matrices $\{A^{(i)}\}$ from

$$A^{(i+1)} = A^{(i)} - \omega_i^{-1} A^{(i)} f^{(i)} g^{(i)\mathrm{T}} A^{(i)} \tag{14.1}$$

for any vectors $f^{(i)} \in \mathbb{R}^{n \times 1}$ and $g^{(i)} \in \mathbb{R}^{m \times 1}$ such that

$$\omega_i = g^{(i)\mathrm{T}} A^{(i)} f^{(i)} \neq 0. \tag{14.2}$$

Then the sequence defined in (14.1) terminates in $\gamma = rank(A)$ steps because each time the rank of the matrix decreases by exactly one. This process is called a *rank-reducing process* and the matrices $A^{(i)}$ are called Wedderburn matrices. For details, see Ref. [1]. This finite terminating process gives a matrix *rank reducing decomposition* that can be defined as

$$A = \widehat{F} \Omega^{-1} \widehat{G}^{\mathrm{T}} \tag{14.3}$$

where

$$\widehat{F} = \left[\widehat{f}_1, \ldots, \widehat{f}_\gamma \right] \in \mathbb{R}^{m \times \gamma} \quad \text{with} \quad \widehat{f}_i = A^{(i)} f^{(i)} \tag{14.4}$$

$$\Omega = \mathrm{diag}(\omega_1, \ldots, \omega_\gamma) \in \mathbb{R}^{\gamma \times \gamma}, \quad \text{and} \tag{14.5}$$

$$\widehat{G} = \left[\widehat{g}_1, \ldots, \widehat{g}_\gamma \right] \in \mathbb{R}^{n \times \gamma} \quad \text{with} \quad \widehat{g}_i = A^{(i)\mathrm{T}} g^{(i)}. \tag{14.6}$$

There are many choices of the vectors $f^{(i)}$ and $g^{(i)}$ that satisfy the condition (14.2). Therefore, various rank reducing decompositions (14.3) are possible. Theorem 14.1 has been used to explain many matrix decompositions [1], and it will be the basis for the rank reduction procedures discussed in this chapter.

Theorem 14.1 can be generalized to the case where the reduction of rank is larger than one, as shown in the next theorem.

Theorem 14.2 (Guttman [4]). *Suppose $A \in \mathbb{R}^{m \times n}$, $F \in \mathbb{R}^{n \times k}$, and $G \in \mathbb{R}^{m \times k}$. Then*

$$rank(A - AFR^{-1}G^{\mathrm{T}}A) = rank(A) - rank(AFR^{-1}G^{\mathrm{T}}A) \tag{14.7}$$

if and only if $R = G^{\mathrm{T}} AF \in \mathbb{R}^{k \times k}$ is nonsingular. ∎

In Ref. [1] the authors discuss rank reduction from the point of view of solving linear systems of equations. It is shown that several standard matrix factorizations in numerical linear algebra are instances of the Wedderburn formula. A complementary view is taken in Ref. [2], where the Wedderburn formula and matrix factorizations are considered as tools for data analysis: "The major purpose of a matrix factorization in this context is to obtain some form of lower rank approximation to A for understanding the structure of the data matrix. . . ."

In this chapter, rank reduction will be discussed mainly from the perspective of data analysis, but it will be seen that data analysis concepts and methods are often the same as those appearing in connection with the solution of linear systems of equations, although different terminology and notation are used. By formulating a specific criterion for the data analysis, a rank reduction is obtained as a particular instance of the matrix rank reduction formula presented in Theorems 14.1 and 14.2. If it is required that the data matrix A be well approximated by a low-rank matrix, then one is led to rank reduction using the SVD or, equivalently, PCA. This is referred to here as rank reduction without constraints. In Section 14.2, some fast methods for approximating the SVD are described. The application of singular value analysis to information retrieval, LSI, is discussed.

In multiple regression, the rank reduction is coupled to the reduction in norm of a residual vector, which is the (stochastic and modeling) error in a linear model. A common method here is principal components regression (PCR), described in Section 14.3.1. An alternative, and often very flexible method is PLS, discussed in Section 14.3.2. In some cases the rank reduction aims at facilitating feature extraction; this is discussed in Section 14.3.3. These procedures can be considered as constrained rank reduction methods.

Throughout this chapter the data matrix is denoted as A with dimension $m \times n$ and it is assumed that A is real. The Euclidean vector norm and its subordinate matrix norm are written as $\| \cdot \|_2$, and $\|A\|_F$ is the Frobenius norm of the matrix A [5, Chapter 2.2–2.3].

14.2   RANK REDUCTION WITHOUT CONSTRAINTS

In this section, the SVD is presented, which gives an *optimal* lower rank approximation to a given matrix, where optimality is measured in the two-norm or the Frobenius norm. Two applications of the SVD, LDI from information retrieval and PCA from statistical pattern recognition, are discussed. Then several methods including orthogonal decomposition with pivoting and the complete orthogonal decomposition are discussed that are often used in practice to approximate the SVD because the SVD is expensive to compute and update.

14.2.1   SINGULAR VALUE DECOMPOSITION

The following theorem defines the SVD. For a proof of the theorem, see Refs. [5, 6], for example. For some early history of the SVD, see Ref. [7].

Theorem 14.3. *For any matrix $A \in \mathbb{R}^{m \times n}$, there exist matrices*

$$U = [u_1, \ldots, u_m] \in \mathbb{R}^{m \times m} \quad and \quad V = [v_1, \ldots, v_n] \in \mathbb{R}^{n \times n},$$

with $U^T U = I$ and $V^T V = I$, such that

$$U^T A V = \Sigma = diag(\sigma_1, \ldots, \sigma_p) \in \mathbb{R}^{m \times n}, \quad p = min(m, n), \tag{14.8}$$

where $\sigma_1 \geq \sigma_2 \geq \cdots \sigma_p \geq 0$. ∎

The σ_i are the singular values, and the columns of U and V are left and right singular vectors, respectively. Suppose for an index $\gamma \leq min(m, n)$, $\sigma_\gamma > 0$ and $\sigma_{\gamma+1} = 0$. Then

$$A = U \Sigma V^T = \begin{pmatrix} U_\gamma & \widehat{U}_\gamma \end{pmatrix} \begin{pmatrix} \Sigma_\gamma & 0 \\ 0 & 0 \end{pmatrix} \begin{pmatrix} V_\gamma^T \\ \widehat{V}_\gamma^T \end{pmatrix} = U_\gamma \Sigma_\gamma V_\gamma^T, \tag{14.9}$$

where $U_\gamma \in \mathbb{R}^{m \times \gamma}$, $\Sigma_\gamma = diag(\sigma_1, \ldots, \sigma_\gamma) \in \mathbb{R}^{\gamma \times \gamma}$, and $V_\gamma \in \mathbb{R}^{n \times \gamma}$.

The SVD provides all four *fundamental* spaces related to the matrix A as follows:

U_γ gives an orthogonal basis for Range(A),
$\widehat{V}_\gamma$ gives an orthogonal basis for Null(A),
V_γ gives an orthogonal basis for Range(A^{T}),
$\widehat{U}_\gamma$ gives an orthogonal basis for Null(A^{T}),

where Range and Null denote the range space and the null space of the matrix, respectively. In addition, the SVD provides the rank of the matrix, which is the number of positive singular values, γ.

However, in floating point arithmetic, these zero singular values may appear as small numbers. In general, a large relative gap between two consecutive singular values is considered to reflect the numerical rank deficiency of a matrix. It is widely known that noise filtering can be achieved via a *truncated* SVD. If trailing *small* diagonal components of Σ are replaced with zeros, then a rank k approximation A_k for A can be obtained as

$$
A = \begin{pmatrix} U_k & \widehat{U}_k \end{pmatrix} \begin{pmatrix} \Sigma_k & 0 \\ 0 & \widehat{\Sigma}_k \end{pmatrix} \begin{pmatrix} V_k^{\mathrm{T}} \\ \widehat{V}_k^{\mathrm{T}} \end{pmatrix}
$$

$$
\approx \begin{pmatrix} U_k & \widehat{U}_k \end{pmatrix} \begin{pmatrix} \Sigma_k & 0 \\ 0 & 0 \end{pmatrix} \begin{pmatrix} V_k^{\mathrm{T}} \\ \widehat{V}_k^{\mathrm{T}} \end{pmatrix} = U_k \Sigma_k V_k^{\mathrm{T}} = A_k, \tag{14.10}
$$

where $\Sigma_k \in \mathbb{R}^{k \times k}$ and $\|\widehat{\Sigma}_k\|_2 < \epsilon$ for a *small* tolerance ϵ. This is summarized in the following theorem, which provides an important application of the SVD, the lower rank approximation. The following theorem [8] is the theoretical foundation of numerous important procedures in science and engineering.

Theorem 14.4. *Let the SVD of $A \in \mathbb{R}^{m \times n}$ be given as in Theorem 14.3, and assume that an integer k is given with $0 < k \le \gamma = \mathrm{rank}(A)$. Then*

$$
\min_{B \in \mathbb{R}^{m \times n}, \mathrm{rank}(B)=k} \|A - B\|_2 = \|A - A_k\|_2 = \sigma_{k+1}
$$

where

$$
A_k = U_k \Sigma_k V_k^{\mathrm{T}} = \sum_{i=1}^{k} \sigma_i u_i v_i^{\mathrm{T}}, \tag{14.11}
$$

U_k consists of the first k columns of U, V_k the first k columns of V, and Σ_k is the $k \times k$ leading principal submatrix of Σ. ∎

A similar result is obtained when the Frobenius norm is used.

Theorem 14.5. *Let the SVD of $A \in \mathbb{R}^{m \times n}$ be given as in Theorem 14.3, and assume that an integer k is given with $0 < k \le \gamma = \mathrm{rank}(A)$. Then*

$$
\min_{B \in \mathbb{R}^{m \times n}, \mathrm{rank}(B)=k} \|A - B\|_F = \|A - A_k\|_F = \sqrt{\sum_{i=k+1}^{p} \sigma_i^2}
$$

where A_k is the same as in Theorem 14.4, and $p = min(m, n)$. ∎

Theorems 14.4 and 14.5 show that the singular values indicate how close a given matrix is to a matrix of lower rank.

The relation between the truncated SVD (14.11) and the Wedderburn matrix rank reduction process (14.7) can be demonstrated as follows. In the rank reduction formula, define the error matrix E as

$$E = A - AF(G^{T}AF)^{-1}G^{T}A, \quad F \in \mathbb{R}^{n \times k}, G \in \mathbb{R}^{m \times k}.$$

Assume that $k \leq \text{rank}(A) = \gamma$, and consider the problem

$$\min \|E\|_2 = \min_{F \in \mathbb{R}^{n \times k}, G \in \mathbb{R}^{m \times k}} \|A - AF(G^{T}AF)^{-1}G^{T}A\|_2.$$

According to Theorem 14.4, the minimum error is obtained when

$$F = V_k \quad \text{and} \quad G = U_k,$$

which gives

$$
\begin{aligned}
(AF)(G^{T}AF)^{-1}(G^{T}A) &= (AV_k)(U_k^{T}AV_k)^{-1}(U_k^{T}A) \\
&= (U_k\Sigma_k)(\Sigma_k)^{-1}(\Sigma_k V_k^{T}) = U_k\Sigma_k V_k^{T}.
\end{aligned}
$$

This same result can be obtained by a stepwise procedure, when k pairs of vectors $f^{(i)}$ and $g^{(i)}$ are to be found, where each pair reduces the matrix rank by 1 in each of k rank reduction problems. Consider the problem

$$\min \|E_1^{(1)}\|_2 = \min_{f^{(1)} \in \mathbb{R}^{n \times 1}, g^{(1)} \in \mathbb{R}^{m \times 1}} \|A - (g^{(1)^{T}}Af^{(1)})^{-1}Af^{(1)}g^{(1)^{T}}A\|_2.$$

Then according to Theorem 14.4, the solution is obtained when $f^{(1)} = v_1$ and $g^{(1)} = u_1$. Now for

$$A^{(2)} = A^{(1)} - A^{(1)}v_1u_1^{T}A^{(1)}/\sigma_1 = A^{(1)} - \sigma_1 u_1 v_1^{T},$$

with $A^{(1)} = A$, the solution for

$$\min \|E_1^{(2)}\|_2 = \min_{f^{(2)} \in \mathbb{R}^{n \times 1}, g^{(2)} \in \mathbb{R}^{m \times 1}} \|A^{(2)} - (g^{(2)^{T}}A^{(2)}f^{(2)})^{-1}A^{(2)}f^{(2)}g^{(2)^{T}}A^{(2)}\|_2$$

is obtained with $f^{(2)} = v_2$ and $g^{(2)} = u_2$, since

$$A^{(1)} - \sigma_1 u_1 v_1^{T} = \sum_{i=2}^{r} \sigma_i u_i v_i^{T}.$$

On the other hand, the rank 2 reduction problem,

$$\min \|E_2\|_2 = \min_{F^{(1)} \in \mathbb{R}^{n \times 2}, G^{(1)} \in \mathbb{R}^{m \times 2}} \|A - AF^{(1)}(G^{(1)^{T}}AF^{1})^{-1}G^{(1)^{T}}A\|_2,$$

has the solution $F^{(1)} = \begin{pmatrix} v_1 & v_2 \end{pmatrix}$ and $G^{(1)} = \begin{pmatrix} u_1 & u_2 \end{pmatrix}$, again due to Theorem 14.4. Therefore, reducing the rank k times by finding the rank-one reducing vectors that give minimum L_2 norm distance between the matrix and the reduced rank matrix is equivalent to reducing the rank by k at once by finding the rank k reducing matrices that satisfy the same criterion.

Similar results can be obtained from Theorem 14.5 when the matrix Frobenius norm is used.

14.2.2 LATENT SEMANTIC INDEXING AND THE SVD

In the vector space-based information retrieval system [9, 10], the document collection is represented as vectors in a matrix. Specifically, a term-document matrix

$$A = \begin{bmatrix} a_1 & a_2 & \cdots & a_n \end{bmatrix} \in \mathbb{R}^{m \times n} \tag{14.12}$$

is formed based on the collection of documents, where m is the total number of terms in the document collection and n is the number of documents. Each column of A represents a document, and in the matrix A the elements a_{ij} are weighted frequencies of each word in a specific document representing the importance of term i in document j. The simplest a_{ij} is binary, but to improve retrieval performance, various weighting methods have been developed [10]. For other related topics such as stemming and removing stop lists, see Refs. [10–12]. The SMART system [10] is one of the most influential test beds where the vector space-based method is successfully implemented. One major advantage of a vector space-based method is that the algebraic structures of the term-document matrix can be exploited using the techniques developed in linear algebra.

For achieving higher efficiency and effectiveness in manipulating these data, it will be necessary to find a lower dimensional representation of the data [13]. A vector space-based information retrieval system needs to solve the following three problems frequently: document retrieval, classification, and clustering. Document retrieval is used to extract relevant documents from a text database given a query. Classification is the process of assigning new data to its proper group. The group is also called class or category [14]. Clustering is the process to find homogeneous groups (clusters) of data based on the values of their vector components and a predefined measure [15]. While the category structure is known in classification, in cluster analysis little or nothing is known about the category structure. All that is available is a collection of data whose category memberships are unknown. The objective is to discover a category structure in the data set.

In document retrieval, classification, and in some clustering algorithms, the major computation involves comparison of two vectors, which will be affected by different weighting schemes and the similarity measures [10]. With dimension reduction of the given text collection, the complexity of subsequent computations involved in these problems can be substantially reduced. To achieve higher efficiency in computation, often it is necessary to reduce the dimension severely, and in the process, one may lose much information that was available in the original data. Therefore, it is important to achieve a *better representation* of data in the lower dimensional space according to specific tasks to be performed after the dimension reduction, such as classification, rather than simply reducing the dimension of the data to best approximate the full term-document matrix. This is discussed later in detail in Section 14.3.3. The significance of this has been recognized by Hubert et al. [2], for example. The difficulty involved is that it is not easy to measure how well a certain dimension reduction method provides a good representation of the original data. It seems that this can only be estimated using experimental results. The dimension reduction/feature extraction methods that will be discussed in this chapter are based on the vector subspace computation in linear algebra.

In LSI [16–19], the SVD of the term-document matrix is utilized for conceptual retrieval and lower dimensional representation. The dimension reduction by the optimal lower rank approximation from the SVD has been successfully applied in numerous other applications, e.g., in signal processing. In these applications, the dimension reduction often achieves the effect of removing noise in the data. In information retrieval, the meaning of *noise* in the text data collection is not as well understood, and it is different from noise in the data in other applications such as signal processing [20] or image processing.

LSI is based on the assumption that there is some underlying latent semantic structure in the data of the term-document matrix that is corrupted by the wide variety of words used in documents and queries for the same objects (the problem of polysemy and synonymy, see Ref. [17]). The basic idea of the LSI/SVD is that if two document vectors represent the same topic, they will share many associating words with a keyword, and they will have very close semantic structures after dimension reduction via the SVD.

In classification, clustering, and document retrieval, the fundamental operation is to compare a document to another document. The choice of similarity measure plays an important role. In vector space-based information retrieval, the most commonly used similarity measures are, L_2 norm (Euclidean distance), inner product, cosine, or variations of these [21]. When the inner product is used as a measure, using the rank k approximation A_k for A, the documents are compared as,

$$A^\mathrm{T} A \approx A_k^\mathrm{T} A_k = V_k \Sigma_k^\mathrm{T} U_k^\mathrm{T} U_k \Sigma_k V_k^\mathrm{T} = (V_k \Sigma_k^\mathrm{T})(\Sigma_k V_k^\mathrm{T}), \tag{14.13}$$

i.e., the inner product between a pair of columns of A can be approximated by the inner product between a pair of columns of $\Sigma_k V_k^\mathrm{T}$. Accordingly, $V_k \Sigma_k^\mathrm{T} \in \mathbb{R}^{k \times n}$ is considered a representation of the document vectors in the reduced k-dimensional space.

Let $q \in \mathbb{R}^{m \times 1}$ represent a query vector, i.e., a document that has been preprocessed in the same way as the documents represented by the columns of A. Consider the inner product between a q and the document vectors in A:

$$q^\mathrm{T} A \approx q^\mathrm{T} A_k = (q^\mathrm{T})(U_k \Sigma_k V_k^\mathrm{T}) = (q^\mathrm{T} U_k)(\Sigma_k V_k). \tag{14.14}$$

Equation (14.14) shows that a new vector $q \in \mathbb{R}^{m \times 1}$ can be represented as

$$\widehat{q} = U_k^\mathrm{T} q \tag{14.15}$$

in the k-dimensional space, because the columns of $\Sigma_k V_k^\mathrm{T}$ represent the columns of A in the k-dimensional space. In LSI/SVD, it has also been proposed that q be reduced to a vector in $\mathbb{R}^{k \times 1}$ as

$$\widehat{q} = \Sigma_k^{-1} U_k q. \tag{14.16}$$

14.2.3 Principal Component Analysis and SVD

A common task in data analysis is to project high-dimensional numerical data on a low-dimensional subspace. The purpose of this transformation is different depending on the application: in some cases feature selection/extraction allows the problem at hand to be solved more effectively and also the low-dimensional representation of the data makes visualization of the data set possible. In other cases the computations using the high-dimensional representation would be too costly (e.g., in a real-time application), so that a dimension reduction is needed for computational efficiency. In data analysis applications this dimensionality reduction is often achieved using PCA. See Refs. [22, 23, Chapter 14.5].

Often in PCA it is assumed that the data are preprocessed so that each column of A has mean zero and Euclidean length one. The reader may assume that this has been done in the sequel (however, it is not essential for the description). PCA is usually presented and its solution is computed in terms of the eigenvalues and eigenvectors of AA^T and A^TA. It turns out that the derivation and computation of solution can also be done, with certain advantages, in terms of the SVD.

From Theorem 14.5 one has

$$\min_{\text{rank}(B)=k} \|A - B\|_F^2 = \|A - A_k\|_F^2 = \|A - U_k \Sigma_k V_k^T\|_F^2 = \left\|A - \sum_{i=1}^{k} \sigma_i u_i v_i^T\right\|_F^2.$$

Defining

$$L_k = \Sigma_k^T V_k^T = \begin{pmatrix} \sigma_1 v_1^T \\ \sigma_2 v_2^T \\ \vdots \\ \sigma_k v_k^T \end{pmatrix} = \begin{pmatrix} l_1^{(k)} & l_2^{(k)} & \cdots & l_n^{(k)} \end{pmatrix},$$

where $l_i^{(k)}$ denotes the ith column vector of L_k, one can write

$$\|A - A_k\|_F^2 = \|A - U_k L_k\|_F^2 = \sum_{i=1}^{n} \|a_i - U_k l_i^{(k)}\|_2^2,$$

where a_i is a column vector of A. This means that there are n linear models for the columns of A,

$$a_i \sim U_k l_i^{(k)}, \qquad i = 1, 2, \ldots, n, \tag{14.17}$$

with the same orthogonal basis vectors $u_1, u_2, \ldots, u_k$.

The SVD also gives linear models for the row vectors. Assume that the ith row of A is denoted as $\hat{a}_i^T$, i.e.,

$$A = \begin{pmatrix} \hat{a}_1^T \\ \hat{a}_2^T \\ \vdots \\ \hat{a}_m^T \end{pmatrix}.$$

Defining

$$\hat{L}_k = \Sigma_k U_k^T = \begin{pmatrix} \hat{l}_1^{(k)} & \hat{l}_2^{(k)} & \cdots & \hat{l}_m^{(k)} \end{pmatrix},$$

one obtains

$$\hat{a}_i \sim V_k \hat{l}_i^{(k)}, \qquad i = 1, 2, \ldots, m. \tag{14.18}$$

The linear models (14.17) and (14.18) are often referred to as PCA of the data matrix; the vectors u_i are called the principal components of the column space, and the v_i the principal components of the row space. It is common to choose $k = 2$ or 3, and plot the coordinates $l_i^{(k)}$ ($\hat{l}_i^{(k)}$) of the columns (rows) in a Cartesian coordinate system, to visualize or cluster the data. According to Theorems 14.4 and 14.5 the principal components are ordered so that the first one reflects the direction of largest variation in the data matrix, the second one the second largest, and so on. In the next example, the principal components are shown to have a concrete interpretation, and the directions of largest variation are illustrated.

Example 14.1 In an application of handwritten digit classification [24], the singular value decomposition of a collection of a couple of hundred handwritten digit "4" was computed. Each image of the digit was discretized on a 16×16 grid (256 pixels), and then made into a vector of length 256. All the vectors were collected in a data matrix, and its SVD was computed. Then the first singular

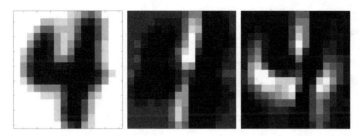

FIGURE 14.1 The first three singular vectors of a collection of handwritten "4."

vectors (each of length 256) were reshaped as 16×16 matrices, which can then be printed as images. In Figure 14.1 the first three left singular vectors are illustrated. The first left singular vector, the dominating direction of variation of the data matrix, represents a rather well written digit "4." The following two illustrate the two most important directions of variation of the set of digits, in the sense that many of the digits in the collection can be represented quite well as linear combinations of these first three basis vectors (principal components). ∎

Sometimes in the literature it is recommended to compute the principal components u_i and v_i as the eigenvectors of AA^{T} and $A^{\mathrm{T}}A$, respectively. However, from the point of view of accuracy it is better to compute the SVD, because the forming (explicitly or implicitly) of AA^{T} ($A^{\mathrm{T}}A$) causes loss of information. There are efficient and accurate algorithms for computing the SVD, implemented in numerical software libraries such as SVDPACK [25] and LAPACK [26], and problem solving environments such as MATLAB [27], Mathematica [28], and statistics software. The SVD of large, sparse matrices is usually computed using some variant of the Lanczos method, see Ref. [29, Chapter 6].

14.2.4 ORTHOGONAL DECOMPOSITION WITH PIVOTING

The power of the SVD lies in the fact that one can determine the rank of the matrix by simply reading off the diagonal elements of the matrix Σ. Due to this fact, orthogonal basis for the four fundamental spaces can also be obtained simply from the columns of the matrices U and V. However, a SVD is rather expensive to compute. Therefore, much research has been devoted to approximating the SVD by computationally less expensive methods. One group of approaches for achieving this is to relax the condition for diagonalization, and compute a triangularization, from which the approximate rank of the matrix is revealed.

Theorem 14.6 (Orthogonal (QR) Decomposition). *For any matrix* $A \in \mathbb{R}^{m \times n}$, *there exists a matrix* $Q \in \mathbb{R}^{m \times m}$ *with* $Q^{\mathrm{T}}Q = I$, *such that*

$$Q^{\mathrm{T}}A = R \in \mathbb{R}^{m \times n} \tag{14.19}$$

where R is upper triangular. ∎

When $m \geq n$ and $\mathrm{rank}(A) = n$, the first n columns of the matrix Q is an orthonormal basis for Range(A). However, this is not true in general. The following example [5] illustrates that the QR decomposition does not provide bases for all four fundamental spaces.

Example 14.2 Consider the QR decomposition of the matrix

$$A = \begin{pmatrix} 1 & 1 & 1 \\ 0 & 0 & 1 \\ 0 & 0 & 1 \end{pmatrix} = \begin{pmatrix} 1 & 0 & 0 & 0 \\ 0 & 1 & 0 & 0 \\ 0 & 0 & 1 & 0 \\ 0 & 0 & 0 & 1 \end{pmatrix} \begin{pmatrix} 1 & 1 & 1 \\ 0 & 0 & 1 \\ 0 & 0 & 1 \end{pmatrix}.$$

Clearly rank$(A) = 2$ but no two columns of the matrix Q form a basis for Range(A). ∎

In addition, the following example [5] shows that one cannot obtain a reliable numerical rank of a triangular matrix based on its diagonal elements. Therefore, the QR decomposition does not provide the rank of the matrix in general.

Example 14.3 Consider the matrix R_n defined as

$$R_n = \begin{pmatrix} 1 & -1 & -1 & \cdots & -1 \\ 0 & 1 & -1 & \cdots & -1 \\ \vdots & \vdots & \ddots & \vdots & \vdots \\ \vdots & & & 1 & -1 \\ 0 & & \cdots & 0 & 1 \end{pmatrix} \in \mathbb{R}^{n \times n}. \tag{14.20}$$

Although there is no small diagonal component, the matrix R_n gets closer to be rank deficient as n becomes larger. For example, $\sigma_{50}(R_{50}) \approx 2.7 \times 10^{-15}$ and $\sigma_{100}(R_{100}) \approx 3.1 \times 10^{-18}$. ∎

In floating point arithmetic, the QR decomposition is guaranteed to provide an orthogonal basis for Range(A) only when the matrix has full column rank. Therefore, the SVD is needed when the matrix is rank deficient or the rank of the matrix is unknown, and any information on the four fundamental subspace is required. One of the simpler methods for approximating the SVD is the QR decomposition with column pivoting, introduced by Golub [30].

Assume that $A \in \mathbb{R}^{m \times n}$ is rank deficient with rank γ. The *exact* QR decomposition with column pivoting of A is then given by

$$Q^T A \Pi = R = \begin{pmatrix} \tilde{R}_\gamma & \tilde{S}_\gamma \\ 0 & 0 \end{pmatrix},$$

where $\tilde{R}_\gamma \in \mathbb{R}^{\gamma \times \gamma}$ is upper triangular with rank$(\tilde{R}_\gamma) = \gamma$, $Q \in \mathbb{R}^{m \times m}$ is orthogonal, and $\Pi \in \mathbb{R}^{n \times n}$ is a permutation matrix. In floating point arithmetic, one may obtain

$$Q^T A \Pi = R = \begin{pmatrix} R_k & S_k \\ 0 & X_k \end{pmatrix}, \tag{14.21}$$

where $R_k \in \mathbb{R}^{k \times k}$, $\|X_k\|$ is small and k is an estimated value for γ.

Assume that the algorithm for computing (14.21) has advanced i steps. The reduced matrix then has the form

$$\begin{pmatrix} R_i & S_i \\ 0 & X_i \end{pmatrix},$$

where $R_i \in \mathbb{R}^{i \times i}$ is upper triangular with rank$(R_i) = i$. The next permutation matrix is chosen so that when it is applied to X_i it moves the column of largest norm to the front of that matrix. Then that column is reduced by an orthogonal transformation so that the dimension of the upper-triangular matrix is increased by one.

Suppose the QR decomposition with column pivoting is terminated after k steps satisfying the stopping criterion that $\|X_k\| \leq \epsilon$, for a given tolerance ϵ. Then since

$$Q^{(k)\mathrm{T}} Q^{(k-1)\mathrm{T}} \cdots A \cdots \Pi_{k-1} \Pi_k = \begin{pmatrix} R_k & S_k \\ 0 & X_k \end{pmatrix},$$

with $Q \equiv Q^{(1)} Q^{(2)} \cdots Q^{(k)} = (Q_1 \quad Q_2)$ where $Q_1 \in \mathbb{R}^{m \times k}$, Q_1 provides an approximate orthogonal basis for Range(A) and Q_2 for Null(A^{T}). Some further computation is needed to obtain the bases for the other two fundamental subspaces.

Although the diagonal values of the resulting matrix R_k reveal gaps in the singular values in most practical problems, the following well-known example shows that the orthogonal decomposition with column pivoting may fail to detect rank deficiency since a matrix may be rank deficient with no small submatrix X_k.

Example 14.4 (Kahan [31]) Let $T_n(c) \in \mathbb{R}^{n \times n}$ be defined as

$$T_n(c) = \mathrm{diag}(1, s, \ldots, s^{n-1}) \begin{pmatrix} 1 & -c & -c & \cdots & -c \\ 0 & 1 & -c & \cdots & -c \\ & \ddots & & \vdots & \vdots \\ \vdots & & & 1 & -c \\ 0 & & \cdots & & 1 \end{pmatrix}, \qquad (14.22)$$

where $c > 0$, $s > 0$, and $c^2 + s^2 = 1$. The QR decomposition with column pivoting applied to $T_n(c)$ does not change the matrix, i.e., $T_n(c) = R$, $Q = I$, and $\Pi = I$. Therefore, $\|X_k\| \geq s^{n-1}$, whatever the value of k is. However, the smallest singular value of $T_n(c)$ is often much smaller than $\|X_k\|$ for any k. For example, for $T_{10}(0.8)$, $\|X_k\|_2 \geq s^9 \approx 8.96 \times 10^{-5}$ but $\sigma_{10}(T_{10}(0.8)) \approx 0.01$. And for $T_{100}(0.2)$, $\|X_k\|_2 \geq s^{99} \approx 0.13$, but $\sigma_{100}(T_{100}(0.2)) \approx 3.68 \times 10^{-9}$. ∎

The singular value gap detecting property of the decomposition can be improved by using decompositions such as the rank revealing QR decomposition [32] and rank revealing URV and ULV decompositions [33, 34]. We first introduce the QLP decomposition [35] because it requires a very simple modification to the QR decomposition with column pivoting without any requirement on estimation of the smallest singular values. For the convergence analysis of the QLP algorithm see Ref. [36]. In the QLP decomposition, the QR decomposition with column pivoting is applied one more time to the matrix R^{T} where R is obtained from the QR decomposition with column pivoting. Specifically, suppose the QR decomposition with column pivoting applied to A gives

$$Q_A^{\mathrm{T}} A \Pi_A = R \qquad (14.23)$$

and applied to R^{T} gives

$$Q_R^{\mathrm{T}} R^{\mathrm{T}} \Pi_R = U. \qquad (14.24)$$

Then combining (14.23) and (14.24),

$$A = (Q_A \Pi_R) U^{\mathrm{T}} (Q_R^{\mathrm{T}} \Pi_A^{\mathrm{T}}) = \mathrm{QLP}. \qquad (14.25)$$

The following example shows that this simple idea of applying the QR decomposition with column pivoting to the triangular matrix R^{T} one more time greatly improves the singular value tracking capability of the algorithm.

Example 14.5 Consider Kahan's example shown in Example 14.4. The matrix $T_{10}(0.8)$ has the diagonal elements (which are the diagonal elements of the R factor in the QR decomposition with column pivoting)

$$[1.00, 0.600, 0.360, 0.216, 0.130, 0.0778, 0.0467, 0.0280, 0.0168, 0.0101],$$

and $T_{10}(0.8)$ has the singular values

$$[2.89, 1.04, 0.608, 0.359, 0.212, 0.124, 0.0722, 0.0412, 0.0225, 8.96 \times 10^{-5}].$$

Therefore, the diagonal elements of R are not very accurate approximations of the singular values. On the other hand, the diagonal elements of the triangular factor L from the QLP algorithm applied to $T_{10}(0.8)$ are

$$[2.60, 1.00, 0.587, 0.376, 0.216, 0.127, 0.0741, 0.0426, 0.0234, 8.96 \times 10^{-5}].$$

Note that all the singular values are now approximated much better. ∎

Example 14.6 Consider the matrix

$$A = \begin{pmatrix} 1 & 1 & 1 \\ 0 & 0 & 1 \\ 0 & 0 & 1 \\ 0 & 0 & 1 \end{pmatrix},$$

which has the singular values

$$[2.18, 1.13, 0].$$

The diagonal elements of the matrix R from its QR decomposition with column pivoting (computed using MATLAB) are

$$[2.00, 0.866, 0]$$

and the diagonal elements from the matrix U from its QLP decomposition are

$$[2.12, 1.15, 0].$$

By allowing the orthogonal matrices applied from both sides of the matrix as in the QLP decomposition instead of simply permuting the columns, one can obtain a better approximation to the SVD. The next section covers this in a more general context.

14.2.5 COMPLETE ORTHOGONAL DECOMPOSITION

As seen earlier, the SVD is of great theoretical and practical importance. Because the computation of the SVD is expensive, there is a need for alternative, cheaper methods. Another important fact is that there are no fast methods to update the SVD when the data matrix is changed slightly, which is a serious drawback when the problem is of recursive nature. For instance, the updating of the SVD when a row is added to A requires $\mathcal{O}(n^3)$ flops (floating point operations) for a matrix of order n, in practice. Although there is an $\mathcal{O}(n^2 \log n)$ algorithm for updating the SVD [37], the constant is machine dependent and can be very large.

Among the most successful methods for approximating the SVD are the rank-revealing URV and ULV decompositions [33, 34]. They are developments of complete orthogonal decompositions [38–40] and are effective in exhibiting the rank and give bases for the range and null spaces, and can be updated in $\mathcal{O}(n^2)$ flops. They are compromises between the SVD and a QR decomposition with some of the virtues of both. Here, for simplicity, only the URVD will be discussed. First the concept of a complete orthogonal decomposition [5, Chapter 5.4.2] is defined.

Let A have rank γ. Then a *complete orthogonal decomposition* is given by

$$U^T A V = \begin{pmatrix} \tilde{R}_\gamma & 0 \\ 0 & 0 \end{pmatrix},$$

where $\tilde{R}_\gamma \in \mathbb{R}^{\gamma \times \gamma}$ has rank γ, and $U \in \mathbb{R}^{m \times m}$ and $V \in \mathbb{R}^{n \times n}$ are orthogonal matrices.

The SVD is a special case of a complete orthogonal decomposition. When the matrix $A \in \mathbb{R}^{m \times n}$, has *numerical rank* k, i.e., σ_k is *large* compared to σ_{k+1}, its URVD is given by

$$U^T A V = \begin{pmatrix} T \\ 0 \end{pmatrix} = \begin{pmatrix} R_k & S_k \\ 0 & \widehat{R}_k \\ 0 & 0 \end{pmatrix}, \tag{14.26}$$

where $U \in \mathbb{R}^{m \times m}$ and $V \in \mathbb{R}^{n \times n}$ are orthogonal matrices, $T \in \mathbb{R}^{n \times n}$, $R_k \in \mathbb{R}^{k \times k}$, and $\widehat{R}_k \in \mathbb{R}^{(n-k) \times (n-k)}$ are upper triangular, and

$$\sigma_k(R_k) \approx \sigma_k, \qquad \| S_k \|_F^2 + \| \widehat{R}_k \|_F^2 \approx \sigma_{k+1}^2 + \cdots + \sigma_n^2.$$

The algorithm for computing the URVD is based on the following observation: Given any upper-triangular matrix $R \in \mathbb{R}^{n \times n}$, let w be a unit norm approximate right singular vector of R corresponding to the smallest singular value σ_n (such an approximation can be obtained using a standard condition estimator, at a cost of $\mathcal{O}(n^2)$ flops). With an orthogonal matrix $Q \in \mathbb{R}^{n \times n}$ such that

$$Q^T w = e_n$$

and the QR decomposition of the product RQ, which is

$$RQ = Q^{(1)} R^{(1)},$$

where $Q^{(1)} \in \mathbb{R}^{n \times n}$ is orthogonal and $R^{(1)} \in \mathbb{R}^{n \times n}$ is upper triangular,

$$\sigma_n \approx \|Rw\|_2 = \|Q^{(1)^T} RQ Q^T w\|_2 = \|R^{(1)} e_n\|_2.$$

Thus the last column of $R^{(1)}$ is small, and can be *deflated* in the sense that it is considered to belong to the right "small part" of the matrix. The computation of $R^{(1)}$ can be organized so that $\mathcal{O}(n^2)$ flops are required. Then the analogous procedure is applied to the $(n-1) \times (n-1)$ principal submatrix. After $n - k$ such steps the rank-revealing URVD is obtained.

When the orthogonal matrix V is partitioned according to the numerical rank as

$$V = (V_1 \; V_2),$$

where $V_1 \in \mathbb{R}^{n \times k}$ and $V_2 \in \mathbb{R}^{n \times (n-k)}$, V_2 is an orthogonal basis for the null space of A. Similarly a basis of the range space can be obtained.

The decompositions shown in this chapter that approximate the SVD, can be used in LSI and PCA to obtain faster methods.

14.3 CONSTRAINED RANK REDUCTION

In Section 14.2 the rank reduction was based on a criterion that gives the directions of largest variation in the data matrix A. In the linear least squares problem

$$\min_x \| Ax - b \|_2, \tag{14.27}$$

where $A \in \mathbb{R}^{m \times n}$, the criterion emphasizes the "covariation" between the data matrix and the right-hand side b. This problem occurs in many problems in data analysis, see Ref. [23]. In statistical terminology this is the *multiple linear regression model*: the right-hand side b contains observations of a response variable, whose variation is to be explained by a linear combination of the explanatory variables in the columns of the data matrix A.

Often the data matrix is large and the columns of A are almost linearly dependent (the explanatory variables are almost collinear). Then one would like to find a hopefully small subset of the explanatory variables that account for as much of the variation in b as possible. In other words, a linear combination of a subset of the columns is to be found such that the corresponding residual is as small as possible. However, this *subset selection problem* [5, Chapter 12.2] is quite difficult, and often one accepts a projection of the solution onto a subspace of small dimension. Such a projection can be computed using the SVD of A or the PLS method. In both methods a rank reduction of A is performed, but here the purpose is not primarily to obtain a low-rank matrix approximation but rather to reduce the norm of the residual as much as possible.

In many cases it is not the least squares solution x that is of interest in data analysis. Instead, x is an auxiliary variable that is used for *prediction:*

$$y_{\text{predicted}} := \widehat{a}^{\mathrm{T}} \widehat{x},$$

where $\widehat{a}^{\mathrm{T}}$ is a new observation of the explanatory variables and $\widehat{x}$ is an estimate of the solution of (14.27).

14.3.1 REGRESSION USING SVD: PRINCIPAL COMPONENT REGRESSION

Using the SVD (14.9) the minimum norm least squares solution of (14.27) can be written

$$x = V \begin{pmatrix} \Sigma_\gamma^{-1} & 0 \\ 0 & 0 \end{pmatrix} U^{\mathrm{T}} b = \sum_{i=1}^{\gamma} \frac{u_i^{\mathrm{T}} b}{\sigma_i} v_i, \tag{14.28}$$

where $\gamma = \operatorname{rank}(A)$. In statistical literature, regression using the SVD goes under the name PCR. The procedure is equivalent to truncating the SVD solution (14.28), i.e., the solution is expressed in terms of the k *first singular vectors*, $k \leq \gamma$,

$$x_{\text{pcr}}^{(k)} = \sum_{i=1}^{k} \frac{u_i^{\mathrm{T}} b}{\sigma_i} v_i. \tag{14.29}$$

This can be interpreted as a rank reduction procedure, where in each step the reduction is made along the direction of largest variation in the data matrix A. PCR is described in Algorithm 1.

Algorithm 1 PCR

Given a data matrix $A \in \mathbb{R}^{m \times n}$ with its SVD, and a right-hand side $b \in \mathbb{R}^m$, it computes a solution projected on the subspace corresponding to the k largest singular values.

1. $x_{\mathrm{pcr}}^{(0)} = 0,$
2. **for** $i = 1, 2, \ldots, k$

$$x_{\mathrm{pcr}}^{(i)} = x_{\mathrm{pcr}}^{(i-1)} + \frac{u_i^{\mathrm{T}} b}{\sigma_i} v_i.$$

The solution (14.29) is also a projection onto a lower dimensional subspace as is seen from the equivalent least squares problem:

$$\min_z \|(A V_k)z - b\|, \qquad x_{pcr}^{(k)} = V_k z, \qquad V_k = \begin{pmatrix} v_1 & v_2 & \cdots & v_k \end{pmatrix}.$$

Thus, the approximate solution of (14.27) is obtained by projecting the solution onto the subspace spanned by the singular vectors $v_1, v_2, \ldots, v_k$. The value of the truncation index k can be chosen, e.g., using cross-validation [41, 42]. The purpose is to choose k such that there is as large a reduction of the norm of the residual for as small a value of k as possible.

It is not uncommon that k must be chosen inconveniently large to obtain the required reduction of the norm of the residual [43, 44]. This happens if the right-hand side b does not have significant components along the first left singular vectors u_i. To explain this, use the SVD to write the residual of the least squares problem,

$$r_k^2 := \|A x_{\mathrm{pcr}}^{(k)} - b\|^2 = \left\| \begin{pmatrix} \Sigma_\gamma & 0 \\ 0 & 0 \end{pmatrix} z^k - U^{\mathrm{T}} b \right\|^2, \qquad z^k = \begin{pmatrix} u_1^{\mathrm{T}} b / \sigma_1 \\ \vdots \\ u_k^{\mathrm{T}} b / \sigma_k \\ 0 \\ \vdots \\ 0 \end{pmatrix}$$

Thus the residual becomes

$$r_k^2 = \sum_{i=k+1}^{m} (u_i^{\mathrm{T}} b)^2.$$

The singular values are ordered in such a way that they reflect the directions of largest variation in the data matrix A. However, if in regression the purpose is to reduce the residual significantly using a projection onto a subspace of as small a dimension as possible, then, naturally, the right-hand side must influence the choice of directions. This can be done simply by choosing the equations to satisfy exactly (i.e., the nonzero components of z) *in the order of the largest magnitudes of the vector* $U^{\mathrm{T}} b$. The rank reduction process, here denoted MPCR (modified PCR), is described in Algorithm 2.

The PLS method, described in the next section, automatically takes into account *both the right-hand side and the data matrix.*

Algorithm 2 MPCR (Modified PCR)

Given a data matrix $A \in \mathbb{R}^{m \times n}$ with its SVD, and a right-hand side $b \in \mathbb{R}^m$, it computes a solution projected on the subspace corresponding to the k largest components of $U^T b$.

1. $x_{\text{mpcr}}^{(0)} = 0, \quad c = U^T b$,
2. **for** $i = 1, 2, \ldots, k$
   (a) Choose j_i such that $|c_{j_i}| = \max_i |c_i|$,
   (b) $x_{\text{mpcr}}^{(i)} = x_{\text{mpcr}}^{(i-1)} + \frac{u_{j_i}^T b}{\sigma_{j_i}} v_{j_i}$,
   (c) $c_{j_i} = 0$.

14.3.2 REGRESSION USING PARTIAL LEAST SQUARES

The PLS method due to Wold [45] is commonly used in data analysis and chemometrics [42]. It has been known for quite some time [43, 46] that PLS is closely related to Lanczos (Golub–Kahan [47]) bidiagonalization. The so-called NIPALS version of the method, given in Algorithm 3, is clearly a rank reduction process.

Defining

$$
\begin{aligned}
W_k &= \begin{bmatrix} w_1 & w_2 & \ldots & w_k \end{bmatrix}, \\
T_k &= \begin{bmatrix} t_1 & t_2 & \ldots & t_k \end{bmatrix}, \\
S_k &= \begin{bmatrix} s_1 & s_2 & \ldots & s_k \end{bmatrix},
\end{aligned}
$$

where the columns w_i, t_i, and s_i are computed as in Algorithm 3, it is straightforward to show that the columns of W_k and T_k are orthonormal (see Refs. [43, 46]). In the NIPALS context the approximate solution $x_{\text{pls}}^{(k)}$ of (14.27) is usually obtained from

$$ x_{\text{pls}}^{(k)} = W_k (S_k^T W_k)^{-1} T_k^T b. \tag{14.30} $$

From the equivalence of PLS and Lanczos bidiagonalization [43, 46], one can easily show that

$$ A W_k = T_k B_k, $$

where B_k is a bidiagonal matrix, the entries of which are normalizing constants in the algorithm [43]. From this identity, it is seen that the PLS solution (after k steps) is the same as the solution of

$$
\begin{aligned}
\min_{x \in \text{span}(W_k)} \| Ax - b \|_2^2 &= \min_z \| A W_k z - b \|_2^2 = \min_z \| T_k B_k z - b \|_2^2 \\
&= \min_z \| B_k z - T_k^T b \|_2^2 + \| (T_k^\perp)^T b \|_2^2,
\end{aligned}
$$

where $T = (T_k \ T_k^\perp) \in \mathbb{R}^{m \times m}$ is an orthogonal matrix. Thus PLS is a projection method, where the solution is expressed as a linear combination of the w_i vectors. The expression for the computation of these vectors,

$$ w_i = \frac{1}{\| (A^{(i)})^T b \|_2} (A^{(i)})^T b, $$

Algorithm 3 PLS

Given a data matrix $A \in \mathbb{R}^{m \times n}$, and a right-hand side $b \in \mathbb{R}^m$, it performs k steps of the partial least squares algorithm.

1. $A^{(1)} = A$,
2. **for** $i = 1, 2, \ldots, k$
   (a) $w_i = \frac{1}{\|(A^{(i)})^{\mathrm{T}} b\|_2} (A^{(i)})^{\mathrm{T}} b$,
   (b) $t_i = \frac{1}{\|A^{(i)} w_i\|_2} A^{(i)} w_i$,
   (c) $s_i = (A^{(i)})^{\mathrm{T}} t_i$,
   (d) $A^{(i+1)} = A^{(i)} - t_i s_i^{\mathrm{T}}$.

---

shows that w_i is the covariance between the data matrix $A^{(i)}$ and the right-hand side. This direction is chosen to *obtain a high correlation between t_i and the right-hand side* [48]. Thus, one can interpret the PLS algorithm as follows:

> In each step, find the new basis vector for the projection of the solution as the co-variance vector between the present data matrix and the right-hand side. After the projection has been computed, remove the corresponding information from the data matrix by performing a Wedderburn rank reduction.

For understanding the PLS, it is essential to know that the columns of W_p and T_p are orthogonal bases for *Krylov subspaces*. Let

$$\mathcal{K}_i(C, b) = \mathrm{span}\{b, Cb, \ldots, C^{i-1}b\}$$

denote the linear subspace spanned by the vectors $b, Cb, \ldots, C^{i-1}b$. The following proposition is well known and is easily proved using the recursion for $A^{(i)}$.

Proposition 14.1. *The vectors $w_1, w_2, \ldots, w_i$ and $t_1, t_2, \ldots, t_i$ constitute orthogonal bases of the Krylov subspaces $\mathcal{K}_i(A^{\mathrm{T}} A, A^{\mathrm{T}} b)$ and $\mathcal{K}_i(AA^{\mathrm{T}}, AA^{\mathrm{T}} b)$, respectively.*

From the above description of PLS one may get the impression that there is always a quick reduction of the residual in the first steps of the algorithm. However, the actual reduction depends in a rather complicated way on the distribution of singular values and the magnitude of the elements of the vector $U^{\mathrm{T}} b$ [43]. The Krylov spaces generated are the key to understanding the behavior of PLS. Consider the space

$$\mathcal{K}_i(A^{\mathrm{T}} A, A^{\mathrm{T}} b) = \mathrm{span}\{A^{\mathrm{T}} b, (A^{\mathrm{T}} A)A^{\mathrm{T}} b, \ldots, (A^{\mathrm{T}} A)^{i-1} A^{\mathrm{T}} b\}.$$

It is easy to show, using the theory of the power method [5, Section 8.2], that for (almost) any vector s,

$$\lim_{j \to \infty} (A^{\mathrm{T}} A)^j s = v_1, \tag{14.31}$$

i.e., the first right singular vector. In view of (14.31), the behavior of the Krylov sequence can be summarized as follows:

1. If the mass of b is concentrated along some intermediate singular vectors, and that mass is large enough (related to the magnitude of the singular values), then the first few Krylov basis vectors will contain large components along those singular vectors. If the mass is

not large enough, then the first basis vectors will be dominated by directions corresponding to the largest singular values.

2. The next several Krylov basis vectors will be dominated by components along the singular vectors corresponding to the largest singular values.
3. Gradually, because a sequence of orthogonal basis vectors is generated, the influence of
   singular vectors corresponding to smaller singular values will be noticed. Unless there is
   a huge component in the right-hand side along a singular vector corresponding to a small
   singular value, it will take long before directions corresponding to the small singular
   values play a role in the approximate solution.

The following example illustrates that the rate of decay of the residual depends on the right-
hand side.

Example 14.7   A problem with a diagonal matrix of dimension 50 was constructed, whose first
22 diagonal elements decay from 1 to 0.1, with the rest decaying from 0.1 to 0.001, approximately.
Two right-hand sides were used as shown in the first two graphs of Figure 14.2. The coordinates
of the right-hand sides in terms of the left singular vectors u_i (here: the standard basis vectors e_i)
are plotted in the first two graphs of Figure 14.2. In the third graph of Figure 14.2, the residuals
are given. For Case 1, PLS automatically reads off the distribution of the $u_i^T b$ (in a way similar to
MPCR), and reduces the residual significantly in the first two steps. On the other hand, in Case 2,
which has a considerably more complicated right-hand side, PLS requires more steps to reduce the
residual.   ∎

14.3.3   FEATURE EXTRACTION OF CLUSTERED DATA

To achieve higher efficiency in manipulating the data represented in a high-dimensional space, it
is often necessary to reduce the dimension *dramatically*. In this section, several dimension reduc-
tion/feature extraction methods that preserve the cluster structure are discussed. The clustering of
data are assumed to be performed already. Each method attempts to choose a projection to the reduced dimension that will capture the cluster structure of the data collection as well as possible. The
dimension reduction is only a preprocessing stage and even if this stage is a little expensive, it may
be worthwhile if it effectively reduces the cost of the postprocessing involved in classification and
query processing, which will dominate computationally. Some experimental results are presented
at the end of this chapter to illustrate the trade-off in effectiveness versus efficiency of the methods,
so that their potential application can be evaluated.

Given a term-document matrix, the problem is to find a transformation that maps each document
vector in the m-dimensional space to a vector in the k-dimensional space for some $k < m$. For this,
either the dimension reducing transformation $T^T \in \mathbb{R}^{k \times m}$ is computed explicitly or the problem
is formulated as that of approximation where the given matrix A is to be decomposed into two
matrices B and Y as

$$A \approx BY \tag{14.32}$$

where $B \in \mathbb{R}^{m \times k}$ with rank$(B) = k$ and $Y \in \mathbb{R}^{k \times n}$ with rank$(Y) = k$. The matrix B is responsible
for the dimension reducing transformation. However, often the dimension reducing transformation
T^T itself is not explicitly needed because the dimension reduced representation Y can be computed
without the transformation T^T. The lower rank approximation (14.32) is not unique since for any
nonsingular matrix $Z \in \mathbb{R}^{k \times k}$,

$$A \approx BY = (BZ)(Z^{-1}Y),$$

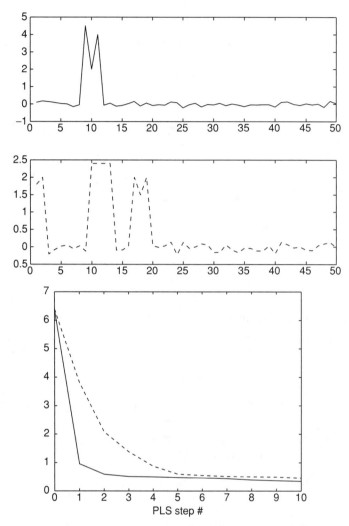

FIGURE 14.2 The upper two graphs show the coordinates of the right-hand sides along the left singular vectors u_i. Cases 1 (top) and 2 (middle). The bottom graph shows the residual as a function of the number of PLS steps. The solid curve is for Case 1 and the dashed for Case 2.

where $\text{rank}(BZ) = k$ and $\text{rank}(Z^{-1}Y) = k$. The decomposition (14.32) is closely related to the rank reducing decomposition presented in (14.3).

The problem of approximate decomposition (14.32) can be recast in two different but related ways. The first is in terms of a matrix rank reduction formula and the second is as a minimization problem

$$\min_{B \in \mathbb{R}^{m \times k}, Y \in \mathbb{R}^{k \times n}} \|A - BY\|_F. \tag{14.33}$$

The incorporation of clustered structure can be translated into choosing the factors F and G in (14.7) or adding a constraint in the minimization problem (14.33). In the next section, ways to compute the factor Y are discussed where knowledge of the clusters from the full dimension is reflected in the dimension reduction. The nonlinear extension of these methods for dimension reduction of nonlinearly separable data (see Refs. [49, 50]).

14.3.3.1 Methods Based on Centroids

For simplicity of discussion, the columns a_j of A are assumed to be grouped into k clusters as

$$A = [\tilde{A}_1 \quad \tilde{A}_2 \quad \cdots \quad \tilde{A}_k] \quad \text{where} \quad \tilde{A}_i \in \mathbb{R}^{m \times n_i}, \quad \text{and} \quad \sum_{i=1}^{k} n_i = n. \quad (14.34)$$

Let N_i denote the set of column indices that belong to cluster i. The centroid $c^{(i)}$ of the ith cluster is the average of the columns in N_i, i.e.,

$$c^{(i)} = \frac{1}{n_i} \sum_{j \in N_i} a_j$$

and the global centroid c is defined as

$$c = \frac{1}{n} \sum_{j=1}^{n} a_j.$$

The centroid vector achieves the minimum variance since

$$\sum_{j=1}^{n} \|a_j - c\|_2^2 = \min_{x \in \mathbb{R}^{n \times 1}} \sum_{j=1}^{n} \|a_j - x\|_2^2 = \min_{x \in \mathbb{R}^{n \times 1}} \|A - xe^{\mathrm{T}}\|_F^2,$$

where $e = (1, \ldots, 1)^{\mathrm{T}} \in \mathbb{R}^{n \times 1}$. In the centroid method for dimension reduction due to Park et al. [51] and summarized in Algorithm 4, the columns of B in the minimization problem (14.33) are chosen as the centroids of the k clusters. Then the k-dimensional representation Y of the matrix A is obtained by solving the least squares problem [6]

$$\min_{Y \in \mathbb{R}^{k \times n}} \|A - CY\|_F, \quad (14.35)$$

where C is the centroid matrix whose ith column is the centroid of the ith cluster. Note that the reduced dimension k is the same as the number of clusters. A k-dimensional representation y of any data $q \in \mathbb{R}^{m \times 1}$ can be found from solving

$$\min_{y \in \mathbb{R}^{k \times 1}} \|q - Cy\|_2. \quad (14.36)$$

The centroid dimension reduction method can be represented using the matrix rank reduction formula. Defining a grouping matrix $H \in \mathbb{R}^{n \times k}$ as

$$H = J \cdot (J^{\mathrm{T}} J)^{-1} \quad (14.37)$$

where $J \in \mathbb{R}^{n \times k}$ and

$$J(i, j) = \begin{cases} 1 & \text{if document } i \text{ belongs to cluster } j, \\ 0 & \text{otherwise,} \end{cases}$$

Algorithm 4 Centroid dimension reduction method

Given a data matrix $A \in \mathbb{R}^{m \times n}$ with k clusters, it computes a k dimensional representation y of any given vector $q \in \mathbb{R}^{m \times 1}$.

    1. Compute the centroid $c^{(i)}$ of the ith cluster, $1 \leq i \leq k$.
    2. Set $C = [c^{(1)} \quad c^{(2)} \quad \cdots \quad c^{(k)}]$.
    3. Solve $\min_Y \|Cy - q\|_2$

---

the centroid matrix C can be written as $C = AH$. The solution for the problem (14.35) is $Y = (C^T C)^{-1} C^T A$, which in turn yields the matrix rank reduction expression

$$
\begin{aligned}
E &= A - CY \\
&= A - (AH)(C^T C)^{-1} C^T A \\
&= A - (AH)(H^T A^T A H)^{-1}(H^T A^T A).
\end{aligned}
$$

This shows that the rank reduction theorem (Theorem 14.2) with $G = H^T A^T$ and $F = H$, produces the centroid method.

In the orthogonal centroid method [51], an orthonormal basis Q_k for the centroid matrix C is chosen as the matrix B and the k dimensional representation of the matrix A is found by solving

$$
\min_{Y \in \mathbb{R}^{k \times n}} \|A - Q_k Y\|_F. \tag{14.38}
$$

Then a k-dimensional representation of any data $q \in \mathbb{R}^{m \times 1}$ can be found by solving

$$
\min_{y \in \mathbb{R}^{k \times 1}} \|q - Q_k y\|_F, \tag{14.39}
$$

where the solution is $y = Q_k^T q$. The matrix Q_k can be obtained from the orthogonal decomposition of the matrix C when it has full rank. This orthogonal centroid method, summarized in Algorithm 5, provides a link between the methods of discriminant analysis and those based on the centroids. More detailed discussions can be found in Ref. [52].

The orthogonal centroid method is mathematically equivalent to the multiple group method known in applied statistics/psychometrics for more than 50 years. In his book on factor analysis [53], Horst attributes the multiple group method to Thurstone [54]. For a description of the multiple group algorithm in modern numerical linear algebra terms, see Ref. [55].

Algorithm 5 Orthogonal centroid dimension reduction method

Given a data matrix $A \in \mathbb{R}^{m \times n}$ with k clusters, it computes a k dimensional representation y of any given vector $q \in \mathbb{R}^{m \times 1}$.

    1. Compute the centroid $c^{(i)}$ of the ith cluster, $1 \leq i \leq k$.
    2. Set $C = [c^{(1)} \quad c^{(2)} \quad \cdots \quad c^{(k)}]$.
    3. Compute the reduced QR decomposition of C, which is $C = Q_k R$.
    4. Solve $\min_Y \|Q_k y - q\|_F$ (in fact, $y = Q_k^T q$).

14.3.3.2 Classical Linear Discriminant Analysis

The goal of linear discriminant analysis is to combine features of the original data in a way that most effectively discriminates between classes. Assuming that the given data are already clustered, a transformation $T^{\mathrm{T}} \in \mathbb{R}^{k \times m}$ is to be found that optimally preserves this cluster structure in the reduced dimensional space. For this purpose, cluster quality needs to be measured. When cluster quality is high, each cluster is tightly grouped, but well-separated from the other clusters. To quantify this, the within-cluster, between-cluster, and mixture scatter matrices are defined [56] as

$$S_{\mathrm{w}} = \sum_{i=1}^{k} \sum_{j \in N_i} (a_j - c^{(i)})(a_j - c^{(i)})^{\mathrm{T}},$$

$$S_{\mathrm{b}} = \sum_{i=1}^{k} \sum_{j \in N_i} (c^{(i)} - c)(c^{(i)} - c)^{\mathrm{T}} = \sum_{i=1}^{k} n_i (c^{(i)} - c)(c^{(i)} - c)^{\mathrm{T}},$$

$$S_{\mathrm{m}} = \sum_{j=1}^{n} (a_j - c)(a_j - c)^{\mathrm{T}},$$

respectively. The scatter matrices have the relationship [15]

$$S_{\mathrm{m}} = S_{\mathrm{w}} + S_{\mathrm{b}}. \tag{14.40}$$

Since

$$\mathrm{trace}(S_{\mathrm{w}}) = \sum_{i=1}^{k} \sum_{j \in N_i} (a_j - c^{(i)})^{\mathrm{T}}(a_j - c^{(i)}) = \sum_{i=1}^{k} \sum_{j \in N_i} \|a_j - c^{(i)}\|_2^2$$

measures the closeness of the columns within the clusters, and

$$\mathrm{trace}(S_{\mathrm{b}}) = \sum_{i=1}^{k} \sum_{j \in N_i} (c^{(i)} - c)^{\mathrm{T}}(c^{(i)} - c) = \sum_{i=1}^{k} \sum_{j \in N_i} \|c^{(i)} - c\|_2^2$$

measures the separation between clusters, an optimal transformation that preserves the given cluster structure would maximize $\mathrm{trace}(T^{\mathrm{T}} S_{\mathrm{b}} T)$ and minimize $\mathrm{trace}(T^{\mathrm{T}} S_{\mathrm{w}} T)$.

One of the most commonly used criteria in discriminant analysis is that of optimizing

$$J_1(T) = \mathrm{trace}((T^{\mathrm{T}} S_2 T)^{-1}(T^{\mathrm{T}} S_1 T)), \tag{14.41}$$

where S_1 and S_2 are chosen from S_{w}, S_{b}, and S_{m}. When S_2 is assumed to be nonsingular, it is symmetric positive definite. According to results from the symmetric-definite generalized eigenvalue problem [5], there exists a nonsingular matrix $X \in \mathbb{R}^{m \times m}$ such that

$$X^{\mathrm{T}} S_1 X = \Lambda = \mathrm{diag}(\lambda_1, \ldots, \lambda_m) \quad \text{and} \quad X^{\mathrm{T}} S_2 X = I_m,$$

where without loss of generality, the following can be assumed:

$$\lambda_1 \geq \cdots \geq \lambda_q > \lambda_{q+1} = \cdots = \lambda_m = 0.$$

Since S_1 is positive semidefinite and $x_i^T S_1 x_i = \lambda_i$, each λ_i is nonnegative and only the largest $q = \text{rank}(S_1)$ λ_i's are nonzero. Letting x_i denote the ith column of X,

$$S_1 x_i = \lambda_i S_2 x_i, \tag{14.42}$$

which means that λ_i and x_i are an eigenvalue–eigenvector pair of $S_2^{-1} S_1$. Then

$$
\begin{aligned}
J_1(T) &= \text{trace}((T^T S_2 T)^{-1}(T^T S_1 T)) \\
&= \text{trace}((T^T X^{-T} X^{-1} T)^{-1} T^T X^{-T} \Lambda X^{-1} T) \\
&= \text{trace}((\tilde{T}^T \tilde{T})^{-1} \tilde{T}^T \Lambda \tilde{T}),
\end{aligned}
$$

where $\tilde{T} = X^{-1} T$. The matrix $\tilde{T}$ has full column rank provided T does, so it has the reduced QR factorization $\tilde{T} = QR$, where $Q \in \mathbb{R}^{m \times l}$ has orthonormal columns and R is nonsingular. Hence

$$
\begin{aligned}
J_1(T) &= \text{trace}((R^T R)^{-1} R^T Q^T \Lambda QR) \\
&= \text{trace}(Q^T \Lambda Q).
\end{aligned}
$$

This shows that

$$\max_T J_1(T) = \max_{Q^T Q = I} \text{trace}(Q^T \Lambda Q) \le \lambda_1 + \cdots + \lambda_q = \text{trace}(S_2^{-1} S_1).$$

(Here only maximization is considered. However, J_1 may need to be minimized for some other choices of S_1 and S_2.) When the reduced dimension k is at least as large as q, i.e., $k \ge q$, this upper bound on $J_1(T)$ is achieved with

$$Q = \begin{pmatrix} I_k \\ 0 \end{pmatrix} \quad \text{or} \quad T = X \begin{pmatrix} I_k \\ 0 \end{pmatrix} R.$$

The transformation T is not unique. That is, J_1 satisfies the invariance property

$$J_1(T) = J_1(TW)$$

for any nonsingular matrix $W \in \mathbb{R}^{k \times k}$, because

$$
\begin{aligned}
J_1(TW) &= \text{trace}((W^T T^T S_2 TW)^{-1}(W^T T^T S_1 TW)) \\
&= J_1(T).
\end{aligned}
$$

Hence, the maximum $J_1(T)$ is also achieved with

$$T = X \begin{pmatrix} I_l \\ 0 \end{pmatrix},$$

and

$$\text{trace}((T^T S_2 T)^{-1} T^T S_1 T) = \text{trace}(S_2^{-1} S_1) \tag{14.43}$$

whenever $T \in \mathbb{R}^{m \times k}$ consists of k eigenvectors of $S_2^{-1} S_1$ corresponding to the l largest eigenvalues.

14.3.3.3 Improved Linear Discriminant Analysis and GSVD

Assume the cluster structure 14.34. Defining the $m \times n$ matrices,

$$H_w = [A_1 - c^{(1)}e^{(1)^T}, A_2 - c^{(2)}e^{(2)^T}, \ldots, A_k - c^{(k)}e^{(k)^T}], \qquad (14.44)$$

$$H_b = [(c^{(1)} - c)e^{(1)^T}, (c^{(2)} - c)e^{(2)^T}, \ldots, (c^{(k)} - c)e^{(k)^T}], \qquad (14.45)$$

$$H_m = [a_1 - c, \ldots, a_n - c] = A - ce^T, \qquad (14.46)$$

where $e^{(i)} = (1, \ldots, 1)^T \in \mathbb{R}^{n_i \times 1}$, the scatter matrices can be expressed as

$$S_w = H_w H_w^T, \quad S_b = H_b H_b^T, \quad \text{and} \quad S_m = H_m H_m^T. \qquad (14.47)$$

Now, a limitation of the J_1 criterion in many applications, including text processing in information retrieval, is that the matrix S_2 must be nonsingular. The matrix S_b is always singular since $\text{rank}(S_b) \leq k - 1$. The matrices S_w and S_m are singular in many applications: when the number of terms in the document collection is larger than the total number of documents (i.e., $m > n$ in the term-document matrix A) and when the data are represented in a high-dimensional space but collecting many samples is expensive. However, a solution based on the GSVD does not impose this restriction, and can be found without explicitly forming S_1 and S_2 from H_w, H_b, and H_m. In addition, the improved LDA/GSVD method is numerically superior to the classical LDA which uses the generalized eigenvalue decomposition, when the matrices involved are ill-conditioned [52]. The GSVD [57, 58], [5, Theorem 8.7.4], as defined in the following theorem is used as a key tool.

Theorem 14.7 (Generalized Singular Value Decomposition [58]). *For any two matrices $K_A \in \mathbb{R}^{n \times m}$ and $K_B \in \mathbb{R}^{p \times m}$ with the same number of columns, there exist orthogonal matrices $U \in \mathbb{R}^{n \times n}$, $V \in \mathbb{R}^{p \times p}$, $W \in \mathbb{R}^{t \times t}$, and $Q \in \mathbb{R}^{m \times m}$ such that*

$$U^T K_A Q = \Sigma_A (\underbrace{W^T R}_{t}, \underbrace{0}_{m-t}) \quad and \quad V^T K_B Q = \Sigma_B (\underbrace{W^T R}_{t}, \underbrace{0}_{m-t}),$$

where

$$K = \begin{pmatrix} K_A \\ K_B \end{pmatrix} \quad and \quad t = \text{rank}(K),$$

$$\Sigma_A = \begin{pmatrix} I_A & & \\ & D_A & \\ & & O_A \end{pmatrix}, \quad \Sigma_B = \begin{pmatrix} O_B & & \\ & D_B & \\ & & I_B \end{pmatrix},$$

and $R \in \mathbb{R}^{t \times t}$ is nonsingular with its singular values equal to the nonzero singular values of K. The matrices

$$I_A \in \mathbb{R}^{r \times r} \quad and \quad I_B \in \mathbb{R}^{(t-r-s) \times (t-r-s)}$$

are identity matrices, where

$$r = \text{rank}\begin{pmatrix} K_A \\ K_B \end{pmatrix} - \text{rank}(K_B) \quad and \quad s = \text{rank}(K_A) + \text{rank}(K_B) - \text{rank}\begin{pmatrix} K_A \\ K_B \end{pmatrix},$$

$$O_A \in \mathbb{R}^{(n-r-s) \times (t-r-s)} \quad and \quad O_B \in \mathbb{R}^{(p-t+r) \times r}$$

are zero matrices with possibly no rows or no columns, and

$$D_A = \text{diag}(\alpha_{r+1}, \ldots, \alpha_{r+s}) \quad and \quad D_B = \text{diag}(\beta_{r+1}, \ldots, \beta_{r+s})$$

satisfy

$$1 > \alpha_{r+1} \geq \cdots \geq \alpha_{r+s} > 0, \qquad 0 < \beta_{r+1} \leq \cdots \leq \beta_{r+s} < 1, \tag{14.48}$$

and $\alpha_i^2 + \beta_i^2 = 1$, *for* $i = r + 1, \ldots, r + s$. ∎

This form of GSVD is related to that of Van Loan [58] as

$$U^T K_A X = (\Sigma_A, 0) \qquad \text{and} \qquad V^T K_B X = (\Sigma_B, 0), \tag{14.49}$$

where $\underset{m \times m}{X} = Q \begin{pmatrix} R^{-1}W & 0 \\ 0 & I \end{pmatrix}$ is a nonsingular matrix. This implies that

$$X^T K_A^T K_A X = \begin{pmatrix} \Sigma_A^T \Sigma_A & 0 \\ 0 & 0 \end{pmatrix} \qquad \text{and} \qquad X^T K_B^T K_B X = \begin{pmatrix} \Sigma_B^T \Sigma_B & 0 \\ 0 & 0 \end{pmatrix}.$$

Defining

$$\alpha_i = 1, \ \beta_i = 0 \quad \text{for } i = 1, \ldots, r$$

and

$$\alpha_i = 0, \ \beta_i = 1 \quad \text{for } i = r + s + 1, \ldots, t,$$

one obtains, for $1 \leq i \leq t$,

$$\beta_i^2 K_A^T K_A x_i = \alpha_i^2 K_B^T K_B x_i, \tag{14.50}$$

where x_i represents the ith column of X. Since

$$K_A^T K_A x_i = 0 \qquad \text{and} \qquad K_B^T K_B x_i = 0$$

for x_i with $t + 1 \leq i \leq m$, the remaining $m - t$ columns of X, (14.50) is satisfied for arbitrary values of α_i and β_i when $t + 1 \leq i \leq m$. The columns of X are the generalized right singular vectors for the matrix pair (K_A, K_B). In terms of the generalized singular values, or the α_i/β_i quotients, r of them are infinite, s are finite and nonzero, and $t - r - s$ are zero.

Expressing λ_i as α_i^2/β_i^2, (14.42) becomes

$$\beta_i^2 S_1 x_i = \alpha_i^2 S_2 x_i, \tag{14.51}$$

which has the form of a problem that can be solved using the GSVD.

Consider the case where

$$(S_1, S_2) = (S_b, S_w).$$

From (14.47) and the definition of H_b given in (14.45), rank$(S_b) \leq k - 1$. To approximate T that satisfies both

$$\max_{T} \text{trace}(T^T S_b T) \qquad \text{and} \qquad \min_{T} \text{trace}(T^T S_w T), \tag{14.52}$$

the x_i's which correspond to the $k - 1$ largest λ_i's are chosen, where $\lambda_i = \alpha_i^2/\beta_i^2$. When the GSVD construction orders the singular value pairs as in (14.48), the α_i/β_i quotients, are in nonincreasing order. Therefore, the first $k - 1$ columns of X are all that is needed. This method based on the GSVD is summarized in Algorithm 6 [55]. In the algorithm, the matrices H_b and H_w from the data matrix A are first computed. Then a limited portion of the GSVD of the matrix pair (H_b^T, H_w^T) is computed.

Algorithm 6 LDA/GSVD

Given a data matrix $A \in \mathbb{R}^{m \times n}$ with k clusters, it computes the columns of the matrix $T \in \mathbb{R}^{m \times (k-1)}$, which preserves the cluster structure in the reduced dimensional space, using

$$J_1(T) = \text{trace}((T^{\mathrm{T}} S_{\mathrm{w}} T)^{-1} T^{\mathrm{T}} S_{\mathrm{b}} T).$$

It also computes the $k-1$ dimensional representation y of any given vector q.

1. Compute H_{b} and H_{w} from A according to

$$H_{\mathrm{b}} = [\sqrt{n_1}(c^{(1)} - c), \sqrt{n_2}(c^{(2)} - c), \ldots, \sqrt{n_k}(c^{(k)} - c)] \in \mathbb{R}^{m \times k},$$

   and (14.44), respectively.
2. Compute the complete orthogonal decomposition of
   $$K = \begin{pmatrix} H_{\mathrm{b}}^{\mathrm{T}} \\ H_{\mathrm{w}}^{\mathrm{T}} \end{pmatrix} \in \mathbb{R}^{(k+n) \times m}, \text{ which is } P^{\mathrm{T}} K Q = \begin{pmatrix} R & 0 \\ 0 & 0 \end{pmatrix}.$$
3. Let $t = \text{rank}(K)$.
4. Compute W from the SVD of $P(1:k, 1:t)$, which is
   $U^{\mathrm{T}} P(1:k, 1:t) W = \Sigma_A$.
5. Compute the first $k-1$ columns of $X = Q \begin{pmatrix} R^{-1} W & 0 \\ 0 & I \end{pmatrix}$,

   and assign them to T.
6. $y = T^{\mathrm{T}} q$

This solution is accomplished by following the construction in the proof of Theorem 1.7 [58]. The major steps are limited to the complete orthogonal decomposition [5, 39, 40] of

$$K = \begin{pmatrix} H_{\mathrm{b}}^{\mathrm{T}} \\ H_{\mathrm{w}}^{\mathrm{T}} \end{pmatrix},$$

which produces orthogonal matrices P and Q and a nonsingular matrix R, followed by the singular value decomposition of a leading principal submatrix of P.

When $m > n$, the scatter matrix S_{w} is singular. Hence, the J_1 criterion cannot be defined, and discriminant analysis fails. Consider a generalized right singular vector x_i that lies in the null space of S_{w}. From Equation (14.51), it is seen that either x_i also lies in the null space of S_{b}, or the corresponding β_i equals zero.

When $x_i \in \text{null}(S_{\mathrm{w}}) \cap \text{null}(S_{\mathrm{b}})$, Equation (14.51) is satisfied for arbitrary values of α_i and β_i. This will be the case for the rightmost $m - t$ columns of X. To determine whether these columns should be included in T, consider

$$\text{trace}(T^{\mathrm{T}} S_{\mathrm{b}} T) = \sum t_j^{\mathrm{T}} S_{\mathrm{b}} t_j \quad \text{and} \quad \text{trace}(T^{\mathrm{T}} S_{\mathrm{w}} T) = \sum t_j^{\mathrm{T}} S_{\mathrm{w}} t_j,$$

where t_j represents the jth column of T. Since $x_i^{\mathrm{T}} S_{\mathrm{w}} x_i = 0$ and $x_i^{\mathrm{T}} S_{\mathrm{b}} x_i = 0$, adding the column x_i to T does not contribute to either maximization or minimization in (14.52). For this reason, these columns of X are not included in the solution.

When $x_i \in \text{null}(S_{\mathrm{w}}) - \text{null}(S_{\mathrm{b}})$, then $\beta_i = 0$. This implies that $\alpha_i = 1$, and hence that the generalized singular value α_i / β_i is infinite. The leftmost columns of X will correspond to these. Including these columns in T increases $\text{trace}(T^{\mathrm{T}} S_{\mathrm{b}} T)$, while leaving $\text{trace}(T^{\mathrm{T}} S_{\mathrm{w}} T)$ unchanged.

It follows that, even when S_w is singular, the rule regarding which columns of X to include in T remains the same as for the nonsingular case. The experiments in Section 14.3.4 demonstrate that Algorithm 6 works well even when S_w is singular, thus extending its applicability beyond that of classical discriminant analysis, and producing more accurate solutions even when S_w is nonsingular and thus the classical LDA can be applied. For equivalence proof of the solutions from various choices of (S_1, S_2), see Ref. [52].

14.3.4 EXPERIMENTS IN CLASSIFICATION

In this section, the effectiveness of the three-dimension reduction methods, centroid, orthogonal centroid, and LDA/GSVD algorithms, are illustrated using experimental results. For LDA/GSVD, its mathematical equivalence to J_1 using an alternative choice of (S_1, S_2) are confirmed. For orthogonal centroid, its preservation of trace(S_b) is shown to be a very effective compromise for the simultaneous optimization of two traces approximated by J_1.

In Tables 14.1 and 14.2, a clustered data set generated by an algorithm adapted from [15, Appendix H] was used. The data consist of 2000 vectors in a space of dimension 150, with $k = 7$ clusters and these data are reclassified in the full space and in the reduced dimensional spaces in Table 14.1. The centroid and orthogonal centroid methods reduce the dimension from 150 to $k = 7$, and the LDA/GSVD reduces the dimension $k - 1 = 6$. In Table 14.1, the LDA/GSVD criterion, $J_1 = \max \text{trace}(S_w^{-1} S_b)$ was compared with the alternative J_1 criterion, $\max \text{trace}(S_w^{-1} S_m)$. The trace values confirm the theoretical finding that the generalized eigenvectors that optimize the alternative J_1 also optimize LDA/GSVDs J_1. The values for trace$(S_w^{-1} S_b)$ and trace$(S_w^{-1} S_m)$ were not computed for the centroid-based dimension reduction methods.

The classification accuracies for a centroid-based classification method [55] and the k nearest neighbor (knn) classification method [59] (see Algorithms 7 and 8), are presented. Note that the classification parameter k of knn differs from the number of clusters k. These are obtained using the L_2 norm similarity measure. In Table 14.2, the results with the same data are shown except that now the 2000 data items are randomly split into 1002 training data items and 998 test data items.

Another set of experiments validates the effectiveness of the dimension reduction methods and the extension of J_1 to the singular case. For this purpose, five categories of abstracts from the MEDLINE database [52] were used (Table 14.3). Each category has 40 documents. There are 7519

TABLE 14.1

Trace Values and Self-Classification Accuracy (in %) with L_2 Norm Similarity on a 150×2000 Artificially Generated Data Set with 7 Clusters

Method	Full	Orthogonal Centroid	Centroid	max trace$(S_w^{-1} S_b)$	max trace$(S_w^{-1} S_m)$
Dim	150×2000	7×2000	7×2000	6×2000	6×2000
trace(S_w)	299705	13885	907.75	1.97	1.48
trace(S_b)	22925	22925	1706.8	4.03	3.04
trace(S_m)	322630	36810	2614.55	6.00	4.52
trace$(S_w^{-1} S_b)$	12.6	-	-	12.6	12.6
trace$(S_w^{-1} S_m)$	162.6	-	-	18.6	18.6
centroid (%)	97.4	97.4	97.1	97.8	98.0
5nn (%)	81.3	96.7	96.8	97.8	97.8
15nn (%)	89.9	97.3	97.2	98.2	98.1

TABLE 14.2
Trace Values and Classification Accuracy (in %) with L_2 Norm
Similarity. The data set is the same as that for Table 14.1 except that out
of 2000 data points, 1002 were used for training and 998 as test data.

Method	Full	Orthogonal Centroid	Centroid	LDA/GSVD
Dim	150×998	7×998	7×998	6×998
trace(S_w)	149,821	6,918.1	436.25	1.81
trace(S_b)	12,025	12,025	855.15	4.19
trace($S_w^{-1} S_b$)	14.358	12.226	12.226	14.358
centroid (%)	95.8	95.9	96.0	94.1
5nn (%)	75.6	95.6	95.6	94.4
15nn (%)	87.3	96.3	96.5	94.3
50nn (%)	92.4	96.1	96.1	93.8

terms after preprocessing with stemming and removal of stop words [60]. Because 7519 exceeds the number of documents (200), S_w is singular and classical discriminant analysis breaks down. However, the LDA/GSVD method circumvents this singularity problem.

The centroid-based dimension reduction algorithms dramatically reduce the dimension 7519 to 5, and the LDA/GSVD reduces the dimension 7519 to 4, which is one less than the number of clusters. Table 14.4 shows classification results using the L_2 norm similarity measure. For the top half of the table where the full problem size is 7519×200, all 200 data items were used for dimension reduction, and the same data items are reclassified in the full space and in the reduced space. Then 200 data points were divided into 100 training data points that generate the dimension reducing transformations, and 100 test data points that were used in prediction. The bottom half of the table shows the prediction accuracy for the test data. Because the J_1 criterion is not defined in this case due to the fact that S_w is singular, the ratio trace(S_b)/trace(S_w) is computed as a rough optimality measure. The ratio is much higher for LDA/GSVD reduction than for the other methods. These experimental results confirm that the LDA/GSVD algorithm effectively extends the applicability of the J_1 criterion to cases that classical discriminant analysis cannot handle. In addition, the

---

Algorithm 7 Centroid-based classification

Given a data matrix A with k clusters and k corresponding centroids, $c^{(i)}$ for $1 \leq i \leq k$, it finds the index j of the cluster to which a vector q belongs.

- Find the index j such that $sim(q, c^{(i)})$, $1 \leq i \leq k$, is minimum (or maximum), where $sim(q, c^{(i)})$ is the similarity measure between q and $c^{(i)}$.

  (For example, $sim(q, c^{(i)}) = \|q - c^{(i)}\|_2$ using the L_2 norm, and the index with the minimum value is taken. Using the cosine measure, $sim(q, c^{(i)}) = \cos(q, c^{(i)}) = \frac{q^T c^{(i)}}{\|q\|_2 \|c^{(i)}\|_2}$, and the index with the maximum value is taken.)

Algorithm 8 k-nearest neighbor (knn) classification

Given a data matrix $A = [a_1 \quad \ldots \quad a_n]$ with k clusters, it finds the cluster to which a vector q belongs.

1. From the similarity measure $sim(q, a_j)$ for $1 \leq j \leq n$, find the k nearest neighbors of q. (k is used to distinguish the algorithm parameter from the number of clusters k.)
2. Among these k vectors, count the number belonging to each cluster.
3. Assign q to the cluster with the greatest count in the previous step.

TABLE 14.3
MEDLINE Data Set, Clusters, and the number of Documents from Each Cluster

Class	Category	No. of documents
1	Heart attack	40
2	Colon cancer	40
3	Diabetes	40
4	Oral cancer	40
5	Tooth decay	40
	Dimension	7519 × 200

TABLE 14.4
Trace Values and Classification Accuracy (in %) with L_2 Norm Similarity on the 7519 × 200 MEDLINE Data. The top half of the table shows the self-classification accuracy, where all 200 data points were used in obtaining the dimension reducing transformations and then reclassified. For the bottom half, 100 data items were used for obtaining the dimension reducing transformations, and the other 100 points were used as a test data set to obtain prediction accuracy.

Method	Full	Orthogonal Centroid	Centroid	LDA/GSVD
Dim	7519 × 200	5 × 200	5 × 200	4 × 200
trace(S_w)	73,048	4,210.1	90.808	0.05
trace(S_b)	6,228.7	6,228.7	160.0	3.95
trace(S_b)/trace(S_w)	0.0853	1.4795	1.7620	79
centroid (%)	95.0	95.0	98.0	99.0
1nn (%)	58.5	96.0	96.5	98.0
Dim	7519 × 100	5 × 100	5 × 100	4 × 100
trace(S_w)	36,599	2,538.5	39.813	6e-16
trace(S_b)	4,329.6	4,329.3	80.0	4.0
trace(S_b)/trace(S_w)	0.1183	1.7056	2.0094	6.7e15
centroid (%)	77.0	77.0	90.0	87.0
1nn (%)	39.0	87.0	85.0	87.0

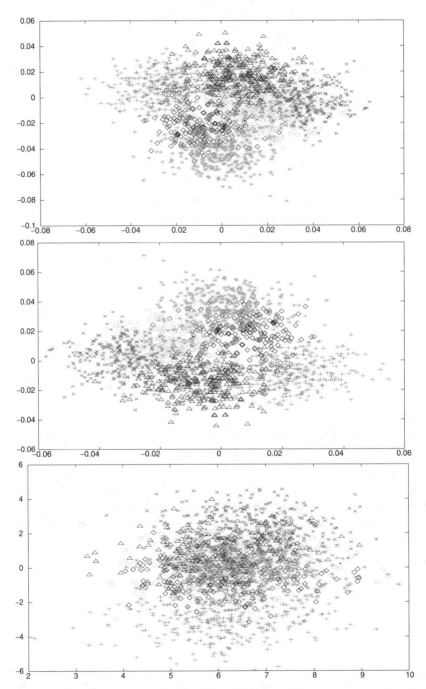

FIGURE 14.3 Two-dimensional representation of 150×2000 data with seven clusters. The first graph is obtained based on the transformation T from max $\mathrm{trace}(T^{\mathrm{T}} S_{\mathrm{w}}^{-1} S_{\mathrm{b}} T)$, the second by max $\mathrm{trace}(T^{\mathrm{T}} S_{\mathrm{w}}^{-1} S_{\mathrm{m}} T)$, and the third by using $\Sigma_2 V_2^{\mathrm{T}}$ from truncated SVD.

Orthogonal Centroid algorithm preserves trace(S_b) from the full dimension without the expense of computing eigenvectors. Taken together, the results for these methods demonstrate the potential for dramatic and efficient dimension reduction without compromising cluster structure.

The experimental results verify that the J_1 criterion, when applicable, effectively optimizes classification in the reduced dimensional space, whereas the LDA/GSVD extends the applicability to cases that classical discriminant analysis cannot handle. In addition, the LDA/GSVD algorithm avoids the numerical problems inherent in explicitly forming the scatter matrices.

In terms of computational complexity, the most costly part of Algorithm 6 is Step 2, where a complete orthogonal decomposition is needed, and a fast algorithm needs to be developed for Step 2. For orthogonal centroid, the most expensive step is the reduced QR decomposition of C [5]. By solving a simpler eigenvalue problem and avoiding the computation of eigenvectors, orthogonal centroid is significantly cheaper than LDA/GSVD. The experiments show it to be a very reasonable compromise.

Finally, it bears repeating that dimension reduction is only a preprocessing stage. Because classification and document retrieval will be the dominating parts computationally, the expense of dimension reduction should be weighed against its effectiveness in reducing the cost involved in those processes.

14.3.5 Two-Dimensional Visualization of Data

Using the same data as in Section 14.3.4, the two-dimensional visualization capability of some dimension reduction methods for clustered data is shown in this section. To illustrate the discriminatory power of LDA/GSVD, the J_1 criterion was used to reduce the dimension from 150 to 2. Even though the optimal reduced dimension is six, J_1 does surprisingly well at discriminating among seven classes, as seen in Figure 14.3. As expected, the alternative J_1 does equally well in Figure 14.3. In contrast, Figure 14.3 shows that the truncated SVD is not the best discriminatory for clustered data sets.

ACKNOWLEDGMENT

The authors are grateful to Prof. Gene Golub for helpful discussions on the Wedderburn rank reduction formula, and to Jimmy Wang for the implementation results presented in the tables. The authors would like to thank the anonymous referee whose comments made it possible to improve this paper greatly.

REFERENCES

[1] M.T. Chu, R.E. Funderlic, and G.H. Golub. A rank-one reduction formula and its applications to matrix factorization. *SIAM Rev.*, 37:512–530, 1995.

[2] L. Hubert, J. Meulman, and W. Heiser. Two purposes for matrix factorization: a historical appraisal. *SIAM Rev.*, 42:68–82, 2000.

[3] J.H.M Wedderburn. *Lectures on Matrices.* Colloquium Publications, American Mathematical Society, New York, 1934.

[4] L. Guttman. A necessary and sufficient formula for matrix factoring. *Psychometrika*, 22:79–81, 1957.

[5] G.H. Golub and C.F. Van Loan. *Matrix Computations. 3rd ed.* Johns Hopkins Press, Baltimore, MD, 1996.

[6] A. Björck. *Numerical Methods for Least Squares Problems.* SIAM, Philadelphia, PA, 1996.

[7] G.W. Stewart. On the early history of the singular value decomposition. *SIAM Rev.*, 35:551–566, 1993.

[8] L. Mirsky. Symmetric gauge functions and unitarily invariant norms. *Quart. J. Math. Oxford*, 11:50–59, 1960.

[9] G. Salton. *The SMART Retrieval System.* Prentice Hall, New York, 1971.

[10] G. Salton and M.J. McGill. *Introduction to Modern Information Retrieval.* McGraw-Hill, New York, 1983.

[11] W.B. Frakes and R. Baeza-Yates. *Information Retrieval: Data Structures and Algorithms.* Prentice Hall, Englewood Cliffs, NJ, 1992.

[12] M.W. Berry and M. Browne. *Understanding Search Engines. Mathematical Modeling and Text Retrieval.* SIAM, Philadelphia, PA, 1999.

[13] H. Kim, P. Howland, and H. Park. Dimension reduction in text classification using support vector machine. *J. Machine Learning Res.*, to appear.

[14] M. R. Anderberg. *Cluster Analysis for Applications.* Academic Press, New York, 1973.

[15] A.K. Jain and R.C. Dubes. *Algorithms for Clustering Data.* Prentice Hall, New York, 1988.

[16] M. Berry, S. Dumais, and G. O'Brien. Using linear algebra for intelligent information retrieval. *SIAM Rev.*, 37:573–595, 1995.

[17] S. Deerwester, S. Dumais, G. Furnas, T. Landauer, and R. Harsman. Indexing by latent semantic analysis. *J. Am. Soc. Inform. Sci.*, 41:391–407, 1990.

[18] M.D. Gordon. Using latent semantic indexing for literature based discovery. *J. Am. Soc. Inform. Sci.*, 49:674–685, 1998.

[19] A.M. Pejtersen. Semantic information retrieval. *Commun. ACM*, 41:90–92, 1998.

[20] J.B. Rosen, H. Park, and J. Glick. Total least norm formulation and solution for structured problems. *SIAM J. Matrix Anal. Appl.*, 17:110–128, 1996.

[21] Y. Jung, H. Park, and D. Du. A balanced term-weighting scheme for improved document comparison and classification. Technical report, Department of Computer Science, University of Minnesota, Minneapolis, 2001.

[22] I. Joliffe. *Principal Component Analysis.* Springer, New York, NY, 1986.

[23] T. Hastie, R. Tibshirani, and J. Friedman. *The Elements of Statistical Learning. Data Mining, Inference and Prediction.* Springer-Verlag, New York, 2001.

[24] B. Savas. Analyses and test of handwritten digit algorithms. Master's thesis, Mathematics Department, Linköping University, Sweden, 2002.

[25] M.W Berry. Large scale singular value computations. *Int. J. Supercomputer Appl.*, 6:13–49, 1992. Also http://www.netlib.org/svdpack.

[26] E. Anderson, Z. Bai, C.H. Bischof, J.W. Demmel, J.J. Dongarra, J.J. Du Croz, A. Greenbaum, S.J. Hammarling, A. McKenney, S. Ostrouchov, and D.C. Sorensen. *LAPACK Users' Guide*, 3rd ed. SIAM, Philadelphia, PA, 1999.

[27] MathWorks. *Matlab User's Guide.* Mathworks, Natick Massachusetts, 1996.

[28] S. Wolfram. *The Mathematica Book, 4th ed.* Cambridge University Press, Cambridge, MA, 1999.

[29] Z. Bai, J. Demmel, J. Dongarra, A. Ruhe, and H. van der Vorst, Eds. *Templates for the Solution of Algebraic Eigenvalue Problems: A Practical Guide.* SIAM, Philadelphia, PA, 2000. http://www.cs.ucdavis.edu/ bai/ET/contents.html.

[30] G.H. Golub. Numerical methods for solving least squares problems. *Numer. Math.*, 7:206–216, 1965.

[31] W. Kahan. Numerical linear algebra. *Can. Math. Bull.*, 9:757–801, 1966.

[32] T.F. Chan. Rank-revealing QR factorizations. *Linear Algebra Appl.*, 88/89:67–82, 1987.

[33] G.W. Stewart. An updating algorithm for subspace tracking. *IEEE Trans. Signal Proc.*, 40:1535–1541, 1992.

[34] F. Stenger. *Numerical Methods Based on Sinc and Analytic Functions.* Springer, New York, 1993.

[35] G.W. Stewart. The QLP approximation to the singular value decomposition. *SIAM J. Sci. Comput.*, 20(4):1336–1348, 1999.

[36] D.A. Huckaby and T.F. Chan. On the convergence of Stewart's QLP algorithm for approximating the SVD. *Numer. Algorithms*, 32:287–316, 2003.

[37] M. Gu and S. Eisenstat. A divide and conquer algorithm fo the bidiagonal SVD. *SIAM J. Sci. Stat. Comput.*, 13:967–993, 1995.

[38] D.K. Faddeev, V.N. Kublanovskaya, and V.N. Faddeeva. Solution of linear algebraic systems with rectangular matrices. *Proc. Steklov Inst. Math.*, 96:93–111, 1968.

[39] R.J. Hanson and C.L. Lawson. Extensions and applications of the Householder algorithm. *Math. Comput.*, 23:787–812, 1969.

[40] C.L. Lawson and R.J. Hanson. *Solving Least Squares Problems*. SIAM, Classics in Applied Mathematics, Philadelphia, PA, 1995.

[41] I.E. Frank and J.H. Friedman. A statistical view of some chemometrics regression tools. *Technometrics*, 35:109–135, 1993.

[42] S. Wold, M. Sjöström, and L. Eriksson. PLS-regression: a basic tool of chemometrics. *Chemometrics Intell. Lab. Syst.*, 58:109–130, 2001.

[43] L. Eldén. Partial least squares vs. Lanczos bidiagonalization I: analysis of a projection method for multiple regression. *Comput. Stat. Data Anal.*, to appear.

[44] O. Svensson, T. Kourti, and J.F. MacGregor. An investigation of orthogonal signal correction algorithms and their characteristics. *J. Chemometrics*, 16:176–188, 2002.

[45] H. Wold. Soft modeling by latent variables; the nonlinear iterative partial least squares approach. In J. Gani, Ed., *Perspectives in Probability and Statistics*, Papers in honour of M.S. Bartlett. Academic Press, London, 1975.

[46] S. Wold, A. Ruhe, H. Wold, and W.J. Dunn. The collinearity problem in linear regression. The partial least squares (PLS) approach to generalized inverses. *SIAM J. Sci. Stat. Comput.*, 5:735–743, 1984.

[47] G.H. Golub and W. Kahan. Calculating the singular values and pseudo-inverse of a matrix. *SIAM J. Numer. Anal. Ser. B*, 2:205–224, 1965.

[48] I.S. Helland. On the structure of partial least squares regression. *Commun. Stat. Simul.*, 17:581–607, 1988.

[49] C. Park and H. Park. Nonlinear feature extraction based on centroids and kernel functions. *Pattern Recog.*, 37:801–810, 2004.

[50] C. Park and H. Park. Kernel discriminant analysis based on the generalized singular value decomposition. Technical report 03-17, Department of Computer Science and Engineering, University of Minnesota, Minneapolis, 2003.

[51] H. Park, M. Jeon, and J.B. Rosen. Lower dimensional representation of text data based on centroids and least squares. *BIT*, 43:427–448, 2003.

[52] P. Howland and H. Park. Generalizing discriminant analysis using the generalized singular value decomposition. *IEEE Trans. Pattern Anal. Machine Learning*, to appear.

[53] P. Horst. *Factor Analysis of Data Matrices*. Holt, New York, 1965.

[54] L.L. Thurstone. A multiple group method of factoring the correlation matrix. *Psychometrika*, 10:73–78, 1945.

[55] P. Howland, M. Jeon, and H. Park. Structure preserving dimension reduction based on the generalized singular value decomposition. *SIAM J. Matrix Anal. Appl.*, 25:165–179, 2003.

[56] K. Fukunaga. *Introduction to Statistical Pattern Recognition*, 2nd ed. Academic Press, New York, 1990.

[57] C.F. Van Loan. Generalizing the singular value decomposition. *SIAM J. Numer. Anal.*, 13:76–83, 1976.

[58] C.C. Paige and M.A. Saunders. Towards a generalized singular value decomposition. *SIAM J. Numer. Anal.*, 18:398–405, 1981.

[59] S. Theodoridis and K. Koutroumbas. *Pattern Recognition*. Academic Press, New York, 1999.

[60] G. Kowalski. *Information Retrieval Systems: Theory and Implementation*. Kluwer Academic Publishers, Dordrecht, 1997.

15 Parallel Computation in Econometrics: A Simplified Approach

Jurgen A. Doornik, Neil Shephard, and David F. Hendry

CONTENTS

ABSTRACT

Parallel computation has a long history in econometric computing but is not at all widespread. A major impediment is the labor cost of coding for parallel architectures. Moreover, programs for specific hardware often become obsolete quite quickly. The proposed approach is to take a popular matrix programming language (Ox) and implement a message-passing interface (MPI). Next, object-oriented programming is used to hide the specific parallelization code, so that a program does not need to be rewritten when it is ported from the desktop to a distributed network of computers.

The focus is on so-called embarrassingly parallel computations, and the issue of parallel random number generation is addressed.

15.1   INTRODUCTION

There can be no doubt that advances in computational power have had a large influence on many, if not all, scientific disciplines. If we compare the current situation with that of 25 years ago, the contrast is striking. Then, electronic computing was almost entirely done on mainframes, and primarily for number crunching purposes. Researchers, particularly in the social sciences, had limited computing budgets (often measured in seconds of computing time) and frequently had to schedule their work at night. Moreover, the available software libraries were limited, and usually much additional programming was required (largely in FORTRAN, and still using punchcards). Today, a simple notebook has more power. In terms of research, the computer is used for a much wider range of tasks, from data management and visualization, to model estimation and simulation, as well as writing papers, to name a few. Most importantly, all access constraints for routine computing have since disappeared.

Regarding high-performance computing (HPC), the situation, until quite recently, is more similar to that of 25 years ago. Of course, systems are very much faster. But they are administered as mainframe systems, largely funded through physical sciences projects, requiring much dedicated programming. The current record holder (see www.top500.org) is the Japanese Earth Simulator. This supercomputer took five years to build, consists of 5120 processors, and is housed in a four-story building with its own small power station. The machine is a parallel vector supercomputer, achieving 36 teraflops (a teraflop is 10^{12} floating point operations per second). Most other machines in the ranking are all massively parallel processor (MPP) and cluster machines using off-the-shelf components: the second fastest is an MPP machine delivering nearly 8 teraflops. The gap is largely due to higher communication overheads in MPP architectures. In contrast, a current desktop reaches 2.5 gigaflops, equivalent to a supercomputer of the mid to late 1980s. The speed difference between an up-to-date desktop and the state-of-the-art supercomputer is of the order 10^4. Moore's law predicts that on-chip transistor count doubles every 18 months. Such an improvement roughly corresponds to a doubling in speed. The "law" describes historical computer developments quite accurately. If Moore's law continues to hold, it will take 20 years for the desktop to match the Earth Simulator.

Our interest lies in HPC for statistical applications in the social sciences. Here, computational budgets are relatively small, and human resources to develop dedicated software is limited (we consider this aspect in more detail in Ref. [1]). A feasible approach uses systems consisting of small clusters of personal computers, which could be called "personal high-performance computing" (PHPC). Examples are:

- A small dedicated compute cluster, consisting of standard PCs
- A connected system of departmental computers that otherwise sit idle most of the time

In both cases, hardware costs are kept to a minimum. Our objective is to also minimize labor costs in terms of software development. A fundamental aspect of these systems is that communication between nodes is slow, so they are most effective for so-called "embarrassingly parallel" problems, where the task is divided into large chunks that can be run independently.

A prime example of an embarrassingly parallel problem is Monte Carlo analysis, where M simulations are executed of the same model, each using a different artificial data set. With P identical processors running one process each, every "slave" can run $M/(P-1)$ experiments, with the "master" running on processor P to control the slaves and collect the result. If interprocess communication is negligible, the speedup is $(P - 1)$-fold. The efficiency is $100(P - 1)/P\%$, i.e., 85% when $P = 7$. In practice, such high efficiency is difficult to achieve, although improved efficiency results when the master process does not need a full processor. In theory, there could be superlinear speedup: e.g., when the distributed program fits in memory, but the serial version does not. Also note that it is usually unnecessary to dedicate a whole processor to the master. Instead, the Pth processor can run the master and a scaled-down slave process, which increases efficiency. We return to this in Section 15.6.2. Many readers will be familiar with the manual equivalent of this design: at the end of the afternoon, go to several machines to start a program, then collect the results the next morning. This procedure is efficient in terms of computer time, but quite error prone, as well as time consuming. Our aim is simply to automate this procedure for certain types of computation.

The hardware developments we described briefly have also resulted in a major transformation of the tools for statistical analysis in the social sciences. Initially, there were quite a few "locally produced" libraries and script-driven packages being used for teaching and research. Today, the default is to look for a readily available package. In econometrics, which is our primary focus, this would be PcGive or EViews for interactive use, and TSP, Stata or RATS for command-driven use. (We do not attempt to provide an exhaustive software listing.) These have been around for a long time and embody many years of academic knowledge. Such canned packages are often too inflexible for research purposes, but a similar shift has occurred in the move from low-level languages like FORTRAN and C, to numerical and statistical programming languages. Examples are MATLAB, GAUSS, Ox, S-plus, and R. The labor savings in development usually outweigh the possible penalty in run-time speed. In some cases, the higher-level language can even be faster than the corresponding user program written in C or FORTRAN (for an Ox and GAUSS example, see Ref. [2]; also see Section 15.4).

Parallel applications in econometrics have quite a long history. For example, one of the authors developed Monte Carlo simulation code for a distributed array processor using DAP FORTRAN [3]. The recent availability of low-cost clusters has resulted in some increase in parallel applications. For example, Swan [4] considers parallel implementation of maximum likelihood estimation; Nagurney and coworkers [5, 6] use massively parallel systems for dynamic systems modeling; Gilli and Pauletto [7] solve large and sparse nonlinear systems; the monograph by Kontoghiorghes [8] treats parallel algorithms for linear econometric models; Ferrall [9] optimizes finite mixture models; examples of financial applications are given in Refs. [10–12]; Kontoghiorghes et al. [13] introduce a special issue of *Parallel Computing* on economic applications.

We believe that more widespread adoption of PHPC in our field of interest depends on two factors. First, only a minimum amount of programming should be required. Second, any method should be built around familiar software tools. Our approach is to take the Ox matrix-programming language (see Ref. [14]) and try to hide the parallelization within the language. This way, the same Ox program can be run unmodified on a stand-alone PC or on a cluster, as discussed in Section 15.6.2. This approach is in the spirit of Wilson [15, p. 497], who argues for the adoption of high-level languages. He also states that a company called Teradata was one of the most successful parallel computer manufacturers of the 1980s because: "in many ways, Teradata's success was achieved by hiding parallelism from the user as much as possible."

Before considering parallel applications for PHPC, we address speed aspects of single-processor implementations. After all, it is pointless gaining a fourfold speed increase by using four computers, but then losing it again because the actual software is four times slower than other programs. This potential drawback has to be balanced against the effort already invested in the slower software: in terms of development time, it may still be beneficial to use a parallel version of the slower software.

The outline of the remainder of the chapter is as follows. The next section discusses the benefits and drawbacks of using a matrix language, followed by a brief introduction to the Ox language. Then, Section 15.4 looks at some relevant single processor performance issues. Next, Section 15.5 has an extensive discussion of how distributed processing is added to Ox using MPI. Important aspects include random number generation (Section 15.5.5) and hiding the parallelism (Section 15.6.2); Section 15.6 provides an application of PHPC. Finally, Section 15.7 concludes.

15.2   USING A MATRIX-PROGRAMMING LANGUAGE

Matrix-programming languages have become the research tool of choice in social sciences, particularly those with a econometric and statistical component, because they make program development substantially easier. Computational expressions are much closer to algebraic statements, thus reducing the amount of coding required. This, in turn, reduces the likelihood of bugs. Moreover, such languages usually come with substantial numerical and statistical libraries built-in.

For example, the very simple Monte Carlo experiment of the next section uses random number generators (RNGs), quantiles of statistical distributions, some basic matrix operations, and graphics functions such as the nonparametrically estimated density. The C or FORTRAN programmer has to find libraries for these and other tasks (many useful libraries, mostly in FORTRAN, can be found at www.netlib.org). The user of a matrix language finds all of these readily presented. Another advantage is that memory allocation and deallocation is automatically handled in the matrix language. This removes a frequent source of bugs.

There can be a speed penalty when using a matrix language. Low-level languages like C, C++, and FORTRAN are translated into machine language by a compiler and saved as an executable image. Matrix languages are interpreted line-by-line as they are read by the interpreter. The latter is similar to a Basic or Java program (provided no "just-in-time" compiler is used). The drawback of an interpreted language is that control-flow statements like loops can be very slow. The advantage is that the edit–run cycle is more convenient.

The aim of this chapter is to conform to the current preference of using a matrix-programming language for research and consider its usability for parallel computation. First, we show that such languages can be sufficiently efficient in a scalar setting, while providing substantial labor savings. Next, we implement a parallel version for embarrassingly parallel problems, which is shown to work well.

We mentioned several matrix languages in the introduction and there are more (although not all are strong on the statistical or econometric side). In the remainder, we restrict ourselves to Ox [14] because we know it well and it is ideally suited for our current objectives.

Efforts for other languages are also underway: there is a PVM package for R, and Caron et al. [16] discusses a parallel version of Scilab. Other projects that consider parallel versions of Ox are ParaDiOx (see rp-40.infra.kth.se/paradiox/, which provides a user-friendly interface, but has now been completed), and using Ox under OpenMosix (ongoing research at the Financial Markets Group of the London School of Economics).

15.3   THE OX MATRIX LANGUAGE

15.3.1   OVERVIEW

Ox has a comprehensive mathematical and statistical library, and, although Ox is a matrix language, it has a syntax that is similar to C and C++ (and Java). Ox is a relatively young language, with a first official release in 1996. Despite this, it has been widely adopted in econometrics and statistics. Two contributing factors are that it is fast, and that there is a free version for academics (but it is not open source). Ox is available on a range of platforms, including Windows and Linux. For a recent review see Ref. [17].

Ox is an interpreted language (more precisely, the code is compiled into an intermediate language, which is then interpreted, similar to Java) with a commensurate speed penalty for unvectorized code such as loops. However, this speed penalty is noticeably less than in other comparable systems. Indeed, several reviewers have noted that their programs, after conversion to Ox, actually run faster than their original FORTRAN or C code; Section 15.4 will explain how this can happen.

Time-critical code sections can be written in FORTRAN or C and added to the language as a dynamic link library. Because of the similarity of Ox to C it is often convenient to do the prototyping of code in Ox before translating it into C for the dynamic link library. All underlying standard library functions of Ox are exported, so can be called from the C code. An example of a library that is implemented this way is SsfPack Ref. [18].

The Ox language is also object-oriented, along the lines of C++, but with a simplicity that bears more resemblance to Java. This is aided by the fact that Ox is implicitly typed. There are several preprogrammed classes; the most important are the Modelbase class and the Simulation class. The latter is of interest here: if we can make it parallel, without affecting its calling signature, we can make existing programs parallel without requiring recoding.

15.3.2 EXAMPLE

As a simple example of a simulation experiment in Ox, we take the asymptotic form of the univariate test for normality, the so-called Bowman–Shenton or Jarque–Bera test (see Refs. [19, 20]; for a small-sample adjusted version, see Ref. [21]). The statistic is the sum of the squared standardized sample skewness and kurtosis. Under the null hypothesis of normality, this has a $\chi^2(2)$ distribution in large samples:

$$N = \frac{T\left(\sqrt{b_1}\right)^2}{6} + \frac{T\left(b_2 - 3\right)^2}{24} \widetilde{a} \; \chi^2\,(2)\,,$$

where T is the sample size and

$$\bar{x} = \frac{1}{T}\sum_{t=1}^{T} x_t, \quad m_i = \frac{1}{T}\sum_{t=1}^{T}(x_t - \bar{x})^i, \quad \sqrt{b_1} = \frac{m_3}{m_2^{3/2}} \quad \text{and} \quad b_2 = \frac{m_4}{m_2^2}.$$

A normal distribution has skewness=0 and kurtosis=3.

The Ox code for this simple simulation example is not reproduced here but would show several similarities with C and C++: these include files; the syntax for functions, loops and indexing is the same; indexing starts at zero; the program has a main function. Comment can be either C or C++ style.

Some differences are: implicit typing of variables (internally, the types are int, double, matrix, string, array, function, object, etc.); variables are declared with `decl`; use of matrices in expressions; matrix constants (`< . . . >`); special matrix operators (`. ^` for element-by-element exponentiation, `|` for vertical concatenation; `. <=` for element-by-element comparison, here used to compare a scalar to a vector); multiple returns; statistical library (random numbers, distributions, quantiles, etc.); and a drawing library.

The text and graphical output of the program is shown together in Figure 15.1. As can be seen, the asymptotic approximation by a $\chi^2(2)$ is sufficiently inaccurate that small-sample corrections are indeed desirable. A comprehensive introduction to Ox is given in Ref. [22].

15.3.3 USING THE SIMULATION CLASS

It is convenient for a Monte Carlo experiment to derive from the preprogrammed simulation class. That way, there is no need to write code to accumulate the results or print the final output. The objective then is to make this class parallel, without affecting its calling signature. Then we can make existing programs parallel without requiring any recoding, just requiring a change of one line of code to switch to the new version.

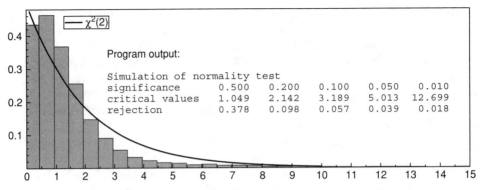

FIGURE 15.1 Empirical distribution of normality test, $T = 50$.

The main aspects of the simulation class are:

- `Simulation` — the constructor function that sets the design parameters such as sample size and number of replications.
- `Generate` — virtual function that is called for each replication. This function should be provided by the derived class, and return 0 if the replication failed.
- `GetTestStatistics` — function that is called after `Generate` to get the values of the test statistics. If coefficients are simulated, or the distribution is known (or conjectured) then `GetCoefficients` and `GetPvalues` are called, respectively.

Converting the program of Section 15.3.2 to use the simulation class does not affect the actual test function but embeds it quite differently. No additional coding is required to get the final results: the Simulation class automatically prints the simulation results. The overhead from switching to the object-oriented version is very small: the new program runs only a fraction slower.

15.4   SINGLE PROCESSOR PERFORMANCE

15.4.1   LOOPS IN MATRIX PROGRAMMING LANGUAGES

Matrix-programming languages tend to be slow when executing loops as a consequence of their interpreted nature. An example that exposes this clearly is the Ox function `vecmulXtY` in Listing A1 (see the Appendix). That function times a loop of `cRep` iterations, which calls a dot product function `vecmul`. This in turn contains a loop to accumulate the inner product:

```
vecmul(const v1, const v2, const c)

{

    decl i, sum;

    for (i = 0, sum = 0.0; i < c; ++i)

        sum += v1[i] * v2[i];

    return sum;

}
```

TABLE 15.1
Number of Iterations per Second for Three Dot Product Implementations

Vector length	10	100	1000	10,000	100,000
Unvectorized Ox code	67,000	8,900	890	89	9
Vectorized Ox code	1,100,000	930,000	300,000	32,000	390
C code	15,000,000	3,200,000	330,000	34,000	380

The corresponding C code is given in Listing A2. Of course, it is better to use matrix expressions in a matrix language, as in the function vecmulXtYvec in Listing A1. The speed comparisons (on a 1.2 GHz Pentium III notebook running Windows XP SP1) are presented in Table 15.1. Clearly, this is a situation where the C code is many orders of magnitudes faster than the unvectorized Ox code. The table also shows that, as the dimension of the vectors grows, the vectorized Ox code catches up and even overtakes the C code.

In the extreme case, the C code is nearly 400 times faster than the unvectorized Ox code. However, this is not a realistic number for practical programs: many expressions do vectorize and even if they do not, they may only form a small part of the running time of the actual program. If that is not the case, it will be beneficial to write the code section in C or FORTRAN and link it to Ox. It is clear that vectorization of matrix-language code is very important.

15.4.2 OPTIMIZATION OF LINEAR ALGEBRA OPERATIONS

Linear algebra operations such as QR and Choleski decompositions underlie many matrix computations in statistics. Therefore, it is important that these are implemented efficiently.

Two important libraries are available from www.netlib.org to the developer:

- BLAS: the basic linear algebra system
- LAPACK: the linear algebra package, see Ref. [23]

LAPACK is built on top of BLAS and offers the main numerical linear algebra functions. Both are in FORTRAN, but C versions are also available. MATLAB was essentially developed to provide easier access to such functionality.

Several vendors of computer hardware provide their own optimized BLAS. There is also an open source project called ATLAS (automatically tuned linear algebra system, www.netlib.org) that provides highly optimized versions for a wide range of processors. ATLAS is used by MATLAB 6 and can also be used in R. Ox has a kernel replacement functionality that allows the use of ATLAS.

Many current processors consist of a computational core that is supplemented with a substantial L2 (level 2) cache (for example 128, 256, or 512 KB for many Pentium processors). Communication between the CPU and the L2 cache is much faster than toward main memory. Therefore, the most important optimization is to keep memory access as localized as possible, to maximize access to the L2 cache, and to reduce transfers from main memory to the L2 cache. For linear algebra operations, doing so implies block partitioning of the matrices involved.

For example, the standard three-level loop to evaluate $C = AB$:

```
for (i = 0; i < m; ++i)

    for (j = 0; j < p; ++j)
```

```
{
    for (k = 0, d = 0.0; k < n; ++k)
        d += a[i][k] * b[k][j];
    c[i][j] = d;
}
```

is fine when A and B are small enough to fit in the L2 cache. If not, it is better to partition A, B into A_{ij} and B_{ij} and use a version that loops over the blocks. Each $A_{ik}B_{kj}$ is then implemented as in the code above. The size of the block depends on the size of the L2 cache.

There are further optimizations possible, but the block partitioning gives the largest speed benefit.

15.4.3 SPEED COMPARISON OF LINEAR ALGEBRA OPERATIONS

We look at the speed of matrix multiplication to show the impact of optimizations, using code similar to that in Listings A1 and A2, but this time, omitting unvectorized Ox code.

We consider the multiplication of two matrices and the cross product of a matrix for four designs as given in Table 15.2. Four implementations are compared on a Mobile Pentium III 1.2 GHz (with 512 KB L2 cache):

1. Vectorized Ox code
   This is the default Ox implementation. Note that Ox does not use BLAS or LAPACK. The Ox version is 3.2.
2. Simple C code
   This is the standard three-loop C code, without any block partitioning. We have seen this being used frequently by C programmers.
3. ATLAS 3.2.1 for PIIISSE1
   An optimized BLAS that is available for many processors. The SSE1 (version 1 streaming SIMD extensions) instructions are not relevant here because they can only be used for single precision. All our computations are double precision.
4. MKL 5.2 (Intel)
   The math kernel library (MKL) is a vendor-optimized BLAS (in this case, with a large part of LAPACK as well). This can be purchased from Intel and is not available for AMD Athlon processors. We used the default version (for all Pentium processors) because it was regularly faster on our benchmarks than the Pentium III specific version.

TABLE 15.2
Designs for Speed Comparison of Linear Algebra Operations

Operation	A	B	Dimensions used
AB	$r \times r$	$r \times r$	$r = 20(10)320$
A'A	$r \times r$		$r = 20(10)320$
AB	$50 \times c$	$c \times 50$	$c = 10, 100, 1,000, 10,000, 100,000$
A'A	$c \times 50$		$c = 10, 100, 1,000, 10,000, 100,000$

Although each implementation is expected to have the same accuracy, the ordering of the summations can be different, resulting in small differences between outcomes.

The Ox, MKL, and ATLAS timings are an average of four runs and are all obtained with the same Ox program (Listing A3). Ox has the facility (as yet undocumented and unsupported) to replace its MKL. This applies to matrix multiplications, eigenvalue and inversion routines, and Choleski, QR, and SVD (singular value) decompositions. Doing so involves compiling a small wrapper around the third-party BLAS/LAPACK libraries. This feature could also be used to implement parallel versions of these operations (such as Ref. [8]) without the need to make any changes to Ox. A command line switch determines the choice of kernel replacement library.

Note that although the ATLAS and MKL timings are obtained from the Ox code, all timed computations take place inside the libraries. Therefore, the same results would be achieved if C were used as the "wrapper" instead of Ox.

Figure 15.2a shows the number of square matrix multiplications per second as a function of dimension r. The left panel, which is for AB, shows that the performance of the C code drops off considerably when the dimension gets to around 220, when inefficient use of the L2 cache starts to result in a severe speed penalty. From then onwards, the Ox implementation is up to 10 times faster than C, ATLAS faster again by about 25%, and finally MKL around 10% faster than ATLAS. Of course, it is the *implementation* that is measured, not the language, as most of the underlying code is C code (Ox is also written in C). In this design ATLAS outperforms Ox at all dimensions, whereas MKL is always faster than ATLAS.

The symmetric case (Figure 15.2b) shows timings for the cross product $A'A$ for the same dimensions. The results are quite similar for large dimensions. At dimensions between 20 and 40,

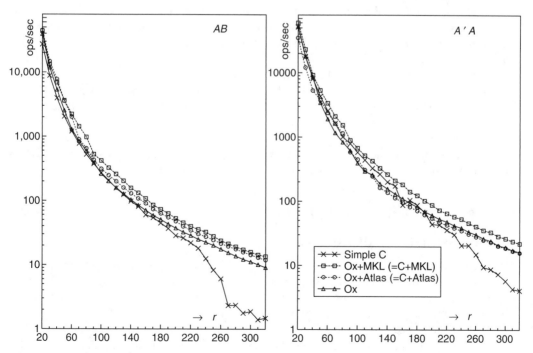

FIGURE 15.2 AB and $A'A$ for square matrices of dimension r (horizontal axis). The number of multiplications per second is on the vertical axis, on a $\log_{10}$ scale.

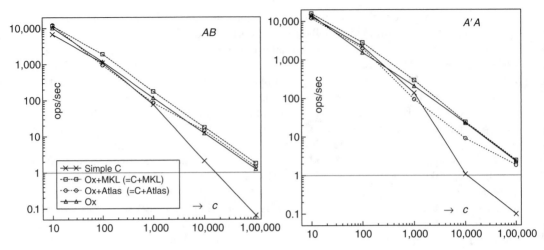

FIGURE 15.3 AB and $A'A$ for matrices of dimension $50 \times c$ (horizontal axis on $\log_{10}$ scale). The number of multiplications per second is on the vertical axis, on a $\log_{10}$ scale.

ATLAS is outperformed by all others, including C. From dimensions 90 to 150, the simple C implementation is faster than ATLAS and Ox, which could be due to overly conservative L2 cache use.

Figure 15.3 presents the cases with $r = 50$ and c varying. At large dimensions, the simple C code is again dominated, by a factor of 20 for $c = 100,000$. In the AB case, Ox, MKL, and ATLAS are quite similar, whereas for $A'A$, Ox, and MKL are twice as fast as ATLAS for $c = 1000, 10,000$.

These examples show that speed depends very much on the application, the hardware, and the developer. For example, on a Pentium 4, we expect ATLAS and MKL to more clearly outperform Ox because they use SSE2 to add additional optimizations.

The results indicate that the C or FORTRAN programmer who uses large matrices is advised to use an optimized BLAS instead of the simple code or standard BLAS. It also shows the possible benefits of a good matrix language, namely that many operations (not just matrix multiplication) have already been implemented and optimized. This reflects the knowledge that is embodied in such systems. It explains why several reviewers and users have found Ox faster than their own low-level program.

The next sections considers what benefits arise for distributed computing using Ox, which has shown itself to be fast enough as a candidate for distributed processing.

15.5 PARALLEL IMPLEMENTATION

15.5.1 PARALLEL MATRIX LANGUAGES

There are several stages at which we can add distributed processing to a matrix language. Like low-level languages, it can be parallelized in the user code. Further, because there is an extensive run-time library, this could be parallelized. Finally, the fact that a matrix language is interpreted creates an additional possibility: to parallelize the interpreter.

Here, we only consider making the parallel functionality directly callable from the matrix language. When a language can be extended through dynamic link libraries in a flexible way, this level can be achieved without changing the language interpreter itself. We would expect optimal performance gains for embarrassingly parallel problems.

When the parallelization is hidden in the run-time library, operations such as matrix multiplication, inversion, etc., will be distributed across the available hardware. Here we could use available libraries for the implementation (such as ScaLAPACK or PLAPACK, also see Ref. [8]). Functions that work on matrix elements (logarithm, loggamma function, etc.) are also easily distributed. The benefits to the user are that the process is completely transparent. The speed benefit will be dependent on the problem: if only small matrices are used, the communication overhead would prevent the effective use of the cluster. The experience of Murphy et al. [24] is relevant at this level: whereas it may be possible to efficiently parallelize a particular operation, the benefit for smaller problems, and, by analogy for a complete program, is likely to be much lower. If a cluster consists of multiprocessor nodes, then this level will be most efficient within the nodes, with the previous user level between nodes. This could be achieved using multithreading, possibly using OpenMP directives (www.openmp.org).

Parallelization of a matrix language at the interpreter level is the most interesting implementation level, and, so far as we are aware, has not been tried successfully before. The basic idea is to run the interpreter on the master, handing elements of expressions to the slaves. The main problem arises when a computation requires a previous result — in that case the process stalls until the result is available. On the other hand, there may be subsequent computations that can already be done. This is an avenue for future research.

15.5.2 AN MPI SUBSET FOR OX

We decided to add distributed processing to Ox using MPI, (see Refs. [25, 26]). This was not based on a strong preference over parallel virtual machine (PVM), (see Ref. [27]; and Ref. [28], for a comparison of PVM and MPI). In PVM, child tasks are spawned from the parent, whereas in MPI the same single program is launched from the command-line, with conditional coding to separate the parent and child. The latter seemed slightly easier to implement within the Ox framework.

Under Windows it was relatively easy to get MPI to work with Ox. The main inconvenience is synchronizing different computers if a network drive is not used for the Ox program and other files that are needed. However, under Linux, MPI uses a different mechanism to launch processes, involving complex command-line argument manipulations. This interferes with the Ox command line unless the Ox driver is recompiled to incorporate the call to MPI_Init. On the other hand, system maintenance is easier.

The objective here is not to provide a full implementation of MPI for Ox, although that can easily be done. Instead we just do what is sufficient to run an example program and then extend this to distributed loops for embarrassingly parallel computations.

The first stage after installation of MPI (we used MPICH 1.2.5 under Windows, and MPICH 1.2.4 under Linux, see Ref. [29]) is to run some example programs. Column one of Listing A4 gives a simplified version of the cpi.c program as provided with MPICH. The simplification is to fix the number of intervals inside the program, instead of asking for this on the command line. The second column of Listing A4 shows that the Ox implementation can be very similar. The MPI functions in their Ox version are:

OxMPI_Init(); Initialize the MPI session. Can be called multiple times, unlike the original.

OxMPI_Comm_size(); Returns the number of processes. Within MPI, communications are associated with communicators which allow processes to be grouped into communication tasks. However, in the current implementation we assume that MPI_COMM_WORLD, i.e., all processes, are the default communicator, and omit the argument to specify the communicator.

`OxMPI_Comm_rank();` Returns the rank of the calling process.

`OxMPI_Get_processor_name();` Returns the name of the current processor.

`OxMPI_Bcast(const aVal, const iRoot);` Broadcasts data from the sending process iRoot to all other processes. So in all other processes than the sender, `OxMPI_Bcast` operates like the receiver. Ox types that can be broadcast are integer, double, matrix, string, and array. `aVal` must be a reference to a variable, which on the sender must hold a value.

`OxMPI_Reduce(const oxval, const iOp, const iRoot);` This performs an action on all the data on the processes other than root, and returns the result to root. In the example, each nonroot holds a part of the approximation to the integral expression for π. The sum is returned to root, to provide an estimate of π. `OxMPI_Reduce` works for integers, doubles, and matrices.

`OxMPI_Wtime();` Returns the time in seconds since a time in the past (which is before the program started running).

`OxMPI_Finalize();` Terminates the MPI session. Unlike the original MPI version, this is an optional call, because `OxMPI_Init` automatically schedules a barrier synchronization and a call to `OxMPI_Finalize` to be executed when the Ox interpreter finishes.

15.5.3 PARALLEL LOOPS

The next stage is to use the MPI facilities to implement a parallel loop in the matrix language that we use. Here we also start using Ox's object-oriented features to facilitate the use of the loop code.

Ox follows a simplified version of the C++ object-oriented paradigm. In its most basic form, a class is a collection of functions and variables (both members of the class), which are grouped together. The variables are accessible to function members, but invisible to the outside world. A new class can derive from an existing class to extend its functionality; virtual functions allow a derived class to override base-class functionality. Finally, an object of a class is created with new for subsequent use (and, as an important improvement over programming with global variables, there can be many objects of the same class). For a more extensive introduction see Ref. [22], whereas a comparison of Ox syntax with C++ and Java is given in Ref. [30].

To implement a parallel library in Ox, we adopt the master–slave model written in MPI: the same program is running on each node, with `if` statements selecting the appropriate code section. Listing A5 gives an outline of the class. In this case, all members have been made static, so there can only be one instance running at a time: more than one would make the program very inefficient anyway. This also implies that there is no need to instantiate the class using new.

The `Init` function in the Loop1 class initializes the MPI library and the class. The number of slaves is determined, whereas the master is arbitrarily set as the process with the highest rank. The `OxMPI_SetMaster` and `OxMPI_IsMaster` functions have been added to the Ox MPI wrapper to facilitate tracking of master–slave status in the Ox code. The second argument of `OxMPI_SetMaster` is used here to switch off all output from the slave processes. Although much of the relevant code has been stripped away, allowance is made for this class to work outside MPI: in that case neither `IsMaster` nor `IsSlave` is true.

The `Run` function is the core of this Loop class and usually the only one called. It initializes MPI and the class, then calls an optional loop initialization function on the slaves. Next, either the master or the slave version is called (or the non-MPI `doLoop`). A final barrier synchronization ensures that all processes get to this point before continuing. `Run` takes three function arguments, in addition to the loop size:

fnInit to initialize the slave loops (optional).

fnRun to run a replication on a slave.

fnProcess is an optional step on the master to process the results as they arrive from the slaves. This is further explained in the next section.

The serial version of the actual loop, see Listing A6, is called doLoop, and simply consists of cRep calls to fnRun, which is expected to return a column vector with results. This is optionally followed by a call to fnProcess. The result is appended to mresult, that is returned when the loop is completed. No allowance is made yet for an iteration to fail, something that could happen, for example, when an iteration involves nonlinear estimation. The getResultSize and storeResult class members have been omitted from Loop1 but would be required for the code to work.

15.5.4 MASTER AND SLAVE LOOP COMPONENTS

To complete the master and slave loops, MPI functions to send and receive data are required, as well as functions to probe if data is ready to be received. The additional MPI functions are:

OxMPI_Barrier(); Initiates a barrier synchronization. This serves as a checkpoint in the code, synchronizing all running processes before continuing.

OxMPI_Recv(const iSource, const iTag); Returns a value received from sender iSource with tag iTag. OxMPI_Recv will wait until the data has been received (so the process is blocked until then).

OxMPI_Send(const val, const iDest, const iTag); Sends the value of val (which may be an integer, double, matrix, string, or array) to process iDest, using send tag iTag. This operation is also blocking: the sending process waits for a matching receive (although the MPI standard allows for a return after buffering the sent data).

OxMPI_Probe(const iSource, const iTag);

OxMPI_Iprobe(const iSource, const iTag); Tests if a message is pending for reception. OxMPI_Probe is blocking, and waits until a message is available, while OxMPI_IProbe is nonblocking: it returns immediately, regardless of the presence of a message. iSource is the message source, which can be MPI_ANY_SOURCE; similarly, iTag could be a specific tag or MPI_ANY_TAG.

The return value of OxMPI_Probe is a vector of three numbers: message source, message tag, and message error status. OxMPI_Iprobe returns an empty matrix if no message is pending, otherwise returns like OxMPI_Probe.

We are now in a position to discuss the slave and master loops. The slave is the simpler code, see Listing A7: there is an outer loop, in which we ask for the number of replications. If it is zero, the loop is exited, and the final (possibly aggregated) results are obtained from the master to synchronize the processes. Otherwise, we run the required number of iterations, accumulate the results by appending the columns of output, and return it to the master. In the simplified implementation, the number of replications is received as a scalar value. At the next stage, this send will be augmented with a seed for random number generation.

The master code in Listing A8 sets the block size as $B = \max[\min(\lfloor M/(P-1)/10 \rfloor, 1000), 1]$, where $P-1$ is the number of slave processes, and M the number of replications. The aim is to achieve a modicum of automatic load balancing: if the processors are not all equal, we would like to allow the more powerful ones to do more, suggesting a fairly small block size. But too small a block size will create excessive communication between processes. On the other hand, if the block size is large, the system may stall to transmit and process large amounts of results. Moreover, processors may sit idle towards the end, while waiting for a large chunk to finish elsewhere,

The master commences by sending each slave the number of replications it should do. The master process will then wait using OxMPI_Probe until a message is available. When that is the case, the message is received from process jnext say. Then, a subloop runs a nonblocking probe over processes jnext+1, ... $P - 2, 0, ...,$ jnext−1, receiving results if available. If a message is received, the master will immediately send the process a new iteration block as long as more work to be done. The intention of the nonblocking subloop is to divide reception of results evenly, and prevent starvation of any process. In principle, this can give some speed improvement, although we have not found any situations on our systems where this is significant. When all the work has been done, the slave sent a message to stop working, together with the overall final result. Afterwards, all the remaining processes will complete their work in isolation. The final synchronization ensures that they subsequently do the same computations. It would be possible for the slave processes to stop altogether, but this has not been implemented.

15.5.5 PARALLEL RANDOM NUMBER GENERATION

Many statistical applications require random numbers. In particular, Monte Carlo experiments, bootstrapping, and simulation estimation are prime candidates for the distributed framework developed here. Scientific replicability is enhanced if computational outcomes are deterministic, i.e., when the same results attain in different settings. This is relevant when using random numbers and is the primary reason why the Ox RNG always starts from the same seed, and not from a time-determined seed. However, in parallel computation doing so would give each process the same seed, and hence be the same Monte Carlo replication, thus a wasted effort.

An early solution, suggested by Smith et al. [31], was to split a linear congruential generator (LCG) into P sequences that are spaced by P (so each process gets a different slice, on processor i: $U_i, U_{i+P}, U_{i+2P}, ...$). As argued by Coddington [32], this leap-frog method can be problematic because LCG numbers tend to have a lattice structure (see Ref., [33], Section 15.2.2 and 15.2.4), so that there can be correlation between streams even when P is large. In particular, the method is likely to be flawed when the number of processors is a power of two. Another method is to split the generation into L sequences, with processor i using $U_{iL}, U_{iL+1}, U_{iL+2}, ...$ This ensures that there is no overlap, but the fixed spacing could still result in correlation between sequences. Both methods produce different results when the number of processors changes.

Recently, methods of parametrization have become popular. The aim is to use a parameter of the underlying recursion that can be varied, resulting in independent streams of random numbers. An example is provided by the SPRNG library (sprng.cs.fsu.edu) (see Refs. [34, 35] for an introduction). This library could be used with Ox because Ox has a mechanism to replace the default internal RNG (once the replacement is made, all Ox random number functions work as normal, using the new uniform generator). The SPRNG library has the ability to achieve the same results independently of the number of processors by assigning a separate parameter to each replication.

Our approach sits half-way between sequence splitting and the method of parametrization. This uses two of Ox's three built-in RNGs. The default generator of Ox is a modified version of Park and Miller [36], which has a period $2^{31} - 1 \approx 2 \times 10^9$. The second is a high-period RNG from L'Ecuyer [37], with period $\approx 4 \times 10^{34}$. Denote the latter as *RanLE*, and the former as *RanPM*. *RanLE* requires four seeds, *RanPM* just one.

We propose to assign seeds to each replication (i.e., each iteration of the loop), instead of each process in the following way. The master process runs *RanPM*, assigning a seed to each replication block. The slaves are told to execute B replications. At the start of the block, they use the received seed to set *RanPM* to the correct state. Then, for each replication, they use *RanPM* to create four random numbers as a seed for *RanLE*. All subsequent random numbers for that replication are then generated with *RanLE*. When the slaves are finished, they receive the final seed of the master, allowing all processes (master and slaves) to switch back to *RanPM*, and finish in the same random number state. On the master, *RanPM* must be advanced by $4B$ after each send, which is simply done by drawing $B \times 4$ random numbers.

Although the replications may arrive at the master in a different order if the number of processes changes, this is still the same set of replications, so the results are independent of the number of processes. Randomizing the seed avoids the problem that is associated with a fixed sequence split. There is a probability of overlap, but, because the period of the generator is so high, this is very small, as we show next.

Let there be M replications, a period of R^*, and L random numbers consumed in each replication. To compute the probability of overlap, the effective period becomes $R = R^*/L$, with overlap if any of the M drawings with replacement are equal. $P\{\text{overlap}\} = 1 - P\{M \text{ different}\} = 1 - P\{M \text{ different}|M-1 \text{ different}\} \cdots P\{1 \text{ different}\} = 1 - \prod_{i=1}^{M}(R-i+1)/R$. For large M and R considerably larger than M: $\log P\{M \text{ different}\} \approx -M^2/(2R)$. Using $x! = x\Gamma(x)$ this is found as follows: $\prod_{i=1}^{M}(R-i+1)/R = [R\Gamma(R)]/[(R-M)\Gamma(R-M)R^M] = q$. Then, for large arguments: $\log \Gamma(x) \approx (x - \frac{1}{2})\log(x) - x + \frac{1}{2}\log(2\pi)$. So $\log q \approx (R - M + \frac{1}{2})[\log R - \log(R-M)] - M$. Finally, using $\log(1-x) \approx -x - \frac{1}{2}x^2$ and dropping some negligible terms gives the approximation.

When $\log P\{M \text{ different}\} \approx -M^2/(2R)$ is close to zero, the probability of overlap is close to zero. In particular, $M < 10^{-m}R^{1/2}$ gives an overlap probability of less than 0.5×10^{-2m}. Selecting $m = 4$ delivers a probability sufficiently close to zero, so $M = 10^7$ allows our procedure to use $L = 10^{16}$ random numbers in each replication. This is conservative, because overlap with very wide spacing is unlikely to be harmful. On the other hand, imperfections in the RNG may make this estimate too optimistic.

15.5.6 PARALLEL RANDOM NUMBER GENERATION: IMPLEMENTATION

The implementation of the random number generation scheme is straightforward. On the master, the first two OxMPI_Send commands are changed to:

```
OxMPI_Send(cb ~ GetMasterSeed(cb), j, TAG_SEED);
```

whereas the final send uses 0 instead of cb. The slave receives two numbers: the first is the number of replications, the second the seed for *RanPM*:

```
repseed = OxMPI_Recv(iMaster, TAG_SEED);

crep = repseed[0];

seed = repseed[1];
```

Then inside the inner loop of doLoopSlave, just before fnRun is called, insert:

```
seed = SetSlaveSeed(seed);
```

Finally, before returning, the slave must switch back to RanPM, and reset the seed to that received last.

The GetMasterSeed function, Listing A9, gets the current seed, and advances it by $4B$; the master is always running *RanPM*. SetSlaveSeed switches to *RanPM*, sets the specified seed, then generates four seeds for *RanLE*, switches to *RanLE* and returns the *RanPM* seed that is to be used next time. The four seeds have to be greater than 1, 7, 15, 127, respectively.

There is one final complicating factor. In some settings, there is random data that is fixed in the experiment. For example, a regression analysis may condition on regressors by keeping them fixed after the first replication. Therefore, the `fnInit` function that is optionally called to initialize the loop is made to use the same seed on each slave.

15.6 SOME APPLICATIONS

15.6.1 TEST ENVIRONMENTS

We have two computational clusters for testing the parallel implementation of the matrix language.

The first is a Linux cluster consisting of four 550 MHz Pentium III nodes, each with 128 MB of memory. The master is a dual processor Pentium Pro at 200 MHz, also with 128 MB of memory. They are connected together using an Intel 100 ethernet switch. The operating system is Linux Mandrake Clic (Phase 1). This facilitated the installation: each slave can be installed via the network. Administration is simplified using networked user accounts. MPICH 1.2.4 is also installed by default (in a non-SMP configuration, so only one processor of the master can be used, giving the cluster $4 \times 1 + 1 = 5$ processors in total). The basic Ox installation has to be mirrored across nodes, but there are commands to facilitate this.

The second "cluster" is a Windows-based collection of machines. In the basic configuration, it consists of one machine with four 500 MHz Intel Xeon processors and 512 MB memory, and a second machine with two 500 MHz Pentium III processors and 384 MB. The four-processor machine runs Windows NT 4.0 Server, whereas the dual processor machine runs Windows 2000 Professional. They are connected via the College network. Installation of MPICH 1.2.5 on these machines is trivial, but synchronization of user programs is a bit more work. The Windows version of MPICH incorporates SMP support by default, so this system is able to use all $4 + 2 = 6$ processors.

15.6.2 NORMALITY TEST

The normality test code has already been rewritten in Section 15.6.2, and only needs to be adjusted to use the new MPI-enabled simulation class by changing the `#import` line at the top of the code. To make the Monte Carlo experiment more involved, we set the sample size to $T = 250$ and the number of replications to $M = 10^6$.

Table 15.3 considers the effect of the block size on program timings. The block size is the number of replications that a slave process is asked to run before returning the result to the master. It is clear that $B = 1$ generates a lot of communication overhead, causing processes to wait for others to be handled. The optimal block size is in the region 100 to 1000, over which the time differences are minimal. A large block size slows the program down, presumably because the overhead it requires in the MPICH communications of large matrices. In the light of these results, we set a maximum block size of 1000 when it is selected automatically.

Table 15.4 investigates the role of the number of processes on program time. The number of processors (the hardware) is fixed, whereas the number of processes (the software) is varied from $P = 1$, when there is no use of MPI, to $P = 8$ (P includes the master process here). It is optimal to run one process more than the number of processors, because the master does not do enough work to keep a processor fully occupied. The efficiency is listed when the hardware is fully employed, and assumes that the speedup could be as high as the number of processors. The efficiency is somewhat limited by the fact that the final report is produced sequentially. Note that the efficiency is higher than reported in [1, Table 1], because we now communicate the seed of *RanPM*, instead of sending the $4B$ *RanLE* seeds. Also, processing and appending results on the master was done

TABLE 15.3
Effect of Blocksize B When T = 250, M = 10^6 for Monte
Carlo Experiment of Normality Test

B	Windows (4+2) with 7 processes		Linux (4×1+1) with 6 processes	
	Time	Relative timing	Time	Relative timing
1	11:54	5.5	13:12	6.1
10	3:15	1.5	3:19	1.5
100	2:29	1.1	2:09	1.0
1,000	2:11	1.0	2:16	1.1
10,000	4:14	1.9	3:25	1.6

Timings in minutes:seconds.

TABLE 15.4
Effect of Number of Processes P When T = 250, M = 10^6 for
Normality Test

P	Windows 4+2 processors, B=1000			Linux 4×1+1 processors, B=100		
	Time	Speedup	Efficiency (%)	Time	Speedup	Efficiency (%)
1	10:30			9:36		
2	11:17	0.9		6:32	1.5	
4	3:30	3.0		3:10	3.0	
6	2:20	4.5	75	2:09	4.5	89
7	2:11	4.8	80	2:09	4.5	89
8	2:32	4.1	68	2:09	4.5	89

Timings in minutes:seconds; speedup is relative to nonparallel code.
Efficiency assumes maximum speedup equals number of processors.

before restarting the slave. Now the send is done first, resulting in better interleaving of the send operation and computation.

The Linux cluster's efficiency is particularly high, especially as the fifth processor is a relatively slow Pentium Pro (indeed, running $P = 1$ on that processor takes more than 21 min). We believe that this gain largely comes from the fact that the cluster is on a dedicated switch. It could also be that the shared memory configuration is a drawback for the Windows system when using MPI.

When adding two more machines to the Windows configuration (another 500 MHz PIII, and a 1.2 MHz PIII notebook), the time goes down to 85 s for $P = 9$ (two of which on the notebook). This is an efficiency of 83%, if the notebook is counted for two processors.

15.7 CONCLUSION

Although computers are getting faster all the time, there are still applications that stretch the capacity of current desktop computers. We focused on econometric applications, with an emphasis on Monte Carlo experiments. Doornik et al. [1] consider simulation-based inference in some detail.

Their example is estimation of a stochastic volatility model, for which the parallel loop class that is discussed here can be used.

We showed how a high-level matrix programming language can be used effectively for parallel computation on small systems of computers. In particular, we managed to hide the parallelization, so that existing code can be used without any additional effort. We believe that this reduction of labor costs and removal of specificity of the final code is essential for wider adoption of parallel computation.

It is outside our remit to consider hardware aspects in any detail. However, it is clear that construction and maintenance of small clusters has become much easier. Here we used both a group of disparate Windows machines on which MPI is installed, as well as a dedicated cluster of Linux machines. As computational grid technologies such as the Globus toolkit (www.globus.org) and Condor (www.cs.wisc.edu/condor) develop, they may also be used to host matrix-programming languages. This could make parallel statistical computation even easier, but requires further research.

ACKNOWLEDGMENTS

We wish to thank Richard Gascoigne and Steve Moyle for invaluable help with the construction of the computational environments. Support from Nuffield College, Oxford, is gratefully acknowledged. The computations were performed using the Ox programming language.

The code developed here, as well as the academic version of Ox Console can be downloaded from www.doornik.com.

Trademarks are: EViews by Quantitative Micro Software, PcGive by Doornik & Hendry, RATS by Estima, Stata by Stata Corp, TSP by TSP Intl., Intel and Pentium by Intel Corp., AMD and Athlon by Advanced Micro Devices Inc.

REFERENCES

[1] Doornik, J.A., Hendry, D.F., and Shephard, N. (2002). Computationally-intensive econometrics using a distributed matrix-programming language. *Philosophical Transactions of the Royal Society of London, Series A*, 360:1245–1266.

[2] Cribari-Neto, F. and Jensen, M.J. (1997). MATLAB as an econometric programming environment. *Journal of Applied Econometrics*, 12:735–744.

[3] Chong Y.Y. and Hendry, D.F. (1986). Econometric evaluation of linear macro-economic models. *Review of Economic Studies*, 53:671–690. Reprinted in Granger, C.W.J. (ed.) (1990), *Modelling Economic Series*. Oxford: Clarendon Press.

[4] Swan, C.A. (2003). Maximum likelihood estimation using parallel computing: an introduction to MPI. *Computational Economics*, 19:145–178.

[5] Nagurney, A., Takayama, T., and Zhang, D. (1995). Massively parallel computation of spatial price equilibrium problems as dynamical systems. *Journal of Economic Dynamics and Control*, 19:3–37.

[6] Nagurney, A. and Zhang, D. (1998). A massively parallel implementation of a discrete-time algorithm for the computation of dynamic elastic demand and traffic problems modeled as projected dynamical systems. *Journal of Economic Dynamics and Control*, 22:1467–1485.

[7] Gilli, M. and Pauletto, G. (1993). Econometric model simulation on parallel computers. *International Journal of Supercomputer Applications*, 7:254–264.

[8] Kontoghiorghes, E.J. (2000). *Parallel Algorithms for Linear Models: Numerical Methods and Estimation Problems*. Boston: Kluwer Academic Publishers.

[9] Ferrall, C. (2001). Solving finite mixture models in parallel. Mimeo, Queens University.

[10] Zenios, S.A. (1999). High-performance computing in finance: the last 10 years and the next. *Parallel Computing*, 25:2149–2175.

[11] Abdelkhalek, A., Bilas, A., and Michaelides, A. (2001). Parallelization, optimization and performance analysis of portfolio choice models. In *Proceedings. of the 2001 International Conference on Parallel Processing (ICPP01)*.

[12] Pauletto, G. (2001). Parallel Monte Carlo methods for derivative security pricing. In L. Vulkov, J. Wasnievski, and P. Yalamov, Eds., *Numerical Analysis and Its Applications*. New York: Springer-Verlag.

[13] Kontoghiorghes, E.J., Nagurney, A., and Rustem, B. (2000). Parallel computing in economics, finance and decision-making. *Parallel Computing*, 26:507–509 [Editorial Introduction to Special Issue].

[14] Doornik, J.A. (2001). *Object-oriented Matrix Programming Using Ox*, 4th ed. London: Timberlake Consultants Press.

[15] Wilson, G.V. (1995). *Practical Parallel Programming*. Cambridge, MA: The MIT Press.

[16] Caron *et al.* (2001). Scilab to scilab//: The ouragan project. *Parallel Computing*, 27:1497–1519.

[17] Cribari-Neto, F. and Zarkos, S.G. (2003). Econometric and statistical computing using Ox. *Computational Economics*, 21:277–295.

[18] Koopman, S.J., Shephard, N., and Doornik, J.A. (1999). Statistical algorithms for models in state space using SsfPack 2.2 (with discussion). *Econometrics Journal*, 2:107–160. Also see: www.ssfpack.com.

[19] Bowman, K.O. and Shenton, L.R. (1975). Omnibus test contours for departures from normality based on $\sqrt{b_1}$ and b_2. *Biometrika*, 62:243–250.

[20] Jarque, C.M. and Bera, A.K. (1987). A test for normality of observations and regression residuals. *International Statistical Review*, 55:163–172.

[21] Doornik, J.A. and Hansen, H. (1994). An omnibus test for univariate and multivariate normality. Discussion paper, Nuffield College.

[22] Doornik, J.A. and Ooms, M. (2001). *Introduction to Ox*. London: Timberlake Consultants Press.

[23] LAPACK. *LAPACK Users' Guide*, 3rd ed. (1999). Philadelphia, PA: SIAM By, Erson, E., Bai, Z., Bischof, C., Blackford, S., Demmel, J., Dongarra, J., Du Croz, J., Greenbaum, A., Hammarling, S., McKenney, A. and Sorensen, D.

[24] Murphy, K., Clint, M., and Perrott, R.H. (1999). Re-engineering statistical software for efficient parallel execution. *Computational Statistics and Data Analysis*, 31:441–456.

[25] Snir, M., Otto, S.W., Hus-Lederman, S., Walker, D.W., and Dongarra, J. (1996). *MPI: The Complete Reference*. Cambridge, MA: The MIT Press.

[26] Gropp, W., Lusk, E., and Skjellum, A. (1999). *Using MPI*, 2nd ed. Cambridge, MA: The MIT Press.

[27] Geist, G.A., Beguelin, A., Dongarra, J., Jiang, W., Manchek, R., and Sunderam, V. (1994). *PVM: Parallel Virtual Machine — A Users' Guide and Tutorial for Networked Parallel Computing*. Cambridge, MA: The MIT Press.

[28] Geist, G.A., Kohl, J.A., and Papadopoulos, P.M. (1996). PVM and MPI: a comparison of features. *Calculateurs Paralleles*, 8. www.csm.ornl.gov/pvm/PVMvsMPI.ps.

[29] Gropp, W. and Lusk, E. (2001). User's guide for mpich, a portable implementation of MPI version 1.2.2. mimeo, Argonne National Laboratory, University of Chicago. www-unix.mcs.anl.gov/mpi/mpich/.

[30] Doornik, J.A. (2002). Object-oriented programming in econometrics and statistics using Ox: a comparison with C++, Java and C#. In S.S. Nielsen, Ed. *Programming Languages and Systems in Computational Economics and Finance*. Dordrecht: Kluwer Academic Publishers, pp 115–147.

[31] Smith, K.A., Reddaway, S.F., and Scott, D.M. (1985). Very high performance pseudo-random number generation on DAP. *Computer Physics Communications*, 37:239–244.

[32] Coddington P.D. (1996). Random number generators for parallel computers. *The NHSE Review*, 2. Version 1.1, April 1997, www.crpc.rice.edu/NHSEreview/RNG/.

[33] Ripley B.D. (1987). *Stochastic Simulation*. John Wiley & Sons, New York.

[34] Mascagni, M. and Srinivasan, A. (2000). Algorithm 806: SPRNG: a scalable library for pseudorandom number generation. *ACM Transactions on Mathematical Software*, 26:436–461.

[35] SCAN. (1999). Parallel random number generators. Scientific computing at NPACI, Vol. 3, March 31, www.npaci.edu/online/v3.7/SCAN1.html.

[36] Park, S. and Miller, K. (1998). Random number generators: good ones are hard to find. *Communications of the ACM*, 31:1192–1201.

[37] L'Ecuyer, P. (1999). Tables of maximally-equidistributed combined LFSR generators. *Mathematics of Computation*, 68:261–269.

APPENDIX

```
#include <oxstd.h>
vecmul(const v1, const v2, const c)
{
    decl i, sum;
    for (i = 0, sum = 0.0; i < c; ++i)
        sum += v1[i] * v2[i];
    return sum;
}
vecmulXtY(const cC, const cRep)
{
    decl i, a, b, c, secs, time;
    a = b = rann(1,cC);
    time = timer();
    for (i = 0; i < cRep; ++i)
    {
        c = vecmul(a, b, cC);
    }
    secs = (timer() - time) / 100;
    println("Vector product x'y:  cC=", "%6d", cC,
            " time=", "%7.4f", secs, " #/sec=", cRep / secs);
}
vecmulXtYvec(const cC, const cRep)
{
    decl i, a, b, c, secs, time;
    a = b = rann(1,cC);
    time = timer();
    for (i = 0; i < cRep; ++i)
    {
        c = a*b';
    }
    secs = (timer() - time) / 100;
     println("Vector product x'y:  cC=", "%6d", cC,
        " time=", "%7.4f", secs, " #/sec=", cRep / secs);
}
main()
{
    decl i;
    for (i = 10; i <= 100000; i *= 10)
        vecmulXtY(i, int(400000 / i));
    for (i = 10; i <= 100000; i *= 10)
        vecmulXtYvec(i, int(100000000 / i));
}
```

Listing A1: Vector multiplication in Ox (vectorized and unvectorized): vecmul.ox

```c
#include <math.h>
#include <stdio.h>
#include <stdlib.h>      /
#include <time.h>
double timer(void)
{
    return clock() * (100.0 / CLOCKS_PER_SEC);
}
typedef double **MATRIX;
typedef double  *VECTOR;
/* code omitted for:
MATRIX matalloc(int cR, int cC)
void  matfree(MATRIX m)
MATRIX matallocRanu(int cR, int cC)
*/
double vecmul(VECTOR v1, VECTOR v2, int c)
{
    int i;   double sum;
    for (i = 0, sum = 0.0; i < c; ++i)
    sum += v1[i] * v2[i];
    return sum;
}
void vecmulXtX(int cC, int cRep)
{
    int i;
        double secs, time;
    MATRIX a, b, c;
    a = matallocRanu(1, cC);
    b = matallocRanu(1, cC);
    c = matalloc(1, 1);
    time = timer();
    for (i = 0, secs = 0.0; i < cRep; ++i)
    {
        c[0][0] = vecmul(a[0], b[0], cC);
    }
    secs = (timer() - time) / 100;
    printf("Product X'X: cC=%7d time=%7.4f #/sec=%g\n",
        cC, secs, cRep / secs);
    matfree(a);  matfree(b);  matfree(c);
}
```

```
int main(int argc, char **argv)
{
    int i, c = 4;
    for (i = 10; i <= 100000; i *= 10)
        vecmulXtX(i, 100000000 / i);
    return 0;
}
```

Listing A2: Vector multiplication in C: vecmul.c

```
#include <oxstd.h>
matmulAB(decl cR, decl cC, decl cRep)
{
    decl i, a, b, c, secs, time;
    a = rann(cR,cC);
    b = a';
    for (i = 0, secs = 0.0; i < cRep; ++i)
    {
        time = timer();
        c = a*b;
        secs += (timer() - time) / 100;
    }
    println("Product AB:  cR=", "%7d", cR, " cC=", "%7d", cC,
        " time=", "%7.4f", secs,
        " log10(#/sec)=", log10(cRep / secs));
}
matmulAtA(decl cR, decl cC, decl cRep)
{
    decl i, a, c, secs, time;
    a = rann(cR,cC);
    for (i = 0, secs = 0.0; i < cRep; ++i)
    {
        time = timer();
        c = a'a;
        secs += (timer() - time) / 100;
    }
        println("Product A'A: cR=", "%7d", cR, " cC=", "%7d", cC,
        " time=", "%7.4f", secs,
        " log10(#/sec)=", log10(cRep / secs));
}
```

Listing A3: Matrix multiplication in Ox (main omitted): matmul.ox

```
#include "mpi.h"
#include <stdio.h>
#include <math.h>
double f(double a)
{
    return (4.0 / (1.0 + a*a));
}
int main(int argc, char *argv[])
{
    int n, myid, numprocs, i;
    double PI25DT = 3.141592653589793238462643;
    double mypi, pi, h, sum, x;
    double startwtime, endwtime;
    char procname[MPI_MAX_PROCESSOR_NAME];
    int   namelen;
    MPI_Init(&argc, &argv);
    MPI_Comm_size(MPI_COMM_WORLD, &numprocs);
    MPI_Comm_rank(MPI_COMM_WORLD, &myid);
    MPI_Get_processor_name(procname, &namelen);
        printf("Process %d of %d on %s\n",
        myid, numprocs, procname);
    if (myid == 0)
    {   n = 1000;
        startwtime = MPI_Wtime();
    }
    MPI_Bcast(&n,1,MPI_INT,0,MPI_COMM_WORLD);
    printf("id: %d n= %d\n", myid, n);
    h = 1.0 / (double) n;
    sum = 0.0;
    for (i = myid + 1; i <= n; i += numprocs)
    {
        x = h * ((double)i - 0.5);
        sum += f(x);
    }
    mypi = h * sum;
    printf("id: %d sum= %g\n", myid, sum);
    MPI_Reduce(&mypi, &pi, 1, MPI_DOUBLE,
    MPI_SUM, 0, MPI_COMM_WORLD);
    if (myid == 0)
    {
        printf("pi=%.16f,Error=%.16f\n",
        pi,
            fabs(pi - PI25DT));
```

```
#include <oxstd.h>
#include <packages/oxmpi/oxmpi.h>
f(const a)
{
    return (4.0 ./ (1.0 + a*a));
}
main()
{
    decl n, myid, numprocs, i;
    decl PI25DT = 3.141592653589793
                        238462643;
    decl mypi, pi, h, sum, x;
    decl startwtime, endwtime;
    decl procname;
    OxMPI_Init();
    numprocs = OxMPI_Comm_size();
    myid = OxMPI_Comm_rank();
     procname = OxMPI_Get_processor_
            name();
      println("Process",myid,"of",
    numprocs," on ", procname);
        if (myid == 0)
    {   n = 1000;
        startwtime = OxMPI_Wtime();
    }
    OxMPI_Bcast(&n, 0);
    println("id: ", myid, " n=", n);
    h = 1.0 / n;
    sum = 0.0;
    for (i = myid + 1; i <= n;
            i += numprocs)
    {
        x = h * (i - 0.5);
        sum += f(x);
    }
     mypi = h * double(sum);
      println("id: ", myid, " sum=",
            sum);
    pi = OxMPI_Reduce(mypi,
                    MPI_SUM, 0);
    if (myid == 0)
    {
      println("pi=", "%.16f", pi,
```

```
        endwtime = MPI_Wtime();                           ", Error=", "%.16f",
        printf("wall clock time = %f\n",                      fabs(pi - PI25DT));
            endwtime - startwtime);                   endwtime = OxMPI_Wtime();
    }                                                 println("wall clock time = ",
    MPI_Finalize();                                       endwtime - startwtime);
    return 0;                                         }
}                                                     OxMPI_Finalize();
                                                  }
```

Listing A4: MPI example program in C (left) and Ox (right)

```
class Loop1
{
    static Run(const const fnInit, const fnRun, const fnProcess, const cRep);
    static doLoopMaster(const cSlaves, const cRep, const fnProcess);
    static doLoopSlave(const iMaster, const fnRun);
    static doLoop(const cRep, const fnRun, const fnProcess);
    static decl m_cSlaves, m_iMaster, m_fInitialized;
    static IsMaster();
    static IsSlave();
    static Init();
};
Loop1::doLoopSlave(const iMaster, const fnRun)
{                  /* code omitted */  }
Loop1::doLoopMaster(const cSlaves, const cRep, const fnProcess)
{                  /* code omitted */  }
Loop1::doLoop(const cRep, const fnRun, const fnProcess)
{                  /* code omitted */  }
Loop1::Run(const fnInit, const fnRun, const fnProcess, const cRep)
{
    decl mresult;
    Init();
    if (isfunction(fnInit) && !IsMaster())
        fnInit();
    if (IsSlave())
        mresult = doLoopSlave(m_iMaster, fnRun);
    else if (IsMaster())
        mresult = doLoopMaster(m_cSlaves, cRep, fnProcess);
```

```
        else
            mresult = doLoop(cRep, fnRun, fnProcess);
        OxMPI_Barrier();
        return mresult;
    }
    Loop1::Init()
    {
        if (m_fInitialized)
            return;
        OxMPI_Init();
        m_cSlaves = OxMPI_Comm_size() - 1;
        m_iMaster = m_cSlaves;
        if (m_cSlaves)
            OxMPI_SetMaster(m_cSlaves && OxMPI_Comm_rank() == m_cSlaves);
        m_fInitialized = TRUE;
    }
    Loop1::IsMaster()
    {
        Init();
        return m_cSlaves ? OxMPI_IsMaster() : 0;
    }
    Loop1::IsSlave()
    {
        Init();
        return m_cSlaves ? !OxMPI_IsMaster() : 0;
    }
```

Listing A5: Loop1: Simplified version of Loop class

```
    Loop1::doLoop(const cRep, const fnRun, const fnProcess)
    {
        decl i, itno, c, mresult = <>, mrep;
        for (itno = i = 0; itno < cRep; ++i)
        {
            mrep = fnRun(i);
            c = getResultSize(mrep);
            storeResult(&mresult, mrep, itno, cRep, fnProcess);
            itno += c;
        }
        return mresult;
    }
```

Listing A6: Loop1: Serial version of loop iteration

```
Loop1::doLoopSlave(const iMaster, const fnRun)
{
    decl itno, i, k, c, iret, mresult, mrep, crep;
    for (k = 0; ; )
    {
        // ask the master for the number of replications
        crep = OxMPI_Recv(iMaster, TAG_SEED);
        if (crep == 0)
            break;
        // run the experiment
        for (i = itno = 0; i < crep; ++i, ++k)
        {
            mrep = fnRun(k);
            c = getResultSize(mrep);
            storeResult(&mresult, mrep, itno, crep, 0);
            itno += c;
        }
        // check the result and send to the master
        checkResult(&mresult, itno, crep);
        OxMPI_Send(mresult, iMaster, TAG_RESULT);
    }
    // receive aggregate and return it
    mresult = OxMPI_Recv(iMaster, TAG_RESULT);
    return mresult;
}
```

Listing A7: Loop1: Slave versions of loop iteration

```
Loop1::doLoopMaster(const cSlaves, const cRep, const fnProcess)
{
    decl itno, i, j, jnext, c, cstarted, status, cb;
    decl mresult = <>, mrep, vcstarted;
    // set the blocksize
    cb = max(min(int((cRep / cSlaves) / 10), 1000), 1);
    // start-up all slaves
    vcstarted = zeros(1, cSlaves);
    for (j = cstarted = 0; j < cSlaves; ++j)
    {
        OxMPI_Send(cb, j, TAG_SEED);
        vcstarted[j] = cb;
        cstarted += cb;
    }
    // replication loop, repeat while any left to do
```

```
for (itno = 0; itno < cRep; )
{
    // wait (hopefully efficiently) until a result comes in
    status = OxMPI_Probe(MPI_ANY_SOURCE, TAG_RESULT);
    jnext = status[0];
    // process and check on others to prevent starvation
    for (i = 0; itno < cRep && i < cSlaves; ++i)
    {
        j = i + jnext;          // first time (i=0): j==jnext
        if (j >= cSlaves) j = 0;
        if (i)                  // next times: non-blocking probe of j
        {   status = OxMPI_Iprobe(j, TAG_RESULT);
            if (!sizerc(status))
                continue;       // try next one if nothing pending
        }
        // now get results from slave j
        mrep = OxMPI_Recv(j, TAG_RESULT);
        // get actual no replications done by j
        c = getResultSize(mrep);
        cstarted += c - vcstarted[j]; // adjust number started
        if (cstarted < cRep) // yes: more work to do
        {   // shrink block size if getting close to finishing
            if (cstarted > cRep - 2 * cb)
                cb = max(int(cb / 2), 1);
            OxMPI_Send(cb, j, TAG_SEED);
            vcstarted[j] = cb;
            cstarted += cb;
        }
        // finally: store the results before probing next
        storeResult(&mresult, mrep, itno, cRep, fnProcess);
        itno += c;
    }
}
// close down slaves
for (j = 0; j < cSlaves; ++j)
{
    OxMPI_Send(0, j, TAG_SEED);
    OxMPI_Send(mresult, j, TAG_RESULT);
}
return mresult;
}
```

Listing A8: Loop1: Master–slave versions of loop iteration

```
Loop::GetMasterSeed(const cBlock)
{
    decl seed = ranseed(0);        // get the current PM seed to return
    ranu(cBlock, 4);               // advance PM seed cBlock * 4 positions
    return seed;                   // return initial PM seed
}


Loop::SetSlaveSeed(const iSeed)
{
    ranseed("PM");                 // reset to PM rng
    ranseed(iSeed);                // and set the seed
                                   // get 4 seeds for LE from PM
     decl aiseed = floor(ranu(1, 4) * INT_MAX);
    aiseed = aiseed .<= <2,8,16,128> .? aiseed + <2,8,16,128> .: aiseed;
    decl seedpm = ranseed(0);      // remember the PM seed for next time
    ranseed("LE");                 // switch to LE, and set seed
    ranseed( {aiseed[0], aiseed[1], aiseed[2], aiseed[3]} );
    return seedpm;
}
```

Listing A9: Communication of seeds for RNGs

16 Parallel Bayesian Computation

Darren J. Wilkinson

CONTENTS

ABSTRACT

The use of Bayesian inference for the analysis of complex statistical models has increased dramatically in recent years, in part due to the increasing availability of computing power. There are a range of techniques available for carrying out Bayesian inference, but the lack of analytic tractability for the vast majority of models of interest means that most of the techniques are numeric, and many are computationally demanding. Indeed, for high-dimensional nonlinear models, the only practical methods for analysis are based on Markov chain Monte Carlo (MCMC) techniques, and these are notoriously computation intensive, with some analyses requiring weeks of CPU time on powerful computers. It is clear therefore that the use of parallel computing technology in the context of Bayesian computation is of great interest to many who analyze complex models using Bayesian techniques.

Of particular interest in the context of Bayesian inference are techniques for parallelization of a computation utilizing the conditional independence structure of the underlying model, as well as strategies for parallelization of Monte Carlo and MCMC algorithms. There are two obvious approaches to parallelization of an MCMC algorithm: one is based on the idea of running different chains in parallel, and the other is based on parallelization of a single chain. It is a subtle and problem-dependent question as to which of these strategies (or combination of the two) is likely to be most appropriate and there are also important issues relating to parallel random number generation which need to be addressed.

16.1 INTRODUCTION

The use of Bayesian inference for the analysis of complex statistical models has increased dramatically in recent years, in part due to the increasing availability of computing power. There are a range of techniques available for carrying out Bayesian inference, but the lack of analytic tractability for the vast majority of models of interest means that most of the techniques are numeric and many are computationally demanding. Indeed, for high-dimensional nonlinear models, the only practical methods for analysis are based on Markov chain Monte Carlo (MCMC) techniques and these are notoriously computation intensive, with some analyzes requiring weeks of CPU time on powerful computers. It is clear therefore that the use of parallel computing technology in the context of Bayesian computation is of great interest to many who analyze complex models using Bayesian techniques.

Section 16.2 considers the key elements of Bayesian inference and the notion of graphical representation of the conditional independence structure underlying a statistical model. This turns out to be key in exploiting the partitioning of computation in a parallel environment. Section 16.3 looks at the issues surrounding Monte Carlo simulation techniques in a parallel environment, laying the foundations for the examination of parallel MCMC in Section 16.4. Standard pseudorandom number generators are not suitable for use in a parallel setting, so this section examines the underlying reasons and the solution to the problem provided by parallel pseudorandom number generators.

Parallel MCMC is the topic of Section 16.4. There are two essentially different strategies that can be used for parallelizing an MCMC scheme (though these may be combined in a variety of ways). One is based on running multiple MCMC chains in parallel and the other is based on parallelization of a single MCMC chain. There are different issues related to the different strategies and each is appropriate in different situations. Indeed, because MCMC in complex models is somewhat of an art form anyway, with a range of different possible algorithms and trade-offs even in the context of a nonparallel computing environment, the use of a parallel computer adds an additional layer of complexity to the MCMC algorithm design process. That is, the trade-offs one would adopt for the design of an efficient MCMC algorithm for the analysis of a given statistical algorithm in a nonparallel environment may well be different in a parallel setting and could depend on the precise nature of the available parallel computing environment. Section 16.4 will attempt to give some background to the kind of trade-offs that are made in the design of an efficient MCMC algorithm and into the reasons why one might design an algorithm differently depending on the available hardware and software.

A wide range of parallel hardware is now available. Dedicated multiprocessor parallel computers offer outstanding interprocessor communication speeds, but networked collections of fairly ordinary PCs running a Unix-derived operating system such as Linux offer the opportunity to get supercomputer performance for relatively little capital outlay. Linux "Beowulf" clusters are rapidly becoming the *de facto* standard for affordable supercomputing. The examples given in the later sections use a small 8-CPU Linux cluster. The cluster was formed by four dual CPU PCs together with a cheap file server linked via switched 100 Mbit network (funded by a small grant from the Royal Society). However, the principles apply also to dedicated parallel computers and to much larger Beowulf clusters using faster (Gigabit or ATM) networking infrastructure. As already mentioned, the design of algorithms will often depend to some extent on the parallel architecture. In particular, the number of processors and the speed (and latency) of interprocessor communication are of particular importance. The examples assume a homogeneous cluster (i.e., all processors run at the same speed linked by a fairly uniform networking infrastructure), but are easily adapted to nonhomogeneous environments.

Parallel computing relies on the ability to write programs that exploit multiple processors and processes and transfer information between them. Rather than developing communication protocols from scratch using low-level network libraries, it is far more efficient to use specially written message-passing libraries specifically designed for the development of parallel applications. The parallel virtual machines (PVM) library and the message passing interface (MPI) are the most popular choices. MPI is a newer more sophisticated library, better tuned to performance-critical environments, and is probably a better choice for people new to parallel computing without an existing PVM code base. The small number of MPI calls necessary for the development of parallel Bayesian inference algorithms will be introduced and illustrated as they are required.

Where necessary, ideas are illustrated with the description of algorithms or simple pieces of code written in ANSI 'C'. Experience of programming will be assumed, but no experience of parallel computing is required, nor is extensive knowledge of ANSI 'C'. Note that the examples adapt easily to other languages, and hence are not ANSI 'C' specific. In particular, the key software libraries used (MPI and SPRNG) both have Fortran interfaces, and hence can be used just as easily from Fortran programs.

16.2   BAYESIAN INFERENCE

16.2.1   INTRODUCTION

This section introduces the key concepts necessary for the development of Bayesian inference algorithms. It also defines the notation used in the subsequent sections. For an introduction to Bayesian inference, O'Hagan [1] covers the theory and Gamerman [2] provides a good introduction to modern computational techniques based on MCMC. In the simplest continuous setting we are interested in inference for the parameter vector ϕ of a probability (density) model $p(y|\phi)$ giving rise to an observed data vector y. If we treat the parameters as uncertain and allocate to them a "prior" probability density $\pi(\phi)$, then Bayes' theorem gives the "posterior" density

$$\pi(\phi|y) = \frac{\pi(\phi)p(y|\phi)}{p(y)},$$

where $p(y)$ is the marginal density for y obtained by integrating over the prior for ϕ. Since $\pi(\phi|y)$ is regarded as a function of ϕ for fixed (observed) y, we can rewrite this as

$$\pi(\phi|y) \propto \pi(\phi)p(y|\phi),$$

so that the posterior is proportional to the prior times the model likelihood function.

For more complex models this simple description hides some important details. If y is the observed data then ϕ represents everything else in the model that is not directly observed. This will no doubt include "conventional" model parameters, but may well include other aspects of a model, such as "nuisance" parameters, random effects and/or missing data. In the context of more complex models it is therefore often helpful to partition $\phi = (\sigma, \theta)$, where σ represents "conventional" model parameters of direct inferential interest and θ represents all other details of the model which are not directly observed. Indeed, it is the ability of the Bayesian framework to handle easily models of this form by treating all aspects of a model in a unified way that makes it so attractive to a broad range of researchers. The prior is usually factored as $\pi(\phi) = \pi(\sigma, \theta) = \pi(\sigma)\pi(\theta|\sigma)$, so that Bayes' theorem now becomes

$$\pi(\sigma, \theta|y) \propto \pi(\sigma)\pi(\theta|\sigma)p(y|\sigma, \theta) \tag{16.1}$$

where the constant of proportionality is independent of both σ and θ. If σ is of primary concern, then real interest will be in the marginal posterior obtained by integrating θ out of (16.1), that is

$$\pi(\sigma|y) \propto \pi(\sigma) \int \pi(\theta|\sigma)p(y|\sigma, \theta)\mathrm{d}\theta,$$

which can be written as

$$\pi(\sigma|y) \propto \pi(\sigma)p(y|\sigma), \tag{16.2}$$

where

$$p(y|\sigma) = \int \pi(\theta|\sigma)p(y|\sigma, \theta)\mathrm{d}\theta \tag{16.3}$$

is the marginal likelihood for θ given the data and will typically be either unavailable or computationally expensive to evaluate. Before moving on, however, it is worth noting that there is an alternative representation of (16.3) that follows immediately from Bayes theorem:

$$p(y|\sigma) = \frac{\pi(\theta|\sigma)p(y|\sigma, \theta)}{\pi(\theta|\sigma, y)}. \tag{16.4}$$

This is sometimes referred to as the basic marginal likelihood identity (BMI) and it is sometimes tractable in situations where it is not at all clear how to tackle the integral in (16.3) directly [3].

The computational difficulties in Bayesian inference arise from the intractability of high-dimensional integrals such as the constants of integration in (16.1) and (16.2). These are typically not only analytically intractable but also difficult to obtain numerically. Although the constant of proportionality in (16.2) appears easier to evaluate, the difficult integration problem has simply been pushed to the evaluation of the marginal likelihood (16.3). Even where the constants of proportionality are known, there are further integration problems associated with obtaining the marginal distributions of various functions of the parameters of interest, as well as computing expectations of summary statistics.

In some situations it will be helpful to further partition $\sigma = (\sigma_1, \sigma_2, \ldots, \sigma_p)$ and/or $\theta = (\theta_1, \theta_2, \ldots, \theta_q)$, but there is no suggestion that the subcomponents are univariate unless explicitly stated.

Example 16.1 In order to make things concrete, consider the simple one-way ANOVA model with random effects:

$$y_{ij} = \mu + \alpha_i + \varepsilon_{ij}, \quad i = 1, 2, \ldots, n, \quad j = 1, 2, \ldots, r_i \tag{16.5}$$

where

$$\alpha_i \sim N(0, \sigma_\alpha^2) \quad \text{and} \quad \varepsilon_{ij} \sim N(0, \sigma_\varepsilon^2)$$

are independent. In fact, the random variables ε_{ij} are redundant and should be removed from the specification by rewriting (16.5) as

$$y_{ij}|\alpha_i \sim N(\mu + \alpha_i, \sigma_\varepsilon^2).$$

Here, one generally regards the α_i to be "missing data," and so the variables would be partitioned as $\sigma = (\mu, \sigma_\alpha, \sigma_\varepsilon)$, $\theta = (\alpha_1, \alpha_2, \ldots, \alpha_n)$, and y as the vector of all y_{ij}. There are other ways of partitioning these parameters, however, depending on the variables of interest and the computational techniques being adopted. For example, the variable μ could be moved from σ to θ if interest is mainly concerned with the variance components σ_α and σ_ε and linear Gaussian techniques are to be used to understand the rest of the model conditional on σ. With the partitioning as initially described, we know that

$$\pi(\theta|\sigma) = \prod_{i=1}^{n} N(\alpha_i; 0, \sigma_\alpha^2) \quad \text{and} \quad \pi(y|\theta, \sigma) = \prod_{i=1}^{n}\prod_{j=1}^{r_i} N(y_{ij}; \mu + \alpha_i, \sigma_\varepsilon),$$

where

$$N(x; \mu, \sigma^2) = \frac{1}{\sigma\sqrt{2\pi}} \exp\left\{ -\frac{1}{2}\left(\frac{x-\mu}{\sigma}\right)^2 \right\}.$$

Note that $\pi(\sigma)$ has not yet been specified, but that this must be done before any Bayesian analysis can be conducted.

16.2.2 GRAPHICAL MODELS AND CONDITIONAL INDEPENDENCE

To exploit a parallel computing environment there must be a way of decomposing the given computational task into parts which may be carried out independently of one another. In the context of complex statistical models the notion of conditional independence of model components provides a natural way of breaking down the model into manageable pieces. Conditional independence structures have a natural representation in terms of graphs, and hence models characterized by their conditional independence structure are often referred to as graphical models.

For a basic introduction to graphical models see Ref. [4] and for a complete account, see Ref. [5]. Here we introduce the essential concepts and notation required for later sections. Consider a finite directed graph $\mathcal{G} = (V, E)$ with vertex set $V = \{X_1, X_2, \ldots, X_n\}$ and finite edge set $E = \{(v, v')|v, v' \in V, v \rightarrow v'\}$. The graph is acyclic if there is no directed sequence of edges starting and finishing at the same vertex. A directed acyclic graph is known as a DAG. The *parents* of $v \in V$ are given by $\text{pa}(v) = \{v' \in V | (v', v) \in E\}$. The density of a random vector $X = (X_1, X_2, \ldots, X_n)$ is said to factor according to $\mathcal{G}$ if the probability density for X can be written as

$$\pi(x) = \prod_{i=1}^{n} \pi(x_i | \text{pa}(x_i)). \tag{16.6}$$

It follows that X_i is conditionally independent of every node that is not a direct descendent given only its parents. Thus for any X_j that is not a descendent of X_i we write $X_i \perp\!\!\!\perp X_j | \text{pa}(X_i)$, where $\perp\!\!\!\perp$ is the notation for conditional independence.

Example 16.2 The basic model factorization

$$\pi(\sigma, \theta, y) = \pi(\sigma)\pi(\theta|\sigma)p(y|\sigma, \theta)$$

has the DAG representation illustrated in Figure 16.1. Note that there are no (nontrivial) conditional independencies implied by this DAG, corresponding to the fact that there are no restrictions placed on the form of the joint density imposed by the chosen factorization. Note also that we could have chosen to factorize the joint density in a different order, and hence there is typically more than one valid DAG corresponding to any given conditional independence structure.

Undirected graphs can also be used to represent conditional independence structures. These graphs have the property that each node is conditionally independent of every other node given only its *neighbors*, defined by $\text{ne}(v) = \{v' \in V | v' \sim v\}$ where $\sim$ here denotes adjacency in the graph. Thus for every X_j not directly joined to X_i we write $X_i \perp\!\!\!\perp X_j | \text{ne}(X_i)$. Such graphs have other important conditional independence properties (and also have implications for the factorization of the joint density) but see Ref. [5] for details. It is worth noting that a valid undirected conditional independence graph can be obtained from a DAG by first "marrying" all pairs of parents of each node and then dropping "arrows." An undirected graph formed in this way is often referred to as a *moral* graph.

16.2.3 LOCAL COMPUTATION IN GRAPHICAL MODELS

Certain classes of statistical model are computationally tractable, at least in principle, but become problematic in higher dimensions. For problems that are purely discrete, for example, Bayesian inference is simply a matter of constructing conditional and marginal probability tables and forming weighted sums of probabilities to obtain posterior probabilities of interest. However, in practice the probability tables involved quickly become infeasibly large and so a direct approach to the problem fails. Fortunately, if the problem has a relatively "sparse" conditional independence structure, represented by a DAG without too many edges, then the conditional independence structure may be exploited so that calculations and computations done at any time involve only "nearby" DAG nodes, yet the results of these "local" computations may be combined to obtain "global" posterior probabilities of inferential interest. Essentially, the DAG is first *moralized* into an undirected graph and then coerced into a tree-shaped conditional independence graph known as a *clique-tree* or *junction tree*. Computations (which involve only adjacent nodes of the tree) are then carried out at each node and passed on to neighbors to solve a given global problem. An excellent introduction to this area can be found in Ref. [6], and a more detailed treatment of the problem is considered in Ref. [7], to which the reader is referred for further information.

There are other classes of models for which local computation is possible. The class of fully specified linear Gaussian models is of particular importance in applied statistics. Again, the conditional independence structure can be exploited to carry out calculations of interest in Bayesian inference based on local information. Wilkinson and Yeung [8] provides a study of this problem from the perspective of simulating unobserved variables in the model conditional on all observations. It turns

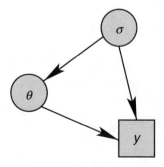

FIGURE 16.1 DAG representing the basic model factorization.

out that exact local computation techniques are also possible for a certain class of models known as conditionally Gaussian (CG) models that involve a mixture of discrete and linear Gaussian random variables; see Refs. [7, 9] for further details.

For pure linear Gaussian systems there is another way of tackling the problem, which utilizes the fact that the joint distribution of all variables in the problem is multivariate normal and that although the variance matrix is typically very large, its inverse (the "precision" matrix) is tractable and very sparse, having nonzero elements only where there are arcs in the associated conditional independence graph. Consequently, computations can be conducted by working with a sparse matrix representation of the problem and applying sparse matrix algorithms to the resulting computational problems to automatically exploit the sparseness and locality of the conditional independence structure. Wilkinson and Yeung [10] give a detailed examination of this approach and shows how to apply it in practice, describing a free software library, GDAGsim, which implements the techniques. If, rather than being DAG based, the linear Gaussian system is derived from a Markov random field, the techniques of Rue [11], implemented in the library GMRFlib, can be used instead.

16.2.4 Parallel Methods for Local Computation

Even using an ordinary serial computer, local computation methods allow the exact analysis of problems that would be difficult to contemplate otherwise. However, the resulting algorithms are often computationally expensive, making exploitation of high-performance parallel computers attractive. It turns out that this is relatively straightforward if the number of available processors is quite small, but the algorithms do not scale well as the number of processors increases. In the context of sending messages around a tree-shaped graph, messages in different branches of the tree can be computed and passed in parallel, allowing a modest speedup if the "diameter" of the graph is relatively small relative to the total number of nodes. The issues of parallel message-passing and "scheduling of flows" are considered in some detail in Refs. [6, 7].

Similar considerations apply to the sparse matrix methods for linear Gaussian systems. Parallel direct sparse matrix algorithms such as provided by the PSPASES library or ScaLAPACK can be used in place of conventional serial sparse matrix algorithms; however, such algorithms tend to scale poorly to large numbers of processors. For certain kinds of computation, parallel *iterative* solvers may be used. These scale much better than direct methods but cannot be used for all computations of interest. See Refs. [12, 13] for further details of the range of techniques that are useful in this context.

16.3 MONTE CARLO SIMULATION

16.3.1 Introduction

Despite the obvious utility of the exact methods discussed in Section 16.2.3, they typically fail to provide a complete solution to the problem of Bayesian inference for the large complex nonlinear statistical models that many statistical researchers are currently concerned with. In such situations, one is faced with the problem of high-dimensional numerical integration and this is a classically hard problem. If the dimensions involved are relatively low, then it may be possible to use standard numerical quadrature techniques. One advantage of such techniques is that they are "embarrassingly" parallel, in that the integration space can be easily partitioned and each processor can integrate the space allocated to it and then the results can be combined to give the final result. This problem is considered in almost every introduction to parallel computing; see Ref. [14] for further details.

Once the dimensions involved are large (say, more than 10), then standard quadrature methods become problematic, even using large parallel computers. However, many statistical problems of interest involve integrating over hundreds or even thousands of dimensions. Clearly a different strategy is required. One way to tackle high-dimensional integrals is to use Monte Carlo simulation methods. Suppose interest lies in the integral

$$I = \mathrm{E}_\Pi(t(\phi)) = \int t(\phi)\mathrm{d}\Pi(\phi) = \int t(\phi)\pi(\phi)\mathrm{d}\phi$$

for some high-dimensional ϕ. If n values of ϕ can be sampled independently from the density $\pi(\phi)$ then I may be approximated by

$$I_1 = \frac{1}{n}\sum_{i=1}^{n}t(\phi^{(i)}).$$

Even if sampling directly from $\pi(\phi)$ is problematic, suppose it is possible to independently sample values of ϕ from a density $f(\phi)$ that has the same support as $\pi(\phi)$, then I may be rewritten

$$I = \int \frac{t(\phi)\pi(\phi)}{f(\phi)}f(\phi)\mathrm{d}\phi = \int \frac{t(\phi)\pi(\phi)}{f(\phi)}\mathrm{d}F(\phi) = \mathrm{E}_F\left(\frac{t(\phi)\pi(\phi)}{f(\phi)}\right).$$

Consequently, given n samples from $f(\phi)$, I may be approximated by

$$I_2 = \frac{1}{n}\sum_{i=1}^{n}\frac{t(\phi^{(i)})\pi(\phi^{(i)})}{f(\phi^{(i)})}.$$

The strong law of large numbers (LLN) assures us that both I_1 and I_2 are strongly consistent unbiased estimators of I, with I_1 preferred to I_2 if available. The quality (precision) of I_2 will depend strongly on how similar $f(\phi)$ is to $\pi(\phi)$. The problem with Monte Carlo methods is that the errors associated with the estimates are proportional to $1/\sqrt{n}$ (assuming the relevant variances exist), so very large samples are required for good estimation. See Ref. [15] for further details of stochastic simulation and Monte Carlo methods, including variance reduction techniques.

Monte Carlo techniques are very computationally intensive but are ideal in the context of parallel computing. For simplicity, consider the evaluation of I_1. Because the samples are independent, it is possible to divide up the n required samples between the available processors. Each processor can then generate and summarize its samples, and once all processors have finished, the summaries can be summarized to give the final result. In a homogeneous environment, it is probably desirable simply to divide up the required number of samples equally between the processors, but in an inhomogeneous environment that is most likely not a good strategy. Strategies for parallel Monte Carlo in a nonhomogeneous environment are examined in Ref. [16], which considers a range of issues including time-limited resources and unreliable processors.

Consider now the problem of evaluating I_1 using N processors where $n = Nm$, for some integer m. The algorithm for parallel simulation can be described as follows.

1. Master program computes $m = n/N$, and passes m to each available processor.
2. Each processor (k):
   (a) simulates m independent realizations of ϕ
   (b) computes $S_k = \sum_{i=1}^{m}t(\phi^{(i)})$
   (c) passes S_k back to master program.
3. Master program collects and sums the S_k to give final sum S.
4. Master program returns S/n.

Section 16.3.6 describes a complete computer program to implement this algorithm but relies on some topics not yet covered.

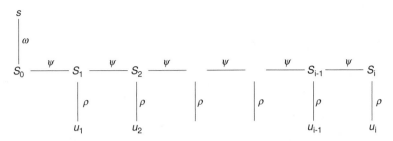

FIGURE 16.2 Illustration of the sequence of deterministic mappings involved in the generation of a pseudo-random number stream.

16.3.2 PSEUDORANDOM NUMBER GENERATION (PRNG)

The above algorithm gives the essential logic, but there are some issues relating to simulation of random quantities on a computer which need to be addressed. The algorithm clearly assumes that all simulated realizations of the random quantity ϕ are independent. This requires not only that the simulated realizations generated on a given processor are independent of one another, but also that the simulated values on different processors are independent of one another. This causes some difficulties with conventional pseudorandom number generation algorithms based on a single underlying "stream." The pseudorandom number generators used by almost all computers are in fact deterministic algorithms carefully designed to give a sequence of numbers which give every appearance of both uniformly distributed and independent of one another. These can then be transformed to give independent realizations of other distributions of interest. Standard pseudorandom number generators are unsatisfactory in a parallel context, but to see why, one must understand a little about how they work.

Pseudorandom number generators start with a seed, s. There may well be a default seed, or in some situations a seed may be chosen automatically based on things such as the computer system clock, but s is always present, and for most random generators, may be explicitly set by the user. The heart of the random number generator is its internal state, S, which is finite, and maintained throughout the time that the generator is used by the program. The initial state is set according to $S_0 = \omega(s)$ for some deterministic function $\omega(\cdot)$. When a uniform random number is required, the state is first updated via $S_i = \psi(S_{i-1})$ and then a uniform random number is computed as $u_i = \rho(S_i)$ for some deterministic functions $\psi(\cdot)$ and $\rho(\cdot)$ (Figure 16.2). Clearly the most important part of the generator is the function $\psi(\cdot)$, which although deterministic, is carefully chosen to give an apparently random sequence of states. The way that $\psi(\cdot)$ is constructed is not of particular concern here, but the interested reader is referred to Ref. [15] for an introduction. Note that the resulting uniform random variate is likely to be further deterministically transformed to give a sample from some nonuniform distribution of interest [17].

16.3.3 PRNG USING THE GNU SCIENTIFIC LIBRARY (GSL)

The GNU Scientific Library (GSL) contains an extensive set of functions for (serial) pseudorandom number generation and transformation. It is perhaps useful to introduce some of these functions at this point. Below is a complete ANSI `C` program to simulate 10 $U(0, 1)$ random variates and print them out to the console.

```
1   #include <gsl/gsl_rng.h>

2   int main(void)
3   {
4     int i; double u; gsl_rng *r;
5     r = gsl_rng_alloc(gsl_rng_mt19937);
6     gsl_rng_set(r,0);
7     for (i=0;i<10;i++) {
8       u = gsl_rng_uniform(r);
9       printf("u(%d) = %f\n",i,u);
10    }
11    return(EXIT_SUCCESS);
12  }
```

Note that the line numbers are not part of the program. The above program will compile with a command like:

```
gcc gsl_rng_demo.c -o gsl_rng_demo -lgsl -lgslcblas
```

The key lines of this program are as follows. Line 1 makes the random number generator functions of the GSL library available to the program. Line 4 allocates the variables used by the program, including a variable r, which contains information about the random number stream being used, including its internal state. Line 5 initializes the generator, setting it to be based on a "Mersenne Twister" generator with a period of $2^{19,937} - 1$. Line 6 sets the seed of the generator to the (default) value of 0. Line 8 requests the next uniform random variate from the stream. Due to the way that the streams are handled via a variable (here r), it is possible to use several streams (possibly of different types) within a single (serial) program.

The GSL has many different kinds of uniform generator, as well as an extensive range of functions for nonuniform random number generation. Importantly, the library is designed so that any kind of uniform stream can be used in conjunction with any nonuniform generator function. Consider the complete program below for simulating standard normal deviates.

```
1   #include <gsl/gsl_rng.h>
2   #include <gsl/gsl_randist.h>

3   int main(void)
4   {
5     int i; double z; gsl_rng *r;
6     r = gsl_rng_alloc(gsl_rng_mt19937);
7     gsl_rng_set(r,0);
8     for (i=0;i<10;i++) {
9       z = gsl_ran_gaussian(r,1.0);
10      printf("z(%d) = %f\n",i,z);
11    }
12    return(EXIT_SUCCESS);
13  }
```

Line 2 makes the nonuniform random number generation functions available to the program, and line 9 requests the generation of a standard normal random variable using the given stream r as a source of randomness. In this case, the uniform stream is based on a "Mersenne Twister," but the gsl_ran_* functions will all happily use any other GSL random number stream that is supplied.

Random number streams of the above form make perfect sense in the context of a serial program, where random variates are read sequentially from a single "stream." However, in a parallel environment it is not at all convenient for all processors to share such a single stream — it involves a prohibitive amount of communication overhead. Consequently, it is highly desirable for each processor to maintain its own stream (state), but this turns out to be problematic.

Clearly if each processor maintains its own random number stream then the random variates generated on that processor will appear to be independent of one another (assuming the generator is good). However, we also require that the variates are independent *across* processors, and this is where the problems lie. Clearly if each processor starts with the *same seed* then the random number streams on each processor will be identical, and each processor will produce exactly the same computations. In this case the results are very clearly not independent across processors. A widely used "solution" to this problem is to use *different seeds* on each processor, but this too presents some difficulties. It is important to realize that the internal state of a pseudorandom number generator is necessarily finite, as it is stored on a digital computer. Given the deterministic nature of the state updating function $\psi(\cdot)$, it follows that the generator will *cycle* through a fixed set of states in a fixed order at some point. The length of the cycle is known as the *period* of the generator, and it is obviously desirable for the period to be larger than the number of random variates that will be required for the simulation problem under consideration (with good modern generators this is rarely a problem). However, suppose that a program is running in a parallel environment and that seed s^A is used on processor A and seed s^B is used on processor B. If one considers the initial states of the random number generators on processors A and B these are $S_0^A = \omega(s^A)$ and $S_0^B = \omega(s^B)$. Now let $\mathcal{S}^A$ denote the *orbit* of S_0^A; that is $\mathcal{S}^A = \{\psi^n(S_0^A)|n \in \mathbb{N}\}$, where $\psi^n(\cdot)$ represents the n-fold composition of the map $\psi(\cdot)$. If it is the case that $S_0^B \in \mathcal{S}^A$ (as is typically the case for random number generators designed with serial computers in mind), then it is clear that for sufficiently long simulation runs the streams on processors A and B will overlap. It follows that unless the seeding algorithms used are extremely sophisticated, then for any substantial simulation there is a nonnegligible probability that the random number streams on parallel processors will have some overlap. This leads to nonindependence of the streams across processors and is obviously undesirable.

16.3.4 PARALLEL PSEUDORANDOM NUMBER GENERATION (PPRNG)

To overcome the difficulties with the use of serial pseudorandom number streams in a parallel environment, a theory of parallel pseudorandom number generation has been developed [18]. There are several different strategies one can employ for adapting standard serial generators into the parallel context, and a detailed examination would not be appropriate here. However, it is useful to have some idea how such generators work, and hence a brief description will now be given of the key idea behind an important class of PPRNG algorithms that are said to be *initial value parameterized*.

Using the notation from Section 16.3.2, the main novelty in this class of PPRNG is the use of an initialization function $\omega(\cdot)$ that depends on the processor number. So for each processor $k = 1, 2, \ldots, N$ we map a single *global* seed, s to N different initial states given by $S_0^k = \omega^k(s)$. Then for each k, a stream is generated in the usual way using $S_i^k = \psi(S_{i-1}^k)$ and $u_i^k = \rho(S_i^k)$ (Figure 16.3). The key to independence of the resulting streams is in the careful choice of updating function $\psi(\cdot)$ and initialization functions $\omega^k(\cdot)$, which ensure that the N resulting orbits, $\mathcal{S}^1, \mathcal{S}^2, \ldots, \mathcal{S}^N$ are nonoverlapping and the same known size irrespective of the choice of global seed s. It is helpful to regard the state space of the PPRNG as forming a toroidal lattice with rows of the lattice forming the streams obtainable on a single processor and the initialization functions $\omega^k(\cdot)$ providing a mechanism for mapping an arbitrary seed onto a given row of the lattice.

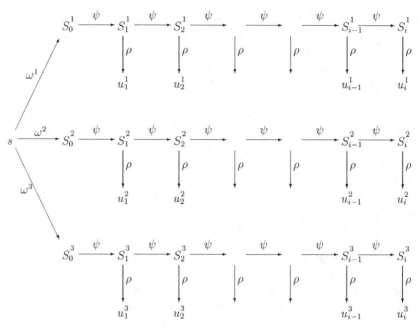

FIGURE 16.3 Illustration of the deterministic mappings involved in the generation of an independent family of pseudorandom number streams based on parameterization of initial values.

16.3.5   The Scalable Parallel Random Number Generator (SPRNG)

A useful family of parameterized parallel random number generators is described in Ref. [19], and implemented in a freely available software library, SPRNG. Being parallel, the library relies on some form of parallel communication (simply to number the different streams on each processor correctly), and the authors of this library have ensured that it is easy to use in conjunction with the MPI; see Ref . [14] for an introduction to MPI, and Ref. [20] for reference. The SPRNG library has a number of sophisticated features, including the ability to use multiple streams on each processor. However, for most Monte Carlo (and MCMC) applications, a single independent stream on each processor is sufficient, and for this there is a simple interface which is very easy to use. Consider the following program that generates 10 $U(0, 1)$ random variates on each available processor.

```
 1 #include <stdio.h>
 2 #include <stdlib.h>
 3 #include <mpi.h>
 4 #define   SIMPLE_SPRNG
 5 #define   USE_MPI
 6 #include "sprng.h"

 7 int main(int argc,char *argv[])
 8 {
 9   double rn; int i,k;
10   MPI_Init(&argc,&argv);
11   MPI_Comm_rank(MPI_COMM_WORLD,&k);
12   init_sprng(DEFAULT_RNG_TYPE,0,SPRNG_DEFAULT);
13   for (i=0;i<10;i++) {
14     rn = sprng();
```

```
15          printf("Process %d, random number %d: %f\n", k, i+1, rn);
16      }
17      MPI_Finalize();
18      return(EXIT_SUCCESS);
19  }
```

This program can be compiled with a command like

```
mpicc sprng_demo.c -o sprng_demo -lsprng
```

and run (on eight processors) with a command like

```
mpirun -np 8 sprng_demo
```

This (and most) MPI programs use the single program multiple data (SPMD) parallel programming paradigm. Exactly the same program is run on every available processor. However, because each process is aware of its "process number," different processes are able to behave differently by using standard conditional directives.

Going through the above code in detail, line 3 makes the MPI functions available to the program. Lines 4 and 5 declare that we are using the "simple" interface to SPRNG and that we are running SPRNG on top of MPI (the definition of these macros affects the function prototypes that are declared when sprng.h is included). Line 6 then makes the relevant SPRNG library functions available to the program. Line 10 initializes MPI, making sure that each copy of the process is started correctly and initialized with any command line arguments. Line 11 records the process number in a variable k, so that it can be used by the program. Line 12 initializes the SPRNG random number generator (on each process), with a common global seed of 0. Line 14 generates a $U(0, 1)$ variable using SPRNG, and line 17 is used to ensure that the MPI program terminates correctly.

Now as previously mentioned, in the context of Monte Carlo integration the uniform random variables generated by SPRNG will typically require transforming to quantities from some nonuniform distribution of interest. The GSL functions of the form gsl_ran_* do exactly this, but only operate on random number streams of type gsl_rng. The author has written a small piece of code, gsl-sprng.h (available from his website), which wraps up the simple SPRNG generator as a GSL PRNG (gsl_rng_sprng20) enabling the development of parallel nonuniform PRNG codes. The following program generates 10 $Pois(2)$ random quantities on each available processor.

```
1   #include <mpi.h>
2   #include <gsl/gsl_rng.h>
3   #include "gsl-sprng.h"
4   #include <gsl/gsl_randist.h>

5   int main(int argc,char *argv[])
6   {
7     int i,k,po; gsl_rng *r;
8     MPI_Init(&argc,&argv);
9     MPI_Comm_rank(MPI_COMM_WORLD,&k);
10    r=gsl_rng_alloc(gsl_rng_sprng20);
11    for (i=0;i<10;i++) {
12      po = gsl_ran_poisson(r,2.0);
13      printf("Process %d, random number %d: %d\n", k, i+1, po);
14    }
15    MPI_Finalize();
16    return(EXIT_SUCCESS);
17  }
```

This combination of ANSI 'C', MPI, SPRNG, and the GSL provides a very effective framework for the development of sophisticated parallel Monte Carlo (and MCMC) algorithms.

16.3.6   A SIMPLE PARALLEL PROGRAM FOR A MONTE CARLO INTEGRAL

This section will finish with a simple example Monte Carlo integral. Suppose we require

$$I = \int_0^1 \exp\{-u^2\}\mathrm{d}u = \mathrm{E}_U\left(\exp\{-U^2\}\right) \simeq \frac{1}{n}\sum_{i=1}^{n}\exp\{-u_i^2\},$$

where $U \sim U(0, 1)$ and $u_i,\ i = 1, 2, \ldots, n$ are independent samples from this distribution. The following program simulates 10,000 variates on each processor, calculates the partial sums, then "gathers" the N partial sums back together on process number 0 where the final result is computed and printed.

```
 1   #include <math.h>
 2   #include <mpi.h>
 3   #include <gsl/gsl_rng.h>
 4   #include "gsl-sprng.h"

 5   int main(int argc,char *argv[])
 6   {
 7     int i,k,N; double u,ksum,Nsum; gsl_rng *r;
 8     MPI_Init(&argc,&argv);
 9     MPI_Comm_size(MPI_COMM_WORLD,&N);
10     MPI_Comm_rank(MPI_COMM_WORLD,&k);
11     r=gsl_rng_alloc(gsl_rng_sprng20);
12     for (i=0;i<10000;i++) {
13       u = gsl_rng_uniform(r);
14       ksum += exp(-u*u);
15     }
16     MPI_Reduce(&ksum,&Nsum,1,MPI_DOUBLE,MPI_SUM,0,MPI_COMM_WORLD);
17     if (k == 0) {
18       printf("Monte carlo estimate is %f\n", (Nsum/10000)/N );
19     }
20     MPI_Finalize();
21     return(EXIT_SUCCESS);
22   }
```

The key new features of this program are the use of MPI_Comm_size (line 9) to obtain the number of processors and the use of MPI_Reduce (line 16) to add together the N partial sums and store the result on process 0. Running the program should give a result close to the true answer of 0.74682.

16.4   MARKOV CHAIN MONTE CARLO (MCMC)

16.4.1   INTRODUCTION

For computations involving high-dimensional posterior distributions, it is usually impractical to implement pure Monte Carlo solutions. Instead, the most effective strategy is often to construct a Markov chain with equilibrium distribution equal to the posterior distribution of interest. There are a variety of strategies that can be used to achieve this; see Refs. [2, 21] for a good introduction. From a practical perspective, the chain is initialized at an essentially arbitrary starting state and successive values are sampled based on the previous value using a PRNG. In principle, after a "burn-in" period the samples should be from the equilibrium distribution, and these can be studied to gain insight into the posterior of interest.

Even in the context of a single serial computer, there is some controversy surrounding the question of whether or not it is better to run one long chain or several shorter ones [22, 23]. The argument in favor of one long run is that the burn-in only happens once, whereas for several shorter chains, each must burn-in, resulting in many wasted samples. The arguments in favor of several chains concern the possibility of being able to better diagnose convergence to the equilibrium distribution. In the context of a serial computer the author tends to side with the one long run camp, but in a parallel environment the situation is rather different. Provided that burn-in times are relatively short, running a chain on each processor can often be the most effective way of utilizing the available hardware.

16.4.2 PARALLEL CHAINS

It is important to bear in mind that for any given posterior distribution, $\pi(\phi|y)$, there are many different MCMC algorithms that can be implemented, all of which have $\pi(\phi|y)$ as the unique equilibrium distribution. When designing an MCMC algorithm for a particular problem there is a range of trade-offs that are made. Often the easiest scheme to implement is based on a Gibbs sampler or simple Metropolis-within-Gibbs scheme that cycles through a large number of low-dimensional components. Although easy to implement, such schemes often have very poor mixing and convergence properties if the dimension of ϕ is high. The output of such a scheme generally has a significant proportion of the run discarded as burn-in, and the iterations left over are often "thinned" (keeping, say, only 1 in 100 iterations) to reduce dependence in the samples used for analysis. The issue of burn-in is of particular concern in a parallel computing environment. The fact that every processor must spend a significant proportion of its time producing samples that will be discarded places serious limitations on the way the performance of the algorithm scales with increasing numbers of processors [16]. Given that there are a number of ways of improving the mixing of MCMC algorithms, including blocking and reparameterization [24], it may be particularly worthwhile spending some time on improving the mixing of the sampler if a parallel chains strategy is to be adopted (see Section 16.4.3). If this is not practical, then parallelization of a single MCMC chain is possible, but this too presents difficulties, and is unlikely to scale well to very large numbers of processors in general (see Section 16.4.4).

Whenever it is felt to be appropriate to use a parallel chain approach, producing code to implement it is extremely straightforward. In this case it is often most sensible to first develop the code for a single processor, and then convert to a parallel program once it is debugged. Debugging MCMC code and parallel code are often both difficult, and so it is best to try to avoid doing the two simultaneously if at all possible!

Example 16.3 Consider generation of a standard normal random quantity using a random walk Metropolis–Hastings sampler with $U(-\alpha, \alpha)$ innovations (α is a pre–chosen fixed "tuning" parameter). Here it is clear that if the chain is currently at x, a proposed new value $x^\star$ should be accepted with probability $\min\{1, \phi(x^\star)/\phi(x)\}$, where $\phi(\cdot)$ is the standard normal density.

The full parallel program to implement the above example is given below.

```
1    #include <gsl/gsl_rng.h>
2    #include "gsl-sprng.h"
3    #include <gsl/gsl_randist.h>
4    #include <mpi.h>
```

```
 5   int main(int argc,char *argv[])
 6   {
 7     int k,i,iters; double x,can,a,alpha; gsl_rng *r;
 8     FILE *s; char filename[15];
 9     MPI_Init(&argc,&argv);
10     MPI_Comm_rank(MPI_COMM_WORLD,&k);
11     if ((argc != 3)) {
12       if (k == 0)
13         fprintf(stderr,"Usage: %s <iters> <alpha>\n",argv[0]);
14       MPI_Finalize(); return(EXIT_FAILURE);
15     }
16     iters=atoi(argv[1]); alpha=atof(argv[2]);
17     r=gsl_rng_alloc(gsl_rng_sprng20);
18     sprintf(filename,"chain-%03d.tab",k);
19     s=fopen(filename,"w");
20     if (s==NULL) {
21       perror("Failed open");
22       MPI_Finalize(); return(EXIT_FAILURE);
23     }
24     x = gsl_ran_flat(r,-20,20);
25     fprintf(s,"Iter X\n");
26     for (i=0;i<iters;i++) {
27       can = x + gsl_ran_flat(r,-alpha,alpha);
28       a = gsl_ran_ugaussian_pdf(can) / gsl_ran_ugaussian_pdf(x);
29       if (gsl_rng_uniform(r) < a)
30         x = can;
31       fprintf(s,"%d %f\n",i,x);
32     }
33     fclose(s);
34     MPI_Finalize(); return(EXIT_SUCCESS);
35   }
```

The first thing to note about the program is that lines 24 to 32 contain the essential part of the algorithm, which may be coded in exactly the same way in the serial context. Lines 1 to 23 set up the parallel processes and create separate MCMC output files for each process. This program also illustrates how to trap errors and exit politely from an MPI program when an error is encountered. Figure 16.4 shows the result of running this program on four processors with parameters 10,000 and 0.1.

An important consideration in the parallel context is that of the starting point of the chain. In principle, provided that a suitable burn-in period is used, the starting point does not matter, and hence any starting point could be used (indeed, all of the chains could be started off at the same point). On the other hand, if each chain could be started with an independent realization from the target distribution, then no burn-in would be necessary, and parallelization would be "perfect." It is suggested in Ref. [16] that wherever possible, an exact sampling technique such as coupling from the past [25] should be used on each processor to generate the first sample, and then the chain run from these starting points, keeping all iterates. Unfortunately, it is difficult to implement exact sampling for many complex models, and so a pragmatic solution is to initialize each chain at a random starting point that is in some sense "overdispersed" relative to the target, and then burn-in until all chains have merged together. Note that the arguments sometimes put forward suggesting that "burn-in is pointless" only hold in the "one-long-run" context; in the context of parallel chains starting points are an issue [16].

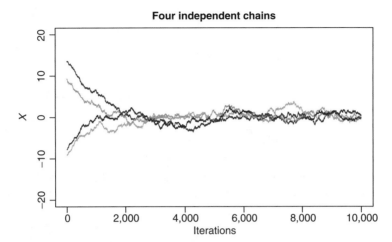

FIGURE 16.4 Trace plot of the output obtained by running the example parallel MCMC chains program for 10,000 iterations on each of four processors with an innovation parameter of $\alpha = 0.1$.

To make things more concrete, consider a problem where preliminary runs have been used to ascertain that a burn-in of b iterations are required, followed by a main monitoring run of n iterations. In the serial context, a chain of length $b + n$ is required. It is usually reasonable to assume that each iteration takes roughly the same amount of processor time to compute, so that iterations may be used as a unit of time. If N processors are available to run parallel chains, then the iterations for the main monitoring run may be divided among them, so that $b + n/N$ iterations are required on each processor. This gives a speedup of

$$\text{SpeedUp}(N) = \frac{b + n}{b + \frac{n}{N}} \xrightarrow[\infty]{N} \frac{b + n}{b},$$

meaning that the potential speed up is limited for $b > 0$. Of course, when N processors are available, then the potential speed up offered in principle for a "perfectly parallel" solution is N (attained here in the special case $b = 0$), so when b is a sizeable proportion of n, the actual speed-up attained falls well short of the potential. As the burn-in time is related to the mixing rate of the chain, and the mixing rate is related to the length of the main monitoring run required, b and n are often related. In many cases, $n = 10b$ is a useful rule-of-thumb. In this case we have

$$\text{SpeedUp}(N) = \frac{11}{1 + \frac{10}{N}} \xrightarrow[\infty]{N} 11,$$

making a parallel chains approach quite attractive on small clusters. In particular, SpeedUp(8) ≈ 5. However, SpeedUp(16) < 7, and so effective utilization of large clusters is not possible (Figure 16.5).

As with any MCMC-based approach to Bayesian computation, processing of the raw output is a vital part of the analysis. The most commonly used systems for output analysis are CODA, a library for the S-PLUS statistical language, and R-CODA, a version of CODA for R, the free S-PLUS alternative. Fortunately, both CODA and R-CODA have excellent support for analysis of parallel chains, and hence can be used without any modification in this context.

Before moving on to the issues that arise when a parallel chains approach is felt not to be appropriate, it is worth considering the popular user-friendly MCMC software, WinBUGS [26]. This

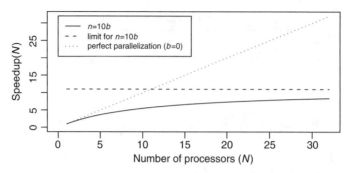

FIGURE 16.5 Graph showing the speedup curve for a parallel chains approach to parallelization in the case $n = 10b$. The limiting speedup is also plotted, along with the curve (a straight line) for a perfect parallelization (attained for $b = 0$).

software allows a user to simply specify a model and prior using a graphical language, and then import data and run an MCMC algorithm for the problem without any programming. Consequently WinBUGS is extremely popular, particularly with applied statisticians. Early versions of WinBUGS offered no facilities for automated control or scripting, and hence were difficult to consider using in a parallel environment. However, as of version 1.4, WinBUGS offers limited remote scripting facilities that now make it realistic to consider using on a parallel cluster. Unfortunately, this approach is not without problems. The most obvious problem is that WinBUGS runs only on Microsoft Windows platforms, whereas many dedicated parallel clusters are Unix-based; on the other hand, many universities have large student PC clusters running Microsoft Windows that stand idle for much of the time, and so a large number of Windows machines are potentially available for parallel computation if configured so to do. Another problem is that WinBUGS currently does not have a parallel random number generator, and hence parallel chains rely on the "different seeds" approach, which does not guarantee independence of the resulting chains. The main problem with using Win-BUGS for parallel chains however (and it is a problem in the serial context too) is that WinBUGS currently uses a simple one-variable-at-a-time Gibbs or Metropolis-within-Gibbs sampler, which leads to very poor mixing for many high-dimensional problems. This in turn leads to very long burn-in times, which make a parallel chains approach to parallelization very unattractive. However, the interested reader is referred to PyBUGS, a Python library that facilitates parallel WinBUGS sessions.

16.4.3 BLOCKING AND MARGINALIZATION

It is tempting to try to separate the issue of parallel computation from that of MCMC algorithm design, but there are at least two reasons why this is misguided. The first is that typically one is interested in parallel strategies because the problem under consideration is of considerable importance and computational complexity. What is the point of parallelizing a bad MCMC algorithm and running it on a powerful computing cluster if a little thought could give an algorithm that performs as well run on a standard desktop PC? Of course if the hardware is available, then one could still parallelize the new algorithm if further speedup is still desirable. The second reason is that different algorithms parallelize in different ways. As has already been demonstrated, for a sophisticated MCMC algorithm with good mixing properties, a parallel chains approach has much to recommend it. However, many simple algorithms with poor mixing properties can often be speeded up by parallelizing a single chain (Section 16.4.4). Consequently the issues of algorithm design and parallelization strategies must be considered together.

If a given Markov chain has poor mixing and long burn-in times, then a straightforward parallel chain approach is unlikely to be sensible. In this case the obvious solution is to work on paralleliz-ing the computations involved in running a single chain (see Section 16.4.4), but another possibility is to develop a more sophisticated MCMC algorithm with better mixing properties. Usually in-dividual iterations are computationally more expensive with better performing algorithms, and so it is important to try and judge the relative performance of the algorithm using a criteria such as "effective sample size" per unit time. CODA and R-CODA contain routines to calculate "effective sample size" from MCMC output. Care must be taken with such an approach, however, as mea-sures of effective sample size based on MCMC output reflect the apparent mixing-rate of the output, and not the true mixing rate of the underlying Markov chain. It is possible for a poor algorithm to apparently mix well, but in fact be only exploring a small part of the true target distribution. Con-sequently, given two algorithms, one of which is very simple, with many components, the other of which is more sophisticated, with fewer components, the more sophisticated algorithm will almost always be preferred, even if the simpler algorithm appears to give slightly more effective indepen-dent samples per unit time. It is clear that such issues are already somewhat difficult to trade-off in the serial context, but things are even less clear in the parallel context, as here it is desirable to construct the algorithm with parallelization in mind.

In this section a brief overview will be given for two techniques for improving the mixing of MCMC algorithms that are often applicable in the context of large complex models, and discuss the use of such techniques in a parallel context. It is helpful to use the notation of Section 16.2.1, con-sidering MCMC algorithms having the posterior $\pi(\sigma, \theta|y)$ as their unique equilibrium distribution. Note that one of the advantages of the MCMC approach is that it renders trivial the marginaliza-tion process; to look at the marginal distribution of any function of σ and θ, simply apply that function to each iterate and study the resulting samples. A simple MCMC algorithm will typi-cally divide up σ and θ into the full collection of univariate components, and sample from each in turn from a kernel that preserves the equilibrium distribution (such as the "full-conditional" or a Metropolis update). However, the large number of components in the sampler typically results in a poorly mixing chain. Usually, there are positive partial correlations between the components of θ, and in this case grouping together elements of θ into "blocks" that are updated simultaneously will invariably improve the mixing of the chain [24]. That is, if the components of the chain are $\sigma_1, \sigma_2, \ldots, \sigma_p, \theta_1, \theta_2, \ldots, \theta_q$, it will typically be beneficial to ensure that p and q are as small as possible while still allowing sampling from the full-conditional of each block directly. Even if di-rect sampling from the full-conditionals is not possible, the above is still valid (at least empirically) if good Metropolis–Hastings jumping rules for the blocks can be found (that is, rules that induce good mixing for the block in question, conditional on all other blocks). In the limiting case of $p = q = 1$, if exact sampling from $\pi(\sigma|\theta, y)$ and $\pi(\theta|\sigma, y)$ are possible, then the result is the classic "data augmentation" technique [27]. For many problems, sampling from $\pi(\sigma|\theta, y)$ will be relatively straightforward, based on semiconjugate updates, or simple Metropolis–Hastings kernels. Updating $\pi(\theta|\sigma, y)$ is typically more problematic. However, in a variety of situations (especially for models where $\theta, y|\sigma$ is linear Gaussian), local computation techniques (see Section 16.2.3) can be used to accomplish this step [8, 10, 11]. In the linear Gaussian case, if the model is DAG based, the free software library GDAGsim can be used to carry out the local computations, and if it is derived from a Markov random field, the library GMRFlib can be used instead.

Markov chains based on data augmentation can still be slow to mix if there is strong posterior dependence between σ and θ. In such cases, a carefully constructed Metropolis–Hastings method is sometimes the only practical way to implement a satisfactory MCMC scheme. See Ref. [28] for an overview of the range of techniques available. One approach is to focus on the marginal posterior density $\pi(\sigma|y)$ (16.2). It is not usually possible to sample from this directly (otherwise MCMC would not be necessary), but a Metropolis–Hastings scheme can be constructed by proposing a new σ^* from a transition density $f(\sigma^*|\sigma)$. The proposal is accepted with probability min$\{1, A\}$, where

$$A = \frac{\pi(\sigma^*)p(y|\sigma^*)f(\sigma|\sigma^*)}{\pi(\sigma)p(y|\sigma)f(\sigma^*|\sigma)}.$$

This involves the marginal likelihood $p(y|\sigma)$, but this is usually tractable using local computation (see Section 16.2.3) and the BMI (16.4) if sampling from $\pi(\theta|\sigma, y)$ is possible [10]. The technique of integrating out the latent process is sometimes referred to as *collapsing* the sampler, and is extremely powerful in a variety of contexts; see for example Ref. [29] for an application in statistical genetics. Application of this technique to lattice-Markov spatiotemporal models is considered in Ref. [30]; the sampler's results have mixing properties that are sufficiently good to make a parallel chains approach to parallelization extremely effective. Although other approaches to parallel MCMC for this model are possible (e.g., parallelization of a single chain for a simple Gibbs sampler), empirical evidence suggests that the combination of sophisticated block sampler and parallel chains results in the most effective use of the available hardware when many processors are available.

The marginal updating schemes are closely related to a class of single-block updating algorithms for $\pi(\sigma, \theta|y)$. Here the proposal is constructed in two stages. First σ^* is sampled from some transition density $f(\sigma^*|\sigma, \theta)$, and then θ^* is sampled from $\pi(\theta^*|\sigma^*, y)$. This proposed joint update is then accepted with probability min$\{1, A\}$, where

$$A = \frac{\pi(\sigma^*)p(y|\sigma^*)f(\sigma|\sigma^*, \theta^*)}{\pi(\sigma)p(y|\sigma)f(\sigma^*|\sigma, \theta)}.$$

Clearly this single-block updating strategy is equivalent to the marginal updating strategy if the proposal density for σ is independent of θ.

Of course in many situations direct sampling from $\pi(\theta|\sigma, y)$ will not be possible, but single-block samplers may still be constructed if this distribution may approximated by one for which sampling is possible. Consider again a proposal constructed in two stages. As before, a candidate σ^* is first sampled from some $f(\sigma^*|\sigma, \theta)$. Next a candidate θ^* is sampled from $g(\theta^*|\sigma^*, \theta)$. The proposed values are then accepted jointly with probability min$\{1, A\}$, where

$$A = \frac{\pi(\sigma^*)\pi(\theta^*|\sigma^*)p(y|\sigma^*, \theta^*)f(\sigma|\sigma^*, \theta^*)g(\theta|\sigma, \theta^*)}{\pi(\sigma)\pi(\theta|\sigma)p(y|\sigma, \theta)f(\sigma^*|\sigma, \theta)g(\theta^*|\sigma^*, \theta)}.$$

In the special case where $g(\theta^*|\sigma^*, \theta) = \pi(\theta^*|\sigma^*, y)$ this reduces to the single-block sampler previously discussed. The above form of the acceptance probability is the form most useful for implementation purposes. However, in order to gain insight into its behavior, it is helpful to rewrite it as:

$$A = \frac{\pi(\sigma^*, y)}{\pi(\sigma, y)} \times \frac{f(\sigma|\sigma^*, \theta^*)}{f(\sigma^*|\sigma, \theta)} \times \frac{\pi(\theta^*|\sigma^*, y)}{g(\theta^*|\sigma^*, \theta)} \times \frac{g(\theta|\sigma, \theta^*)}{\pi(\theta|\sigma, y)}.$$

It is then clear that for relatively small proposed changes to σ, A depends mainly on how well $g(\theta^*|\sigma^*, \theta)$ approximates $\pi(\theta^*|\sigma^*, y)$. A good example of a situation where this technique is useful arises when $\pi(\theta^*|\sigma^*, y)$ is not tractable, but is well-approximated by a linear Gaussian system which is [31]. See also Section 16.5 for a related technique.

16.4.4   PARALLELIZING SINGLE CHAINS

In the context of a parallel computing environment, the strategies outlined in the previous section may lead to a sampler that mixes sufficiently well when a parallel chains approach to parallelization is quite adequate. Given that this is very simple to implement and scales well with the number of available processors (provided burn-in times are short), this is usually the strategy to strive for in the first instance. However, there are many situations (e.g., intrinsically nonlinear high-dimensional latent process models), where the best practically constructible samplers mix poorly and have long burn-in times. In such cases it is desirable to take an algorithm for generating a single chain and run it on a parallel computer to speed up the rate at which iterations are produced. It is worth noting, however, that even in this situation, it may be possible to improve the sampler by constructing an improved algorithm based on a noncentered parameterization (NCP) of the problem [32]. It turns out that NCPs can be parallelized in the same way as other algorithms, so we will not give further details here.

Clearly Markov chain simulation is an iterative process; simulation of the next value of the chain cannot begin until the current value has been simulated.[1] However, for high-dimensional problems every iteration is a computationally intensive task and there is often scope for parallelization of the computation required for each.

If the MCMC algorithm is based on one of the sophisticated samplers from the previous section, which at each stage involve the updating of only one or two very large blocks, then parallelization strategies must focus on speeding up the computations involved in those updates. Parallelization of local computation algorithms has already been discussed in Section 16.2.4. For example, if θ is sampled using a local computation algorithm, then a parallel updating strategy must be adopted; see Refs. [6, 7] for further details, but again note that the speedup provided by graph-propagation algorithms in the statistical context tends not to scale well as the number of available processors increases. If the system is linear Gaussian conditional on σ, then an approach based on large (probably sparse) matrices has probably been used for sampling and/or the computation of marginal likelihoods required in the acceptance probabilities. These computations can be speeded up by using parallel matrix algorithms. There is an issue of whether or not direct or iterative matrix algorithms must be used; generally speaking, direct algorithms must be used if samples from θ are required, but iterative algorithms will suffice if θ has been marginalized out of the problem completely. Note that parallel matrix algorithms are covered in detail elsewhere in this volume, to which the reader is referred for further details.

For high-dimensional nonlinear problems, a sampler based on many components is more likely to be effective. Once again it will be assumed that the components of the chain are $\sigma_1, \sigma_2, \ldots, \sigma_p, \theta_1,$ $\theta_2, \ldots, \theta_q$, and that each block is updated once per iteration using a kernel which preserves the desired target distribution $\pi(\sigma, \theta | y)$. Typically the updates of σ will be very fast (given some sufficient statistics regarding the current state of θ), and the updates of θ will be very slow. Consequently, a serial MCMC algorithm will spend almost all of its time carrying out the updating of θ. If a parallel computing environment is available, it is therefore natural to try and speedup the algorithm by parallelizing the θ update step.

Parallelization of the update of θ depends crucially on the conditional independence structure of the model (Section 16.2.2). The structure of interest is the undirected conditional independence

---

[1] There is actually an exception to this; where a pure Metropolis–Hastings *independence sampler* is being used, the next proposed value may be simulated and work may start on evaluation of the next acceptance probability before the current iteration is complete. In practice, however, this is not much help, as pure independence samplers tend to perform poorly for large nonlinear models.

FIGURE 16.6 Undirected conditional independence graph for the case of serial dependence between the θs (conditional on σ and y).

graph for $\theta_1, \theta_2, \ldots, \theta_q$ conditional on both σ and y. Assume first the simplest possible case, where $\theta_i \perp\!\!\!\perp \theta_j | \sigma, y$, $i \neq j$. If this is the case then the update of any particular θ_i will not depend on the state of any other θ_j ($j \neq i$). Then clearly each θ_i can be updated in parallel, so the q blocks of θ can be evenly distributed out to the N available processors to be updated in parallel. The essential details of each iteration of the MCMC algorithm can then be described as follows.

1. Each processor (k):
   (a) sequentially updates the q_k θs that have been assigned to it, using the current value of σ
   (b) computes summary statistics for the new θs that will be required to update σ
   (c) passes the summary statistics back to the root process (master program)
2. The root process combines the summary statistics to obtain an explicit form for the updating of σ.
3. A new σ is sampled and output.
4. The new σ is distributed out to each process.

In Section 16.3.6 it has already been demonstrated how to gather and summarize statistics to a root processor, using `MPI_Reduce`. The only new feature required for the implementation of the above algorithm is the ability to "broadcast" a statistic from the root process to the slave processes. This is accomplished with the function `MPI_Bcast`, which has a similar syntax to `MPI_Reduce`. For example, the command

```
MPI_Bcast(sigma,p,MPI_DOUBLE,0,MPI_COMM_WORLD);
```

will take a p-vector of "doubles" called `sigma` defined on the root process 0 (but also defined and allocated on every other process), and distribute the current values of `sigma` on process 0 to every other process.

The above strategy will often work quite well when the θs are all independent of one another. Unfortunately, this is rarely the case in practice. Consider next the case of serial dependence between the blocks, as depicted in Figure 16.6. Here we have the property $\theta_i \perp\!\!\!\perp \theta_j | \theta_{i-1}, \theta_{i+1}$, $i \neq j$. The update of θ_i therefore depends on its neighbors θ_{i-1} and θ_{i+1} meaning that the θ cannot be updated in parallel. However, all of the odd-numbered blocks are independent of one another given the even-numbered blocks (together with σ and y). So, breaking the updating of θ into two stages, all of the odd-numbered blocks may be updated in parallel conditional on the even-numbered blocks. When this is complete, the even-numbered blocks may be updated in parallel conditional on the odd-numbered blocks. See Section 16.5 for an example of this approach in practice.

The same basic strategy can be applied when the conditional independence structure is more complex. Using the terminology of Winkler [33] (to which the reader is referred for more details), let $T = \{T_1, T_2, \ldots, T_c\}$ be a partition of $\{\theta_1, \theta_2, \ldots, \theta_q\}$, where each T_i is a set of blocks that are mutually independent given the remaining blocks (e.g., the odd-numbered blocks from the previous example). The smallest value of c for which it is possible to construct such a partition is known as the *chromatic number* of the graph (e.g., for the example of serial dependence, the chromatic number of the graph is 2). Incidentally, this is the minimum number of colors required to paint the

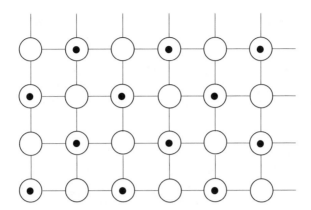

FIGURE 16.7 Undirected conditional independence graph for a four-neighbor Markov random field on a regular square lattice.

nodes of the graph so that no two adjacent nodes are of the same color. It is clear that the blocks allocated to each T_i can be updated in parallel, and so it is clearly desirable to make each T_i as large as possible, and c as small as possible. It should be noted that for completely general conditional independence graphs, finding good partitions is an NP-hard combinatorial optimization problem; however, for most statistical applications, it tends to be quite straightforward to construct good partitions "by hand." Consider, for example, a four-neighbor Markov random field model based on a two-dimensional regular square lattice (Figure 16.7). Here again the chromatic number is just 2; if the graph is painted as a "chequer-board", the black nodes are mutually independent given the white nodes and vice versa.

To implement the parallel updating in practice, the blocks in each T_i are distributed evenly among the available processors so that they can be updated in parallel. In general, then the computations carried out at each iteration of the parallel MCMC algorithm will be as follows.

1. Each processor (k):
   (a) For each i in 1 to c:
       (i) sequentially updates all the blocks in T_i allocated to it
       (ii) distributes necessary state information regarding updates to adjacent processors
       (iii) receives such information from adjacent processors
   (b) Computes summary statistics for the new θs that will be required to update σ.
   (c) Passes the summary statistics back to the root process (master program).
2. The root process combines the summary statistics to obtain an explicit form for the updating of σ.
3. A new σ is sampled and output.
4. The new σ is distributed out to each process.

The only new features required to implement this algorithm concern the exchange of information between processors. Here, the functions `MPI_Send` and `MPI_Recv` can be used to send on one processor and receive on another. The function `MPI_Sendrecv` is also useful for exchanging information between a pair of processors. See Ref. [20] for precise syntax and further details.

In fact the above algorithms have made a simplifying assumption regarding the update of the parameter set σ. The above algorithms work as described if σ is univariate, or if it is convenient to update directly all of σ as a single block. Note that if the elements of $\sigma | \theta, y$ are conditionally independent of one another, then updating each in turn and updating in a single block are equivalent, so this case occurs quite often in practice. However, it is also the case that often the p components of σ are not conditionally independent, and are most conveniently updated separately. Here the above algorithm is not adequate, as the summaries required for the update of σ_j may depend on the result of the update for σ_i ($i < j$). In this situation a revised algorithm is required.

1. Each processor (k):
   (a) For each i in 1 to c:
       (i) sequentially updates all the blocks in T_i allocated to it
       (ii) distributes necessary state information regarding updates to adjacent processors
       (iii) receives such information from adjacent processors
2. For each i in 1 to p:
   (a) each processor (k):
       (i) computes summary statistics that will be required to update σ_i
       (ii) passes the summary statistics back to the root process
   (b) the root process combines the summary statistics to obtain an explicit form for the updating of σ_i
   (c) a new σ_i is sampled and output
   (d) the new σ_i is distributed out to each processor

This algorithm involves a greater number of passed messages than the previous algorithm, so the latency associated with each passed message will make this algorithm less efficient than the previous when both are valid. No additional MPI commands are required to implement this scheme.

Note that a fundamental limitation of these so-called "partially parallel" algorithms is that they cannot provide an N-fold speedup on N processors even for the θ-updating step due to the synchronization requirements. There are various "check points" where further progress cannot be made by one processor if every other processor has not reached the same point. The very tight synchronization requirements make careful load-balancing key. Even then, because not all processors are not carrying out *exactly* the same computations (even if they are executing essentially the same sequence of function calls), processors stand idle until the last one "catches up." This has serious implications for the extent to which such algorithms will scale up to large numbers of processors. Also note that communication overheads can become an issue if processors need to exchange a lot of information regarding the current state of θ. Obviously it is desirable to construct the partition T and the allocation of blocks to processors in such a way as to minimize the interprocessor communication. See Ref. [34] for a thorough investigation of these issues in the context of spatial latent Gaussian models with Poisson count data.

16.5   CASE STUDY: STOCHASTIC VOLATILITY MODELING

16.5.1   Background

This section is concerned with parallelization strategies for Bayesian inference in a discrete-time univariate stochastic volatility model. This problem is high-dimensional and intrinsically nonlinear, so even "good" MCMC algorithms tend to exhibit poor mixing and long burn-in times, making a parallel chains approach apparently unattractive. Therefore parallelization of a single chain will be explored, and its performance relative to the serial and parallel chains implementations examined as a function of the number of available processors.

The log-Gaussian stochastic volatility model is considered, as examined in Refs. [35, 36], though here it is presented using the notation of Ref. [37]. The block-updating algorithm used is similar to that of Ref. [38], which uses a simulation smoother [39] to construct proposed updates for blocks that are then accepted/rejected using a Metropolis–Hastings step. It is worth noting that there is another strategy for constructing an efficient sampler for this model, as described in Ref. [40], but such an approach will not be considered here. The use of parallel computing for estimation of stochastic volatility models in a non-Bayesian and non-MCMC context is discussed in Ref. [41].

The basic model describes a Gaussian noise process y, which has zero mean, but has variance (volatility) varying through time according to a mean-reverting stationary stochastic process. More precisely

$$y_t = \varepsilon_t \exp\{\alpha_t/2\}, \quad t = 1, 2, \ldots, n \tag{16.7}$$

$$\alpha_t = \mu + \phi(\alpha_{t-1} - \mu) + \eta_t, \tag{16.8}$$

where ε and η are mean zero Gaussian white noise processes with variances 1 and σ_η^2, respectively. This model is written $y \sim ISV_n(\phi; \sigma_\eta; \mu)$. It is common practice to initialize the (log) volatility process α with the stationary distribution: $\alpha_0 \sim N(\mu, \sigma_\eta^2/\{1 - \phi^2\})$. In terms of the notation from Section 16.2, we have $\sigma = (\phi, \sigma_\eta, \mu)$ and $\theta = (\alpha_1, \alpha_2, \ldots, \alpha_n)$ with data y as before. Interest is in the joint posterior $\pi(\sigma, \theta|y)$, but for this to be completely specified, $\pi(\sigma)$ is also required. For illustrative purposes, an independent semiconjugate specification is adopted:

$$\mu \sim N(\mu_\mu, \sigma_\mu^2), \quad \phi \sim N(\mu_\phi, \sigma_\phi^2), \quad \sigma_\eta^{-2} \sim \Gamma(\alpha_\sigma, \beta_\sigma).$$

Note, however, that this is not necessarily optimal. In particular, a Beta prior for ϕ is very often used in practice to describe an appropriate informative prior on this parameter.

16.5.2 BASIC MCMC SCHEME

A simple one-variable-at-a-time sampler for this model has $3 + n$ components, and exhibits very poor mixing for large n. It is therefore desirable to block together sequences of successive α_ts to reduce the number of components in the sampler to $3 + q$ ($q \ll n$), where then $\theta = (\theta_1, \theta_2, \ldots, \theta_q)$, and hence improve mixing. This is only likely to be beneficial, however, if a good updating mechanism can be found for the update of each of the components θ_i. This is achieved by approximating the joint distribution of θ and y as linear Gaussian, and then using linear Gaussian local computation techniques to sample an approximation to the full conditional for each θ_i, which is then corrected using a Metropolis–Hastings step. The linear Gaussian approximation is constructed following Ref. [35] by rewriting (16.7) as

$$\log(y_t^2) = \alpha_t + \log(\varepsilon_t^2). \tag{16.9}$$

The random quantity $\log(\varepsilon_t^2)$ is not Gaussian, but its mean and variance may be computed as -1.27 and $\pi^2/2$, respectively. Now if $\log(\varepsilon_t^2)$ is approximated by a $N(-1.27, \pi^2/2)$ random quantity, then (16.9) and (16.8) together form a linear Gaussian system in the form of a dynamic linear model [42]. Local computation algorithms applied to dynamic linear models are generally referred to as Kalman filters. Here the simulation smoother [39] will be used to sample from each block

θ_i based on the linear Gaussian approximation. The log-likelihood for a sampled block under this approximate model is

$$l_G(\theta_i) = -\frac{n}{2}\log\pi^3 - \frac{1}{\pi^2}\sum(\log y_t^2 + 1.27 - \alpha_t)^2,$$

where the sum is over those α_t in block θ_i. Under the true model, however, the log-likelihood is

$$l(\theta_i) = -\frac{n}{2}\log(2\pi) - \frac{1}{2}\sum(\alpha_t + y_t^2\exp\{-\alpha_t\}).$$

Therefore the proposed new block $\theta_i^\star$ should be accepted as a replacement of the current θ_i with probability $\min\{1, A\}$, where

$$\log A = [l(\theta_i^\star) - l_G(\theta_i^\star)] - [l(\theta_i) - l_G(\theta_i)].$$

This step corrects the linear Gaussian approximation, keeping the exact posterior distribution as the equilibrium of the Markov chain. This algorithm was given in Ref. [43] in the special case of blocks of size one. Fixed block sizes of m will be assumed, and it will be assumed that $n = 2rmN$, where N is the number of processors and r is an integer. In the serial context $N = 1$.

Updating the elements of σ is particularly straightforward due to the semiconjugate prior specification. Note that while for many models, the elements of $\sigma|\theta, y$ are conditionally independent, this is not the case here. One consequence of this is that there is potential for improving mixing of the chain by using a block update for the elements of σ (though this is not considered here). Another consequence is that necessary sufficient statistics required for the updating of the elements of σ must be computed as they are required. They cannot all be computed and passed back to the root process in one go after the update of the latent process (which would be desirable, due to the latency associated with interprocess communication). Standard Bayesian calculations lead to the following full-conditionals for the elements of σ:

$$\mu|\cdot \sim N\left(\frac{\frac{\mu_\mu}{\sigma_\mu^2} + \frac{1-\phi}{\sigma_\eta^2}\sum_{i=2}^n(\alpha_i - \phi\alpha_{i-1})}{\frac{1}{\sigma_\mu^2} + \frac{(n-1)(1-\phi)^2}{\sigma_\eta^2}}, \frac{1}{\frac{1}{\sigma_\mu^2} + \frac{(n-1)(1-\phi)^2}{\sigma_\eta^2}}\right) \tag{16.10}$$

$$\phi|\cdot \sim N\left(\frac{\frac{\mu_\phi}{\sigma_\phi^2} + \frac{1}{\sigma_\eta^2}\sum_{i=2}^n(\alpha_{i-1} - \mu)(\alpha_i - \mu)}{\frac{1}{\sigma_\phi^2} + \frac{1}{\sigma_\eta^2}\sum_{i=2}^n(\alpha_i - \mu)^2}, \frac{1}{\frac{1}{\sigma_\phi^2} + \frac{1}{\sigma_\eta^2}\sum_{i=2}^n(\alpha_i - \mu)^2}\right) \tag{16.11}$$

$$\sigma_\eta^{-2} \sim \Gamma\left(\alpha_\sigma + \frac{1}{2}(n-1), \beta_\sigma + \frac{1}{2}\sum_{i=2}^n\{(\alpha_i - \mu) - \phi(\alpha_{i-1} - \mu)\}^2\right). \tag{16.12}$$

In each case the (very small) contribution to the likelihood due to α_1 has been neglected.

16.5.3 PARALLEL ALGORITHM

As always, adaptation of the serial algorithm to parallel chains is trivial. Construction of a parallel algorithm for a single chain requires a little more work. However, there is a simple serial dependence for the blocks of θ, as shown in Figure 16.6. We therefore use the assumed relationship $n = 2rmN$ giving $2rN$ blocks for θ. The first $2r$ blocks are assigned to the first processor, the next $2r$ to the second processor, etc., so that r "odd" and r "even" blocks are assigned to each processor. The basic algorithm is then as described in Section 16.4.4.

As well as storing the blocks allocated to it, each processor also needs to know something of the blocks adjacent to the blocks allocated to it. In this case, all that is required is the very last α value assigned to the previous processor, and the very first α value assigned to the subsequent processor, keeping communication overheads fairly minimal. Assuming that the processes are all initialized in a self-consistent way, each processor passes its first α value down to the previous processor after updating the "odd" blocks, and passes its last α value up to the subsequent processor after updating the "even" blocks. The situation is illustrated in Figure 16.8.

It is clear from the form of the full-conditionals for the model parameters (16.10)–(16.12) that the summary statistics required for parameter updating are

$$\text{Summ}_1 = \sum_{i=2}^{n} (\alpha_i - \phi \alpha_{i-1})$$

$$\text{Summ}_2 = \sum_{i=2}^{n} (\alpha_i - \mu)^2$$

$$\text{Summ}_3 = \sum_{i=2}^{n} (\alpha_{i-1} - \mu)(\alpha_i - \mu)$$

$$\text{Summ}_4 = \sum_{i=2}^{n} \left(\{\alpha_i - \mu\} - \phi\{\alpha_{i-1} - \mu\} \right)^2 .$$

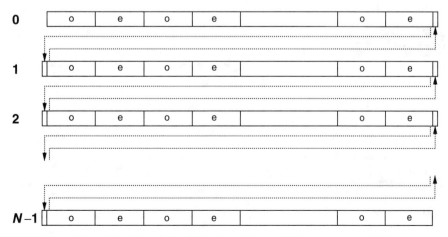

FIGURE 16.8 Plan of the allocation of the θ blocks to the available processes, and the messages passed between them.

Each processor therefore calculates these four summaries for the part of the θ process allocated to it, and passes them back to the root process where they are combined and then used to sample new parameters. The new parameters are then broadcast back to all processors, and the update of the θ process starts again. Note again, however, that the summaries are not all computed at the same time. Summ$_1$ is computed for the update of μ, then Summ$_2$ and Summ$_3$ are computed for the update of ϕ (using the new value of μ), then Summ$_4$ is computed for the update of σ_η (using the new values of μ and ϕ). This additional message-passing overhead is undesirable, but a necessary consequence of the lack of conditional independence of the elements of σ.

16.5.4 RESULTS

To compare the performance of the serial and parallel algorithms, some data simulated from the true model will be used. A total of 3200 observations were simulated using parameters $\mu = 1$, $\phi = 0.8$, $\sigma_\eta = 0.1$. The hyperparameters used to complete the prior specification were

$$\mu_\mu = 0, \quad \sigma_\mu = 100, \quad \mu_\phi = 0.8, \quad \sigma_\phi = 0.1, \quad \alpha_\sigma = 0.001, \quad \beta_\sigma = 0.001.$$

This follows common practice for stochastic volatility modelling, where a fairly informative prior is used for ϕ, but weak priors are adopted for μ and σ_η. A block size of $m = 50$ was used for the MCMC algorithm. Initial values must be chosen for σ and θ. Although these are essentially arbitrary in principle, they actually have important implications for burn-in times, and hence are quite important in practice. Setting σ can be done either by setting the components to plausible values or by sampling from a range of plausible values. Once this has been done, θ can be set by

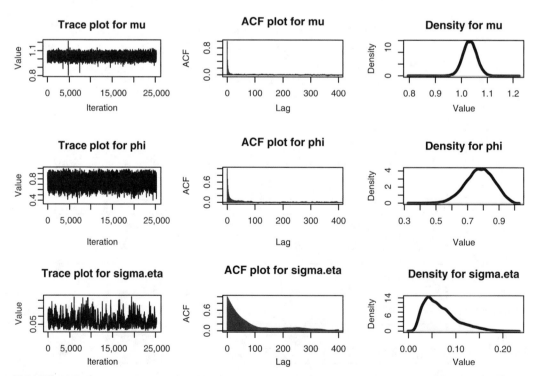

FIGURE 16.9 Analysis of output from the serial algorithm (thinned by a factor of 40), showing trace plots, autocorrelation plots, and marginal density estimates for the three key parameters.

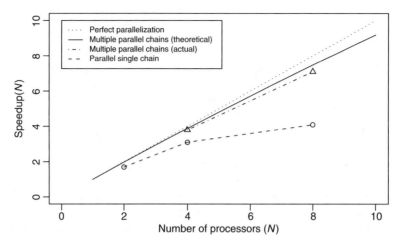

FIGURE 16.10 Graph showing how the speedup varies with the number of processors for the two competing parallel algorithms in this particular example.

sampling from the linear Gaussian approximation to $\theta|\sigma, y$. This ensures that θ is consistent with σ, and hence will not immediately "drag" σ away from the range of plausible values. In cases where it is less straightforward to initialize θ to be consistent with σ, burn-in times can be very much longer.

Preliminary runs suggest that a burn-in of $b = 10,000$ iterations should be used, followed by a monitoring run of $n = 1,000,000$ iterations. The results for running the serial algorithm are illustrated in Figure 16.9. In general the model does a good job of uncovering the true parameters; for example, they all are contained within 95% equitailed posterior probability intervals. It is clear, however, that σ_η is slightly underestimated and μ is a little overestimated, which is a common experience using such models. There are various alternative prior specifications and model parameterizations that can be used to reduce this effect, but such issues are somewhat tangential to the current discussion, and will not be considered further here.

Despite the relatively long burn-in time, the fact that a such a long monitoring run is required means that the burn-in time is a fairly small proportion of the total running time. This means that despite initial appearances to the contrary, a parallel chains approach to this problem is well worth considering. Here the limiting speedup is 101, which is very good (though a very large number of processors would be required to get close to this). For $N = 8$ processors, the theoretical speedup is roughly 7.5, which represents close to perfect utilization of the available hardware.

Running the parallel version of the algorithm for 1,010,000 iterations resulted in speedups 1.7, 3.1, and 4.1 on 2, 4, and 8 processors, respectively. This is plotted together with the (theoretical and actual) parallel chains performance in Figure 16.10. The actual speedups obtained by executing a parallel-chains program on four and eight processors were 3.8 and 7.1, respectively; close to (but less than) the theoretical values of 3.9 and 7.5. It is clear therefore, that in this example, even though the mixing is very poor, by initializing the chain carefully to minimize burn-in times, the parallel chains approach appears to be superior to an approach based on parallelizing a single chain (and is much easier to program!). All of the ANSI 'C' source code for the analysis of this example is available (together with the other simple example programs) from the author's website.

16.6 SUMMARY AND CONCLUSIONS

A number of techniques for utilizing parallel computing hardware in the context of Bayesian inference have been examined. For conventional (non MCMC) methodology, techniques based on parallelization of local computation algorithms are most promising. In particular, parallel algorithms for large matrices are especially valuable in the context of large linear models (such techniques are explored in detail elsewhere in this volume). If a pure Monte Carlo solution to the problem can be found, then parallelization of computations is very straightforward provided a good parallel random number generator library (such as SPRNG) is used. The same applies to MCMC algorithms if a parallel chains approach is felt to be appropriate.

Poorly mixing MCMC algorithms with long burn-in times are not ideally suited to a parallel chains approach, because the burn-in must be repeated on every available processor. However, as the case study illustrated, even where this is the case it often turns out that the length of monitoring run required is such that the burn-in time is a relatively small proportion making a parallel chains approach attractive. In the case study the burn-in time was small relative to the length of monitoring run because it was relatively easy to initialize the latent process (θ) in a sensible way (using a sample from the linear Gaussian approximation). In cases where initializing the latent process well is less straightforward, burn-in times may not be such a small proportion of the required monitoring run. In such situations it is desirable to redesign the sampler if possible. If it is not possible so to do, it may be desirable to consider strategies for parallelizing a single chain. If the algorithm contains large blocks of computation (such as calculations involving large matrices), then it may be possible to parallelize the sampler by using parallel libraries (such as a parallel matrix library). Alternatively, if the sampler is constructed by cycling through a large number of blocks of variables, the sampler may be parallelized by sampling carefully selected sets of blocks in parallel. Unfortunately, this requires considerable modification of the serial algorithm, and does not always scale well as the number of processors increases, due to the very tight synchronization requirements of MCMC algorithms.

MCMC algorithms have revolutionized Bayesian inference (and statistical inference for complex models more generally), rendering tractable a huge range of inference problems previously considered to be of impossible complexity. Of course statisticians are becoming ever more ambitious in the range of models they consider. While MCMC algorithms for large complex stochastic models are now relatively straightforward to implement in principle, many require enormous amounts of computing power to execute. Consequently, effective exploitation of parallel hardware is of clear relevance to many Bayesian statisticians.

ACKNOWLEDGMENTS

This research is supported by a grant from the UK Royal Society (Grant Ref. 22873).

REFERENCES

[1] A. O'Hagan. *Bayesian Inference*, volume 2B of *Kendall's Advanced Theory of Statistics*. Arnold, London, 1994.
[2] D. Gamerman. *Markov Chain Monte Carlo. Texts in Statistical Science*. Chapman & Hall, 1997.
[3] S. Chib. Marginal likelihood from the Gibbs output. *Journal of the American Statistical Association*, 90(432):1313–1321, 1995.
[4] J. Whittaker. *Graphical Models in Applied Multivariate Statistics*. Wiley, Chichester, 1990.
[5] S.L. Lauritzen. *Graphical Models*. Oxford Science Publications, Oxford, 1996.
[6] J. Pearl. *Probabilistic Reasoning in Intelligent Systems*. Morgan Kaufmann San Francisco, CA, 1988.
[7] R.G. Cowell, A.P. Dawid, S.L. Lauritzen, and D.J. Speigelhalter. *Probabilistic Networks and Expert Systems*. Springer, New York, 1999.
[8] D.J. Wilkinson and S.K.H. Yeung. Conditional simulation from highly structured Gaussian systems, with application to blocking-MCMC for the Bayesian analysis of very large linear models. *Statistics and Computing*, 12:287–300, 2002.

[9] S.L. Lauritzen. Propagation of probabilities, means, and variances in mixed graphical association models. *Journal of the American Statistical Association*, 87(420):1098–1108, 1992.

[10] D.J. Wilkinson and S.K.H. Yeung. A sparse matrix approach to Bayesian computation in large linear models. *Computational Statistics and Data Analysis*, 44:493–516, 2004.

[11] H. Rue. Fast sampling of Gaussian Markov random fields. *Journal of the Royal Statistical Society*, B:63(2):325–338, 2001.

[12] A. George and J.W. Liu. *Computer Solution of Large Sparse Positive Definite Systems*. Series in Computational Mathematics. Prentice Hall, Englewood Cliffs, 1981.

[13] E.J. Kontoghiorghes. *Parallel Algorithms for Linear Models: Numerical Methods and Estimation Problems*, volume 15 of *Advances in Computational Economics*. Kluwer, Boston, 2000.

[14] P.S. Pacheco. *Parallel Programming with MPI*. Morgan Kaufmann, San Francisco, CA, 1997.

[15] B.D. Ripley. *Stochastic Simulation*. Wiley, New York, 1987.

[16] J.S. Rosenthal. Parallel computing and Monte Carlo algorithms. *Far East Journal of Theoretical Statistics*, 4:207–236, 2000.

[17] L. Devroye. *Non-uniform Random Variate Generation*. Springer-Verlag, New York, 1986.

[18] A. Srinivasan, D.M. Ceperley, and M. Mascagni. Random number generators for parallel applications. *Advances in Chemical Physics*, 105:13–36, 1999. Monte Carlo Methods in Chemical Physics.

[19] M. Mascagni and A. Srinivasan. SPRNG: A scalable library for pseudorandom number generation. *ACM Transactions on Mathematical Software*, 23(3):436–461, 2000.

[20] M. Snir, S. Otto, S. Huss-Lederman, D. Walker, and J. Dongarra. *MPI — The Complete Reference: Volume 1, The MPI Core*, 2nd ed. MIT Press, Cambridge, MA, 1998.

[21] S.P. Brooks. Markov chain Monte Carlo method and its application. *The Statistician*, 47(1):69–100, 1998.

[22] A. Gelman and D. Rubin. Inference from iterative simulation using multiple sequences. *Statistical Science*, 7:457–511, 1992.

[23] C.J. Geyer. Practical Markov chain Monte Carlo. *Statistical Science*, 7:473–511, 1992.

[24] G.O. Roberts and S.K. Sahu. Updating schemes, correlation structure, blocking and parameterisation for the Gibbs sampler. *Journal of the Royal Statistical Society*, B:59(2):291–317, 1997.

[25] J.G. Propp and D.B. Wilson. Exact sampling with coupled Markov chains and applications to statistical mechanics. *Random Structures and Algorithms*, 9:223–252, 1996.

[26] D.J. Lunn, A. Thomas, N. Best, and D.J. Spiegelhalter. WinBUGS — a Bayesian modelling framework: concepts, structure, and extensibility. *Statistics and Computing*, 10(4):325–337, 2000.

[27] M.A. Tanner and W.H. Wong. The calculation of posterior distributions by data augmentation. *Journal of the American Statistical Association*, 82(398):528–540, 1987.

[28] L. Tierney. Markov chains for exploring posterior distributions (with discussion). *Annals of Statistics*, 21:1701–1762, 1994.

[29] L.A. García-Cortés and D. Sorensen. On a multivariate implementation of the Gibbs sampler. *Genetics Selection Evolution*, 28:121–126, 1996.

[30] L.M. Garside and D.J. Wilkinson. Dynamic lattice-Markov spatio-temporal models for environmental data. In J.M. Bernardo et al., eds., *Bayesian Statistics*, vol. 7. Oxford University Press, Tenerife, 2003, pp. 535–542.

[31] L. Knorr-Held and H. Rue. On block updating in Markov random field models for disease mapping. *Scandanavian Journal of Statistics*, 29(4):597–614.

[32] O. Papaspiliopoulos, G.O. Roberts, and M. Sköld. Non-centered parameterisations for hierarchical models and data augmentation (with discussion). In J.M. Bernardo et al., eds., *Bayesian Statistics*, vol. 7, Oxford University Press, Tenerife, 2003, pp. 307–326.

[33] G. Winkler. *Image Analysis, Random Fields and Dynamic Monte Carlo Methods*. Springer, Heidelberg, 1995.

[34] M. Whiley and S.P. Wilson. Parallel algorithms for Markov chain Monte Carlo methods in latent Gaussian models. Technical report, Department of Statistics, Trinity College Dublin, 2004. 14: pp 171–179.

[35] A. Harvey, E. Ruiz, and N. Shephard. Multivariate stochastic variance models. *The Review of Economic Studies*, 61(2):247–264, 1994.

[36] E. Jaquier, N.G. Polson, and P.E. Rossi. Bayesian analysis of stochastic volatility models. *Journal of Business and Economic Statistics*, 12:371–417, 1994.

[37] M. Pitt and N. Shephard. Time varying covariances: a factor stochastic volatility approach. In J.-M. Bernardo et al., eds., *Bayesian Statistics* vol. 6, Oxford University Press, Oxford, 1999, pp. 3–26.

[38] N. Shephard and M. K. Pitt. Likelihood analysis of non-Gaussian measurement time series. *Biometrika*, 84(3):653–667, 1997.

[39] P. de Jong and N. Shephard. The simulation smoother for time-series models. *Biometrika*, 82:339–350, 1995.

[40] S. Kim, N. Shephard, and S. Chib. Stochastic volatility: likelihood inference and comparison with ARCH models. *Reviews Economic Studies*, 65:361–393, 1998.

[41] J.A. Doornik, D.F. Hendry, and N. Shephard. Computationally-intensive econometrics using a distributed matrix-programming language. *Philosophical Transactions of the Royal Society of London, Series A*, 360:1245–1266, 2002.

[42] M. West and J. Harrison. *Bayesian Forecasting and Dynamic Models*, 2nd ed. Springer, New York, 1997.

[43] N. Shephard. Fitting nonlinear time-series models with applications to stochastic variance models. *Journal of Applied Econometrics*, 8(S):S135–S152, 1993.

WEB REFERENCES

Name	URL	Comment
Beowulf	http://www.beowulf.org/	Cluster computing
GDAGsim	http://www.staff.ncl.ac.uk/d.j.wilkinson/software/gdagsim/	
GMRFlib	http://www.math.ntnu.no/~hrue/GMRFLib/	Spatial library
GSL	http://www.gnu.org/software/gsl/	GNU Scientific library
GSL-SPRNG	http://www.staff.ncl.ac.uk/d.j.wilkinson/software/	
LA Software	http://www.netlib.org/utk/people/JackDongarra/la-sw.html	
LAM	http://www.lam-mpi.org/	MPI implementation
LDP	http://www.tldp.org/	Linux documentation
MPI	http://www-unix.mcs.anl.gov/mpi/	Message-passing interface
MPICH	http://www-unix.mcs.anl.gov/mpi/mpich/	MPI implementation
PSPASES	http://www-users.cs.umn.edu/~mjoshi/pspases/	Parallel direct solver
PVM	http://www.epm.ornl.gov/pvm/	Parallel virtual machines
PyBugs	http://www.simonfrost.com/	Parallel WinBUGS
Python	http://www.python.org/	Scripting language
R Project	http://www.r-project.org/	Statistical language
R-CODA	http://www-fis.iarc.fr/coda/	MCMC Output analysis
ScaLAPACK	http://www.netlib.org/scalapack/	Parallel solver
SPRNG	http://sprng.cs.fsu.edu/	PPRNG library
WinBUGS	http://www.mrc-bsu.cam.ac.uk/bugs/	MCMC software

Index